Computational Linear Algebra
With Models

Computational Linear Algebra
With Models

SECOND EDITION

Gareth Williams
Stetson University

ALLYN AND BACON, INC.
Boston,   London,   Sydney,   Toronto

To Jeff

10 9 8 7 6   85 84 83 82

Library of Congress Cataloging in Publication Data

Williams, Gareth.
    Computational linear algebra with models.

    Includes index.
    1. Algebras, Linear. I. Title.
QA184.W53   1978   512'.5   77-27240
ISBN 0-205-05998-8

Printed in the United States of America.

Contents

Preface   *ix*
Overall Discussion of the Text   *xiii*

CHAPTER 1   **Vectors and Matrices**   1

1-1.   Introduction to Vectors   1
1-2.   Addition and Scalar Multiplication of Vectors   5
1-3.   Introduction to Matrices   17
1-4.   Multiplication of Matrices   27
1-5.   An Analysis of Matrix Multiplication   37
1-6.   Properties of Matrices   48
1-7.*   A Population Movement Model   60
1-8.*   Communication Model and Group Relationships   73

CHAPTER 2   **Systems of Linear Equations—A Quantitative Discussion**   99

2-1.   Systems of Two Linear Equations in Two Variables   99
2-2.   General Systems of Linear Equations   104
2-3.*   Models Involving Systems of Linear Equations   120
2-4.*   Use of Pivots   136
2-5.   Inverses of Matrices   143
2-6.*   Leontief Input-Output Models in Economics   151
2-7.*   Iterative Methods for Linear Systems   157

*Asterisked chapters, sections, and examples are optional. Many of these areas contain additional applications, some further theory. The instructor can use these sections to build around the core course giving the course the desired flavor.

Contents

CHAPTER 3   **Vector Spaces**   163

     3-1.   Subspaces   163
     3-2.   Linear Combinations of Vectors   167
     3-3.   Linear Dependence of Vectors   173
     3-4.   Bases   177
     3-5.   Inner Products, Norms, Angles and Distances   186
     3-6.   Projections and the Gram-Schmidt Orthogonalization Process   198
     3-7.*   Function Spaces   207
     3-8.*   The Inner Product of Special Relativity   213

CHAPTER 4   **Mappings**   219
     4-1.   Mappings Defined by Square Matrices   219
     4-2.   Linear Mappings   232
     4-3.   Mappings and Systems of Equations   244
     4-4.   Matrix Representations of Linear Mappings   251

CHAPTER 5   **Determinants**   259

     5-1.   Introduction to Determinants   259
     5-2.   Properties of Determinants   267
     5-3.   The Evaluation of a Determinant   273
     5-4.*   Vector Products   277

CHAPTER 6   **Systems of Linear Equations—A Qualitative Discussion**   283

     6-1.   The Rank of a Matrix   283
     6-2.   Existence and Uniqueness of Solutions   289
     6-3.   Cramer's Rule   293
     6-4.   A Formula for the Inverse of a Matrix   297
     6-5.   Eigenvalues and Eigenvectors   302
     6-6.*   An Application of Eigenvalues and Eigenvectors   308
     6-7.*   Eigenvalues and Eigenvectors by Iteration   319
     6-8.   Similarity Transformations—Diagonalization of Symmetric Matrices   323
     6-9.   Coordinate Transformations   331
     6-10.*   Normal Modes of Oscillating Systems   353
     6-11.*   The Concept of a Coordinate System   363

CHAPTER 7*   **Linear Programming (Prerequisite—Section 2-2)**   369

     7-1.   Systems of Linear Inequalities   370
     7-2.   Linear Programming—A Geometrical Introduction   376
     7-3.   The Simplex Method   389
     7-4.   Geometrical Explanation of the Simplex Method   398

APPENDIX   **Computing***   **409**

      A.   Introduction to Computing   409
      B.   Introduction to BASIC   412
      C.   Matrices and Vectors (Section 1-3)   424
      D.   Multiplication of a Matrix by a Scalar (Section 1-3)   429
      E.   Addition of Matrices (Section 1-3)   432
      F.   Multiplication of Matrices (Section 1-4)   434
      G.   Significant Figures   436
      H.   Programs for Systems Without Built-In Matrix Commands   438
      I.   Echelon Form of a Matrix (Section 2-2)   443
      J.   Inverse of a Matrix using Gauss-Jordan Elimination (Section 2-5)   446
      K.   Determinant (Section 5-3)   447
      L.   Eigenvalues and Eigenvectors (Section 6-7*)   449
      M.   Simplex Method (Section 7-3*)   451

Answers to Selected Exercises   **455**

Index   **477**

Preface

This book is designed for an introductory course in linear algebra. Its aims are two-fold. One is to teach the reader to understand and perform computations involving vectors, matrices, and systems of linear equations. The other is to communicate an understanding of when and how to apply these tools and an appreciation of the vital role played by this branch of mathematics in many areas of life.

The main change in this second edition is to move the discussion of vector spaces back to Chapter 3, thereby bringing forward matrix algebra to Chapter 1 and systems of linear equations to Chapter 2. This enables us to discuss the concepts of linear independence and bases within the framework of subspaces in more depth than previously, having the tools of systems of equations available. The applications have been revised and more linear programming has been included.

I have included an unusually large number of applications based on real data; these examples bridge the gaps to other fields and show students the outside relevance of the mathematics they are learning. A discussion of computer methods in linear algebra is included as an optional appendix; the use of the computer can add a further dimension to the course. The appendix and numerous exercises provide guidelines for use of the computer in the development of models.

A growing number of schools are encouraging students in mathematics, the sciences, and business to take a solid course in linear algebra as early as practically possible. Linear algebra has long been important in the physical and engineering sciences; it is becoming increasingly necessary in the behavioral, social, and management sciences. It is, for example, the foundation of factor analysis, a field that was formerly used in psychology, and that is now being used in economics, biology, physiology, and medicine as well. It is the basis of linear programming and is a tool used in such areas as differential equations and numerical analysis. This book is intended for a course designed to expose the student to the necessary fundamentals of linear algebra early in his studies. With this purpose in mind, a number of theorems have been stated and illustrated without rigorous proofs. Many theorems in linear algebra can be readily understood and applied; however, if one attempts to verify them without sophisticated concepts, the proofs often become long, cumbersome, and meaningless to the student. Furthermore, it is a working knowledge of such results—not necessarily the proofs—that is needed by some of the groups mentioned above. For example, some proofs have been omitted in the discussion of the properties of determinants.

Calculus is not an essential prerequisite. A knowledge of linear algebra is becoming necessary for many students who do not have a background in calculus. This book is designed to be used by these students, as well as by students of mathematics, the sciences, and engineering who do have a working knowledge of calculus. With this aim in mind, calculus has been omitted from the main body of the text. Optional sections and examples requiring calculus can be included in classes where students have the necessary background. For example, there is an optional section on integral inner products on a function space that would be appropriate for mathematics and engineering students.

The many and varied applications help motivate the reader. Not only is linear algebra intellectually stimulating from a theoretical viewpoint, but also it is a subject of immense breadth. Its spectrum ranges from the abstract through numerical techniques to innumerable applications. Here I have attempted to give the reader a flavor of all these aspects while neither neglecting theory nor being too rigorous. Numerical techniques are presented; for example, errors in Gaussian elimination are discussed and the method of Gaussian elimination is compared to the Gauss-Seidel method. Many applications are given, including the classification of artifacts in archaeology, an analysis of electrical circuits and bridge structures, a mathematical model of population movement between city and suburbia, and an explanation of the behavior of time in Einstein's special relativity model of space-time. The applications are as diverse as one can imagine. These examples vary in degree of difficulty, in completeness of discussion, and in background required for their understanding. The instructor should select those applications that are most appropriate for his class. Other examples can be given as browsing assignments in order to acquaint the students with the breadth of application of the field.

Many examples are designed to give the reader an understanding of the concept of the mathematical model. (The book could be used for a course in modeling at the undergraduate level.) I believe that an understanding of mathematical modeling is absolutely fundamental for nonmathematicians who wish to use mathematics, and is most desirable for mathematicians. For many readers, mathematics will be a mode of expression, a language, and as such it should not be taught in complete isolation from other subjects, but rather in relation to the world around us. The demand in industry and commerce is for mathematicians with a sound foundation in mathematics but also with an appreciation of the wide applicability of the subject—for versatile mathematicians who know when and how to apply the tools of their trade. There is a need for mathematicians who can converse knowledgeably with technologists and sociologists, translate a problem presented in nonmathematical terms into a mathematical model, arrive at a solution or decision, and then communicate the result in a form that is meaningful to nonmathematical colleagues. The analyses of population movements, traffic flow in a road network, and linear programming are among those that illustrate the idea of a mathematical model. The reader will learn to appreciate the various stages: formulating the problem, setting up the mathematical model, developing the mathematical solution, and interpreting the mathematical solution. Some of the models, such as the analysis of the gasoline shortage in Section 2-3 and the comparison of supermarket food prices in Section 1-4, can lead to individual or class projects.

The computer is often used in practice, when linear algebra is applied. One of the most remarkable features of the change in mathematics over the past decade is the birth and growth of the computer. This advent has opened up entirely new fields of mathematical activity, and it has changed the ways in which many traditional areas of mathematics are applied. The computer has also wrought changes in mathematics education. Time-sharing systems are being introduced in ever-greater numbers; this kind of computing facility has proven ideal for mathematics education. I feel that a course in linear algebra should capitalize on the availability of the computer. The computer approach allows the students to learn what the computer can do while, at the same time, it drives home the concepts of linear algebra because the programming process requires a thorough understanding of the concept or problem. Many computational techniques, such as iterative procedures, can be presented in their true framework if the students have access to a computer. The use of the computer widens the scope of the applications in depth, breadth, and realism. Furthermore, when students actually go to apply linear algebra, many will be solving problems involving matrices on the computer.

Because of its diverse applications, I feel that a computer-oriented course in linear algebra will become part of the regular curriculum of most universities and colleges in the near future. This text, with its appendix, is well-suited for such a course. The language is BASIC. A brief introduction to BASIC is included for those who have had no previous computing experience. Further programs are related to sections in the main text where appropriate methods have been developed. (The main text was written with this appendix in mind.) The programs increase in sophistication; new programming concepts are explained when they are first introduced. Thus the student can use the appendix and the main text simultaneously. BASIC (or APL) is very appropriate for a course such as this one because it can be introduced quickly; the programming aspect does not require extensive explanations.

The book has been arranged to permit the instructor a great deal of flexibility in choice of material. I recognize that time is a deciding factor in the choice of material for a course, particularly when applications are to be introduced. One has to balance the pros and cons of introducing certain applications at the expense of theory. The book consists of twenty-nine sections which form the core mathematical content. Other sections, marked optional, contain further theory, such as an explanation of function spaces or discussions of numerical techniques. The applications fall into two categories: brief and in-depth discussions. From within this framework the instructor can cover the core material and select the optional material that gives the desired perspective to the course.

It is a pleasure to acknowledge the help that made this book possible. I would like to thank my former colleagues at the University of Denver: Professors Herbert J. Greenberg, Loren J. Haskins, David L. Hector, Michael S. Martin, and John E. Skelton of the Mathematics Department; Gaylen A. Thurston of the Mechanical Sciences and Environmental Engineering Department; Dennis Peacock of the Physics Department; and William Hildred of the Industrial Economics Division of the Denver Research Institute. To colleagues from the Mathematics and Sociology Departments who participated in a joint seminar, I express my gratitude not only for insights into mathematical modeling, but for stimulating this type of approach toward mathematics. For illuminating discussions on traffic flow problems I thank Greg G. Henk of the Colorado Highway Department and Jim O'Grady, Transportation Planner for the City of Arvada, Colorado.

I thank current colleagues Elizabeth A. Magarian, Bruce Bradford, Raymond J. Cannon, and Gene W. Medlin for many discussions on modeling and on the use of the computer as a tool in the classroom.

I would also like to express my appreciation to the reviewers of this text for their constructive comments: Professor Jean Bevis of Georgia State University, Professor David Buchthal of the University of Akron, Professor Richard Faber of Boston College, Professor Charles Freifeld of Northeastern University, Professor Garret Etgen of the University of Houston, Professor Harvey Keynes of the University of Minnesota, and Professor Berrien Moore of the University of New Hampshire.

My thanks go to David Dahlbacka, Carl Harris, and Garen Wickham of Allyn and Bacon, Inc., for their many suggestions that made this text a better product.

I am grateful to my wife Donna for many valuable discussions on this subject and for her typing contribution. I would like to thank Kathy McCormick, who did most of the typing, and also Sally Lokey, Kathleen Russell, Loretta Norris, and Anita Johnson for their role in the typing.

DeLand, Florida G. Williams

Overall Discussion of the Text

This discussion is intended to give the instructor an overall view of the course so that he or she may see the course as a whole and the continuity of the material.

The first six chapters lead in a natural manner from one to the other. Chapter 7 on Linear Programming has been included to give a fairly in-depth example of an application of linear algebra and can be covered any time after Section 2-2.

For the most part the approach adopted is to develop mathematics, followed by examples of applications. This, we feel, makes for the clearest text presentation. However, instructors can also look ahead with the class to an application and use the application to motivate the mathematics.

Chapter 1.   We start with the geometrical interpretation of vectors in $\mathbf{R}^2$ and $\mathbf{R}^3$. $\mathbf{R}^n$ is introduced as a generalization of these sets. Operations of addition and scalar multiplication are defined on $\mathbf{R}^n$, making it into a vector space. Examples are given that illustrate geometrical interpretations of these operations and also their importance in physical applications and as tools in computations. Example 6, Section 1-2, shows the importance of these operations in the understanding of solutions to systems of linear equations. Throughout the text, we relate the theory developed to systems of linear equations. Matrices and their operations are introduced as natural extensions of vectors to rectangular arrays. Many discussions of applications, such as the highway-model use of the origin-destination matrix in Section 1-3, have been included in a nontechnical manner. These are of interest to all readers and can be left as reading assignments. Some discussions, such as that of the stress matrix in Section 1-6, are more specialized and can be omitted in certain classes. Even then, my hope is that the reader, in passing over these examples, will be left with some impression of the diverse applications of linear algebra. The first chapter closes with two optional sections, 1-7 and 1-8, that should have broad appeal. Section 1-7 introduces the reader to Markov chain models in the fields of demography and genetics. Section 1-8 uses graph theory in models in communication, economics, geography, history, archaeology, and ecology. These sections not only illustrate the usefulness of vectors and matrices, but also introduce the reader to two very important branches of mathematics, Markov Processes and Graph Theory. Instructors who cannot fit these sections into their formal class schedule should encourage the students to browse through them.

Chapter 2. With the tools of vectors and matrices available, we enter on a computational discussion of systems of linear equations. Gaussian elimination is presented in Section 2-2 with specific examples of applications—electrical network analysis, traffic flow, and supply and demand models—in the following section. The use of pivots in reducing round-off errors is discussed. It is shown how matrix inverses can be used to solve certain equations. This leads into a discussion of the Leontieff model for analyzing the interdependence of industries in economics; the matrix inverse method is used there. Finally iterative methods are discussed and compared with the method of Gaussian elimination. After completing this chapter, the reader should have a good grasp of various techniques for solving systems of linear equations, their relative merits, and a feel for when and how such systems can arise in applications.

Chapter 3. At this point the ground work has been laid for a more theoretical turn. We introduce the concepts of subspaces, linear dependence, bases, and dimension. These concepts are given geometrical interpretations whenever possible. The dot product, norms, angles, and distances follow. Numerous proofs have been included in this chapter. Instructors desiring less rigor can choose to omit any of them. The function space section is optional. It can be presented as an example of another kind of vector space and can be used to bridge the gap into more advanced linear algebra courses for many readers. The optional section on special relativity is self-contained, is of broad appeal, and represents in my opinion an important aspect of mathematics—mathematics as developed, modified, and molded by human beings to suit their needs, mathematics as their servant! Further, this section illustrates how mathematical tools can lead to intellectual insights of the deepest kind.

Chapter 4. Mappings are introduced geometrically. The concepts of kernels and ranges are discussed. The use of these tools leads to insights concerning the behavior of systems of linear equations. It is shown, for example, how the set of solutions to a system of nonhomogeneous equations can be decomposed into a particular solution and the set of solutions to the corresponding homogeneous system. This decomposition leads to the geometrical interpretation of the solutions. Its significance to differential equations is discussed.

Chapter 5. I have introduced determinants and their properties as quickly and as painlessly as I can in the first two sections. The numerical technique of evaluating a determinant is then covered. An optional section on vector products follows.

Chapter 6.   This chapter brings together many of the topics introduced in previous sections.   The reader by this time fully realizes the importance of linear systems of equations.   In this chapter he gains deeper insight into their behavior.   Section 6-1 introduces the concept of rank.   In Section 6-2 the existence and uniqueness of solutions to systems of linear equations are discussed in terms of the ranks of the matrix of coefficients and the augmented matrix.   In Section 6-3 the tie-in is made between determinant and rank, leading to the use of the determinant in the analyses of square systems of equations.   The reader is thus prepared for Section 6-5 on eigenvalues and eigenvectors.   He or she is instructed in the role of eigenvalues and eigenvectors in similarity transformations.   Two optional sections on the applications of eigenvalues and eigenvectors are given.   Section 6-6 illustrates their importance in the social sciences.   Section 6-10 illustrates their role in the physical sciences; here the reader also sees the linkup between linear algebra and differential equations.

Chapter 7.   This chapter gives a fairly in-depth example of the application of linear algebra.   Readers who desire just a brief introduction to linear programming can get it in the first two sections.   Those who want to be able to use the simplex method should include Section 7-3 and those who desire to know why the simplex method works should go on to Section 7-4.

Appendix.   A brief introduction to BASIC is included in Appendix B for those who have no previous experience with this language.   It covers essentials and can easily be covered in two 50-minute class periods.   The computer can be used in either one of two ways: (1) Simultaneously with the main text (programs are related to sections in the main text and computer exercises are included in the main text); or (2) A block of classes can be devoted to the computing aspect at the end of the course.

Vectors and Matrices

INTRODUCTION

Mathematics is, of course, a discipline in its own right. It is, however, more than that—it is also a tool used in many other fields. Linear algebra is a branch of mathematics that plays a central role in modern mathematics and is also of increasing importance to engineers, and physical, social and behavioral scientists. In this course the reader will learn mathematics and also be instructed in the art of applying mathematics; the course is a blend of theory, numerical techniques, and applications. He will develop mathematical models, mathematical "pictures" of certain aspects of situations, and see how such descriptions lead to a deeper understanding of the situation. Actual applications, from the aerospace industry to archaeology, will be discussed. The aim of the course is to develop an overall competence in linear algebra as well as an appreciation of the role this subject plays in many other fields.

1–1. INTRODUCTION TO VECTORS

The relative locations of points in a plane are usually discussed in terms of a coordinate system. For example, in Figure 1-1 the location of each point in a plane can be described relative to a rectangular coordinate system. The point A in Figure 1-1 is the point $(2, 1)$.

Furthermore, A is a certain distance in a certain direction from $(0, 0)$; the distance and direction are characterized by the length and direction of the line segment OA. We call such a line segment a *position*

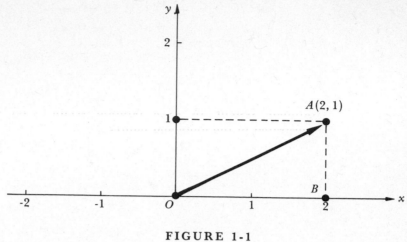

FIGURE 1-1

vector and indicate it by means of an arrow, as shown. There are thus two ways of interpreting $(2, 1)$; it defines the location of point A in the plane, and it also defines the position vector OA.

Example 1

In Figure 1-2 we diagram the position vectors $(0, 4)$, $(-2, 3)$, and $(4, 4)$, labeling them **u**, **v**, and **w**, respectively.

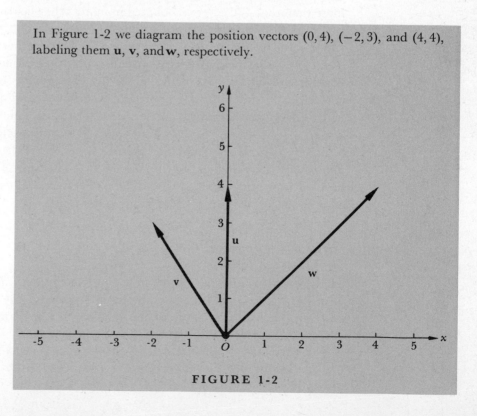

FIGURE 1-2

2

Denote the collection of all ordered pairs of real numbers by $\mathbf{R}^2$. Note the significance of "ordered" here; $(2, 1)$ is not the same vector as $(1, 2)$. The order is significant.

These concepts can be extended to apply to arrays consisting of three real numbers, such as $(1, 2, 3)$. $(1, 2, 3)$ can be interpreted in two ways—as the location of a point in three-space relative to an *xyz* coordinate system, or as a position vector. These interpretations are illustrated in the following figures.

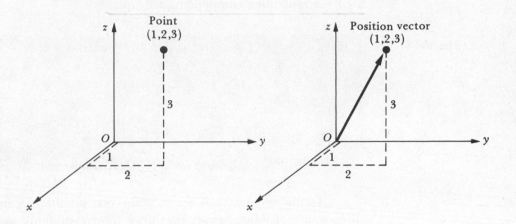

We shall denote the set of all ordered triples of real numbers by $\mathbf{R}^3$.

We now generalize these concepts mathematically. $\mathbf{R}^n$ is defined to be the set of all ordered collections of n real numbers. Thus $(1, 2, ..., n)$ would be an element of $\mathbf{R}^n$. We shall call such an element a *vector*. The numbers that make up the vector are called its *components*. Many of the results and techniques that we develop for $\mathbf{R}^n$ with $n > 3$ will be useful mathematical tools without direct geometrical significance. The elements of $\mathbf{R}^n$ can, however, be interpreted as locations of points in n-space, or as position vectors in n-space. It is difficult to visualize an n-space for $n > 3$, but the reader is encouraged to try to form an intuitive picture. A geometrical "feel" for what is taking place often makes an algebraic discussion easier to follow. The mathematical structures that we shall place on $\mathbf{R}^n$ will be motivated by structures on $\mathbf{R}^2$ and $\mathbf{R}^3$ which have direct geometrical interpretation.

In keeping with this scheme, let us denote the set of real numbers by $\mathbf{R}$.

3

Example 2

$\mathbf{R}^4$ is the collection of all sets of four ordered real numbers.   For example, $(1, 2, 3, 4)$ and $(-1, 0, \frac{1}{2}, 3)$ are elements of $\mathbf{R}^4$.

$\mathbf{R}^5$ is the collection of all sets of five ordered real numbers. $(-1, 0, 3, 4, 2)$ and $(\frac{1}{2}, \frac{1}{2}, 2, 0, 3)$ are in this set.

We complete this section by defining a concept of *equality* for vectors. Two vectors $(x_1, x_2, ..., x_n)$ and $(y_1, y_2, ..., y_n)$ are said to be equal if and only if $x_1 = y_1, x_2 = y_2, ..., x_n = y_n$, that is, if and only if their corresponding components are equal.

Example 3

If the vector $(a+2, a-b)$ is equal to $(3, 4)$, determine a and b.
We know that

$$(a+2, a-b) = (3, 4)$$

This implies that

$$a + 2 = 3 \quad \text{and} \quad a - b = 4$$

Thus $a = 1$ and $b = -3$.

In this text we shall use only the real number system. However, other number systems exist; the other most commonly used one is the complex number system. Most of the theory developed in this text is also appropriate for complex numbers. The term *scalar*, which we shall use, is often used by mathematicians for a number when they do not want to commit themselves to a particular number system.

EXERCISES

1. Diagram the position vectors $(1, 0)$ and $(0, 1)$. Label these vectors $\mathbf{i}$ and $\mathbf{j}$. This notation is often used for these vectors.

2. Diagram the position vectors $(1, 0, 0)$, $(0, 1, 0)$, and $(0, 0, 1)$. Label these vectors $\mathbf{i}$, $\mathbf{j}$, and $\mathbf{k}$.

3. Diagram the position vectors $(5, 6)$, $(-3, -2)$, and $(\frac{1}{2}, -1)$. Label these vectors $\mathbf{u}$, $\mathbf{v}$, and $\mathbf{w}$.

4. Diagram the position vectors $(2, 3, 1)$, $(0, 5, -1)$, and $(-\frac{3}{2}, 2, 4)$. Label these vectors $\mathbf{u}$, $\mathbf{v}$, and $\mathbf{w}$.

5. $\mathbf{R}$ has been defined to be the set of real numbers. Illustrate how the elements of $\mathbf{R}$ can be interpreted as points or position vectors in one-space.

6. The vector $(a+b, b)$ is equal to $(2, 1)$. Determine a and b.

7. Solve the vector equation $(a+c, b, b+c) = (1, -1, 2)$ for a, b, and c.

8. Prove that there are many values of a and b and a unique c that satisfy the vector equation $(a+b, 1, c) = (3, 1, 4)$.

9. Prove that there are no solutions to the vector equation $(a-b, a, b+c, c) = (3, 2, 1, -1)$.

10. Solve, if possible, the following vector equations:

   a) $(a^2, b, c^2) = (4, 3, 1)$ **b)** $(a+b+c, b, b+c) = (4, 1, 3)$

   c) $(a+c, b+c) = (2, 4)$

1–2. ADDITION AND SCALAR MULTIPLICATION OF VECTORS

In this section we add further structure to $\mathbf{R}^n$. We define operations of addition and scalar multiplication. Let $(x_1 \ldots x_n)$, $(y_1 \ldots y_n)$ be arbitrary elements of $\mathbf{R}^n$, and let a be an arbitrary scalar. Define

Addition: $(x_1, \ldots, x_n) + (y_1, \ldots, y_n) = (x_1 + y_1, \ldots, x_n + y_n)$

Scalar Multiplication: $a(x_1, \ldots, x_n) = (ax_1, \ldots, ax_n)$.

To add two vectors in $\mathbf{R}^n$ we add corresponding components, and to multiply a vector by a scalar we multiply every component by that scalar; the resulting vectors are in $\mathbf{R}^n$.

$\mathbf{R}^n$, with these two operations, addition and scalar multiplication, is called a *vector space*. We shall interpret $\mathbf{R}^n$ to be a vector space.

Example 1

$(-1, 2, 3, 4)$ and $(0, -3, 2, 3)$ are elements of $\mathbf{R}^4$. Adding these vectors,

$$(-1, 2, 3, 4) + (0, -3, 2, 3) = (-1, -1, 5, 7).$$

To illustrate scalar multiplication we multiply $(-1, 2, 3, 4)$ by 4.

$$4(-1, 2, 3, 4) = (-4, 8, 12, 16).$$

Observe that the resulting vector under each operation is in the original vector space, $\mathbf{R}^4$ here.

These operations lead to rich mathematical developments and are also useful in applications. We now give examples to illustrate geometrical interpretations of these operations and some applications.

Example 2

This example illustrates a geometrical interpretation of vector addition. Consider the two vectors $(3, 1)$ and $(2, 5)$. On adding,

$$(3, 1) + (2, 5) = (5, 6).$$

Let us interpret the vectors as position vectors:

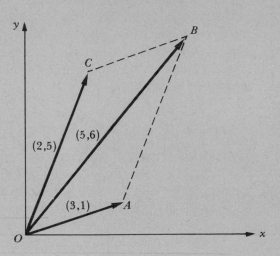

Construct the parallelogram $OBAC$ having the vectors $(3, 1)$ and $(2, 5)$ as adjacent sides.   Then the vector $(5, 6)$, the sum, will be the diagonal of this parallelogram.

In general, if **u** and **v** are vectors in the same vector space then we can geometrically interpret **u** + **v** as the diagonal of the parallelogram defined by **u** and **v**:

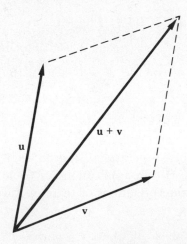

This way of visualizing vector addition is useful in all vector spaces.

Example 3

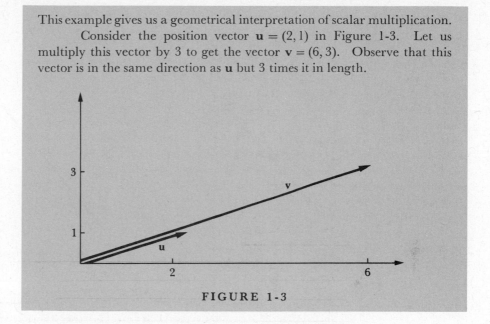

This example gives us a geometrical interpretation of scalar multiplication.

Consider the position vector $\mathbf{u} = (2, 1)$ in Figure 1-3. Let us multiply this vector by 3 to get the vector $\mathbf{v} = (6, 3)$. Observe that this vector is in the same direction as $\mathbf{u}$ but 3 times it in length.

FIGURE 1-3

In general, if $\mathbf{u}$ is an arbitrary vector and a a scalar, there are the following geometrical interpretations for $a\mathbf{u}$, depending on a:

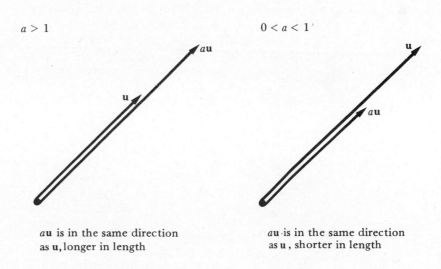

$a > 1$

$0 < a < 1$

$a\mathbf{u}$ is in the same direction as $\mathbf{u}$, longer in length

$a\mathbf{u}$ is in the same direction as $\mathbf{u}$, shorter in length

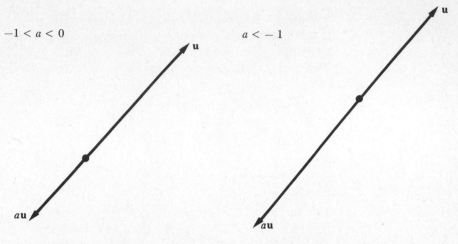

$-1 < a < 0$

$a < -1$

*a*u is in the opposite direction
to u, shorter in length

*a*u is in the opposite direction
to u, longer in length

Example 4*

Physical quantities such as velocities and forces that have both magnitude and direction are called vectors.   They can be represented mathematically by elements of $\mathbf{R}^2$ or $\mathbf{R}^3$ as this example illustrates.

A force of two pounds tends to pull a body in a direction $N\,60°\,E$. Represent this force by an element of $\mathbf{R}^2$.   Let the body be at O with axes OE and ON representing easterly and northerly directions from O, respectively (Figure 1-4).   To obtain the direction $N\,60°\,E$ one turns 60° in an easterly direction from N.   This gives the direction OA.   The length of the vector OA is 2, this represents the magnitude of the force.   The lengths of OB and AB will then be $\sqrt{3}$ and 1.   Hence the force is represented by the vector $OA\left(\sqrt{3},1\right)$, an element of $\mathbf{R}^2$.

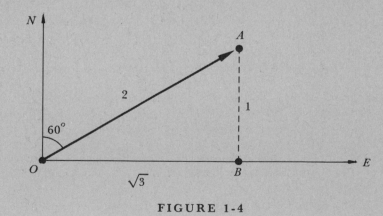

FIGURE 1-4

* Starred examples are optional.

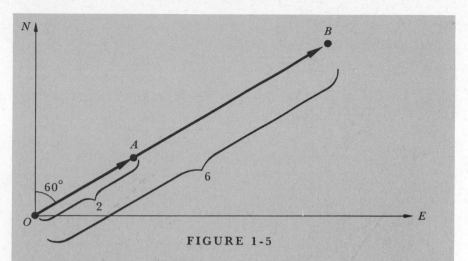

FIGURE 1-5

Using this same example, we can see how multiplication of a vector by a scalar has physical interpretation. The vector $3(\sqrt{3,1})$, that is, the vector $(3\sqrt{3,3})$, represents a force three times the magnitude of OA, acting in the same direction as OA. It is thus a force of six pounds acting in the same direction as OA (Figure 1-5).

Example 5* We now look at a physical interpretation of vector addition. Two forces acting simultaneously on a body are equivalent to a single force, called the *resultant force*. Experiments show that if two such forces are represented by vectors in $\mathbf{R^2}$ or $\mathbf{R^3}$ the resultant force is represented by the sum of these vectors. This law is called the *Parallelogram of Forces*.

Let $(1,3)$ and $(2,1)$ represent two forces acting on a body situated at O below. On adding $(1,3)$ and $(2,1)$ we find that the resultant force is $(3,4)$.

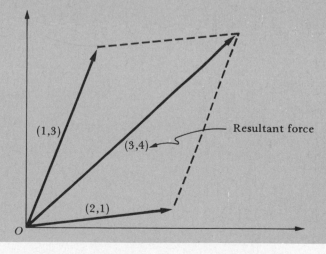

From our previous discussion we know that the resultant force will be defined by the diagonal of the parallelogram having the given vectors as adjacent sides—hence the term parallelogram of forces.

We now give a practical example of addition and scalar multiplication of the elements of $\mathbf{R}^4$.

Example 6

Consider the following system of equations.

$$x - 2y + 3z + w = 0$$
$$2x - 3y + 2z - w = 0$$
$$3x - 5y + 5z = 0$$
$$x - y - z - 2w = 0$$

There are four equations in four variables, x, y, z, and w. There will be certain values of x, y, z, and w that satisfy all of these equations simultaneously. Any such set is called a *solution* to the system. We can interpret a solution, such as x_1, y_1, z_1, w_1, as a vector (x_1, y_1, z_1, w_1), an element of $\mathbf{R}^4$. For example, $x_1 = 10$, $y_1 = 7$, $z_1 = 1$, $w_1 = 1$ is a solution that can be written as the vector $(10, 7, 1, 1)$. (Substitute these values into the equations to see that every equation is satisfied. The reader will see how this solution is arrived at in the following chapter.)

Another solution is $(0, 1, 1, -1)$. Such a set of equations may have only a single solution or many solutions. Note that $(0, 0, 0, 0)$, called the *zero vector*, will always be a solution to such a system.

Let (x_1, y_1, z_1, w_1) and (x_2, y_2, z_2, w_2) represent two arbitrary solutions. Thus,

$$x_1 - 2y_1 + 3z_1 + w_1 = 0$$
$$2x_1 - 3y_1 + 2z_1 - w_1 = 0$$
$$3x_1 - 5y_1 + 5z_1 = 0$$
$$x_1 - y_1 - z_1 - 2w_1 = 0$$

and

$$x_2 - 2y_2 + 3z_2 + w_2 = 0$$
$$2x_2 - 3y_2 + 2z_2 - w_2 = 0$$
$$3x_2 - 5y_2 + 5z_2 = 0$$
$$x_2 - y_2 - z_2 - 2w_2 = 0$$

Adding the two first equations, the two second equations, etc., we get

$$(x_1 + x_2) - 2(y_1 + y_2) + 3(z_1 + z_2) + (w_1 + w_2) = 0$$

$$2(x_1 + x_2) - 3(y_1 + y_2) + 2(z_1 + z_2) - (w_1 + w_2) = 0$$

$$3(x_1 + x_2) - 5(y_1 + y_2) + 5(z_1 + z_2) \qquad\qquad = 0$$

$$(x_1 + x_2) - (y_1 + y_2) - (z_1 + z_2) - 2(w_1 + w_2) = 0$$

implying that $(x_1 + x_2, y_1 + y_2, z_1 + z_2, w_1 + w_2)$ also represents a solution; that is, $(x_1, y_1, z_1, w_1) + (x_2, y_2, z_2, w_2)$, the vector sum of the solutions, is a solution.

In our example, $(10, 7, 1, 1)$ and $(0, 1, 1, -1)$ were solutions. According to our theory, $(10, 7, 1, 1) + (0, 1, 1, -1)$, or $(10, 8, 2, 0)$, should also be a solution. Substitute $x = 10$, $y = 8$, $z = 2$, and $w = 0$ into the equations to see that they are satisfied, proving that this is a solution.

If c is any scalar, then $c(x_1, y_1, z_1, w_1)$ will also be a solution. (The proof is left as an exercise.)

We call a system of equations with zeros to the right of each equal sign a system of *homogeneous equations*. We have seen that if any two solutions are written in vector form, their vector sum is a solution, and so also is a scalar multiple of any vector solution. New solutions can be generated in this manner from known solutions.

In the following chapter we discuss solutions to systems of linear equations in detail. We shall see the importance of these systems of equations. The mathematical analysis of many situations reduces to examining such systems of equations.

Up to this point we have defined only *row vectors*; that is, the components of a vector were written in row form. We shall find that it is more suitable sometimes to use vectors in the form of columns, both in theoretical discussions and in applications. We define addition and scalar multiplication in $\mathbf{R}^n$, when the elements are written in column form, as

$$\begin{pmatrix} x_1 \\ \vdots \\ x_n \end{pmatrix} + \begin{pmatrix} y_1 \\ \vdots \\ y_n \end{pmatrix} = \begin{pmatrix} x_1 + y_1 \\ \vdots \\ x_n + y_n \end{pmatrix} \quad \text{and} \quad a\begin{pmatrix} x_1 \\ \vdots \\ x_n \end{pmatrix} = \begin{pmatrix} ax_1 \\ \vdots \\ ax_n \end{pmatrix}.$$

For example in $\mathbf{R}^2$,

$$\begin{pmatrix} 1 \\ 2 \end{pmatrix} + \begin{pmatrix} 0 \\ 1 \end{pmatrix} = \begin{pmatrix} 1 \\ 3 \end{pmatrix} \quad \text{and} \quad 4\begin{pmatrix} 1 \\ 2 \end{pmatrix} = \begin{pmatrix} 4 \\ 8 \end{pmatrix}.$$

It is appropriate to use column vectors in the following applications.

Data is often analyzed on computers using concepts of linear algebra. Computers can be programmed to handle vectors and to perform operations of addition and scalar multiplication. The following examples illustrate how certain analyses can be set up in terms of vectors.

Example 7* A class of ten students has had five tests during the quarter. A perfect score on each of the tests is 50. The scores are listed in Table 1-1.

Table 1-1

	Test 1	Test 2	Test 3	Test 4	Test 5
Anderson	40	45	30	48	42
Boggs	20	15	30	25	10
Chittar	40	35	25	45	46
Diessner	25	40	45	40	38
Farnam	35	35	38	37	39
Gill	50	46	45	48	47
Homes	22	24	30	32	29
Johnson	35	27	20	41	30
Schomer	28	31	25	27	31
Wong	40	35	36	32	38

We can express these scores as column vectors

$$
\begin{pmatrix} 40 \\ 20 \\ 40 \\ 25 \\ 35 \\ 50 \\ 22 \\ 35 \\ 28 \\ 40 \end{pmatrix}
\begin{pmatrix} 45 \\ 15 \\ 35 \\ 40 \\ 35 \\ 46 \\ 24 \\ 27 \\ 31 \\ 35 \end{pmatrix}
\begin{pmatrix} 30 \\ 30 \\ 25 \\ 45 \\ 38 \\ 45 \\ 30 \\ 20 \\ 25 \\ 36 \end{pmatrix}
\begin{pmatrix} 48 \\ 25 \\ 45 \\ 40 \\ 37 \\ 48 \\ 32 \\ 41 \\ 27 \\ 32 \end{pmatrix}
\begin{pmatrix} 42 \\ 10 \\ 46 \\ 38 \\ 39 \\ 47 \\ 29 \\ 30 \\ 31 \\ 38 \end{pmatrix}
$$

where each vector is an element of $\mathbf{R}^{10}$. To obtain each person's average we use vector addition to add the vectors, and then scalar multiplication to multiply by $\frac{1}{5}$ (dividing by the number of tests). We get

$$
\frac{1}{5}
\begin{pmatrix} 205 \\ 100 \\ 191 \\ 188 \\ 184 \\ 236 \\ 137 \\ 153 \\ 142 \\ 181 \end{pmatrix}
=
\begin{pmatrix} 41 \\ 20 \\ 38.2 \\ 37.6 \\ 36.8 \\ 47.2 \\ 27.4 \\ 30.6 \\ 28.4 \\ 36.2 \end{pmatrix}
\qquad
\text{\textit{Column vector giving each}}
$$
person's average score

Row vectors are also useful; a person's complete set of scores corresponds to a row vector. For example,

$$(25, 40, 45, 40, 38)$$

is a row vector giving Diessner's scores.

To determine the average score on each test, we add all the row vectors using vector addition and then multiply this vector by $\frac{1}{10}$ using scalar multiplication. This gives

$$\tfrac{1}{10}(335, 333, 324, 375, 350)$$

$$= (33.5, 33.3, 32.4, 37.5, 35) \quad \textit{Row vector giving average}$$
$$\textit{score on each test}$$

Example 8*   Consider Table 1-2, a mileage chart between certain cities in the state of Florida.

Table 1-2

	1	2	3	4	5	6	7	8	9	10	11
1. Clearwater	0	160	256	132	200	399	271	106	430	235	228
2. Daytona Beach	160	0	232	97	91	416	258	55	436	237	192
3. Fort Lauderdale	256	232	0	321	323	183	25	214	650	455	40
4. Gainesville	132	97	321	0	68	476	336	109	339	143	283
5. Jacksonville	200	91	323	68	0	505	348	136	368	168	283
6. Key West	399	416	183	476	505	0	164	378	805	609	223
7. Miami Beach	271	258	25	336	348	164	0	238	665	469	65
8. Orlando	106	55	214	109	136	378	238	0	438	243	174
9. Pensacola	430	436	650	339	368	805	665	438	0	199	612
10. Tallahassee	235	237	455	143	168	609	469	243	199	0	417
11. West Palm Beach	228	192	40	283	283	223	65	174	612	417	0
				↑						↑	
				a						**b**	

Each column can be interpreted as a vector with eleven components. Note that, for mathematical reasons, we have put zeros into the locations that give the mileages from the places to themselves. These are usually left blank.

The University of Florida is at Gainesville. To find the distances from Gainesville to various cities look up the column vector **a**. Florida State University is at Tallahassee. We can examine column vector **b** to find the distances from Tallahassee to various cities.   To find the round-trip mileage from Gainesville to various cities, one would scalar multiply the column vector **a** by 2.

Subtract vector **a** from vector **b** to get column vector **c**.

Clearwater	103
Daytona Beach	140
Fort Lauderdale	134
Gainesville	143
Jacksonville	100
Key West	133
Miami Beach	133
Orlando	134
Pensacola	−140
Tallahassee	−143
West Palm Beach	134

column vector **c** = **b** − **a**

This column vector gives an indication of the relative distances of Gainesville and Tallahassee from the various cities.   A positive component x indicates that Gainesville is x miles nearer than Tallahassee.   A negative component y indicates that Gainesville is y miles further than Tallahassee. Thus, Gainesville is 134 miles nearer to Fort Lauderdale than Tallahassee is, but it is 140 miles further from Pensacola.   Can you deduce the distance between Gainesville and Tallahassee from this column vector in two distinct ways?

Because of the preponderance of positive components, one could conclude from column vector **c** that Gainesville was more centrally situated in the state than Tallahassee.   If a salesman for the state of Florida had to visit each of these cities in a certain period, returning to his place of residence each time, then mileage-wise he would be better off living in Gainesville than in Tallahassee.   We can actually be more specific; we can calculate the extra miles he would have to travel if he lived in Tallahassee.   The vector 2**c**, a scalar multiple of **c**, gives the round-trip differences in miles from the various cities to Gainesville and to Tallahassee. The sum of the components of 2**c**, namely 1,742, is the extra number of miles the salesman would have to travel in this period if he lived in Tallahassee.

This type of problem can become much more sophisticated when there are many cities involved and certain cities are covered en route to others.   Such situations have been studied extensively and good algorithms for their analysis have been devised.   This class of problems includes a famous traveling salesman problem: A traveling salesman has a girlfriend in the capital city of each of the 50 states.   He feels it his duty to arrange his tour in order to be able to visit each one of his girlfriends.   Find the shortest route.   One branch of mathematics that deals with such problems is *graph theory*.

EXERCISES

1. Multiply the following vectors by the given scalars. Interpret the results for **a)**–**e)** geometrically.

   a) $(1, 4)$ by 3.   **b)** $(-1, 3)$ by -2.

   c) $(2, 6)$ by $\frac{1}{2}$.   **d)** $(2, 4, 2)$ by $-\frac{1}{2}$.

   e) $(-1, 2, 3)$ by 3.   **f)** $(-1, 2, 3, -2)$ by 4.

   g) $(1, -4, 3, -2, 5)$ by -5.

2. $(1, 2)$ and $(3, 4)$ are elements of $\mathbf{R}^2$. Add these elements and interpret the addition geometrically.

3. $(-1, 3)$ and $(-1, -2)$ are elements of $\mathbf{R}^2$. Add these elements and interpret the addition geometrically. Prove that

   $$(-1, 3) + (-1, -2) = (-1, -2) + (-1, 3);$$

   that is, prove that it does not matter in which order the vectors are added.

4. Let (x_1, x_2) and (y_1, y_2) be arbitrary elements of $\mathbf{R}^2$. Prove that

   $$(x_1, x_2) + (y_1, y_2) = (y_1, y_2) + (x_1, x_2).$$

5. $(1, 2, 4)$ and $(2, 3, 1)$ are elements of $\mathbf{R}^3$. Determine their sum. Prove that $(2, 3, 1) + (1, 2, 4)$ leads to the same element of $\mathbf{R}^3$.

6. $(1, 1, -1, 0, 2)$ and $(4, 1, 5, -1, 3)$ are elements of $\mathbf{R}^5$. Add these vectors.

7. $(2, 1, 4, 1, 5, -1)$ and $(3, 1, 5, 6, -1, 4)$ are elements of $\mathbf{R}^6$. Add these vectors.

8. Let $(x_1, x_2, \ldots, x_n)$ and $(y_1, y_2, \ldots, y_n)$ be arbitrary vectors in $\mathbf{R}^n$. Prove that

   $$(x_1, x_2, \ldots, x_n) + (y_1, y_2, \ldots, y_n) = (y_1, y_2, \ldots, y_n) + (x_1, x_2, \ldots, x_n);$$

   that is, prove that the order in which they are added is immaterial. We shall see later that this is not a trivial property of vector addition on $\mathbf{R}^n$. We call this property a *commutative property*.

9. $(1, 4)$, $(0, 1)$, and $(5, 6)$ are elements of $\mathbf{R}^2$. Add these elements in a meaningful manner.

10. $(3, 1)$, $(-1, 2)$, and $(2, 3)$ are elements of $\mathbf{R}^2$. Prove that it does not matter in which order these vectors are grouped for adding, in the sense that $[(3, 1) + (-1, 2)] + (2, 3) = (3, 1) + [(-1, 2) + (2, 3)]$.

11. Motivated by the previous question, give a rule for adding more than two elements of $\mathbf{R}^n$. Prove that the addition may be performed by grouping the elements together in any order, as in the previous question. We call this property an *associative property*. The elements of $\mathbf{R}^n$ are said to be associative under vector addition. Thus, if $\mathbf{v}_1, \ldots, \mathbf{v}_m$ are m elements of $\mathbf{R}^n$, they have a unique sum $\mathbf{v}_1 + \cdots + \mathbf{v}_m$; we do not have to include brackets in this expression to indicate the order of addition.

12. Show that $(2, 3) = (1, 2) + (3, 3) - \frac{1}{2}(4, 4)$. We say that $(2, 3)$ is a *linear*

combination of the vectors $(1, 2)$, $(3, 3)$, and $(4, 4)$. We discuss this concept in greater depth in Chapter 3.

13. Show that $(-1, 1, 3) = 2(1, 0, 2) + (4, 3, 1) - (7, 2, 2)$. $(-1, 1, 3)$ is a linear combination of $(1, 0, 2)$, $(4, 3, 1)$, and $(7, 2, 2)$.

14. Show that $(-1, 2, 3) + 3(1, 2, -1) - (2, 8, 0) = (0, 0, 0)$. The vector $(0, 0, 0)$ is called the *zero vector* of $\mathbf{R}^3$.

15. Consider the following system of homogeneous equations.

$$x + 2y - z - 2w = 0$$
$$2x + 5y \quad\;\; - 2w = 0$$
$$4x + 9y - 2z - 6w = 0$$
$$x + 3y + z \quad\quad = 0$$

You are given two solutions, $(5, -2, 1, 0)$ and $(11, -4, 1, 1)$. Using the operations of $\mathbf{R}^4$, generate five other solutions.

16. $(0, -1, 1, -1)$ and $(3, 7, 2, 1)$ are solutions of the system

$$x - y + z + 2w = 0$$
$$3x - 2y + z + 3w = 0$$
$$5x - 4y + 3z + 7w = 0$$
$$2x - y \quad\;\; + w = 0$$

Use the operations of $\mathbf{R}^4$ to generate five further solutions.

17. If $\mathbf{a}$, $\mathbf{b}$, and $\mathbf{c}$ are the column vectors $\begin{pmatrix} 1 \\ 2 \\ -1 \end{pmatrix}$, $\begin{pmatrix} 3 \\ 0 \\ 1 \end{pmatrix}$, and $\begin{pmatrix} -1 \\ 0 \\ 5 \end{pmatrix}$, respectively, all elements of $\mathbf{R}^3$, determine

    a) $\mathbf{a} + \mathbf{b}$ **b)** $\mathbf{a} + \mathbf{b} - \mathbf{c}$

    c) $3\mathbf{a}$ **d)** $2\mathbf{a} + \mathbf{b} - 3\mathbf{c}$

18. **a)** If $(2, -1) + (a, b) = (3, 4)$, determine the vector (a, b).

    b) If $(1, 0, -3) + (b, q, r) + (2, 1, -3) = (1, 5, -1)$, determine the vector (b, q, r).

    c) If $\mathbf{a}$, $\mathbf{b}$, and $\mathbf{x}$ are all elements of $\mathbf{R}^n$ and $\mathbf{a} + \mathbf{x} = \mathbf{b}$, prove that $\mathbf{x} = \mathbf{b} - \mathbf{a}$.

19. Which of the following physical quantities are scalars and which are vectors?

    a) temperature   **b)** acceleration   **c)** pressure

    d) frequency   **e)** gravity   **f)** position

    g) time   **h)** sound   **i)** cost

20. Determine the resultant of the following forces:

    a) a force of two pounds acting in a northerly direction, and a force of three pounds in a westerly direction.

    b) forces of two pounds in the direction N, two pounds in the direction $N\,60°\,E$, and three pounds S. (The resultant is given by vector addition.)

21. The following forces act on a body simultaneously: four pounds pulling it *N*, two pounds pulling it *W*, and two pounds pulling it *N* 120° *E*. Represent these forces graphically as elements of $\mathbf{R}^2$. Using vector addition, determine the resultant force.

22. In Example 7 of this section:
    a) Suppose the final grade corresponds to a comprehensive examination and is worth two tests. Using vector operations, work out the average grade of each student.
    b) Using vector operations, express the final grades as percentages.

23. Using Table 1-2 of Example 8:
    a) Determine the column vector that gives the round-trip mileages from Orlando to various cities.
    b) A salesman who has to visit each of these cities in turn and return to his place of residence after each visit has a choice of living in Daytona Beach or in Orlando. From the mileage viewpoint, where should he live?

1–3. INTRODUCTION TO MATRICES

In the previous sections we developed the vector space $\mathbf{R}^n$. The vectors were made up of real numbers and could be written in either row or column form. In this section we extend these concepts to rectangular arrays called matrices.

A *matrix* is a rectangular array of scalars. Examples of matrices, in standard notation are

$$\begin{pmatrix} 1 & 2 & 3 \\ 0 & -1 & 1 \end{pmatrix}, \begin{pmatrix} 2 & 3 \\ 1 & 1 \\ 4 & 1 \end{pmatrix}, \begin{pmatrix} 1 & 2 & 3 \\ 4 & 5 & 6 \\ 0 & 1 & 2 \end{pmatrix}.$$

Each matrix has a certain number of rows and a certain number of columns. For example, the matrix

$$\begin{pmatrix} 1 & 2 & 3 \\ 0 & -1 & 1 \end{pmatrix}$$

has 2 rows and 3 columns. We call such a matrix a 2×3 matrix; the first number indicates the number of rows in the matrix, the second indicates the number of columns. When the number of columns and rows are equal the matrix is said to be a *square matrix*.

$$\begin{pmatrix} 2 & 3 \\ 1 & 1 \\ 4 & 1 \end{pmatrix} \text{ is a } 3 \times 2 \text{ matrix.}$$

$$\begin{pmatrix} 1 & 2 & 3 \\ 4 & 5 & 6 \\ 0 & 1 & 2 \end{pmatrix}$$ is a 3×3 matrix, a square matrix.

The array $(\ 1 \quad 2 \quad 3 \quad 4\ )$ is a row vector; it can also be interpreted as a 1×4 matrix. Every vector can be interpreted as a matrix. The vector

$$\begin{pmatrix} 1 \\ 2 \\ 3 \\ 0 \end{pmatrix}$$

is a 4×1 matrix.

Matrices are used in many fields. The following examples illustrate two areas.

Example 1 *

Highway departments use origin-destination data in mathematical models to analyze traffic patterns. These data are based on door-to-door questionnaires and are drawn up in the form of origin-destination matrices (O–D matrices). The region of interest is divided into zones and an estimate of the daily traffic between zones is made. An example of such a matrix for passenger cars in the city of Kyoto, Japan for 1962 is given below.† The city is divided into nine zones.

	1	2	3	4	5	6	7	8	9
1. Kita-Ku	2,119	2,113	1,429	2,813	309	1,772	439	810	80
2. Kamigyo-Ku	1,887	4,708	2,770	4,441	1,220	3,931	375	827	207
3. Sakyo-Ku	1,334	2,435	6,975	3,967	3,051	3,423	110	212	22
4. Nakagyo-Ku	2,807	4,936	3,935	11,822	4,446	10,864	1,473	1,374	768
5. Higashiyama-Ku	450	1,257	2,709	4,601	8,230	4,987	809	395	568
6. Shimogyo-Ku	1,724	3,653	2,937	10,259	5,441	16,480	3,224	1,754	945
7. Minami-Ku	307	294	310	1,379	681	3,499	2,930	227	870
8. Ukyo-Ku	905	736	330	1,955	169	1,391	411	2,173	22
9. Fushimi-Ku	118	297	222	542	616	922	780	73	2,234

To interpret the matrix, one reads the origin of interest in the row and the destination of interest in the column. The element lying in the relevant row and column represents the total daily car traffic from the origin to that destination. For example, according to the matrix, the daily car traffic from Sakyo-Ku to Ukyo-Ku is 212. Between Shimogyo-Ku and Kamigyo-Ku it is 3,653.

Such matrices are used in a mathematical model, called the *gravity model*, by highway departments to analyze traffic patterns and predict

---

† From E. Kometani, "The Estimation of Origin-Destination Trips Using a Transition Matrix Method," in L. Edie, R. Herman, and R. Rothery, eds. *Vehicular Traffic Science* (New York: American Elsevier, 1967).

future highway needs. The city of Chicago has, for example, used the gravity model to analyze traffic flow.

Ideally, the highway division works with the local planning authority in applying such a model. Currently, a regional transit system is being planned for the Denver metropolitan area. It is to be completed by 1983 and will include a computerized rapid transit network, a regional bus service, and a local bus service. The planners of this system are working with the Denver Regional Council of Governments and the Colorado Division of Highways. The gravity model is being used to help determine the various routes operated by the transit system.†

(The gravity model takes its name from Newton's law of gravity, from which it was developed. Newton's law states that the force of attraction between two bodies is inversely proportional to the square of the distance between them. The gravity model assumes that the number of vehicles traveling between zones is inversely proportional to some power of the distance between the zones.)

Example 2*

The following type of matrix is used in the Leontief INPUT-OUTPUT model in economics. This model is used to analyze interdependence of industries.

Consider an economic situation involving three interdependent industries, each industry producing a single commodity. The output of any one industry is needed as input by the other industries, and possibly also by the industry itself. This interdependence can be described by a matrix such as

$$
\begin{array}{c}
 \\
1 \\
2 \\
3
\end{array}
\begin{array}{ccc}
1 & 2 & 3 \\
\left(\begin{array}{ccc}
0.25 & 0.40 & 0.50 \\
0.35 & 0.10 & 0.20 \\
0.20 & 0.30 & 0.10
\end{array} \right).
\end{array}
$$

We have labeled the industries as 1, 2 and 3. The interpretation of the elements of the matrix is as follows:

Consider the 0.20 in row 2, column 3 of this matrix (circled). The implication is that 20 cent's worth of commodity 2 is required to produce a dollar's worth of commodity 3. The 0.40 in row 1, column 2 implies that 40 cents' worth of commodity 1 is required to produce a dollar's worth of commodity 2, etc.

We further discuss the Leontief INPUT-OUTPUT model in Section 2-6.

† A text that gives a discussion of the gravity model is *New Perspectives in Urban Transportation Research*, edited by Anthony J. Catanese, Lexington Books, 1972.

We extend the definition of equality of $1 \times n$ matrices (vectors) to $m \times n$ matrices by saying that two $m \times n$ matrices are equal if and only if their corresponding elements are equal.

We next define operations of scalar multiplication and addition for matrices.

Scalar Multiplication   In the case of $1 \times n$ matrices, we multiplied the matrix by a scalar by multiplying every element in the matrix by that scalar. We extend this rule to apply to all matrices. Scalar multiplication is performed on a matrix by multiplying every element of the matrix by the scalar.

For example,

$$3 \begin{pmatrix} 1 & 2 & 3 \\ 0 & -1 & 2 \\ 4 & 1 & 0 \end{pmatrix} = \begin{pmatrix} 3 & 6 & 9 \\ 0 & -3 & 6 \\ 12 & 3 & 0 \end{pmatrix}$$

Addition   To define addition we again generalize the definition of vector addition. Note that one can add two vectors such as $(1, 2)$ and $(3, 4)$ to get $(4, 6)$, but one cannot add two vectors such as $(1, 2)$ and $(3, 4, 5)$. Vector addition is defined between vectors having the same number of components, resulting in a vector with that same number of components. This operation is extended to matrices having the same number of rows and the same number of columns. We call such matrices *matrices of the same kind*. Addition is performed on matrices of the same kind by adding corresponding elements, resulting in a matrix of that same kind.

For example,

$$\begin{pmatrix} 1 & 2 & 3 \\ 0 & 4 & 1 \\ 2 & 1 & 0 \end{pmatrix} + \begin{pmatrix} -1 & 0 & 1 \\ 0 & 1 & 1 \\ 2 & 3 & 0 \end{pmatrix} = \begin{pmatrix} 0 & 2 & 4 \\ 0 & 5 & 2 \\ 4 & 4 & 0 \end{pmatrix}$$

These are all 3×3 matrices.

However, we cannot add the matrices

$$\begin{pmatrix} 1 & 2 & 3 \\ 0 & 4 & 1 \\ 2 & 1 & 0 \end{pmatrix} \text{ and } \begin{pmatrix} 0 & 1 \\ 0 & -1 \\ 2 & 3 \end{pmatrix}.$$

The former is a 3×3 matrix, and the latter is a 3×2 matrix; they are not matrices of the same kind.

Subtraction is performed on matrices of the same kind by subtracting corresponding elements.

A convenient notation has been developed to handle matrices.

Let us denote a general $m \times n$ matrix A

$$\begin{pmatrix} a_{11} & \cdots & a_{1n} \\ & \vdots & \\ a_{m1} & \cdots & a_{mn} \end{pmatrix}$$

It has m rows and n columns. The subscripts of the elements in the matrix indicate the locations of those elements. For example, the element a_{11} lies in the first row and first column; the element a_{32} lies in the third row and second column.

Example 3

Consider the matrix A,

$$\begin{pmatrix} 1 & -2 & -1 & 3 \\ 3 & 4 & 2 & 1 \\ 5 & 1 & 0 & 6 \end{pmatrix}.$$

A is a 3×4 matrix. $a_{11} = 1, a_{13} = -1, a_{23} = 2, a_{34} = 6$, etc.

An arbitrary matrix A can be referred to in terms of an arbitrary element a_{ij}, the element in the ith row and jth column. If one knows what is happening to an arbitrary element of the matrix, one knows what is happening to every element. Thus we say that $A = (a_{ij})$. Scalar multiplication can be expressed

$$c(a_{ij}) = (ca_{ij})$$

One multiplies the matrix with arbitrary element a_{ij} by c to get the matrix with the arbitrary element ca_{ij}.

Let $A = (a_{ij})$ and $B = (b_{pq})$ be matrices of the same kind. The sum $A + B$ is defined to be the matrix obtained by adding corresponding elements. Using this notation we write

$$A + B = (a_{ij}) + (b_{pq})$$
$$= (a_{ij} + b_{ij})$$

In the first line we introduced different letters for the subscripts of a and b in order to indicate that these elements represent independent arbitrary elements of A and B. However, applying the rule of addition, $a_{ij} + b_{ij}$ represents an arbitrary element in $A + B$. Since it is obtained by adding elements in corresponding rows and columns of A and B, the subscripts of a and b have to be identical.

We illustrate the power of this notation in the proofs of the following two important algebraic results.

⁕<u>Theorem 1-1</u>   *If A and B are matrices of the same kind, then $A + B = B + A$. The order in which we add the matrices is immaterial. Matrices of the same kind are said to be* commutative under addition. *They have this property in common with real numbers.*

⁕*Proof:*   We shall prove this result first using the longhand notation, then using the compact notation.

Let

$$A = \begin{pmatrix} a_{11} & \cdots & a_{1n} \\ & \vdots & \\ a_{m1} & \cdots & a_{mn} \end{pmatrix} \quad \text{and} \quad B = \begin{pmatrix} b_{11} & \cdots & b_{1n} \\ & \vdots & \\ b_{m1} & \cdots & b_{mn} \end{pmatrix}.$$

Then

$$A + B = \begin{pmatrix} a_{11} & \cdots & a_{1n} \\ & \vdots & \\ a_{m1} & \cdots & a_{mn} \end{pmatrix} + \begin{pmatrix} b_{11} & \cdots & b_{1n} \\ & \vdots & \\ b_{m1} & \cdots & b_{mn} \end{pmatrix}$$

$$= \begin{pmatrix} a_{11} + b_{11} & \cdots & a_{1n} + b_{1n} \\ & \vdots & \\ a_{m1} + b_{m1} & \cdots & a_{mn} + b_{mn} \end{pmatrix}$$

$$= \begin{pmatrix} b_{11} + a_{11} & \cdots & b_{1n} + a_{1n} \\ & \vdots & \\ b_{m1} + a_{m1} & \cdots & b_{mn} + a_{mn} \end{pmatrix}, \quad \begin{array}{l} \text{since real numbers are} \\ \text{commutative under} \\ \text{addition,} \end{array}$$

$$= B + A$$

This proof is correct, but all the indices make it tedious. The reasoning used in this proof can be applied to arbitrary elements of A and B. Let $A = (a_{ij})$ and $B = (b_{pq})$. Then

$$A + B = (a_{ij}) + (b_{pq})$$

$$= (a_{ij} + b_{ij})$$

$$= (b_{ij} + a_{ij})$$

$$= (b_{pq}) + (a_{ij})$$

$$= B + A$$

⁕ <u>Theorem 1-2</u>   *If A, B and C are matrices of the same kind*

$$(A + B) + C = A + (B + C)$$

Either pair can be added first.

We say that matrices of the same kind are associative *under addition.*

WHEN ADDING — HAVE TO BE SAME CORRESPONDENTS

✱ *Proof:*   Let $A = (a_{ij})$, $B = (b_{pq})$ and $C = (c_{rs})$.   Then

$$(A + B) + C = (a_{ij} + b_{ij}) + (c_{rs}) = (a_{ij} + b_{ij} + c_{ij}).$$

$$A + (B + C) = (a_{ij}) + (b_{pq} + c_{pq}) = (a_{ij} + b_{ij} + c_{ij})$$

$$= (A + B) + C.$$

We denote this sum $A + B + C$.   Thus

$$A + B + C = (a_{ij} + b_{ij} + c_{ij}).$$

This result can be extended to add any finite number of matrices.   The sum is obtained by adding corresponding elements.

$$A + B + C + \cdots + Z = (a_{ij} + b_{ij} + c_{ij} + \cdots + z_{ij}).$$

The following examples illustrate the uses of matrices in correlating data.   We see the need for matrix operations.

Example 4

The following matrix represents total trade figures between certain countries for the five year period 1969–1973.   The numbers are values in million U.S. dollars.†   We use this matrix to illustrate a manner in which scalar multiplication of a matrix can arise.

$$\begin{array}{l} \\ \textit{From} \\ \text{Canada} \\ \text{E.E.C.‡} \\ \text{Japan} \\ \text{U.S.A.} \end{array} \quad \begin{array}{cccc} \textit{To}\ \text{Canada} & \text{E.E.C.} & \text{Japan} & \text{U.S.A.} \\ \left(\begin{array}{cccc} 0 & 12{,}600 & 4{,}890 & 63{,}690 \\ 9{,}230 & 0 & 8{,}280 & 56{,}510 \\ 4{,}020 & 13{,}230 & 0 & 37{,}170 \\ 54{,}750 & 59{,}690 & 25{,}190 & 0 \end{array}\right) = A. \end{array}$$

The annual average over this period is $\frac{1}{5}$ of the five year figure.   We can get the annual figures by scalar multiplying the above matrix by $\frac{1}{5}$.

Annual Figures

$$\begin{array}{l} \\ \textit{From} \\ \text{Canada} \\ \text{E.E.C.} \\ \text{Japan} \\ \text{U.S.A.} \end{array} \quad \begin{array}{cccc} \textit{To}\ \text{Canada} & \text{E.E.C.} & \text{Japan} & \text{U.S.A.} \\ \left(\begin{array}{cccc} 0 & 2{,}520 & 978 & 12{,}738 \\ 1{,}846 & 0 & 1{,}656 & 11{,}302 \\ 804 & 2{,}646 & 0 & 7{,}434 \\ 10{,}950 & 11{,}938 & 5{,}038 & 0 \end{array}\right) = \frac{1}{5}A. \end{array}$$

Example 5

The following matrix A represents energy production statistics for certain

---

† Reproduced from U.N. Statistical Year Book 1974.

‡ European Economic Community Figures include Denmark, Ireland and United Kingdom, for the whole of this five year period. These countries actually became part of the E.E.C. in 1973.

countries for the period 1970–73.   The units are million metric tons of coal equivalent.†

	1970	1971	1972	1973
Canada	206.03	223.25	257.25	285.35
U.S.A.	2062.53	2029.17	2065.22	2052.26
U.K.	163.73	176.16	159.71	175.00
Japan	58.82	49.47	44.92	39.73

$= A.$

The following matrix B gives energy consumption for the same countries.

	1970	1971	1972	1973
Canada	201.07	209.68	235.01	248.61
U.S.A.	2269.49	2317.04	2426.07	2516.44
U.K.	300.20	309.21	301.83	323.91
Japan	332.37	342.02	344.55	390.20

$= B.$

The matrix $A - B$ gives the difference between production and consumption in these countries.

	1970	1971	1972	1973
Canada	4.96	13.57	22.24	36.74
U.S.A.	− 206.96	− 287.87	− 360.85	− 464.18
U.K.	− 136.47	− 133.05	− 142.12	− 148.91
Japan	− 273.55	− 292.55	− 299.63	− 350.47

$= A - B$

The negative elements indicate a deficit in energy.   Note the increasing energy deficit in the United States over this period while Canada has an increasing surplus.

Example 6

A clinic has three doctors, each with his own speciality.   Patients attending the clinic see more than one doctor.   The accounts are drawn up monthly and handled systematically through the use of matrices. We illustrate this accounting with four patients.   This illustration can easily be extended to accommodate any number of patients.

	Doctor		
	I	II	III
A	10	25	0
B	0	20	40
Patient C	15	0	25
D	10	10	0

† Reproduced from U.N. Statistical Year Book 1974.

The entries in the above matrix are in dollars and represent the bill from each doctor to each patient for a certain month.

During a given quarter there would be three such matrices. To determine the quarterly bill to each patient from each doctor one adds the three matrices. Assume the monthly bills are as follows:

$$
\begin{array}{c}
\begin{array}{cccc} & \text{First month} \\ & \text{I} & \text{II} & \text{III} \end{array} \\
\begin{array}{c} A \\ B \\ C \\ D \end{array}
\left(\begin{array}{ccc}
10 & 25 & 0 \\
0 & 20 & 40 \\
15 & 0 & 25 \\
10 & 10 & 0
\end{array}\right)
\end{array}
+
\begin{array}{c}
\begin{array}{ccc} \text{Second month} \\ \text{I} & \text{II} & \text{III} \end{array} \\
\left(\begin{array}{ccc}
0 & 0 & 0 \\
0 & 100 & 0 \\
10 & 0 & 0 \\
20 & 0 & 0
\end{array}\right)
\end{array}
$$

$$
+
\begin{array}{c}
\begin{array}{ccc} \text{Third month} \\ \text{I} & \text{II} & \text{III} \end{array} \\
\left(\begin{array}{ccc}
20 & 0 & 0 \\
0 & 0 & 0 \\
0 & 0 & 20 \\
15 & 15 & 0
\end{array}\right)
\end{array}
=
\begin{array}{c}
\begin{array}{ccc} \text{Quarter} \\ \text{I} & \text{II} & \text{III} \end{array} \\
\left(\begin{array}{ccc}
30 & 25 & 0 \\
0 & 120 & 40 \\
25 & 0 & 45 \\
45 & 25 & 0
\end{array}\right)
\end{array}
$$

Thus A's bill to doctor I during the quarter would total \$30. B's quarterly bill to doctor III would total \$40, and so on.

Although this analysis, and others like it in this text, can theoretically be carried out without the use of matrices, the handling of large quantities of data is often most efficiently done on computers using matrix techniques and storing facilities.

EXERCISES

1. If $A = \begin{pmatrix} 1 & 2 & -3 & 5 & 7 \\ 0 & 4 & 6 & -1 & 5 \\ -1 & 2 & -7 & 3 & 4 \end{pmatrix}$, determine a_{11}, a_{24}, a_{34}, and a_{25}.

2. If

$$
A = \begin{pmatrix} 1 & 2 \\ 3 & 0 \end{pmatrix}, \qquad B = \begin{pmatrix} -1 & 2 \\ 1 & 1 \end{pmatrix},
$$

and

$$
C = \begin{pmatrix} 0 & 1 \\ 1 & 4 \end{pmatrix},
$$

determine the matrices:

a) $2A, 3B, -2C$

b) $A+B, B+A, A+C, B+C$

c) $A+2B, 3A+C, 2A+B-C$ (These are said to be linear combinations of the matrices A, B, and C.)

3. If

$$A = \begin{pmatrix} -1 & 4 & 5 \\ 0 & 1 & 2 \end{pmatrix}, \qquad B = \begin{pmatrix} 1 & 1 & 0 \\ 2 & 3 & 1 \end{pmatrix},$$

and

$$C = \begin{pmatrix} 4 & 5 & 0 \\ 2 & -1 & -1 \end{pmatrix},$$

determine:

a) $A - B, B + 3C, 2B + C$

b) $4A, -B, 3C$

c) $A + 3B, 2A - B + C, A + 2B - 2C$

4. Let A be an arbitrary 2×2 matrix. The columns of A may be viewed as column vectors. Define the second column vector.

5. Let B be an arbitrary 3×3 matrix. The rows of B may be viewed as row vectors. Define the third row vector.

6. Let C be an $m \times n$ matrix. Its rows and columns may be viewed as vectors. Define the ith row vector and jth column vector.

7. Let a and b be scalars and let C be a matrix. Prove, using techniques similar to those used in proofs of Theorems 1-1 and 1-2, that

a) $(a+b)C = aC + bC$.   **b)** $(ab)C = a(bC)$.

8. In the Exercises of Section 1-1 we introduced vector equations. Here we extend the concept to matrix equations. Solve, if possible, the following matrix equations:

a) $\begin{pmatrix} a & a+b \\ c & c-d \end{pmatrix} = \begin{pmatrix} 1 & 3 \\ 2 & 5 \end{pmatrix}$

b) $\begin{pmatrix} a & a-b \\ a+c & b+c \end{pmatrix} = \begin{pmatrix} 3 & 5 \\ 4 & 6 \end{pmatrix}$

c) $\begin{pmatrix} a & a+b \\ c & c-d \end{pmatrix} + \begin{pmatrix} a & 3 \\ b & 4 \end{pmatrix} = \begin{pmatrix} -4 & 0 \\ 2 & 6 \end{pmatrix}$

d) $\begin{pmatrix} a \\ a-b \\ c+d \end{pmatrix} = \begin{pmatrix} 3 \\ 4 \\ 7 \end{pmatrix}$

9. Observe that each of the above matrix equations is equivalent to a set of equations. The matrix equation $\begin{pmatrix} a & a+b \\ c & c-d \end{pmatrix} = \begin{pmatrix} 1 & 3 \\ 2 & 5 \end{pmatrix}$ is equivalent to the system of equations $a = 1$, $a+b = 3$, $c = 2$, $c-d = 5$. In fact, the matrix equations above were solved by solving the component systems. It is useful to be able not only to decompose matrix equations into component equations but to do the reverse—combine equations into a single matrix equation. Combine each of the following systems of equations into a single matrix equation of the type indicated.

a) $a+b = 3$
$a = -4$
$a+b+c = 6$
$d = 3$
Use 2×2 matrices

b) $a = 5$
$a+b = 6$
$b+c = 7$
Use 3×1 matrices

c) $a = 3$
$a+b = 4$
$c-d = 3$
$p = -1$
$q = 7$
$p-r = 8.$
Use 2×3 matrices

In general relativity Einstein described gravity by means of geometry. In this theory the relation between matter and geometry is described by the single matrix equation

$$T_{ij} = R_{ij} - \tfrac{1}{2}RG_{ij}.$$

Here T_{ij} is a 4×4 matrix that represents matter. R_{ij} and G_{ij} are 4×4 matrices, R a scalar. The right side of the above equation describes geometry. Finding the appropriate geometry to describe a certain gravitational field involves solving the above matrix equation.

Readers who are interested in relativity should read Section 3-8 at the appropriate time.

10. The reader has been presented with a number of examples of matrices arising in applications in this section: an origin-destination matrix, an input-output matrix, a matrix representing trade figures, and matrices representing energy production and consumption. Give your own further example of an application of matrices.

1–4. MULTIPLICATION OF MATRICES

Having introduced addition and scalar multiplication of matrices, the next natural step in the development of a theory of matrices is multiplication of matrices. Here again we have complete freedom in defining multiplication. The most natural way may seem to be to multiply matrices in a component-wise manner, as follows:

$$\begin{pmatrix} 1 & 2 & 3 \\ 0 & 1 & -1 \end{pmatrix} \begin{pmatrix} 2 & 4 & 6 \\ 1 & 3 & 0 \end{pmatrix} = \begin{pmatrix} 1 \times 2 & 2 \times 4 & 3 \times 6 \\ 0 \times 1 & 1 \times 3 & -1 \times 0 \end{pmatrix}$$

$$= \begin{pmatrix} 2 & 8 & 18 \\ 0 & 3 & 0 \end{pmatrix}$$

It has been found that this is not the most useful way of multiplying matrices. Mathematicians have introduced a different rule of multiplication.

To lead up to this rule we ask the reader to think about matrices in terms of rows and columns.   For example, consider the matrix

$$\begin{pmatrix} 1 & 2 & -1 & 3 \\ 2 & 1 & 4 & 6 \\ 0 & 1 & 3 & 2 \end{pmatrix}$$

The first row is $(\, 1 \quad 2 \quad -1 \quad 3 \,)$, the second column is $\begin{pmatrix} 2 \\ 1 \\ 1 \end{pmatrix}$, and

the fourth column is $\begin{pmatrix} 3 \\ 6 \\ 2 \end{pmatrix}$, etc.   One interprets the matrices in terms

of rows and columns when multiplying.

In the multiplication of two matrices such as

$$\overset{A}{\begin{pmatrix} 1 & 2 \\ -1 & 3 \end{pmatrix}} \overset{B}{\begin{pmatrix} -2 & 4 \\ 5 & -3 \end{pmatrix}},$$

we shall interpret the first matrix in terms of its rows and the second in terms of its columns.

To get the first row of the product C multiply the first row of A by each column of B in turn, in an appropriate manner.

$$\overset{A}{\begin{pmatrix} 1 & 2 \\ -1 & 3 \end{pmatrix}} \overset{B}{\begin{pmatrix} -2 & 4 \\ 5 & -3 \end{pmatrix}}$$

$$= \overset{C}{\left(\begin{array}{c|c} (1 \;\; 2) \times \begin{pmatrix} -2 \\ 5 \end{pmatrix} & (1 \;\; 2) \times \begin{pmatrix} 4 \\ -3 \end{pmatrix} \\ & \end{array} \right)}$$

First row of C. (*Dotted lines will be included in examples to clarify components of the matrices.*)

To get the second row of the product multiply the second row of A by each column of B.

$$= \overset{C}{\left(\begin{array}{c|c} (1 \;\; 2) \times \begin{pmatrix} -2 \\ 5 \end{pmatrix} & (1 \;\; 2) \times \begin{pmatrix} 4 \\ -3 \end{pmatrix} \\ \hline (-1 \;\; 3) \times \begin{pmatrix} -2 \\ 5 \end{pmatrix} & (-1 \;\; 3) \times \begin{pmatrix} 4 \\ -3 \end{pmatrix} \end{array} \right)}$$

It remains to define the special multiplication of the rows by the columns; one multiplies corresponding components and adds. In the above matrix C,

$$(\ 1 \quad 2 \ ) \times \begin{pmatrix} -2 \\ 5 \end{pmatrix} = (1 \times (-2)) + (2 \times 5) = 8,$$

$$(\ 1 \quad 2 \ ) \times \begin{pmatrix} 4 \\ -3 \end{pmatrix} = (1 \times 4) + (2 \times (-3)) = -2, \text{ etc.}$$

Thus

$$\begin{pmatrix} 1 & 2 \\ -1 & 3 \end{pmatrix} \begin{pmatrix} -2 & 4 \\ 5 & -3 \end{pmatrix} = \begin{pmatrix} 8 & -2 \\ 17 & -13 \end{pmatrix}.$$

We now illustrate this rule further with examples.

Example 1

If $A = \begin{pmatrix} 0 & 1 & 2 \\ -1 & 0 & 3 \\ 0 & 4 & 1 \end{pmatrix}$ and $B = \begin{pmatrix} -1 & 3 \\ 2 & 1 \\ 0 & 1 \end{pmatrix}$, determine AB.

$$AB = \begin{pmatrix} 0 & 1 & 2 \\ -1 & 0 & 3 \\ 0 & 4 & 1 \end{pmatrix} \begin{pmatrix} -1 & 3 \\ 2 & 1 \\ 0 & 1 \end{pmatrix}$$

$$= \begin{pmatrix} (\ 0 \ 1 \ 2 \ ) \times \begin{pmatrix} -1 \\ 2 \\ 0 \end{pmatrix} & (\ 0 \ 1 \ 2 \ ) \times \begin{pmatrix} 3 \\ 1 \\ 1 \end{pmatrix} \\ (-1 \ 0 \ 3 \ ) \times \begin{pmatrix} -1 \\ 2 \\ 0 \end{pmatrix} & (-1 \ 0 \ 3 \ ) \times \begin{pmatrix} 3 \\ 1 \\ 1 \end{pmatrix} \\ (\ 0 \ 4 \ 1 \ ) \times \begin{pmatrix} -1 \\ 2 \\ 0 \end{pmatrix} & (\ 0 \ 4 \ 1 \ ) \times \begin{pmatrix} 3 \\ 1 \\ 1 \end{pmatrix} \end{pmatrix}$$

$$= \begin{pmatrix} (0 \times -1) + (1 \times 2) + (2 \times 0) & (0 \times 3) + (1 \times 1) + (2 \times 1) \\ (-1 \times -1) + (0 \times 2) + (3 \times 0) & (-1 \times 3) + (0 \times 1) + (3 \times 1) \\ (0 \times -1) + (4 \times 2) + (1 \times 0) & (0 \times 3) + (4 \times 1) + (1 \times 1) \end{pmatrix}$$

$$= \begin{pmatrix} 2 & 3 \\ 1 & 0 \\ 8 & 5 \end{pmatrix}$$

Example 2

If $A = \begin{pmatrix} 2 & -1 & 1 \\ 3 & 1 & 2 \\ -2 & 3 & -1 \end{pmatrix}$ and $B = \begin{pmatrix} 3 \\ 4 \\ 5 \end{pmatrix}$, determine AB.

$$AB = \begin{pmatrix} 2 & -1 & 1 \\ 3 & 1 & 2 \\ 2 & 3 & -1 \end{pmatrix} \begin{pmatrix} 3 \\ 4 \\ 5 \end{pmatrix}$$

$$= \begin{pmatrix} (2 \quad -1 \quad 1) \times \begin{pmatrix} 3 \\ 4 \\ 5 \end{pmatrix} \\ \overline{} \\ (3 \quad 1 \quad 2) \times \begin{pmatrix} 3 \\ 4 \\ 5 \end{pmatrix} \\ \overline{} \\ (-2 \quad 3 \quad -1) \times \begin{pmatrix} 3 \\ 4 \\ 5 \end{pmatrix} \end{pmatrix}$$

$$= \begin{pmatrix} (2 \times 3) + (-1 \times 4) + (1 \times 5) \\ \overline{} \\ (3 \times 3) + (1 \times 4) + (2 \times 5) \\ \overline{} \\ (-2 \times 3) + (3 \times 4) + (-1 \times 5) \end{pmatrix}$$

$$= \begin{pmatrix} 6 - 4 + 5 \\ 9 + 4 + 10 \\ -6 + 12 - 5 \end{pmatrix} = \begin{pmatrix} 7 \\ 23 \\ 1 \end{pmatrix}$$

We now illustrate how matrices can be used in a mathematical model to compare food bills from various supermarkets. Here we demonstrate the art of setting up the model in terms of matrices in a manner that makes use of the matrix multiplication operation.

Example 3

Data taken from three supermarkets are conveniently represented in the following 4×3 matrix, A.

	Store 1	Store 2	Store 3
Sugar (per lb)	54¢	52¢	58¢
Peaches (per can)	48	53	56
Chicken (per lb)	65	64	62
Bread (per loaf)	60	62	58

For our model we have taken a "blue chip" sample of items—sugar, peaches, chicken, and bread. They represent a cross-section of items. Suppose we purchase five pounds of sugar, three cans of peaches, three pounds of chicken, and two loaves of bread during an average week.

Our aim is to set these amounts up as the components of a matrix B which will, on multiplication with A, result in a matrix giving the total grocery bill at each store.   Note that a bill is obtained by multiplying the amount of an item by its price per unit for all items and then adding.

If we let

$$B = (\; 5 \quad 3 \quad 3 \quad 2\;)$$

then we see that

$$BA = (\; 5 \quad 3 \quad 3 \quad 2\;) \begin{pmatrix} 54 & 52 & 58 \\ 48 & 53 & 56 \\ 65 & 64 & 62 \\ 60 & 62 & 58 \end{pmatrix}$$

$$= ((5 \times 54) + (3 \times 48) + (3 \times 65) + (2 \times 60)$$
$$(5 \times 52) + (3 \times 53) + (3 \times 64) + (2 \times 62)$$
$$(5 \times 58) + (3 \times 56) + (3 \times 62) + (2 \times 58))$$

$$= \quad (\$7.29 \qquad \$7.35 \qquad \$7.60)$$

| *total bill at* | *total bill at* | *total bill at* |
| *store* 1 | *store* 2 | *store* 3 |

giving a 1×3 matrix whose components are the bills at each store.

A more accurate model to assess the comparative costs of shopping at the different stores would include many more items and an analysis carried out weekly over a period.

In this type of model the aim is to set up given data in an appropriate manner, in terms of matrices, so that matrix operations can be used to lead to desired information.   Large amounts of data can be controlled in this manner, often on a computer.   Trial and error can be involved in constructing such a model; the first attempt may not lead to an appropriate model.   Exercises 11, 12, 13, and 15 have been included to help develop these instincts for constructing models.   In Section 1-7 the reader will meet a population movement model based on such a use of matrices.

The rule of multiplication of two matrices A and B involves taking the products of rows of A with columns of B.   The rule can only be applied if the number of elements in any row of A is equal to the number of elements in any column of B.   Hence *the product AB only exists if the number of columns in A is equal to the number of rows in B*.

Example 4

Let $A = \begin{pmatrix} 1 & 2 & 3 \\ 4 & 1 & 2 \\ 3 & -1 & 0 \end{pmatrix}$ and $B = \begin{pmatrix} 1 & 2 \\ 0 & 1 \end{pmatrix}$.   Then AB does not exist, since A has three columns and B has two rows.   (Try multiplying these matrices!)

In the previous section we saw that matrices of the same kind were commutative under addition, $A + B = B + A$.   Matrices are not, however, commutative under multiplication as the following example illustrates.

Example 5

Let $A = \begin{pmatrix} 1 & 2 \\ -1 & 0 \end{pmatrix}$ and $B = \begin{pmatrix} 3 & 1 \\ 1 & 2 \end{pmatrix}$.   Determine AB and BA

and compare these products.

$$AB = \begin{pmatrix} 1 & 2 \\ -1 & 0 \end{pmatrix} \begin{pmatrix} 3 & 1 \\ 1 & 2 \end{pmatrix} = \begin{pmatrix} 5 & 5 \\ -3 & -1 \end{pmatrix}$$

$$BA = \begin{pmatrix} 3 & 1 \\ 1 & 2 \end{pmatrix} \begin{pmatrix} 1 & 2 \\ -1 & 0 \end{pmatrix} = \begin{pmatrix} 2 & 6 \\ -1 & 2 \end{pmatrix}.$$

Thus, for the above matrices, $AB \neq BA$.   This one *counterexample* proves that matrices are not commutative under multiplication; the order of multiplication *is* important, unlike real numbers.   [There are special cases of matrices A and B where $AB = BA$, but in general this rule does not hold.]

For some matrices A and B, AB exists but BA does not even exist. The reader is asked to construct such an example in Exercise 6.

We now investigate the kind of matrix that results from a product. Since the product of two matrices A and B, AB exists only if the number of columns in A is equal to the number of rows in B, let A be an $m \times n$ matrix and B an $n \times r$ matrix; AB then exists.

The first row of AB is obtained by multiplying the first row of A by each column of B in turn in the appropriate manner.   Thus the number of columns in AB is equal to the number of columns in B.   The first column of AB results from multiplying the first column of B by each row of A in turn.   Thus the number of rows in AB is equal to the number of rows in A.

Therefore, *if A is an $m \times n$ matrix and B is an $n \times r$ matrix, then AB will be an $m \times r$ matrix.*

Example 6

If A is a 5×6 matrix and B is a 6×7 matrix, what kind of matrix is AB?

First we note that A has six columns and B has six rows; thus AB exists.   It will be a 5×7 matrix.

Example 7

If A is a 4×3 matrix, B is a 3×7 matrix, and C is a 7×4 matrix, determine whether AB, BC, CA, and CB exist. What kinds of matrices are those that exist?

AB exists since A has three columns and B has three rows. It will be a 4×7 matrix. BC exists and is a 3×4 matrix. CA exists and is a 7×3 matrix. CB does not exist; the number of columns in C is 4 and the number of rows in B is 3.

Any desired element in a product matrix can be computed without calculating all the elements in the product. Let A and B be matrices for which the product AB exists. *To obtain the element in row i and column j of AB, one multiplies the ith row of A with the jth column of B in the appropriate manner.*

Example 8

If $A = \begin{pmatrix} -1 & 2 \\ 3 & 4 \\ 1 & 2 \end{pmatrix}$ and $B = \begin{pmatrix} 1 & 2 & 4 & -1 \\ 3 & 6 & 1 & 2 \end{pmatrix}$, determine the element in row 2, column 3 of AB.

The product AB will be a 3×4 matrix, since A has three rows and B has four columns.

$$\begin{pmatrix} -1 & 2 \\ 3 & 4 \\ 1 & 2 \end{pmatrix} \begin{pmatrix} 1 & 2 & 4 & -1 \\ 3 & 6 & 1 & 2 \end{pmatrix} = \begin{pmatrix} x & x & x & x \\ x & x & \textcircled{x} & x \\ x & x & x & x \end{pmatrix}$$

We desire
this element

We take the second row of A and the third column of B, and we get the required element.

$$(\; 3 \quad 4 \;) \times \begin{pmatrix} 4 \\ 1 \end{pmatrix} = 12 + 4 = 16$$

Example 9

Let $A = \begin{pmatrix} -1 & 2 & 0 \\ 3 & 4 & 5 \\ 2 & 1 & 1 \end{pmatrix}$, $B = \begin{pmatrix} 0 & -2 & 1 \\ 1 & 3 & 1 \\ 2 & 4 & 1 \end{pmatrix}$, and $C = AB$.

Writing

$$C = \begin{pmatrix} c_{11} & c_{12} & c_{13} \\ c_{21} & c_{22} & c_{23} \\ c_{31} & c_{32} & c_{33} \end{pmatrix}$$

determine the elements c_{31} and c_{23}.

To obtain c_{31} we take the third row of A and first column of B.

$$c_{31} = (\; 2 \quad 1 \quad 1 \;) \times \begin{pmatrix} 0 \\ 1 \\ 2 \end{pmatrix} = (2 \times 0) + (1 \times 1) + (1 \times 2) = 3$$

Similarly, c_{23} is obtained using the second row of A and third column of B.

$$c_{23} = (\,3 \quad 4 \quad 5\,) \times \begin{pmatrix} 1 \\ 1 \\ 1 \end{pmatrix} = (3 \times 1) + (4 \times 1) + (5 \times 1) = 12$$

EXERCISES

1. Let

$$A = \begin{pmatrix} 0 & 1 \\ 2 & 3 \end{pmatrix}, \quad B = \begin{pmatrix} 4 \\ 5 \end{pmatrix}, \quad C = \begin{pmatrix} -1 & 3 \\ 2 & 1 \end{pmatrix},$$

and

$$D = \begin{pmatrix} 1 & 3 & -1 \\ 2 & 4 & 1 \end{pmatrix}.$$

Determine the following products if they exist: $AB, AC, AD, BA, DB, BD,$ $A^2, C^3, AC+CA.$ $(A^2 = AA,$ etc.$)$ Predict the kind of matrix that will result, beforehand, in each case.

2. Let $A = \begin{pmatrix} -1 & 0 \\ 2 & 1 \\ 3 & 4 \end{pmatrix}$, $B = \begin{pmatrix} 1 \\ 2 \\ 3 \end{pmatrix}$, $C = (-1, 2, 3)$, and $D = (1, 3)$.

Determine the following products if they exist: $AB, BC, CB, AD, DA, CA, BD.$ Predict the kind of matrix that will result, beforehand, in each case.

3. If $A = \begin{pmatrix} -1 & 0 & 2 \\ 3 & 4 & 5 \\ 1 & 2 & 0 \end{pmatrix}$ and $B = \begin{pmatrix} 1 & -1 \\ 3 & -2 \\ 3 & -2 \end{pmatrix}$, determine the elements

a) in the third row and first column of the product matrix,
b) in the second row and second column of the product matrix.

4. If

$$A = \begin{pmatrix} 1 & 3 & 0 \\ 2 & 2 & 1 \\ 3 & 1 & 0 \end{pmatrix}, \quad B = \begin{pmatrix} -1 & 0 & -1 \\ -2 & 1 & 1 \\ 0 & 1 & 2 \end{pmatrix},$$

and

$$C = \begin{pmatrix} c_{11} & c_{12} & c_{13} \\ c_{21} & c_{22} & c_{23} \\ c_{31} & c_{32} & c_{33} \end{pmatrix} = AB,$$

determine $c_{21}, c_{22},$ and c_{33} without evaluating the whole product matrix.

5. Let A be a 3×5 matrix, B a 5×2 matrix, C a 3×4 matrix, D a 4×2 matrix, and E a 4×5 matrix. Determine which of the following exist and give the size of the resulting matrix when it does exists.

a) $AB + C$ b) $AB + CD$ c) $3EB + 4D$

d) $CD - 2(CE)B$ e) $2EB + DA$

6. Give examples of
   a) two matrices, A and B, for which AB exists but BA does not,
   b) two matrices, P and Q, for which PQ and QP exist, but are not equal. (Other than the matrices of Example 5.) Each of these examples proves that the matrices are not commutative under multiplication.

7. Let A be an $m \times n$ matrix, B and $n \times s$ matrix having columns $v_1, ..., v_s$. Show that the product AB can be interpreted as the matrix having columns $Av_1, ..., Av_s$; that is $AB = (Av_1 ... Av_s)$. [We shall from time to time have occasion to use this interpretation of AB.]

8. a) Let A be an $n \times n$ matrix. Prove that A^2 is $n \times n$.
   b) Let A be an $m \times n$ matrix with $m \neq n$. Prove that A^2 does not exist. Thus one can only talk about powers of a square matrix.

9. Consider two matrices, P and Q, such that $PQ = 0$, the zero matrix of the appropriate kind. (A zero matrix is one whose elements are all zero.) Does this imply that either $P = 0$ or $Q = 0$? (*Hint*: Construct an example.)

10. Consider three matrices A, B and C such that $AC = BC$. Does this imply that $A = B$?

11. What modifications would you make in the model in Example 3 of this section to make it more reliable?

12. In Example 3, determine a vector C that could be multiplied by the matrix A to give a vector whose components would be the average price of sugar, the average price of peaches, the average price of a pound of chicken, and the average price of a loaf of bread. (*Hint*: C would be a column vector with three identical components; AC would give the average prices.)

13. Construct a model similar to the one in Example 3 to analyze the increase in food prices per quarter. Recorded below are the prices of four different items—canned peas, peaches, chicken, and bread—in one store during four successive quarters. Base your model on buying two cans of peas, one can of peaches, three pounds of chicken and two loaves of bread.

	Beginning of 1st quart.	Beginning of 2nd quart.	Beginning of 3rd quart.	Beginning of 4th quart.	End of 4th quart.
Canned peas (per can)	44¢	44¢	47¢	47¢	52¢
Peaches (per can)	49	52	52	52	51
Chicken (per lb)	65	65	65	67	67
Bread (per loaf)	58	60	60	61	63

14. †Write a computer program that can be used in your model for Exercise 13. It could include many items, leading to an accurate analysis of the increase in the cost of living.

15. In 1958 the European Common Market was formed. This was a European trading block consisting of six countries, France, West Germany, Italy, Belgium, Luxemburg, and The Netherlands. At the time, Great Britain did not favor entering the Common Market for a variety of political reasons, including factors concerning its trading relationships with the Commonwealth countries. However, during the years that followed, the climate of opinion in Great Britain shifted somewhat; many people concluded that Britain's trading future lay within Europe. In 1963 an unsuccessful application was made by the Conservative government; in 1967 an equally unsuccessful one was made by the Labour government. Both attempts were vetoed by De Gaulle's French government. However, in 1972 Britain's application was accepted, and in 1973 Britain, Eire, and Denmark officially became members of the Common Market. During the period from 1958 to 1972 (and afterwards), the pros and cons of entering the Market were widely debated. One of the main arguments put forward for entering the Market was that, in the long run, Britain could not afford to remain out—the Common Market had vast potential as a trading block, and Britain, a small island trading nation on its fringe, would suffer. Another argument was that in a world dominated by superpowers, the Common Market would be a platform from which Britain could again play an important role in international affairs. The antimarketeers, on the other hand, thought that Britain was already in a trading group of Commonwealth countries. These ties would eventually have to be severed if Britain entered the Common Market. Here the matter of loyalty entered in. Various statistics were presented by both sides—the split was not along party lines. The statistics in Table 1-3 put

Table 1-3

	London	Frankfurt	Paris	Rome
Bread (per loaf)	21¢	40¢	25¢	11¢
Milk (per pt)	12	12	14	13
Pork chops (per lb)	94	99	79	129
Eggs (per doz)	54	69	42	56
Soap powder (per large pkt)	34	70	59	32
Tomatoes (per lb)	42	40	32	31
Fruit (per large tin)	39	37	62	60

---

† Exercises on the computer have been included. These are all optional.

out by the anti-Common Market group were aimed at the British house-wife. They gave a comparison of grocery prices in various European cities in April, 1970.

a) Use the statistics in a manner similar to that of Example 3 in order to estimate the relative cost of groceries in the various European cities and London.

b) Extend the model to include the U.S. These statistics were for prices in Europe in 1970.

c) A more reliable model of relative costs of living would also include earnings in the various cities. The following row matrix represents average weekly wages.

$$
\begin{array}{cccc}
\text{London} & \text{Frankfurt} & \text{Paris} & \text{Rome} \\
(\$80 & \$110 & \$92 & \$98)
\end{array}
$$

Include these figures in a mathematical model giving the relative costs of living in the various European cities.

1–5. AN ANALYSIS OF MATRIX MULTIPLICATION

Let A be an $m \times n$ matrix and B an $n \times t$ matrix. The number of columns in A is n, and the number of rows in B is also n; hence the product $C = AB$ exists and is given by

$$
C = AB = \begin{pmatrix} a_{11} & \cdots & a_{1n} \\ & \vdots & \\ a_{m1} & \cdots & a_{mn} \end{pmatrix} \begin{pmatrix} b_{11} & \cdots & b_{1t} \\ & \vdots & \\ b_{n1} & \cdots & b_{nt} \end{pmatrix}
$$

$$
= \begin{pmatrix} (a_{11}b_{11} + \cdots + a_{1n}b_{n1}) & \cdots & (a_{11}b_{1t} + \cdots + a_{1n}b_{nt}) \\ & \vdots & \\ (a_{m1}b_{11} + \cdots + a_{mn}b_{n1}) & \cdots & (a_{m1}b_{1t} + \cdots + a_{mn}b_{nt}) \end{pmatrix}
$$

The general element c_{ij} in the ith row and jth column of C is obtained by taking the product of the ith row vector of A, $(a_{i1} \cdots a_{in})$, and the jth column vector of B, $\begin{pmatrix} b_{1j} \\ \vdots \\ b_{nj} \end{pmatrix}$.

Thus

$$
c_{ij} = a_{i1}b_{1j} + a_{i2}b_{2j} + \cdots + a_{in}b_{nj}
$$

We can write

$$
AB = (a_{rs})(b_{pq}) = (a_{i1}b_{1j} + \cdots + a_{in}b_{nj})
$$

We can represent this arbitrary element of AB using sigma notation:

$$
a_{i1}b_{1j} + \cdots + a_{in}b_{nj} = \sum_{k=1}^{n} a_{ik}b_{kj}
$$

Here k takes on all the values from 1 to n and the terms are summed.

Thus we can define matrix multiplication as follows.

$$AB = (a_{rs})(b_{pq}) = \left(\sum_{k=1}^{n} a_{ik} b_{kj} \right)$$

Example 1

Write $a_{11}b_{11} + a_{12}b_{21} + a_{13}b_{31} + a_{14}b_{41}$ in sigma notation.

The first subscript of a, 1, remains constant in all the terms, and so does the second subscript of b, 1. The other two subscripts range from 1 to 4 and are identical for each term. Hence we can indicate these two subscripts by a letter k and sum over k from 1 to 4.

$$a_{11}b_{11} + a_{12}b_{21} + a_{13}b_{31} + a_{14}b_{41} = \sum_{k=1}^{4} a_{1k} b_{k1}$$

Example 2

In sigma notation,

$$a_{21}b_{13} + a_{22}b_{23} + a_{23}b_{33} = \sum_{k=1}^{3} a_{2k} b_{k3}$$

Example 3

Expand $\sum_{k=1}^{5} a_{3k} b_{k2}$.

Here k ranges from 1 to 5. The expansion will consist of five terms, each term corresponding to a value of k.

$$\sum_{k=1}^{5} a_{3k} b_{k2} = a_{31}b_{12} + a_{32}b_{22} + a_{33}b_{32} + a_{34}b_{42} + a_{35}b_{52}$$

Example 4

If A is a 5×6 matrix (a_{rs}) and B is the 6×7 matrix (b_{pq}), write down the element in the third row and fourth column of AB and the element in the fourth row and second column of AB.

We know that the element in the ith row and jth column of AB is

$$\sum_{k=1}^{6} a_{ik} b_{kj}$$

The summation here is from 1 to 6, as this is the number of columns in A and rows in B.

For the element in the third row and fourth column of AB, $i = 3$ and $j = 4$. Hence this element is

$$\sum_{k=1}^{6} a_{3k} b_{k4}$$

Similarly, the element in the fourth row and the second column is

$$\sum_{k=1}^{6} a_{4k} b_{k2}$$

Example 5

If A is a 4×4 matrix (a_{rs}), B is a 4×6 matrix (b_{pq}), and C, or (c_{ij}), is AB, determine c_{35} and c_{24}. Then determine the location in C of the elements $\sum_{k=1}^{4} a_{2k} b_{k5}$ and $\sum_{k=1}^{4} a_{3k} b_{k2}$.

We know that the arbitrary element c_{ij} of C is given by

$$c_{ij} = \sum_{k=1}^{4} a_{ik} b_{kj}$$

Hence

$$c_{35} = \sum_{k=1}^{4} a_{3k} b_{k5}$$

and

$$c_{24} = \sum_{k=1}^{4} a_{2k} b_{k4}$$

Further,

$$\sum_{k=1}^{4} a_{2k} b_{k5} = c_{25}$$

and

$$\sum_{k=1}^{4} a_{3k} b_{k2} = c_{32}$$

Example 6   This example illustrates a property that we shall use to form a general result for matrices in the following theorem.

Prove that

$$\sum_{j=1}^{3} \left(\sum_{k=1}^{2} a_{ik} b_{kj} \right) c_{js} = \sum_{k=1}^{2} a_{ik} \left(\sum_{j=1}^{3} b_{kj} c_{js} \right)$$

The significance of this result is that it does not matter which of the finite summations, the one over j or the one over k, is carried out first.

We have

$$\sum_{j=1}^{3} \left(\sum_{k=1}^{2} a_{ik} b_{kj} \right) c_{js} = \sum_{j=1}^{3} (a_{i1} b_{1j} + a_{i2} b_{2j}) c_{js}$$

$$= (a_{i1} b_{11} + a_{i2} b_{21}) c_{1s} + (a_{i1} b_{12} + a_{i2} b_{22}) c_{2s}$$

$$+ (a_{i1} b_{13} + a_{i2} b_{23}) c_{3s}$$

$$= a_{i1} b_{11} c_{1s} + a_{i2} b_{21} c_{1s} + a_{i1} b_{12} c_{2s}$$

$$+ a_{i2} b_{22} c_{2s} + a_{i1} b_{13} c_{3s} + a_{i2} b_{23} c_{3s}$$

The secret is now to regroup the terms, keeping an eye on the desired expression.

$$= a_{i1} (b_{11} c_{1s} + b_{12} c_{2s} + b_{13} c_{3s})$$

$$+ a_{i2} (b_{21} c_{1s} + b_{22} c_{2s} + b_{23} c_{3s})$$

$$= a_{i1} \left(\sum_{j=1}^{3} b_{1j} c_{js} \right) + a_{i2} \left(\sum_{j=1}^{3} b_{2j} c_{js} \right)$$

$$= \sum_{k=1}^{2} a_{ik} \left(\sum_{j=1}^{3} b_{kj} c_{js} \right)$$

We shall now see the usefulness of sigma notation in formal proofs of identities involving the multiplication of matrices.

THEOREM 1-3   *If A, B and C are matrices for which AB and BC exist, then $(AB)C = A(BC)$; that is, it does not matter which multiplication is performed first.   This is called the associative property of matrix multiplication.*

Thus matrices have this property in common with real numbers.

A significance of this result is that an expression such as ABC (without brackets) can have meaning.   We define the product of three matrices by $ABC = (AB)C = A(BC)$.   This concept can be extended to the product of any number of matrices such as $ABCD$; this is well defined without having to indicate with brackets the order of multiplication.

Proof:

$$AB = \left(\sum_k a_{ik} b_{kj} \right)$$

and

$$(AB)C = \left[\sum_j \left(\sum_k a_{ik} b_{kj} \right) c_{js} \right]$$

If we expand both sums, this expression would be an addition of a finite string of terms of the type $a_{i2} b_{24} c_{4s}$.   As in the previous example, we can sum all these terms to get

$$(AB)C = \sum_k a_{ik} \left(\sum_j b_{kj} c_{js} \right)$$

which gives

$$(AB)C = A(BC)$$

proving the result.

Note that you cannot prove this general result by showing that it works for specific matrices.   This is a general result that holds for all matrices where $(AB)C$ exists, and thus it has to be proved in terms of arbitrary matrices.

Example 7*

Here we analyze a *two-port* in an electrical circuit.   The analysis illustrates the importance of matrix algebra in the theory of electrical circuits.

Many networks are designed to accept signals at certain points and to deliver a modified version of the signals.   The usual arrangement at such locations is illustrated in Figure 1-6.   A current i_1 at voltage V_1 is delivered into the black box, and it in some way determines the current i_2 at voltage V_2.   The term *black box* is used to imply that very often one is ignorant of the actual structure of the interior of the two-port.   The two-port may, for example, be a transistor.   The only assumption we shall make concerning the terminal is that it behaves in a linear manner—the

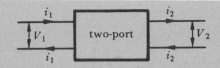

FIGURE 1-6

input and output currents and voltages are related by the expression

$$\begin{pmatrix} V_2 \\ i_2 \end{pmatrix} = \begin{pmatrix} a_{11} & a_{12} \\ a_{21} & a_{22} \end{pmatrix} \begin{pmatrix} V_1 \\ i_1 \end{pmatrix}$$

This is very often the case in practice.

The matrix $\begin{pmatrix} a_{11} & a_{12} \\ a_{21} & a_{22} \end{pmatrix}$ is called the *transmission matrix*.

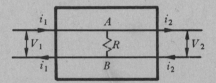

FIGURE 1-7

Figure 1-7 is an example of a two-port. The interior consists of a resistance R connected as shown. We shall illustrate that the currents and voltages do indeed behave in the linear manner described above. *Ohm's law* describes the behavior of currents in such circuits; it tells us that the voltage drop across any portion of a circuit is equal to the current multiplied by the resistance of that part. $V_1 = V_2$, since the terminals are connected directly. Applying Ohm's law to the part AB of the circuit: The voltage drop across AB must be V_1 (voltage drops occur across resistances.) The current through AB is $i_1 - i_2$ in the direction A to B. Thus $V_1 = (i_1 - i_2) R$.

We have the two equations

$$V_1 = V_2,$$

$$V_1 = (i_1 - i_2) R.$$

These two equations can be rearranged in the form

$$V_2 = V_1 + 0i_1$$

$$i_2 = -\frac{1}{R} V_1 + i_1$$

to give the matrix equation

$$\begin{pmatrix} V_2 \\ i_2 \end{pmatrix} = \begin{pmatrix} 1 & 0 \\ -\dfrac{1}{R} & 1 \end{pmatrix} \begin{pmatrix} V_1 \\ i_1 \end{pmatrix}$$

Thus the transmission matrix is $\begin{pmatrix} 1 & 0 \\ -\dfrac{1}{R} & 1 \end{pmatrix}$

If a number of two-ports are placed in series as in Figure 1-8, matrix operations make the dependence of the output current and voltage on the input immediately apparent.

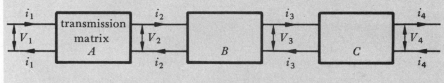

FIGURE 1-8

Let the transmission matrices of the two-ports be A, B, and C. Considering each port separately, we have

$$\begin{pmatrix} V_2 \\ i_2 \end{pmatrix} = A \begin{pmatrix} V_1 \\ i_1 \end{pmatrix}, \quad \begin{pmatrix} V_3 \\ i_3 \end{pmatrix} = B \begin{pmatrix} V_2 \\ i_2 \end{pmatrix}, \quad \begin{pmatrix} V_4 \\ i_4 \end{pmatrix} = C \begin{pmatrix} V_3 \\ i_3 \end{pmatrix}$$

Combining the first two equations by substituting the value of $\begin{pmatrix} V_2 \\ i_2 \end{pmatrix}$ from the first equation into the second equation gives $\begin{pmatrix} V_3 \\ i_3 \end{pmatrix} = BA \begin{pmatrix} V_1 \\ i_1 \end{pmatrix}$. Then substituting this value of $\begin{pmatrix} V_3 \\ i_3 \end{pmatrix}$ into the third equation gives

$$\begin{pmatrix} V_4 \\ i_4 \end{pmatrix} = CBA \begin{pmatrix} V_1 \\ i_1 \end{pmatrix}$$

Note that the order of matrix multiplication is important, for matrices are not commutative under multiplication. The physical interpretation of this is that, in general, the order of the two-ports in the circuit is important. Since CBA is a matrix, we see that the three individual two-ports have been combined to give one two-port.

There are many possible circuits within two-ports. The name *two-port* arises from the fact that the terminal is characterized by external measurements of the two ports. The four elements of the transmission matrix can be determined by measuring two separate input and output currents and voltages. (The reader is asked to determine the transmission matrix from such data in the exercises.) Three-ports and, in general, n-ports are analyzed similarly in network theory. The transmission matrix for an n-port will be an $n \times n$ matrix. Readers who are interested in pursuing this analysis further can consult *Linear Active Networks*, by Robert Spence, Wiley-Interscience, 1970.

PARTITIONING OF MATRICES*

A matrix may be subdivided into a number of *sub-matrices* by means of horizontal and vertical lines drawn through the array of elements. For example the following matrix A can be subdivided into the sub-matrices P, Q, R, S.

$$A = \left(\begin{array}{c|cc} 0 & -1 & 2 \\ 3 & 1 & 4 \\ \hline -2 & 5 & -3 \end{array} \right) = \left(\begin{array}{cc} P & Q \\ R & S \end{array} \right).$$

where $P = \begin{pmatrix} 0 \\ 3 \end{pmatrix}$, $Q = \begin{pmatrix} -1 & 2 \\ 1 & 4 \end{pmatrix}$, $R = (-2)$ and $S = (5, -3)$.

Such subdivisions are used both in computations and theoretical discussions involving matrices.

Provided appropriate rules are followed, the matrix operations previously described can be applied to sub-matrices as if they were elements of an ordinary matrix.

Let us look at addition. Let A and B be matrices of the same kind. If A and B are partitioned in the same way, their sum is the sum of the corresponding sub-matrices. For example,

$$A + B = \left(\begin{array}{cc} P & Q \\ R & S \end{array} \right) + \left(\begin{array}{cc} H & I \\ J & K \end{array} \right) = \left(\begin{array}{cc} P+H & Q+I \\ R+J & S+K \end{array} \right)$$

In multiplication, any partition of the first matrix in a product determines the row partition of the second matrix. For example, let us consider the product AB of the matrices

$$A = \begin{pmatrix} 1 & 2 & -1 \\ 3 & 0 & -2 \\ 4 & -3 & 2 \end{pmatrix} \quad \text{and} \quad B = \begin{pmatrix} -1 & 0 \\ 2 & 1 \\ 5 & 4 \end{pmatrix}$$

Let A be subdivided

$$\left(\begin{array}{cc|c} 1 & 2 & -1 \\ \hline 3 & 0 & -2 \\ 4 & -3 & 2 \end{array}\right) = \left(\begin{array}{cc} P & Q \\ R & S \end{array}\right)$$

A is interpreted as having two columns in this form. B must be subdivided into a suitable form having two rows for multiplication to be possible.

$$B = \left(\begin{array}{cc} -1 & 0 \\ 2 & 1 \\ \hline 5 & 4 \end{array}\right) = \left(\begin{array}{c} M \\ N \end{array}\right)$$

would be a suitable partition for B.

$$AB = \left(\begin{array}{cc} P & Q \\ R & S \end{array}\right)\left(\begin{array}{c} M \\ N \end{array}\right) = \left(\begin{array}{c} PM+QN \\ RM+SN \end{array}\right).$$

If A is subdivided into a form having q columns, to obtain a product AB, B must be subdivided into an appropriate form having q rows. There is freedom in the column subdivision of B. If the partition of A is a $p \times q$ form, and B is a $q \times s$ form, then AB will be partitioned into p rows and s columns.

Example 8

Let $A = \left(\begin{array}{cc} 1 & -1 \\ 3 & 0 \\ 2 & 4 \end{array}\right)$ and $B = \left(\begin{array}{ccc} 1 & 2 & -1 \\ 1 & 3 & 1 \end{array}\right)$.

Consider the partition $\left(\begin{array}{c|c} 1 & -1 \\ 3 & 0 \\ \hline 2 & 4 \end{array}\right) = \left(\begin{array}{cc} P & Q \\ R & S \end{array}\right)$ of A.

A, under this partition is interpreted as a 2×2 matrix. For the product to exist B must be appropriately partitioned into a matrix having 2 rows, $\left(\begin{array}{c} H \\ J \end{array}\right)$. The number of columns of P and Q determine the number of rows of H and J, since PH and QJ must exist:

$$AB = \left(\begin{array}{cc} P & Q \\ R & S \end{array}\right)\left(\begin{array}{c} H \\ J \end{array}\right) = \left(\begin{array}{c} PH + QJ \\ RH + SJ \end{array}\right)$$

Thus an appropriate partition for B is $\left(\begin{array}{c|cc} 1 & 2 & -1 \\ \hline 1 & 3 & 1 \end{array}\right)$, B is a 2×1 matrix.

Under these partitions, A is 2×2, B is 2×1, and AB is 2×1.

There are other possible partitions of B, can you give them, and also the corresponding partitions of AB in each case?

EXERCISES

1. Write $a_{11}b_{11} + a_{12}b_{21} + a_{13}b_{31}$ in sigma notation.

2. Write $a_{31}b_{12} + a_{32}b_{22} + a_{33}b_{32} + a_{34}b_{42}$ in sigma notation.

3. Write out $\sum_{k=1}^{3} a_{2k}b_{k3}$ in full.

4. Expand $\sum_{k=1}^{4} a_{4k}b_{k3}$.

5. Expand $\sum_{k=1}^{5} a_{ik}b_{kj}$.

6. Prove that $\sum_{j=1}^{2} \left(\sum_{k=1}^{3} a_{ik}b_{kj} \right) c_{js} = \sum_{k=1}^{3} a_{ik} \left(\sum_{j=1}^{2} b_{kj}c_{js} \right)$ by expanding the left side and then rearranging the terms to get the right side.

7. If A and B are both 5×5 matrices, use sigma notation to write the elements in
   a) the first row and first column of AB
   b) the second row and third column of AB
   c) the fourth row and fifth column of AB

8. If L, M, and R are matrices for which $L+M$ and LR exist, prove that $(L+M)R = LR+MR$. In a similar manner, $A(B+C) = AB + AC$ for matrices A, B, and C, where AB and $B+C$ exist. This property of matrix multiplication is called the *distributive property*.

9. Discuss the significance of the condition in Exercise 8 that $L+M$ and LR exist. Does it necessarily follow that MR exists?

10. Let A and B be square matrices of the same kind. Prove that, in general,
$$(A+B)^2 \neq A^2 + 2AB + B^2.$$
Under what condition is $(A+B)^2 = A^2 + 2AB + B^2$?

11. If A and B are square matrices of the same kind, and are such that $AB = BA$, prove that
$$(AB)^2 = A^2 B^2.$$
Show that this identity does not hold in general, for all square matrices of the same kind, by constructing a counter example.

12. Let A be an arbitrary square matrix. Prove that $A^r A^s = A^{r+s}$ and $(A^r)^s = A^{rs}$ for all nonnegative integer values of r and s.

13. Prove that if A, B and C are $m \times n$, $n \times p$, and $p \times q$ matrices, respectively, then ABC is an $m \times q$ matrix.

14. Let $A = \begin{pmatrix} 1 & 2 & 3 \\ -1 & 1 & 4 \\ 0 & 1 & 2 \end{pmatrix}$ and $B = \begin{pmatrix} -1 & -2 \\ 0 & 3 \\ 4 & 1 \end{pmatrix}$.

For each partition of A given below find all partitions of B that can be used for calculating AB. Give the corresponding dimensions of AB in each case.

a) $\begin{pmatrix} 1 & \vdots & 2 & 3 \\ \hline -1 & \vdots & 1 & 4 \\ 0 & \vdots & 1 & 2 \end{pmatrix}$ **b)** $\begin{pmatrix} 1 & \vdots & 2 & 3 \\ -1 & \vdots & 1 & 4 \\ 0 & \vdots & 1 & 2 \end{pmatrix}$

c) $\begin{pmatrix} 1 & \vdots & 2 & \vdots & 3 \\ -1 & \vdots & 1 & \vdots & 4 \\ \hline 0 & \vdots & 1 & \vdots & 2 \end{pmatrix}$

15. Let A and B be $n \times n$ matrices.
   a) Partition A into n row matrices $A_1, ..., A_n$, keeping B as it is. Determine the product AB in terms of submatrices.
   b) Partition B into n column matrices $B^1, ..., B$ keeping A unchanged. Determine AB in terms of submatrices.
   c) Partition A into n row matrices $A_1, ..., A_n$ and B into n column matrices $B^1, ..., B^n$. Determine AB.
   d) Partition A into n column matrices $A^1, ..., A^n$ and B into n row matrices $B_1, ..., B_n$. Does AB exist for these partitions?

   Later, we find that partitions of matrices into row and column matrices (or vectors) are very useful.

16. Determine the transmission matrices for the two-ports in Figure 1-9.

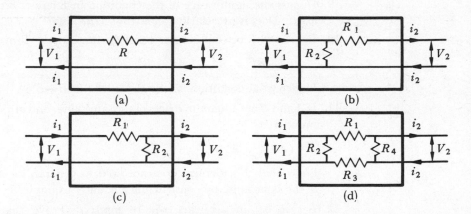

(a)   (b)

(c)   (d)

FIGURE 1-9

17. Determine the transmission matrices for the two-ports in Figure 1-10. They are combinations of the two-ports in Figure 1-9. Note that the transmission matrices differ—this is a direct consequence of the fact that matrices are not commutative under multiplication.

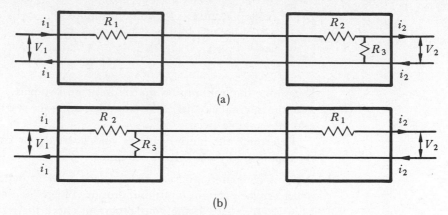

(a)

(b)

FIGURE 1-10

18. The two-port in Figure 1-11 consists of three two-ports placed in a series. The transmission matrices are indicated. What is the transmission matrix of the composite two-port?

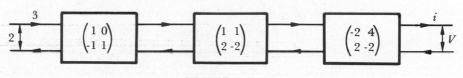

FIGURE 1-11

If the input voltage is two volts and current is three amps, determine the output voltage and current.

19. Write a computer program that can be used for evaluating the transmission matrix of a composite two-port such as the one in Exercise 18 and for determining output voltage and current given certain inputs.

20. The internal circuits of the two-ports in Figure 1-12 are not known.

FIGURE 1-12

External measurements of certain currents and voltages are

a)

V_1 volts	i_1 amps	V_2 volts	i_2 amps
3	2	3	1
6	5	6	3

b)

V_1 volts	i_1 amps	V_2 volts	i_2 amps
4	1	6	1
3	2	7	2

Determine the transmission matrix for each two-port. Can you speculate as to the internal circuits of the two-ports using the results from Exercise 16 and the example given?

Determine the output voltage and currents for each two-port when the inputs are $V_1 = 1$ volt, $i_1 = 3$ amps.

21. Write a computer program that can be used to determine transmission matrices when certain input and output voltages and currents are known, as in Exercise 20. Using your program, check the results of Exercise 20.

1–6. PROPERTIES OF MATRICES

We now examine certain relations between matrices and define special types of matrices.

If A is an $m \times n$ matrix, then the *transpose* of A, denoted by A^t, is the $n \times m$ matrix obtained by writing the rows of A as columns. The first row of A becomes the first column of A^t, the second row of A becomes the second column of A^t, etc.

Example 1

Determine the transposes of the following matrices A, B, and C:

$$A = \begin{pmatrix} 0 & -2 \\ 1 & 3 \end{pmatrix}, \quad B = \begin{pmatrix} 1 & 2 & 1 \\ 3 & -4 & 1 \end{pmatrix}, \quad C = \begin{pmatrix} 1 \\ -2 \\ 3 \end{pmatrix}$$

A is a 2×2 matrix; A^t is the 2×2 matrix $\begin{pmatrix} 0 & 1 \\ -2 & 3 \end{pmatrix}$.

B is a 2×3 matrix; B^t is the 3×2 matrix $\begin{pmatrix} 1 & 3 \\ 2 & -4 \\ 1 & 1 \end{pmatrix}$.

C is a 3×1 matrix; C^t is the 1×3 matrix $\begin{pmatrix} 1 & -2 & 3 \end{pmatrix}$.

In determining the transpose of a matrix A the element that was in row i and column j of A goes to row j and column i of A^t. We illustrate this property, which is used in theoretical work involving transposes, with the following example.

48

Example 2

Consider the matrix $A = \begin{pmatrix} 2 & 1 \\ 3 & 4 \\ -1 & 0 \end{pmatrix}$

Let $B = A^t = \begin{pmatrix} 2 & 3 & -1 \\ 1 & 4 & 0 \end{pmatrix}$.

Then $a_{21} = 3 = b_{12}$; $a_{32} = 0 = b_{23}$, etc.
In general, if $B = A^t$ then $a_{ij} = b_{ji}$.

We now look at the properties of transpose. There are three operations that have been defined for certain matrices: addition, matrix multiplication, and scalar multiplication. Let us see how transpose works in conjunction with these operations. We examine $(A+B)^t$, $(AB)^t$, and $(cA)^t$

THEOREM 1-4 $(A+B)^t = A^t + B^t$

Proof: We shall have need to work with the element in the ith row, and jth column of an arbitrary matrix. Let us adopt the terminology (i, j)th element for this element.

Let $A = (a_{mn})$ and $B = (b_{st})$. $(A+B)^t$ and $A^t + B^t$ will both, on carrying out the necessary additions and transpositions, be matrices. We prove that these matrices are equal by proving that corresponding arbitrary elements are equal.

$$\text{The } (i, j)\text{th element of } (A+B)^t = (j,i)\text{th element of } A + B$$
$$= a_{ji} + b_{ji}$$

$$\text{The } (i, j)\text{th element of } A^t + B^t = (i,j)\text{th element of } A^t$$
$$+ (i,j)\text{th element of } B^t$$

$$= (j,i)\text{th element of } A$$
$$+ (j,i)\text{th element of } B$$

$$= a_{ji} + b_{ji}$$

Thus corresponding arbitrary elements of $(A+B)^t$ and $A^t + B^t$ are equal, proving that these matrices are equal.

THEOREM 1-5 $(AB)^t = B^t A^t$.

Note the reversal of the order of the matrices on the right side.

Proof:

$$\text{The } (i, j)\text{th element of } (AB)^t = (j,i)\text{th element of } AB$$
$$= \sum_k a_{jk} b_{ki}$$

49

The (i, j)th element of $B^t A^t$ = product of ith row of B^t with jth column of A^t

= product of ith column of B with jth row of A

$$= \sum_j a_{jk} b_{ki}$$

Corresponding elements of $(AB)^t$ and $B^t A^t$ are equal, proving the equality of these matrices.

THEOREM 1-6 $(cA)^t = cA^t$, for a scalar c.

Proof: We leave the proof of this theorem and the following theorem as exercises.   (Exercise 7 p. 58.)

THEOREM 1-7 $(A^t)^t = A$.

The transpose of the transposed matrix is, as we might expect, the original matrix.

A matrix A is *symmetric* if it is equal to its transpose; that is, if $A = A^t$.   This condition can also be expressed $a_{ij} = a_{ji}$, for every element a_{ij}.   A symmetric matrix is necessarily square.   (Why?)   Examples of symmetric matrices are

$$\begin{pmatrix} 1 & -2 \\ -2 & 3 \end{pmatrix}, \begin{pmatrix} 1 & 3 & 0 \\ 3 & -1 & 2 \\ 0 & 2 & 4 \end{pmatrix}, \begin{pmatrix} 1 & 2 & 3 & -1 \\ 2 & 5 & 4 & -2 \\ 3 & 4 & 0 & -3 \\ -1 & -2 & -3 & 6 \end{pmatrix}.$$

Symmetric matrices form an important class.   They arise frequently in applications, as we shall see, and they have special properties.   They have been studied in great depth by mathematicians.   We shall look at their special properties from time to time.

Example 3

Consider the following matrix, representing the distances between various American cities.

		1	2	3	4	5	6
1.	Atlanta	0	702	1,411	2,215	667	862
2.	Chicago	702	0	1,013	2,092	1,374	841
3.	Denver	1,411	1,013	0	1,157	2,074	1,866
4.	Los Angeles	2,215	2,092	1,157	0	2,733	2,797
5.	Miami	667	1,374	2,074	2,733	0	1,336
6.	New York	862	841	1,866	2,797	1,336	0

To determine the mileage between any two of the cities listed, one looks along the row of one city and the column of the other. The element that lies at the intersection of the row and the column is the required mileage. For example, the distance between Miami and Denver is 2,074 miles.

Notice that we can look up the mileage between any two cities in two distinct ways. For example, we can look up Miami in the row and Denver in the column to arrive at 2,074. We arrive at the same number if we look up Denver in the row and Miami in the column. Each number appears twice in the matrix; there is a symmetry about the matrix. It is, in fact, symmetric.

Every such mileage matrix will be symmetric.

The *main diagonal* of a square matrix is the set of elements $a_{11}, a_{22}, ..., a_{nn}$, where the two subscripts are equal. The main diagonal of the matrix $\begin{pmatrix} 3 & 4 & 1 \\ 2 & 2 & -1 \\ 0 & 2 & -1 \end{pmatrix}$ is the set $3, 2, -1$.

A *diagonal matrix* is a square matrix with nonzero elements only on the main diagonal. The matrix $\begin{pmatrix} 1 & 0 & 0 \\ 0 & -1 & 0 \\ 0 & 0 & 3 \end{pmatrix}$ is a diagonal matrix.

The square $n \times n$ matrix

$$\begin{pmatrix} 1 & 0 & \cdots & 0 \\ 0 & 1 & \cdots & 0 \\ & & \vdots & \\ 0 & 0 & \cdots & 1 \end{pmatrix}$$

which has ones on the main diagonal and zeros elsewhere, is called the *unit $n \times n$ matrix*; we shall denote it by I_n.

Example 4

$$I_2 = \begin{pmatrix} 1 & 0 \\ 0 & 1 \end{pmatrix}, \quad I_3 = \begin{pmatrix} 1 & 0 & 0 \\ 0 & 1 & 0 \\ 0 & 0 & 1 \end{pmatrix}, \quad \text{and} \quad I_4 = \begin{pmatrix} 1 & 0 & 0 & 0 \\ 0 & 1 & 0 & 0 \\ 0 & 0 & 1 & 0 \\ 0 & 0 & 0 & 1 \end{pmatrix}.$$

THEOREM 1-8   *If A is an $n \times n$ matrix, then $I_n A = A I_n = A$.*

Thus I_n plays a role in the multiplication of $n \times n$ matrices similar to the role of 1 in the multiplication of real numbers. Hence the terminology *unit $n \times n$ matrix*.

Proof: To prove this result we employ a very useful tool in the manipulation of matrices called the *kronecker delta*. It is defined as follows:

$$\delta_{ab} \begin{cases} = 1, & \text{if } a = b \\ = 0, & \text{if } a \neq b \end{cases}$$

Hence, $\delta_{11} = 1, \delta_{12} = 0, \delta_{22} = 1, \delta_{24} = 0$, etc.

I_n may be represented by (δ_{ij}). This gives

$$I_n A = \left(\sum_{k=1}^{n} \delta_{ik} a_{kj} \right)$$

$$= (\delta_{i1} a_{1j} + \delta_{i2} a_{2j} + \cdots + \delta_{ii} a_{ij} + \cdots + \delta_{in} a_{nj})$$

The only nonzero term in this expansion is the term $\delta_{ii} a_{ij}$, or a_{ij}. Hence

$$I_n A = (a_{ij})$$

$$= A$$

The proof of $AI_n = A$ is similar; we leave it as an exercise for the reader.

Our final definition in this section is that of the *trace* of a matrix. If $A = (a_{ij})$ is a $n \times n$ matrix, then the trace of A, denoted $tr(A)$, is the sum of the diagonal elements,

$$tr(A) = a_{11} + a_{22} + \cdots + a_{nn}$$

$$= \sum_{i=1}^{n} a_{ii}.$$

Example 5

Determine the trace of the matrix $A = \begin{pmatrix} 0 & 1 & 2 & 3 \\ -1 & 4 & 2 & 1 \\ 3 & 4 & -1 & 2 \\ 0 & 1 & 2 & -4 \end{pmatrix}$.

$$tr(A) = 0 + 4 - 1 - 4 = -1$$

The trace of a matrix plays an important role in matrix theory and matrix applications because of its properties and the ease with which it can be evaluated. It is important in such fields as statistical mechanics, general relativity, and quantum mechanics, where it has physical significance.

There are results concerning the trace of the sum of two matrices, the trace of the product of two matrices and the trace of the scalar multiple of a matrix, analogous to those for transpose.

THEOREM 1-9   *If A and B are two $n \times n$ matrices, then*

$$\mathrm{tr}(A+B) = \mathrm{tr}(A) + \mathrm{tr}(B).$$

Proof:

$$\mathrm{tr}(A+B) = \sum_{i=1}^{n} (a_{ii} + b_{ii})$$

$$= \sum_{i=1}^{n} a_{ii} + \sum_{i=1}^{n} b_{ii}$$

$$= \mathrm{tr}(A) + \mathrm{tr}(B)$$

THEOREM 1-10 $\mathrm{tr}(AB) = \mathrm{tr}(BA)$

Proof: The proof of this theorem and the following one are left as exercises.   (Exercise 8 p. 58.)

THEOREM 1-11 $\mathrm{tr}(cA) = c\,\mathrm{tr}(A)$ *for scalar c.*

The following example illustrates how scientists are using some of these techniques of linear algebra in archaeology.

Example 6*

A problem confronting archaeologists is that of placing sites and artifacts in proper chronological order.   Let us look at this general problem in terms of graves and varieties of pottery found in graves.   This approach to *sequence dating* or *seriation* in archaeology began with the work of Flinders Petrie in the late nineteenth century.   Petrie studied graves in the cemeteries of Nagada, Ballas, Abadiyeh and Hu, all located in what was prehistoric Egypt.   (Recent radiocarbon dating shows that the graves ranged from 6000 B.C. to 2500 B.C.)   Petrie used the data from approximately 900 graves to order them and assign a time period or sequence to each type of pottery found.

An assumption usually made in archaeology is that two graves which lie close together in temporal order will be more likely to have similar contents than would be the case for two graves which lie further apart. The model we now construct leads to information concerning the relative common contents of graves.   We construct a matrix A, all of whose elements are either 1 or 0, that relates graves to pottery.   Label the graves $1, 2, \ldots$ and the types of pottery $1, 2, \ldots$.   Let the matrix A be defined by

$$a_{ij} = \begin{cases} 1, & \text{if grave } i \text{ contains pottery type } j \\ 0, & \text{if grave } i \text{ does not contain pottery } j. \end{cases}$$

The matrix A contains all the information about the pottery content of the various graves. The following result tells us how information is now extracted from A.

The element g_{ij} of the matrix $G = AA^t$ is equal to the number of varieties of pottery found in both graves i and j. Let us verify this result:

$$g_{ij} = \sum_k a_{ik} b_{kj} \qquad \text{where} \quad (b_{kj}) = A^t.$$

But $b_{kj} = a_{jk}$. Thus

$$g_{ij} = \sum_k a_{ik} a_{jk},$$

$$g_{ij} = a_{i1} a_{j1} + a_{i2} a_{j2} + a_{i3} a_{j3} + \cdots + a_{in} a_{jn}.$$

Each term in this sum, such as $a_{i3} a_{j3}$, will be either 0 or 1. It will be 1 if and only if a_{i3} and a_{j3} are both 1, that is, if and only if variety 3 pottery is common to both graves i and j. Thus g_{ij} will give the number of varieties of pottery found in both graves. The relative magnitudes of the elements of the matrix G thus leads scientists to the relative temporal proximity of the various graves.

The matrices A and A^t also lead to information concerning the sequence dating of the varieties of pottery. The element v_{ij} of the matrix $V = A^t A$ is equal to the number of graves in which the ith and jth varieties of pottery both appear. The assumption is made that the larger the number of graves in which two varieties of pottery appear, the closer they are chronologically. The reader is asked to verify this result in Exercise 21, p. 60.

Furthermore, the matrices $G(=AA^t)$ and $V(=A^t A)$ are symmetric. (See Exercise 21, p. 60.)

We illustrate the method for a situation involving four graves and three types of pottery. Let the following matrix A represent the pottery contents of the various graves.

$$A = \begin{pmatrix} 1 & 0 & 1 \\ 1 & 0 & 0 \\ 0 & 1 & 1 \\ 0 & 1 & 0 \end{pmatrix}$$

Thus, for example, $a_{32} = 1$ implies that grave 3 contains pottery type 2; $a_{43} = 0$ implies that grave 4 does not contain pottery type 3. G is determined:

$$G = AA^t = \begin{pmatrix} 1 & 0 & 1 \\ 1 & 0 & 0 \\ 0 & 1 & 1 \\ 0 & 1 & 0 \end{pmatrix} \begin{pmatrix} 1 & 1 & 0 & 0 \\ 0 & 0 & 1 & 1 \\ 1 & 0 & 1 & 0 \end{pmatrix} = \begin{pmatrix} 2 & 1 & 1 & 0 \\ 1 & 1 & 0 & 0 \\ 1 & 0 & 2 & 1 \\ 0 & 0 & 1 & 1 \end{pmatrix}$$

Since G is symmetric, the information contained in the elements above the main diagonal is duplicated in the elements below it. We systematically look at the elements above the diagonal.

$g_{12} = 1 \implies$ graves 1 and 2 have one type of pottery in common.

$g_{13} = 1 \implies$ graves 1 and 3 have one type of pottery in common.

$g_{14} = 0 \implies$ graves 1 and 4 do not have common content.

$g_{23} = 0 \implies$ graves 2 and 3 do not have common content.

$g_{24} = 0 \implies$ graves 2 and 4 do not have common content.

$g_{34} = 1 \implies$ graves 3 and 4 have one type of pottery in common.

Thus graves 1 and 2 are close together in time, having common content, and so are graves 1 and 3, 3 and 4. The other graves, having no common content are further apart. This information leads to the following possible orderings of the graves.

$$\text{Time}$$
$$\rightarrow$$
$$2 \rightarrow 1 \rightarrow 3 \rightarrow 4$$
$$\text{or}$$
$$4 \rightarrow 3 \rightarrow 1 \rightarrow 2$$

We see that the mathematical model does not in general lead to a unique ordering—it does however narrow down the possibilities. The archaeologist is then in a position to use other information to possibly further pinpoint the ordering.

The matrices G and V, through the relative magnitudes of their elements, contain information about the chronological order of the graves and pottery. These matrices are in practice large and the information cannot be extracted as easily as it was in the above illustration. For example, Petrie examined 900 graves; his matrix G would be a 900×900 matrix. Special mathematical techniques have been developed for extracting information from these matrices, these methods being executed on the computer. Readers who are interested in pursuing this topic further should consult "Some Problems and Methods in Statistical Archaeology" by David G. Kendall, in *World Archaeology*, **1**, 61–76, 1969. A further discussion of the role of mathematics in archaeology will also be presented in Section 1–8 of this text.

We now discuss another area of application of these techniques, the analysis of stresses within bodies in engineering.

Example 7*

The *stress matrix*, used in engineering.
Consider a body subject to external forces. The external forces

cause forces distributed throughout the body. These *internal forces* transmit the effects of the external forces. Knowledge of their distribution is necessary in order that the body may be so designed that the forces do not exceed its strength. This example illustrates how the analysis of such forces is carried out in terms of matrices.

Let O be an arbitrary point in the body. Consider a cross section of the body through O. The material on one side of this section exerts forces on the material on the other side. It is customary to analyze the effect of such forces relative to the plane, at the point, in terms of *stresses* or forces per unit area at the point. One examines the stress exerted by the portion on one side of the plane on the portion on the other side of the plane. A convention is that pull is positive and push is negative.

Consider the body in Figure 1-13, under the action of external forces.

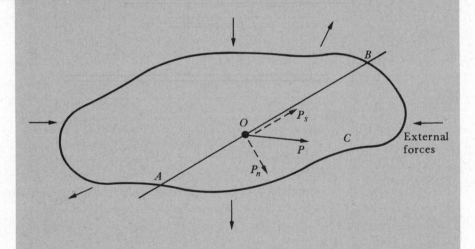

FIGURE 1-13

Let O be the point of interest in the body, and let AB be a cross section through O. Consider the stress exerted on the side C. Let it be P. This stress is analyzed by decomposing it into two components: one perpendicular to the plane P_n, called *normal stress*, and one parallel to the plane, P_s, called *shearing stress* $(P = P_n + P_s)$. The normal stress tends to elongate the body, and the shearing stress tends to cause the plane to slide on an adjacent plane.

To completely specify the forces at a point O the stresses for three mutually perpendicular planes passing through the point are given. The approach taken is to construct rectangular coordinate axes at the

point (the relevant point becomes the origin), and to let the three perpendicular planes be the *xy*, *xz*, and *yz* planes. We now introduce the
notation that is usually used. Consider the *xz* plane. The normal stress
is denoted T_y (it is in the direction of the *y* axis). The shearing stress
is considered in terms of its two components, in the *x* and *z* directions,
denoted T_{yx} and T_{yz}, respectively (see Fig. 1-14). Thus, this stress is
$T_{yx}i + T_y j + T_{yz}k$.

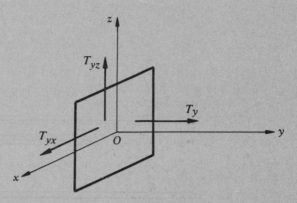

FIGURE 1-14

Similarly the components of the stress on the *xy* plane are T_z,
T_{zx}, and T_{zy}. On the *xz* plane they are T_x, T_{xy}, and T_{xz}.

Nine quantities completely specify the stresses at *O*; they are
usually written in the form of a matrix called a *stress matrix*.

$$\begin{pmatrix} T_x & T_{xy} & T_{xz} \\ T_{yx} & T_y & T_{yz} \\ T_{zx} & T_{zy} & T_z \end{pmatrix}$$

For physical reasons (which we shall not go into here) this stress matrix
is such that $T_{yx} = T_{xy}$, $T_{zx} = T_{xz}$, and $T_{zy} = T_{yz}$. Thus, in actual fact,
the six quantities T_x, T_y, T_z, T_{xy}, T_{xz}, and T_{yz} completely characterize
the stresses at a point. Such a matrix is a *symmetric matrix*. In Section 6-9
we return to this model and further analyze stresses in a body. The special
properties of symmetric matrices turn out to be crucial to this analysis.

For an introduction to stress analysis see *Introduction To Solid
Mechanics*, James McD. Baxter Brown, John Wiley & Sons, 1973.

EXERCISES

1. Determine the transpose of each of the following matrices. Indicate whether
or not the matrix is symmetric.

a) $\begin{pmatrix} -1 & 2 \\ 2 & -3 \end{pmatrix}$ b) $\begin{pmatrix} 1 & 2 \\ 0 & 3 \end{pmatrix}$ c) $\begin{pmatrix} 3 & -1 \\ 2 & 4 \end{pmatrix}$

d) $\begin{pmatrix} 4 & 5 & 6 \\ -1 & 2 & 3 \\ 0 & 1 & 2 \end{pmatrix}$ e) $\begin{pmatrix} 2 & -1 & 3 \\ 4 & 5 & 6 \\ 7 & 8 & 9 \end{pmatrix}$ f) $\begin{pmatrix} 1 & 2 \\ 3 & 4 \\ 1 & 0 \end{pmatrix}$

g) $\begin{pmatrix} 1 & -1 & 3 \\ -1 & 2 & 0 \\ 3 & 0 & 4 \end{pmatrix}$ h) $\begin{pmatrix} 3 & 1 & 4 & 5 \\ 2 & 1 & 0 & -1 \end{pmatrix}$

i) $\begin{pmatrix} 1 \\ 2 \\ 3 \end{pmatrix}$ j) $(-1 \quad 4 \quad 7 \quad 0)$

2. a) If A is an $n \times n$ matrix, prove that $AI_n = A$. [This is the second part of Theorem 1-8.]
   b) If A is an $m \times n$ matrix, prove that $AI_n = A$.

3. Determine the traces of the following matrices:

   a) $\begin{pmatrix} 1 & 2 \\ -1 & 3 \end{pmatrix}$ b) $\begin{pmatrix} 0 & -1 \\ 2 & 3 \end{pmatrix}$

   c) $\begin{pmatrix} 1 & 2 & 3 \\ 0 & -1 & 1 \\ 4 & 1 & 0 \end{pmatrix}$ d) $\begin{pmatrix} -1 & 2 & 3 & 4 \\ 0 & 0 & 0 & 0 \\ 1 & 2 & 3 & 4 \\ 1 & 2 & 3 & 1 \end{pmatrix}$

4. Consider two matrices of the same kind, A and B, such that $A^t = B^t$. Prove that $A = B$.

5. Consider two matrices of the same kind, A and B, such that $\operatorname{tr}(A) = \operatorname{tr}(B)$. Does this imply that $A = B$?

6. Let A be a diagonal matrix. Prove that $A = A^t$.

7. Prove the following results for transpose.
   a) $(cA)^t = cA^t$, c being a scalar, A a matrix.
   b) $(A^t)^t = A$.
   c) $(A+B+C)^t = A^t+B^t+C^t$
   d) $(ABC)^t = C^t B^t A^t$.

8. Prove the following results for trace.
   a) $\operatorname{tr}(cA) = c\operatorname{tr}(A)$, c being a scalar, A a square matrix.
   b) $\operatorname{tr}(AB) = \operatorname{tr}(BA)$, A and B being square matrices of the same kind.
   c) $\operatorname{tr}(A) = \operatorname{tr}(A^t)$, A being a square matrix.
   d) $\operatorname{tr}(A+B+C) = \operatorname{tr}(A)+\operatorname{tr}(B)+\operatorname{tr}(C)$, A, B, and C being square matrices of the same kind.

9. Prove that the product of two diagonal matrices of the same kind is a diagonal matrix.

10. **a**) Prove that the sum of two symmetric matrices of the same kind is symmetric.

    b) Prove that the scalar multiple of a symmetric matrix is symmetric.

11. Prove that a symmetric matrix is necessarily square.

12. A matrix A is said to be antisymmetric if $A = -A^t$. Give an example of an antisymmetric matrix. Prove that such a matrix is a square matrix having diagonal elements all zero. Prove that the sum of two antisymmetric matrices is itself antisymmetric. Further, prove that a scalar multiple of an antisymmetric matrix is antisymmetric.

13. If A is a symmetric matrix, prove that **a**) $A + A^t$ is symmetric, **b**) $A - A^t$ is antisymmetric.

14. Prove that any square matrix A can be decomposed into the sum of a symmetric matrix B and antisymmetric matrix C. $A = B + C$. (*Hint*: Use Exercise 13.)

15. Prove that $(I_3)^2 = (I_3)^3 = I_3$. Generalize this result by proving that $(I_n)^m = I_n$, where m is any positive integer.

16. An *upper triangular matrix* is a square matrix of the type

$$\begin{pmatrix} a_{11} & a_{12} & \cdots & a_{1n} \\ & a_{22} & & \\ 0 & & \ddots & \\ & & & a_{nn} \end{pmatrix},$$

having zeros below the main diagonal. Prove that the product of two upper triangular matrices of the same kind is an upper triangular matrix. [A lower triangular matrix is a square matrix having zeros above the main diagonal.]

17. **a**) Let A be a 4×4 matrix defined by $a_{ij} = \delta_{i\ j-1}$. Write down A.

    b) B is a 5×5 matrix defined by $b_{ij} = \delta_{i+1\ j-1}$. Write down B.

18. Expand the sums

    a) $\sum_{k=1}^{4} \delta_{2k} a_{k4}$ **b**) $\sum_{k=1}^{3} b_{2k} \delta_{k3}$

19. Let A be an arbitrary matrix. Prove that the products AA^t and A^tA always exist. [This result has been assumed in the archaeology model.]

20. The following matrices describe the pottery contents of various graves. For each situation determine possible chronological orderings of (i) the graves, and (ii) the pottery types.

    a) $A = \begin{pmatrix} 1 & 0 \\ 0 & 1 \\ 1 & 1 \end{pmatrix}$ **b**) $A = \begin{pmatrix} 0 & 0 & 1 \\ 1 & 1 & 0 \\ 1 & 0 & 1 \\ 0 & 1 & 0 \end{pmatrix}$

c) $A = \begin{pmatrix} 1 & 0 & 1 & 0 \\ 0 & 1 & 1 & 1 \\ 1 & 1 & 1 & 1 \end{pmatrix}$ d) $A = \begin{pmatrix} 0 & 0 & 0 & 1 \\ 1 & 1 & 0 & 0 \\ 0 & 0 & 1 & 1 \\ 1 & 0 & 1 & 0 \end{pmatrix}$

e) $A = \begin{pmatrix} 1 & 0 & 1 & 0 \\ 1 & 0 & 0 & 0 \\ 0 & 1 & 0 & 1 \\ 0 & 1 & 1 & 0 \end{pmatrix}$.

21. Let A be a matrix of 1s and 0s that defines the relationship between pottery varieties and graves, as in example 4. Let $V = A^t A$. Prove that v_{ij} gives the number of graves in which the ith and jth varieties of pottery are both found. This leads to information concerning the temporal proximities of various varieties of pottery: the assumption is made that the larger the number of graves in which two varieties of pottery appear, the closer they are chronologically.

    Prove that $G(= AA^t)$ and $V(= A^t A)$ are symmetric matrices.
    [*Hint*: Use Theorems 1–5 and 1–7.]

22. Let A be the matrix that defines the relationship between pottery and graves in archaeology. Write a computer program for determining $G(= AA^t)$ and $V(= A^t A)$.

23. The model used here in archaeology can be used in sociology to analyze relationships within a group of people. Consider a group of people and let us look at the relationship of friendship. Let us assume that all friendships are mutual. Label the people 1 to n and define a matrix A:

    $a_{ii} = 0$ for all i (diagonal element zero)

    $a_{ij} = \begin{cases} 1 \text{ if } i \text{ and } j \text{ are friends} \\ 0 \text{ if } i \text{ and } j \text{ are not friends.} \end{cases}$

    a) Prove that if $G = AA^t$ then g_{ij} is the number of friends that i and j have in common.

    b) Suppose we drop the condition of mutual friendship, i.e., it is possible for i to interpret j as a friend but that j might not consider i a friend. How will this possibly affect the model?

1–7.* A POPULATION MOVEMENT MODEL

Certain types of matrices, called *stochastic matrices*, are of great use in the study of random phenomena where the exact outcome is not known but probabilities can be determined. In this section we introduce stochastic matrices, derive some of their properties, and give examples of their application. One example is an analysis of population movement between a city and its suburbs. The trend from the city to suburbia is one that should be of concern not only to sociologists and politicians, but to all of us. This example shows how a mathematical model can be constructed to analyze this important problem.

If the outcome of an event is certain to occur, we say that its probability is 1. On the other hand, if it will not occur, its probability is said to be 0. Other probabilities can be represented by fractions between 0 and 1: the larger the fraction, the greater the probability p of that outcome occurring. Thus we have the restriction $0 \leq p \leq 1$ for a probability p.

If any one of n completely independent outcomes is equally likely and if m of these are of interest to us, then the probability p that one of these outcomes will occur is defined to be the fraction m/n.

As an example, let us work out the probability that, when a coin is flipped, it will land heads. There are two possible outcomes, one of which is of interest (landing heads). The probability p of this occurrence is $\frac{1}{2}$.

As another example, consider the event of drawing a single card from a deck of 52 playing cards. What is the probability that the card will be either an ace or a king? First of all, we see that there are 52 possible outcomes. There are four aces and four kings in the pack. Since there are eight outcomes of interest to us, the probability is $\frac{8}{52}$, or $\frac{2}{13}$.

Let an event have n outcomes with probabilities $p_1, \ldots, p_n$. One of the outcomes is sure to occur. This leads to a second restriction on the probabilities: $p_1 + \cdots + p_n = 1$.

We can represent such probabilities by the elements of vectors and matrices.

A *stochastic vector* is a vector whose components can be interpreted as the probabilities of all the possible outcomes of a given event. Thus (p_1, p_2) is a stochastic vector if and only if $0 \leq p_1 \leq 1$, $0 \leq p_2 \leq 1$, and $p_1 + p_2 = 1$. For example, $(\frac{1}{2}, \frac{1}{2})$ would be a stochastic vector whose components correspond to the probabilities of the various outcomes when a coin is flipped.

A square matrix is called a *stochastic matrix* if its rows are stochastic vectors. The following matrices are stochastic matrices.

$$\begin{pmatrix} \frac{1}{2} & \frac{1}{2} \\ \frac{1}{3} & \frac{2}{3} \end{pmatrix} \begin{pmatrix} 0 & 1 \\ \frac{3}{4} & \frac{1}{4} \end{pmatrix} \begin{pmatrix} 1 & 0 & 0 \\ 0 & \frac{1}{2} & \frac{1}{2} \\ \frac{3}{4} & \frac{1}{8} & \frac{1}{8} \end{pmatrix}$$

The matrix $\begin{pmatrix} \frac{1}{2} & 1 \\ \frac{3}{4} & \frac{1}{4} \end{pmatrix}$ would not be a stochastic matrix—its first row is not a stochastic vector because the sum of its elements is not 1.

A general 2×2 stochastic matrix can be represented by

$$\begin{pmatrix} x & 1-x \\ y & 1-y \end{pmatrix}$$

where $0 \le x \le 1, 0 \le y \le 1$.

Stochastic matrices have certain properties. The following theorem illustrates one of these properties.

THEOREM 1-12   *If A and B are stochastic matrices of the same kind, then AB is a stochastic matrix.*

Proof: We shall prove this theorem for 2×2 stochastic matrices. Let

$$\begin{pmatrix} x & 1-x \\ y & 1-y \end{pmatrix} \quad \text{and} \quad \begin{pmatrix} z & 1-z \\ m & 1-m \end{pmatrix}$$

represent two 2×2 arbitrary stochastic matrices. We know that their product is

$$\begin{pmatrix} xz + (1-x)m & x(1-z) + (1-x)(1-m) \\ yz + (1-y)m & y(1-z) + (1-y)(1-m) \end{pmatrix}$$

Consider the first row vector. The sum of its components is

$$xz + (1-x)m + x(1-z) + (1-x)(1-m)$$

$$= xz + m - xm + x - xz + 1 - x - m + xm = 1$$

Next consider the $(1, 1)$ element, $xz + (1-x)m$. $x \ge 0, z \ge 0$, and $(1-x)m \ge 0$, since $x \le 1$ and $m \ge 0$. Hence this element is ≥ 0. Similarly, the $(1, 2)$ element $x(1-z) + (1-x)(1-m)$ is ≥ 0. The elements must both be ≤ 1 since their sum is 1. Hence the first row vector is a stochastic vector.

It can be shown in a similar manner that the second row vector is a stochastic vector, proving that the product matrix is a stochastic matrix.

It follows from this result that if A is a stochastic matrix, then so are A^2 and all A^n, where n is any positive integer.

Now let us look at applications of stochastic matrices in the social sciences.

Example 1

This example gives an analysis of land use succession using available statistics for center-city Toronto for the period 1952–1962.

The data are represented in the form of a stochastic matrix p_{ij}, where each element p_{ij} represents the probability of succession from use i to use j.   In this context, the stochastic matrix is called a *matrix of transition probabilities*.   In this model, change is defined to include all new construction, demolitions, or major structural adjustments in the building stock.   The matrix in Table 1-4 is stochastic; each element satisfies $0 \leq p_{ij} \leq 1$ and the sum of the elements in each row is 1.

Table 1-4. *Land Use Succession and Transition Probabilities for the Total City*, 1952–1962

1952 Existing use	Terminal use									
	1	2	3	4	5	6	7	8	9	10
1. Low-density residential	0.13	0.34	0.10	0.04	0.04	0.22	0.03	0.02	0.00	0.08
2. High-density residential	0.02	0.41	0.05	0.04	0.00	0.04	0.00	0.00	0.00	0.44
3. Office	0.00	0.07	0.43	0.05	0.01	0.28	0.14	0.00	0.00	0.02
4. General commercial	0.02	0.01	0.09	0.30	0.09	0.27	0.05	0.08	0.01	0.08
5. Auto commercial	0.00	0.00	0.11	0.07	0.70	0.06	0.00	0.01	0.00	0.05
6. Parking	0.08	0.05	0.14	0.08	0.12	0.39	0.04	0.00	0.01	0.09
7. Warehousing	0.01	0.03	0.02	0.12	0.03	0.11	0.38	0.21	0.01	0.08
8. Industry	0.01	0.02	0.02	0.03	0.03	0.08	0.18	0.61	0.00	0.02
9. Transportation	0.01	0.18	0.14	0.04	0.10	0.39	0.03	0.03	0.08	0.00
10. Vacant	0.25	0.08	0.03	0.03	0.05	0.15	0.22	0.13	0.00	0.06

Let us now interpret some of the information contained in this matrix.

For example, $p_{36} = 0.28$.   This represents the probability that what is an office in 1952 will have become a parking space by 1962.   The sixth column gives the probabilities that various areas will have become parking areas by 1962.   These relatively large figures reveal the increasingly dominant role of parking in land use.   Relatively high proportions of all types of land were converted into parking.

The diagonal elements represent the probabilities that land use will remain in the same category.   For example, $p_{77} = 0.38$ represents the probability that warehousing land will remain warehousing land.   The relatively high figures of these diagonal numbers reveal the marked tendency for land to remain in the same broad category.   The exceptions are land used for transportation and vacant land.

It is interesting that $p_{2\ 10}$ is equal to 0.44.   This is the probability that area that is high-density residential in 1952 will become vacant by 1962.   Note that this is an area considered to be in center-city Toronto.

The reader who is interested in following up this type of analysis is encouraged to read the article from which these data have been abstracted: Larry S. Bourne, "Physical Adjustment Processes and Land Use Succession: Toronto," in *Economic Geography*, **47**, No. 1, January,

1971, pp. 1–15.   Another article by the same author which outlines the use of such matrices is "A Spatial Allocation—Land Use Conversion Model of Urban Growth," in *Journal of Regional Science*, **9**, August, 1969, pp. 261–272.   References to other similar works are found in both articles.

Example 2

This example is an analysis of the population movement between cities and their surrounding suburbs in the United States.   The numbers given are based on statistics in *Statistical Abstracts of the U.S.A.*, 1969–1971.

The number of people (in thousands of persons 1 year old or over) who lived in cities in the U.S. during 1971 was 57,633.   The number of people who lived in the surrounding suburbs was 71,549.   We represent this information by the vector $\mathbf{x}_0 = (57{,}633,\ 71{,}549)$.

Consider the population flow from city centers to suburbs. During the period 1969–1971, the average probability per year of a person staying in the city was 0.96.   Thus the probability of moving to the suburbs was 0.04 (assuming that all those moving went to the suburbs). We can represent this by the stochastic vector $(0.96, 0.04)$.

Consider now the reverse population flow from suburbia to city. The probability of a person moving to the city was 0.01, the probability of a person remaining in suburbia was 0.99.   Represent these data by the stochastic vector $(0.01, 0.99)$.

These probabilities can be written in the form of a stochastic matrix

$$P = (p_{ij}) = \begin{array}{c} \\ \text{Initial} \end{array} \begin{array}{c} \\ \begin{array}{c} \text{City} \\ \text{Suburb} \end{array} \end{array} \overset{\begin{array}{c} \text{Final} \\ \text{City}\quad\text{Suburb} \end{array}}{\begin{pmatrix} 0.96 & 0.04 \\ 0.01 & 0.99 \end{pmatrix}}$$

To determine the probability of moving from location A to location B, look up row A and column B.

Consider the population of the city centers in 1972—one year later. 0.96 is the probability of a person living in the city staying there; it represents the fraction of those living in the city who stay there.   We have that

City population in 1972

= people who remained from 1971 + people who moved in from the suburbs

$= (57{,}633 \times 0.96) + (71{,}549 \times 0.01)$

$= 56{,}043.2$

Similarly,

Suburban population in 1972

= people who moved in from cities + people who stayed, from 1971

$$= (57{,}633 \times 0.04) + (71{,}549 \times 0.99)$$

$$= 73{,}138.8$$

Note that we can arrive at these two numbers using matrix multiplication.

$$(\ 57{,}633 \quad 71{,}549 \ ) \begin{pmatrix} 0.96 & 0.04 \\ 0.01 & 0.99 \end{pmatrix} = (\ 56{,}043.2 \quad 73{,}139.8 \ )$$

or

$$\mathbf{x}_1 = \mathbf{x}_0 P$$

where $\mathbf{x}_0$ is the population vector in 1971, P is the stochastic matrix giving probability distributions, and $\mathbf{x}_1$ is the population vector one year later.

Assuming that the population flow represented by the matrix P is unchanged, the population distribution $\mathbf{x}_2$ after 2 years is given by

$$\mathbf{x}_2 = \mathbf{x}_1 P,$$

that is

$$\mathbf{x}_2 = \mathbf{x}_0 P^2.$$

After 3 years the population distribution is given by

$$\mathbf{x}_3 = \mathbf{x}_2 P,$$
$$= \mathbf{x}_0 P^3.$$

Assuming that the matrix of transition probabilities does not vary, we can predict the distribution any number of years later, from the general result,

$$\mathbf{x}_n = \mathbf{x}_0 P^n$$

Observe that the matrix P^n is a stochastic matrix and that it does "take $\mathbf{x}_0$ into $\mathbf{x}_n$", n stages later. This result can be generalized to

$$\mathbf{x}_{i+n} = \mathbf{x}_i P^n.$$

That is P^n can be used to predict the distribution n stages later from any given distribution. P^n is called the *n-step transition matrix*.

The (i, j)th element of P^n gives the probability of going from state S_i to state S_j in n steps.

The predictions of this model are

	City	Suburb
$\mathbf{x}_1 =$	(56043.2	73138.8)
$\mathbf{x}_2 =$	(54532.8	74649.2)
$\mathbf{x}_3 =$	(53098	76084)

$$\mathbf{x}_4 = (51734.9 \quad 77447.1)$$

$$\mathbf{x}_5 = (50440 \quad 78742)$$

etc.

Observe how the city population is decreasing annually while that of the suburbs is increasing. We return to this model in Section 6-6. There we find that the sequence $\mathbf{x}_1$, $\mathbf{x}_2$, $\mathbf{x}_3$ $\cdots$ approaches (25836.4 103345.6).

The probabilities in this model depend only on the current status of a person—whether he is living in the city or suburbia. This type of model, where the probability of going from one state to another depends only on the current state rather than on a more complete historical description, is called a *Markov chain*.

These concepts can be extended to Markov processes involving more than two states. The reader meets a model involving three states in a following example.

A further modification of the model would give an improved estimate of future population distributions. Let us take into account the fact that the population of the United States increased by 1% per annum during the period 1969–1971. We assume that the population will increase by the same amount annually during the years immediately following 1971.

Thus, if the population was b during any year, the population the following year would be $b+1\%$ of b.

$$b + b/100 = \tfrac{101}{100} b$$

Our model now becomes

$$\mathbf{x}_1 = \tfrac{101}{100}\mathbf{x}_0 P$$

Note that since $\tfrac{101}{100}$ is a scalar, we are multiplying a matrix by a scalar and so the multiplying sequence does not matter.

$$\mathbf{x}_1 = \tfrac{101}{100}(\; 57,633 \quad 71,549 \;)\begin{pmatrix} 0.96 & 0.04 \\ 0.01 & 0.99 \end{pmatrix}$$

$$= (\; 56603.632 \quad 73870.188 \;)$$

The population of city centers is now 56603.632 thousand and that of suburbia is 73870.188 thousand.

For 1973, future population distribution based on a uniform 1% annual population increase would be

$$\mathbf{x}_2 = (\tfrac{101}{100})^2 \mathbf{x}_0 P^2$$

After n years, it would be

$$\mathbf{x}_n = (\tfrac{101}{100})^n \mathbf{x}_0 P^n$$

Note that in this model we have made the assumption that the increase in population is 1% in both cities and suburbia since it is 1% nationally. The increase might be slightly higher than 1% in the cities and correspondingly less than 1% in suburbia. A more accurate model would incorporate a breakdown of this growth statistic. The reader is asked to carry out this modification in Exercise 11, p. 71.

Example 3

This example illustrates the use of stochastic matrices in analyzing intergenerational mobility.

In 1953, N. Rogoff collected data on occupations listed on marriage license applications in Marion County, Indiana. The licenses describe the occupations of both male applicant and his father. The data collected refer to two time periods—1905 through 1912, and 1938 through the first half of 1941. These will be abbreviated as 1910 and 1940, respectively. This example is particularly interesting because we can compare the matrices of transition probabilities of two different time periods. We distinguish three broad occupational categories: (1) nonmanual nonfarm, (2) manual nonfarm, and (3) farming. The matrices for 1910 and 1940 are, respectively,

$$P_1 = \begin{pmatrix} 0.594 & 0.396 & 0.009 \\ 0.211 & 0.782 & 0.007 \\ 0.252 & 0.641 & 0.108 \end{pmatrix} P_2 = \begin{pmatrix} 0.622 & 0.375 & 0.003 \\ 0.274 & 0.721 & 0.005 \\ 0.265 & 0.694 & 0.042 \end{pmatrix}$$

These matrices are fairly similar, but the probabilities of transition to the nonmanual category (first column) have all increased in the years between 1910 and 1940, whereas the probabilities of transition to farming (third column) have all decreased.

The observed occupational distribution in 1910 is 31% : 65. 6% : 3.4%. Represent these statistics as the components of a vector: $(31, 65.6, 3.4)$. Postmultiplying this vector by P_1 gives $(33.1, 65.8, 1.1)$ for the distribution of the next generation.

These predictions are confirmed—at least qualitatively—by the observed occupational distribution of 1940, which is $(37.3, 61.6, 1.1)$. There was a decrease of farming, as predicted by our model, but the increase of the nonmanual category was beyond the levels predicted by our model.

Using the 1940 occupational distribution $(37.3, 61.6, 1.1)$ and the matrix P_2, the occupational distribution for the present should, according to our model, be $(40.4, 59.1, 0.5)$. It would be interesting to compare this prediction with the current situation.

Readers who desire further knowledge of these mathematical methods for analyzing social mobility should refer to *Statistical Decomposition Analysis with Applications in the Social and Administrative Sciences*, Henri

Theil, American Elsevier Publishing Company, 1972, Chap. 5. (This example is a modification of an example that appears in the text.) For an introduction to probability theory and stochastic matrices see *An Introduction to Probability Theory and Its Applications*, William Feller, John Wiley & Sons, 1971.

Example 4

Markov chains are useful tools for scientists in many fields. We now discuss the role of Markov chains in the field of *genetics*.

Genetics is the branch of biology that deals with heredity. It is the study of units called genes which determine the characteristics living things inherit from their parents. The inheritance of such traits as sex, height, eye color, and hair color of human beings, and such traits as petal color and leaf shape of plants are governed by genes. Because many diseases are inherited, genetics is important in medicine. In agriculture, breeding methods based on genetic principles led to important advances in both plant and animal breeding. High-yield hybrid corn ranks as one of the most important contributions of genetics to increasing food production. We shall consider a model developed for analyzing the behavior of simple traits involving a pair of genes. We illustrate the concepts involved in terms of the crossing of guinea pigs.

The traits that we shall study in the guinea pigs are the traits of long hair and short hair. The length of hair is governed by a pair of genes which we shall denote A and a. A guinea pig may have any one of the combinations AA, Aa (which is genetically the same as aA), or aa. Each of these classes is called a *genotype*. The AA type guinea pig is indistinguishable in appearance from the Aa type—both have long hair— while the aa type has short hair. We say that the A gene *dominates* the a gene. An animal is called *dominant* if it has AA genes, *hybrid* with Aa genes, and *recessive* with aa genes.

In crossing two guinea pigs, the offspring inherits one gene from each parent in a random manner. Given the genotypes of the parents we can determine the probabilities of the genotype of the offspring. Consider a given population of guinea pigs. Let us perform a series of experiments in which *we keep crossing offspring with dominant animals only*. Thus we keep crossing AA, Aa, and aa with AA. What are the probabilities of the offspring being AA, Aa, or aa in each of these cases?

Consider the crossing of AA with AA. The offspring will have one gene from each parent, thus it will be of type AA. Thus the probabilities of AA, Aa, and aa resulting are 1, 0, and 0, respectively. All offspring will have long hair.

Next consider the crossing of Aa with AA. Taking one gene from each parent, there are the possibilities AA, AA (taking the A from the first parent with each A in turn from the second parent), aA, and aA (taking the a from the first parent with each A in turn from the second parent).

Thus the probabilities of *AA*, *Aa*, and *aa* resulting are $\frac{1}{2}$, $\frac{1}{2}$, and 0, respectively. All offspring again have long hair.

Finally, on crossing *aa* with *AA* we get the possibilities *aA*, *aA* (taking the first *a* in *aa* with each *A* in *AA* in turn), *aA*, and *aA*. Thus the probabilities of *AA*, *Aa* and *aa* are 0, 1, 0 respectively.

All offspring resulting in these experiments have long hair. This series of experiments is a Markov chain having transition matrix

$$
\begin{array}{c}
\quad\; AA\; Aa\; aa \\
P = \begin{array}{c} AA \\ Aa \\ aa \end{array}\!\left(\begin{array}{ccc} 1 & 0 & 0 \\ \frac{1}{2} & \frac{1}{2} & 0 \\ 0 & 1 & 0 \end{array}\right)
\end{array}
$$

The genotype *AA* is called an *absorbing* state in this model. It is not possible to go from it into another state. Further, observe that the other genotypes will gradually become a smaller and smaller fraction of the population.

Here we have considered the case of crossing offspring with a dominant animal. The reader is asked to construct a similar model that describes the crossing of offspring with a hybrid in Exercise 13 following. Some of the offspring will have long hair, some short hair in that series of experiments.

EXERCISES

1. State which of the following matrices are stochastic and which are not.

   a) $\begin{pmatrix} \frac{1}{4} & \frac{3}{4} \\ 0 & 1 \end{pmatrix}$ **b)** $\begin{pmatrix} \frac{1}{2} & \frac{1}{2} \\ -1 & 0 \end{pmatrix}$ **c)** $\begin{pmatrix} \frac{1}{2} & \frac{1}{4} & \frac{1}{4} \\ 1 & 0 & 0 \\ \frac{1}{8} & \frac{1}{8} & \frac{3}{4} \end{pmatrix}$

2. If **x** is a 1×2 stochastic vector and A is a 2×2 stochastic matrix, prove that **x**A is a stochastic vector.

3. A stochastic matrix the sum of whose columns is 1 is called *doubly stochastic*. Each column of such a matrix can be interpreted as giving the complete set of probabilities of the outcomes of an event. Give an example of a doubly stochastic matrix. Is the product of two doubly stochastic matrices of the same kind doubly stochastic?

4. Use the matrix of transition probabilities in Example 1 to answer the following questions:
   a) What is the probability that land used for industry in 1952 will be used for offices in 1962?
   b) What is the probability that land used for parking in 1952 will be a high-density residential area in 1962?

c) Vacant land in 1952 has the highest probability of becoming what kind of land by 1962?

d) Which is the most stable usage of land over the period from 1952–1962?

5. Assuming the same matrix of transition probabilities for land usage in Toronto for the period 1962–1972 as for the period 1952–1962, determine the matrix of transition probabilities for the period 1952–1972. Generalize this result.

6. In the original model in Example 2, determine
   a) the probability of moving from center city to the suburbs in two years
   b) the probability of moving from the suburbs to center city in three years

7. Draw up a model of population flow between metropolitan areas and non-metropolitan areas given that their respective populations in 1971 were 129,182 and 69,723 (in thousands of persons one year old or over). The probabilities are given by the matrix

$$
\begin{array}{cc}
 & \text{Final} \\
 & \text{Metro \quad Nonmetro}
\end{array}
$$

$$
\text{Initial} \quad
\begin{array}{c}
\text{Metro} \\
\text{Nonmetro}
\end{array}
\left(
\begin{array}{cc}
0.99 & 0.01 \\
0.02 & 0.98
\end{array}
\right)
$$

8. Construct a model of population flow between cities, suburbs, and non-metropolitan areas. Their respective populations in 1971 were 57,633, 71,549, and 69,723 (in thousands of persons one year old or over). The stochastic matrix giving the probabilities of the moves is

$$
\begin{array}{ccc}
 & \text{Final} & \\
\text{City} & \text{Suburb} & \text{Nonmetro}
\end{array}
$$

$$
\text{Initial} \quad
\begin{array}{c}
\text{City} \\
\text{Suburb} \\
\text{Nonmetro}
\end{array}
\left(
\begin{array}{ccc}
0.96 & 0.03 & 0.01 \\
0.01 & 0.98 & 0.01 \\
0.015 & 0.005 & 0.98
\end{array}
\right)
$$

This model is a refinement on the model of Exercise 7 in that the Metro population is broken down into City and Suburb. It is also a more complete model than that of Example 2 of this section, leading to more accurate predictions than that model.

9. The following stochastic matrix gives occupational transition probabilities.

$$
\begin{array}{cc}
 & \text{Following generation} \\
 & \text{White collar \quad Manual}
\end{array}
$$

$$
\begin{array}{c}
\text{Initial} \\
\text{generation}
\end{array}
\quad
\begin{array}{c}
\text{White collar} \\
\text{Manual}
\end{array}
\left(
\begin{array}{cc}
1 & 0 \\
0.2 & 0.8
\end{array}
\right)
$$

a) If the father is a manual worker, what is the probability that the son will be a white collar worker?

b) If there are 10,000 in the white collar category and 20,000 in the manual category, what will be the distribution one generation later? three generations later?

10. The following stochastic matrix gives occupational transition probabilities.

$$
\begin{array}{cc}
 & \text{Following generation} \\
 & \begin{array}{cc} \text{Nonfarming} & \text{Farming} \end{array}
\end{array}
$$

$$
\begin{array}{cc}
\text{Initial} & \text{Nonfarming} \\
\text{generation} & \text{Farming}
\end{array}
\begin{pmatrix} 1 & 0 \\ 0.4 & 0.6 \end{pmatrix}
$$

a) If the father is a farmer, what is the probability that the son will be a farmer?

b) If there are 10,000 in the nonfarming category and 1,000 in the farming category at a certain time, what will the distribution be one generation later? four generations later?

c) If the father is a farmer, what is the probability that the grandson (two generations later) will be a farmer?

11. Carry out the proposed modification of the model in Example 2. Assume that the population increase (that allows for births and deaths) per annum in the cities is 1.2% and in suburbia is 0.8%. [*Hint*: Represent this information in the form of a 2×2 matrix A. Then $\mathbf{x}_1 = \mathbf{x}_0\, PA$, $\mathbf{x}_2 = \mathbf{x}_0\, (PA)\, (PA)$, ..., $\mathbf{x}_n = \mathbf{x}_0\, (PA)^n$.] Note that the order of multiplication does matter here since matrices are not commutative (unlike the situation in the previous model, where the scalar $\frac{101}{100}$ was used).

12. A certain society is made up of two groups, a large majority group and a small minority group. During a certain period in the history of the society, it is fashionable to become part of the majority group by marrying into this group, by adopting the culture of the group, etc. Initially, the populations of a single generation of the two groups are 50 million and $\frac{1}{2}$ million. The following is a matrix of transition probabilities.

$$
\begin{array}{cc}
 & \text{Following Generation} \\
 & \overline{\begin{array}{cc} \text{Majority} & \text{Minority} \\ \text{Group} & \text{Group} \end{array}}
\end{array}
$$

$$
\begin{array}{cc}
\text{Initial} & \left\{ \begin{array}{l} \text{Majority group} \\ \text{Minority group} \end{array} \right.
\end{array}
\begin{pmatrix} 1 & 0 \\ 0.2 & 0.8 \end{pmatrix}
$$

The matrix represents the probabilities of a following generation being in a certain group.

a) Assuming that the total population of each successive group remains the same, construct a model that gives the populations after n years.

b) Allow for a 1% uniform population increase in each successive generation.

c) Assume that in each generation there is a 2% increase in population of the majority group and a 3% decrease in population of the minority group due to the births and deaths. Allow for these figures in your model.

A state is said to be an *absorbing state* if it is not possible to leave it. The majority group is an absorbing state in all the models of this exercise, the probability of leaving the majority group being 0. The minority group is being absorbed into the majority group. The minority culture will be lost to future generations unless the society acts.

13. Construct a model, similar to that of Example 4, that describes the crossing of offspring of guinea pigs with hybrids only. [We shall pursue this model further in Exercise 5, Section 6-6.]

14. **a**) Write a computer program for the model of Example 2 that gives the city and suburb populations annually from 1971 to 1981.

    b) Modify the program to allow for a uniform 1% annual increase in population. Use your program to observe the effect of varying the uniform increase. This increase is called a *parameter* of the model.

    c) Modify the above program to allow for a 1.2% annual increase in city population and 0.8% increase in suburbia (Exercise 11) due to births and deaths. Use your program to observe the effect of varying these parameters.

15. **a**) Given a vector $\mathbf{x} = (x_1, \ldots, x_n)$ having non-negative components, there exists a stochastic vector $\mathbf{y} = (y_1, \ldots, y_n)$ corresponding to $\mathbf{x}$, in the sense that its components are in the same ratio as those of $\mathbf{x}$, that is $x_i/x_j = (y_i/y_j)$. Determine a method for computing such a vector $\mathbf{y}$ from a given vector $\mathbf{x}$.

    Determine the stochastic vector corresponding to each of the following vectors: (i) $(1, 2)$, (ii) $(1, 0, 3)$, and (iii) $(1, 2, 3, 4)$.

    b) Given a square matrix A, having nonnegative elements, there exists a stochastic matrix B corresponding to A in the sense that the components of each row vector of B are in the same ratio as the components of the corresponding row vector of A. Determine a method for computing such a stochastic matrix B that corresponds to a given matrix A.

    Determine the stochastic matrix corresponding to each of the following matrices:

$$\text{(i)} \begin{pmatrix} 0 & 2 \\ 0 & 2 \end{pmatrix} \qquad \text{(ii)} \begin{pmatrix} 1 & 2 \\ 3 & 4 \end{pmatrix} \qquad \text{(iii)} \begin{pmatrix} 1 & 2 & 3 \\ 1 & 2 & 1 \\ 4 & 1 & 5 \end{pmatrix}$$

    c) Write a computer program that can be used for deriving stochastic vectors and matrices from given appropriate vectors and matrices.

16. A market analysis group studying car purchasing trends in a certain region has concluded that a new car is purchased once every 3 years on an average. The buying patterns are described by the matrix

$$A = \begin{array}{c} \\ \text{Small} \\ \text{Large} \end{array} \begin{array}{c} \text{Small Large} \\ \begin{pmatrix} 80\% & 20\% \\ 40\% & 60\% \end{pmatrix} \end{array}$$

The elements of A are to be interpreted as follows. The first row indicates that of the current small cars, 80% will be replaced with a small car, 20% with a large car. The second row implies that 40% of the current large cars will be replaced with small cars while 60% will be replaced by large cars.

Construct a stochastic matrix P from A that will define a Markov chain model of these buying trends.

If there are currently 40,000 small cars and 50,000 large cars in the region what is your prediction of the distribution in 12 years time?

17. The conclusions of an analysis of voting trends in a certain state is that the voting patterns of successive generations are described by the following matrix A.

$$
\begin{array}{cc}
& \begin{array}{ccc} \text{Dem.} & \text{Rep.} & \text{Ind.} \end{array} \\
A = \begin{array}{c} \text{Democrat} \\ \text{Republican} \\ \text{Independent} \end{array} & \left(\begin{array}{ccc} 80\% & 15\% & 5\% \\ 20\% & 70\% & 10\% \\ 60\% & 30\% & 10\% \end{array} \right)
\end{array}
$$

Among the democrats of one generation, 80% of their next generation are democrats, 15% are republican, 5% are independent, etc.

Construct a stochastic matrix P from A that defines a Markov chain model of these voting patterns.

If there are 2.5 million registered democrats, 1.5 million registered republicans and .25 million registered independents at a certain period, what is the distribution likely to be in the next generation?

1–8.* COMMUNICATION MODEL AND GROUP RELATIONSHIPS

Many branches of the social sciences and business use models from an area of mathematics called *graph theory* to analyze relationships among elements of a group. The group could consist of people within a certain company; for example, the techniques of graph theory can be used to analyze the hierarchy within a company. Sociologists apply the results of graph theory to analyze the relations among people that make up a group. Principles of graph theory are also used to improve the efficiency of computer networks. In this section we introduce the reader to this increasingly important field which uses matrix algebra. To motivate the concepts involved let us look at the following communication system.

Consider a communication system involving five stations labeled $P_1, ..., P_5$. The communication links could be roads, phone lines, etc. Certain stations are linked by two-way communication links, others by one-way links. Still others have only indirect communication by way of intermediate stations. Suppose the network of interest is described by the diagram in Figure 1-15. Curves joining stations represent links; the arrows give the directions of those links. Stations P_1 and P_2 have

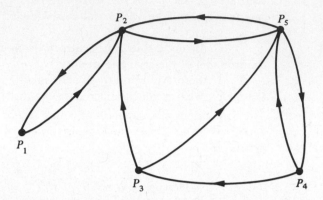

FIGURE 1-15

two-way direct communication.   Station P_3 can send a message to station P_5 directly, whereas station P_5 can only send a message to P_3 by way of P_4.   Station P_4 can send a message to P_1 by way of stations P_3 and P_2 or by way of P_5 and P_2.

Such communication networks can be vast, involving many stations.   The diagrams of vast systems become complicated and difficult to interpret in this geometrical manner.   The mathematical vocabulary and theory we now present can be used to analyze such networks.   This theory can be used, for example, to give the minimum number of links needed to send a message from one station to another.

A *graph*† is a finite collection of vertices $P_1, P_2, ..., P_n$ together with a finite number of edges joining certain pairs of vertices.   Below we give some examples of graphs.   Observe that it is possible to have an edge from a vertex to itself (P_1 in the second example).

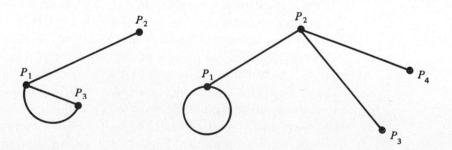

---

† In this context the term graph has a meaning different to that of the graph of a function.

In the communication network there is a sense of direction in the edges. This leads to a special class of graphs called digraphs. We now focus on this class.

A *digraph* (for "directed graph") is a graph in which each edge carries a direction, with not more than one edge in any one direction between any pair of vertices. The graph of the communication network is a digraph. Graphs having more than a single edge in a direction between two vertices are called *multidigraphs*. We shall not be concerned with such graphs here.

Example 1

Organization of a Company

In 1921 E. I. duPont de Nemours and Company Ltd. became the first company to establish committees as a permanent part of its top management structure. The following digraph describes the current internal structure of DuPont.

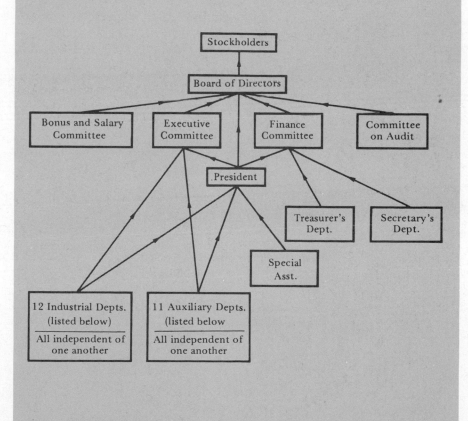

Industrial Depts Elastomer Chemicals, Electrochemicals, Explosives, Fabrics and Finishes, Film, Industrial and Biochemicals, International,

Organic Chemicals, Photo Products, Pigments, Polychemicals, Textile Fibers. (Each of these departments is under a General Manager.)

Auxiliary Depts Advertising, Central Research, Development, Employee Relations, General Services, Legal, Public Relations, Purchasing, Traffic. (Each of these departments is under a Director.) Engineering Dept. (under Chief Engineer.) Economist.

Readers who desire further discussion of the DuPont structure should consult *Management and Organization*, 2nd Ed., Henry L. Sisk, South-Western Publishing Co., 1973. The text contains further references to the internal organization of DuPont and other companies.

It is convenient to represent digraphs by matrices. These representations, as we shall see, can be used to analyze digraphs. The *adjacency matrix A* of a digraph consists of 1's and 0's and is defined by

$$a_{ij} = \begin{cases} 1 & \text{If there is an edge from vertex} \\ & P_i \text{ to vertex } P_j. \\ 0 & \text{Otherwise} \end{cases}$$

The adjacency matrix of the communication digraph is

$$A = \begin{pmatrix} 0 & 1 & 0 & 0 & 0 \\ 1 & 0 & 0 & 0 & 1 \\ 0 & 1 & 0 & 0 & 1 \\ 0 & 0 & 1 & 0 & 1 \\ 0 & 1 & 0 & 1 & 0 \end{pmatrix}$$

For example, $a_{32} = 1$, since there is an edge from P_3 to P_2; $a_{34} = 0$, there is no edge from P_3 to P_4.

Example 2

The network in Figure 1-16 describes a system of one-way streets in a downtown area. Let us interpret the network as a digraph and give the adjacency matrix of this network.

Label the intersections $P_1, ..., P_6$; these are the vertices of the digraph. The adjacency matrix is

$$A = \begin{pmatrix} 0 & 1 & 0 & 0 & 0 & 0 \\ 0 & 0 & 1 & 0 & 1 & 0 \\ 0 & 0 & 0 & 1 & 0 & 0 \\ 0 & 0 & 0 & 0 & 1 & 0 \\ 0 & 1 & 0 & 0 & 0 & 1 \\ 1 & 0 & 0 & 0 & 0 & 0 \end{pmatrix}$$

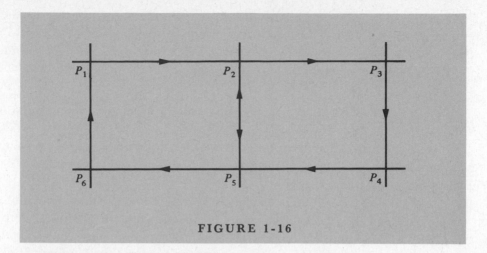

FIGURE 1-16

A (*chain* or *path*) between two vertices of a digraph is a sequence of directed edges that allows one to proceed in a continuous manner from one of the vertices to the other. For example, in the communication digraph (digraph and adjacency matrix duplicated below) there is a chain from P_1 to P_4; this is denoted $P_1 \rightarrow P_2 \rightarrow P_5 \rightarrow P_4$.

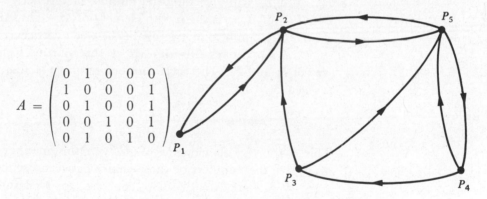

$$A = \begin{pmatrix} 0 & 1 & 0 & 0 & 0 \\ 1 & 0 & 0 & 0 & 1 \\ 0 & 1 & 0 & 0 & 1 \\ 0 & 0 & 1 & 0 & 1 \\ 0 & 1 & 0 & 1 & 0 \end{pmatrix}$$

Let us look at certain chains. There are various chains from P_3 to P_1; let us look at $P_3 \rightarrow P_2 \rightarrow P_1$ and $P_3 \rightarrow P_5 \rightarrow P_2 \rightarrow P_1$. These chains differ in length. We call $P_3 \rightarrow P_2 \rightarrow P_1$ a two-chain, and $P_3 \rightarrow P_5 \rightarrow P_2 \rightarrow P_1$ a three-chain, according to the number of edges traversed. In general, a chain consisting of $n+1$ vertices is called an *n-chain*. When the same vertex occurs more than once in an *n*-chain, the *n*-chain is said to be *redundant*. In the above network, $P_4 \rightarrow P_5 \rightarrow P_4 \rightarrow P_3$ would be a redundant three-chain. We are primarily interested in *n*-chains that are not redundant.

77

In many instances one is interested in the shortest chains (if any exist) between two vertices. In the communication network, for example, the routes $P_4 \to P_5 \to P_2 \to P_1$ and $P_4 \to P_3 \to P_2 \to P_1$ would be of interest, whereas the $P_4 \to P_3 \to P_5 \to P_2 \to P_1$ route would not be of interest.

The following theorem from graph theory gives information about chains.

THEOREM 1-13   *Let A be an adjacency matrix representing a digraph. The (i, j)th element of A^n gives the number of n-chains from P_i to P_j.*

Thus, in a given graph, to find the number of 4-chains from vertex P_2 to P_5 for example, one would compute A^4; the $(2, 5)$th element of this matrix would give the number of 4-chains. In particular, we shall see that this theorem can be used to reveal the number of links in the shortest chains between a given pair of vertices.

Proof: a_{ik} will be the number of one-chains from P_i to P_k, and a_{kj} will be the number of one-chains from P_k to P_j. Thus $a_{ik} a_{kj}$ is the number of distinct two chains from P_i to P_j passing through P_k. $\sum_k a_{ik} a_{kj}$ will give the total number of distinct two-chains from P_i to P_j. $\sum_k a_{ik} a_{kj}$ is the (i, j)th element of A^2. Thus the elements of A^2 give the numbers of two-chains between vertices.

$a_{ik} a_{ks} a_{sj}$ gives the number of three-chains from P_i to P_j by way of P_k and P_s. The total number of three-chains from P_i to P_j is given by

$$\sum_k \sum_s a_{ik} a_{ks} a_{sj}$$

This is the (i, j)th element of A^3. In this manner the elements of A^3 give the numbers of three-chains between vertices.

We can extend these concepts to n-chains. $a_{ik} a_{ks} \cdots a_{rj}$ would give the number of distinct n-chains from P_i to P_j by way of $P_k, P_s, \dots P_r$. $\sum_k \sum_s \cdots \sum_r a_{ik} a_{ks} \cdots a_{rj}$ would give the total number of n-chains between P_i and P_j. This is the (i, j)th element of A^n. Thus A^n represents the n-chains between vertices, proving the theorem.

Let us use the notation $a_{ij}{}^{(n)}$ for the (i, j)th element of A^n. We have seen that A^n represents the n-chains between vertices. These chains can include redundant chains. However, if $a_{ij}{}^{(n)} = c (\neq 0)$ and $a_{ij}{}^{(m)} = 0$ for all $m < n$ then these c distinct n-chains must be chains of minimum length linking P_i to P_j—there can be no shorter chains. Furthermore, there can be no redundant chains among these, for, otherwise, they would not be the chains of minimum length linking P_i to P_j.

Let us check the theorem and illustrate its application with our communication network. In particular we know that there are two three-chains linking station P_4 to station P_1 and that these are the chains of minimum length. Successive powers of the adjacency matrix A are determined.

$$A = \begin{pmatrix} 0 & 1 & 0 & 0 & 0 \\ 1 & 0 & 0 & 0 & 1 \\ 0 & 1 & 0 & 0 & 1 \\ 0 & 0 & 1 & 0 & 1 \\ 0 & 1 & 0 & 1 & 0 \end{pmatrix}, \quad A^2 = \begin{pmatrix} 1 & 0 & 0 & 0 & 1 \\ 0 & 2 & 0 & 1 & 0 \\ 1 & 1 & 0 & 1 & 1 \\ 0 & 2 & 0 & 1 & 1 \\ 1 & 0 & 1 & 0 & 2 \end{pmatrix},$$

$$A^3 = \begin{pmatrix} 0 & 2 & 0 & 1 & 0 \\ 2 & 0 & 1 & 0 & 3 \\ 1 & 2 & 1 & 1 & 2 \\ 2 & 1 & 1 & 1 & 3 \\ 0 & 4 & 0 & 2 & 1 \end{pmatrix}, \quad A^4 = \begin{pmatrix} 2 & 0 & 1 & 0 & 3 \\ 0 & 6 & 0 & 3 & 1 \\ 2 & 4 & 1 & 2 & 4 \\ 1 & 6 & 1 & 3 & 3 \\ 4 & 1 & 2 & 1 & 6 \end{pmatrix}, \quad A^5 = \cdots$$

Let us focus our attention on communication between the two stations of interest, namely communication from station P_4 to station P_1.

A gives $a_{41} = 0$; no direct link.

A^2 gives $a_{41}^{(2)} = 0$; no two-chain communication.

A^3 gives $a_{41}^{(3)} = 2$; two distinct three-chains from P_4 to P_1.

These are the chains of minimum length. The matrix approach is confirmed for this test case.

As a second example, let us determine the minimum number of links for communication from station P_1 to station P_3.

A gives $a_{13} = 0$; no direct link.

A^2 gives $a_{13}^{(2)} = 0$; no two-chain communication.

A^3 gives $a_{13}^{(3)} = 0$; no three-chain communication.

A^4 gives $a_{13}^{(4)} = 1$; a single four-chain from P_1 to P_3.

This result is confirmed when we examine the graph. The quickest way of sending a message from station P_1 to station P_3 is by using the four-chain $P_1 \rightarrow P_2 \rightarrow P_5 \rightarrow P_4 \rightarrow P_3$. The model that we have discussed is somewhat limited; although it gives us the shortest linkage possible, it does not tell us which intermediate stations to use. A more complete model that would yield this information has yet to be developed by mathematicians.

In the remainder of this section we discuss the role of graph theory in the fields of sociology, economic geography, history, archaeology, and ecology.

Sociology   Graph theory is used by sociologists to analyze group relationships.

Consider a group of six people. A sociologist is interested in determining which one of the six has most influence over, or dominates, the others. The group is asked to fill out the following questionnaire:

Your name	Person whose opinion you value most

These answers are then tabulated. Let us, for convenience, label the group members $M_1, M_2, ..., M_6$. The results are given in Table 1-5. The sociologist at this point makes the assumption that the person whose opinion a member values most is the person who influences that member most. This seems to be a natural assumption to make. The question the group was asked is more tactful than, "Who influences you most?"

Table 1-5

Group member	Person whose opinion he values most
M_1	M_5
M_2	M_4
M_3	own
M_4	M_3
M_5	M_3
M_6	M_5

Let us represent these results by means of the digraph in Figure 1-17. The edges correspond to direct influence; the arrows indicate direction of influence. M_1 values the opinion of M_5 most; thus M_1 is influenced by M_5. The arrows go from right to left in the table.

In this type of graph one-chains correspond to direct influence. Chains involving two, three, or more links correspond to indirect influences; the longer the chain, the more remote and presumably the smaller the influence. From Figure 1-17 it becomes apparent that M_3 is the person who exerts most influence over the group.

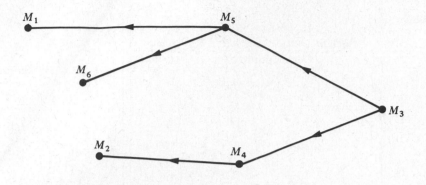

FIGURE 1-17

As groups become larger, the analysis becomes more complex. A matrix description of the graph is used, its powers revealing the indirect influences.

Readers who are interested in further applications of this theory in sociology are referred to "A Method of Matrix Analysis of Group Structure," R. Duncan Luce and Albert D. Perry in *Readings in Mathematical Social Science*, Paul F. Lazarsfeld and Neil W. Henry, MIT Press, 1968.

Economic Geography Geographers use graph theory to analyze networks. The most common types of networks looked at in this field are airways, sea routes, roads, railways, navigable rivers and canals, pipelines, and electricity transmission cables. Here we illustrate a road network model.

Consider a network of cities connected by roads. Let us represent the cities as vertices of a digraph and the roads as edges. For a given number of cities, the more routes there are, the greater the *connectivity* of the network. The *shortest-chain* matrix is used to analyze the connectivity of a region. Here one calculates the chains of minimum length between every pair of cities and represents this information in the form of a matrix, the shortest-chain matrix. By totalling the shortest chains across the rows of this matrix, the city with the lowest total is revealed as the place with the best connectivity. Comparisons of such matrices for various regions lead to information concerning the development of the regions.

We compare road networks in two regions of England, the Lake District and Lincolnshire, using these techniques.

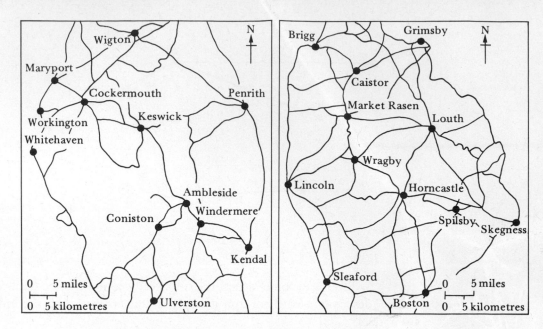

Road network in the Lake District   Road network in Lincolnshire

Twelve towns were taken in each region. The numbers of edges in the shortest chains from each town to each of the other eleven towns were determined and written as matrices, the shortest-chain matrices for these regions. Observe that a shortest-chain matrix is symmetric.

MATRIX FOR THE LAKE DISTRICT

	Maryport	Workington	Whitehaven	Wigton	Cockermouth	Keswick	Coniston	Ambleside	Windermere	Ulverston	Kendal	Penrith	Line Total
Maryport	0	1	2	1	1	1	2	2	2	2	2	1	17
Workington	1	0	1	2	1	2	2	3	2	2	2	2	20
Whitehaven	2	1	0	2	1	2	1	2	1	1	1	2	16
Wigton	1	2	2	0	1	1	2	2	2	2	2	1	18
Cockermouth	1	1	1	1	0	1	1	2	1	1	1	1	(12)
Keswick	1	2	2	1	1	0	2	1	1	2	2	1	16
Coniston	2	2	1	2	1	2	0	1	1	1	1	2	16
Ambleside	2	3	2	2	2	1	1	0	1	2	2	1	19
Windermere	2	2	1	2	1	1	1	1	0	1	1	1	14
Ulverston	2	2	1	2	1	2	1	2	1	0	1	2	17
Kendal	2	2	1	2	1	2	1	2	1	1	0	1	16
Penrith	1	2	2	1	1	1	2	1	1	2	1	0	15

Total 196

MATRIX FOR LINCOLNSHIRE

	Brigg	Caistor	Grimsby	Lincoln	Market Rasen	Wragby	Louth	Horncastle	Sleaford	Spilsby	Boston	Skegness	Line Total
Brigg	0	1	1	1	1	2	1	1	2	1	2	1	14
Caistor	1	0	1	1	1	2	1	1	2	1	2	1	14
Grimsby	1	1	0	2	1	2	1	1	2	1	2	1	15
Lincoln	1	1	2	0	1	1	2	1	1	1	1	2	14
Market Rasen	1	1	1	1	0	1	1	1	2	1	2	1	13
Wragby	2	2	2	1	1	0	1	1	1	1	1	2	15
Louth	1	1	1	2	1	1	0	1	2	1	2	1	14
Horncastle	1	1	1	1	1	1	1	0	1	1	1	1	(11)
Sleaford	2	2	2	1	2	1	2	1	0	1	1	2	17
Spilsby	1	1	1	1	1	1	1	1	1	0	1	1	(11)
Boston	2	2	2	1	2	1	2	1	1	1	0	1	16
Skegness	1	1	1	2	1	2	1	1	2	1	1	0	14

Total 168

We see that Cockermouth is the best connected place in the Lake District, it has the lowest figure, 12. The towns of the Lake District, arranged in a hierarchy based on their connectivity, are:

1 = Cockermouth	4 = Keswick	8 = Ulverston
2 = Windermere	4 = Coniston	10 = Wigton
3 = Penrith	4 = Kendal	11 = Ambleside
4 = Whitehaven	8 = Maryport	12 = Workington

In the area of Lincolnshire, Spilsby and Horncastle both have values of 11, these are the towns of greatest connectivity.

The Lake District network has a total value of 196 while the Lincolnshire total is 168. The Lincolnshire network therefore has greater connectivity than the Lake District network.

The smaller range of values from the Lincolnshire matrix (11–17), compared with the Lake District matrix (12–20), demonstrates the greater uniformity of the road network in the relatively low and level parts of eastern England, and the almost equal connectivity of these places.

The shortest-chain matrix is a powerful tool for a precise and detailed analysis of an area, making possible a much better appreciation of the accessibility of each town.

In this section we have introduced the reader to an application of graph theory in economic geography. There are many others. For example, graphs are used to analyze road plan efficiency using what is known as Smeed's Index. The connectivity of central business districts of cities are compared using this index.

Readers who are interested in an introduction to networks in economic geography should read *Networks in Geography*, Roger Dalton, Joan Garlick, Roger Minshull, and Alan Robinson; George Philip and Son, Ltd., London, 1973. The shortest-chain matrix example presented here was adapted from this text.

History We shall now initiate an algebraic analysis of the Houses of York and Lancaster!

From William the Conqueror to Richard II, hereditary right to the English throne was indisputable. In 1399 Henry IV overthrew Richard II. This resulted in the Wars of the Roses, and it was not until 1485 when the marriage of Henry VII of the House of Lancaster and

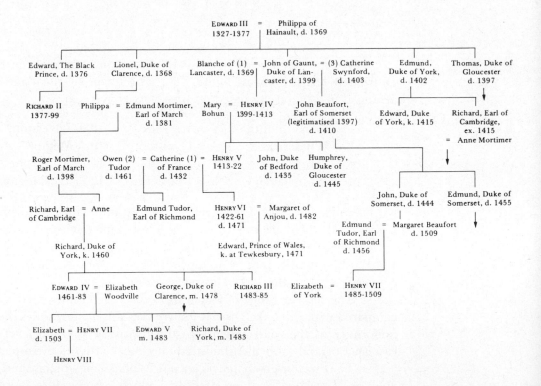

FIGURE 1-18

Elizabeth of the House of York united both the Lancastrian and Yorkist claims to the throne that the hereditary right once again became undisputed. Their son Henry VIII was the undisputed heir. Figure 1-18 is the genealogical table.

For this analysis we make the above genealogical table into a digraph. We focus on certain relationships between people, namely the following relationships: father, mother, son, and daughter. If a person P is related in any of these ways to a person Q, we draw a directed edge from P to Q. For example, if P is the father of Q, there will be a directed edge from P to Q. Observe that in this case Q will be the son or daughter of P; there is also a directed edge from Q to P. This symmetry enables us to start at a point in the table and work down (into the future) or up the table (into the past). The adjacency matrix A is, in fact, a symmetric matrix.

Let us investigate how Elizabeth and Henry VII are each related to Edward III. By calculating the powers of A, we find that the first matrix to have a nonzero element in the $(1, 8)$ location is A^5. This number is 1, implying that there is one five-chain between Edward III and Elizabeth. The interpretation of this five-chain is that Elizabeth is

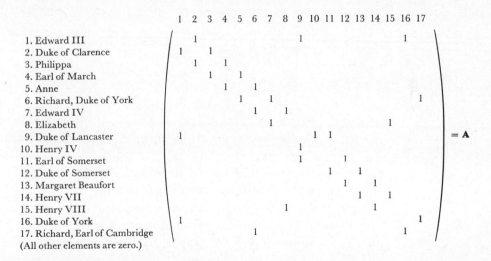

directly descended from Edward III through five generations and that this is the shortest link between them. (Can you find this single five-chain in the genealogical table?)

A^5 also provides the shortest link between Henry VII and Edward III. The 1 in its $(1, 14)$ location implies that Henry VII is descended from Edward III through five generations.

$$
\mathbf{A}^5 = \begin{array}{c} \text{Edward III} \\ \\ \\ \\ \\ \\ \\ \\ \\ \\ \\ \end{array}
\begin{pmatrix}
0 & 18 & 0 & 7 & 0 & 8 & 0 & \textcircled{1} & 24 & 0 & 0 & 7 & 0 & \textcircled{1} & 0 & 18 & 0 \\
18 & 0 & 11 & 0 & 6 & 0 & 2 & 0 & 0 & 6 & 7 & 0 & 1 & 0 & 0 & 10 & 7 \\
0 & 11 & 0 & 10 & 0 & 7 & 0 & 1 & 7 & 0 & 0 & 1 & 0 & 0 & 0 & 7 & 0 \\
7 & 0 \\
0 & 6 \\
8 & 0 \\
0 & 2 \\
1 & 0 \\
24 & 0 \\
\vdots & \vdots
\end{pmatrix}
$$

Column headings above: **Elizabeth** ... **Henry VII**

The first power to have a nonzero element in the $(1, 15)$ location is A^6. This gives a linkage between Henry VIII and Edward III. A 2 in this location implies two distinct six-chains. These two chains arise from the distinct York and Lancastrian lines which date back to Edward III. Henry VIII combined both these claims to the throne.

$$
\mathbf{A}^6 = \begin{array}{c} \text{Edward III} \\ \\ \\ \\ \\ \\ \\ \\ \\ \\ \end{array}
\begin{pmatrix}
60 & 0 & 25 & 0 & 15 & 0 & 9 & 0 & 0 & 24 & 31 & 0 & 8 & 0 & \textcircled{2} & 0 & 26 \\
0 & 29 & 0 & 17 & 0 & 15 & 0 & 2 & 31 & 0 & 0 & 8 & 0 & 1 & 0 & 25 & 0 \\
25 & 0 & 21 & 0 & 17 & 0 & 8 & 0 & 0 & 7 & 8 & 0 & 1 & 0 & 1 & 0 & 14 \\
0 & 17 \\
15 & 0 \\
0 & 15 \\
9 & 0 \\
0 & 2 \\
\vdots & \vdots
\end{pmatrix}
$$

Column heading above: **Henry VIII**

Links between various members of each line can be found in a similar manner.

In this analysis we have omitted the line of Henry IV, which could be included in a larger matrix for a more complete analysis. For example, using such matrix techniques, the mathematical claim of Edward IV to the throne could be compared with that of Henry VI.

Three types of graphs have already been mentioned—graphs, digraphs, and multidigraphs. We now introduce the reader to another type of graph, an *interval graph*. The field of interval graphs and their applications is currently a very active research area. We shall discuss applications of these graphs in archaeology and ecology.

Consider a number of intervals $p_1, \ldots, p_n$ on the real line, either all open (i.e., of the form (a, b)), or all closed (i.e., of the form $[a, b]$). We can construct a graph corresponding to these intervals called an *interval graph*.

Let the vertices of the graph, labelled $P_1, ..., P_n$, represent the intervals, P_i representing p_i. We draw an edge between P_i and P_j, for $i \neq j$, if and only if $p_i \cap p_j \neq \varnothing$.

Example 3

Let $p_1 = [-8, 2]$, $p_2 = [-4, 3]$, $p_3 = [1, 5]$, and $p_4 = [4, 8]$ be intervals on the real line. The corresponding interval graph is

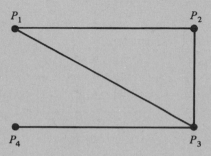

There is, for example, an edge between P_1 and P_2 since $p_1 \cap p_2 = [-4, 2] \neq \varnothing$. There is no edge between P_1 and P_4 since $p_1 \cap p_4 = \varnothing$.

Not every graph is an interval graph. (This fact will be important when we use interval graphs in archaeology.) For example, the following graph is not an interval graph.

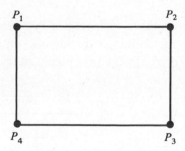

If it were an interval graph, representing the intervals p_1, p_2, p_3, and p_4, then we would have to have $p_1 \cap p_2 \neq \varnothing$, $p_1 \cap p_3 = \varnothing$, $p_1 \cap p_4 \neq \varnothing$, $p_2 \cap p_3 \neq \varnothing$, $p_2 \cap p_4 = \varnothing$, and $p_3 \cap p_4 \neq \varnothing$. By considering all possibilities, it can be shown that four such intervals cannot exist on the real line.

Archaeology   We now illustrate the role of interval graphs in archaeology. Several types of pottery (or other artifacts) are found in different prehistoric graves. Archaeologists attempt to place the pottery in proper chronological order, assigning to each type of pottery a time period. Seriation in archaeology, as we discussed previously, began with the work of Flinders Petrie in the late nineteenth century. Here we formulate the problem of sequence dating, or seriation, in terms of graph theory.

The approach is to assign intervals to various pieces of pottery according to three conditions. This assignment is called an *admissable chronological ordering*.

(1) If type p_2 pottery *strictly followed* type p_1 pottery, i.e., type p_1 pottery disappeared before type p_2 appeared, then p_2 is represented by an interval strictly to the right of p_1:

p_1 p_2

(2) If type p_2 pottery *weakly followed* type p_1, i.e., p_2 appeared when type p_1 was still around and did not disappear until after p_1 had, then p_1 and p_2 are represented by intervals that overlap:

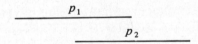

(3) If type p_2 *over-reached* type p_1, i.e., type p_2 appeared before p_1 was present, remained during the entire period that p_1 was in use, and disappeared after p_1, then the intervals are as follows:

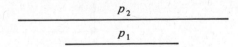

Any pair of pieces of pottery p_i and p_j will fall into one of the categories (1), (2), or (3) above, the category being determined by an examination of the contents of the graves. For example, if p_i and p_j are found in a common grave, we can assume that the time periods of the two types of pottery must have overlapped. Conversely, if the selection of graves is extensive enough, it can be assumed that if two types of pottery appeared in overlapping periods they will appear in some common grave.

Thus a given admissable chronological ordering leads to a unique interval graph.

Example 4

Consider five pieces of pottery $p_1, ..., p_5$. Suppose that examination of graves leads to the following admissable chronological ordering:

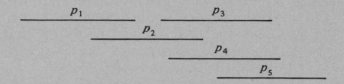

The interval graph corresponding to this ordering is

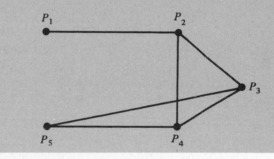

Conversely, suppose we are given n pieces of pottery $p_1, ..., p_n$, their chronological ordering not being completely known, can we use graphs to represent various *possible chronological ordering*? A knowledge of these representations could be useful in narrowing down the alternatives. Let us represent the pieces of pottery by the vertices of a graph $P_1, ..., P_n$. Such a graph would have to be an interval graph. The partial knowledge of the chronological orderings available could reduce the number of interval graphs under consideration. However, each interval graph leads to a number of possible chronological orderings. For example, this last interval graph also corresponds to the chronological ordering:

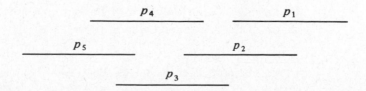

It is currently an open question in graph theory to be able to determine from a given interval graph how many different chronological orderings are possible and what they are. Answers to these questions could lead to useful archaeological techniques.

Ecology Graph theoretical tools are currently used in mathematical models being developed to enable us to better understand the delicate balance of nature. The *food web* of an ecological community is a digraph whose vertex set is the set of all species† in the community. A directed edge is drawn from species P_1 to P_2 if P_1 preys on P_2. The following figure illustrates a small oversimplified food web with five species.

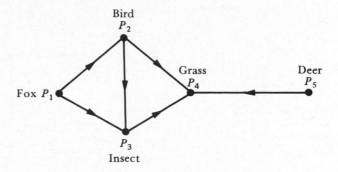

The adjacency matrix of this food web is

$$\begin{pmatrix} 0 & 1 & 1 & 0 & 0 \\ 0 & 0 & 1 & 1 & 0 \\ 0 & 0 & 0 & 1 & 0 \\ 0 & 0 & 0 & 0 & 0 \\ 0 & 0 & 0 & 1 & 0 \end{pmatrix}$$

From the food web, we can define competition: two species *compete* if and only if they have a common prey. This leads to the *competition graph*. This graph has a vertex set the set of all species and two species are joined by an edge if and only if they have a common prey. The competition graph of the above food web is

---

† The word "species" will be used rather loosely here.

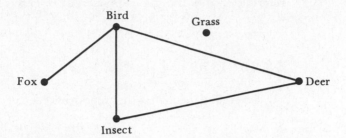

Applied mathematicians are currently investigating food webs of various ecological communities and their corresponding competition graphs. Strangely enough, inspection of a number of food webs has turned up the same result—their competition graphs are interval graphs. It is not known whether this is a general law. An interval assignment for the above competition graph is

In these examples we have seen how applied mathematicians are attempting to construct models that may lead to archaeological and ecological insights, while simultaneously stimulating development of new mathematical concepts and results. It often happens that the mathematical theory developed in conjunction with one application turns out to be relevant in quite another field. The theory of interval graphs provides a very good illustration of this point. Interval graphs first arose in connection with a problem in genetics called Benzer's problem. This involved investigating the structure inside a gene. Further, the archaeological model discussed here is also applicable in developmental psychology when various traits or characteristics are being studied. However, we shall not delve into this here. Readers who are interested in pursuing this further should consult *Discrete Mathematical Models*, Fred S. Roberts, Prentice-Hall, Englewood Cliffs, NJ, 1976. The models in archaeology and ecology were adapted from this text. The book gives an excellent introduction to graph theory, many examples of applications, numerous references, and also discusses current research in the field.

EXERCISES

1. Determine the matrix that represents each of the digraphs below.

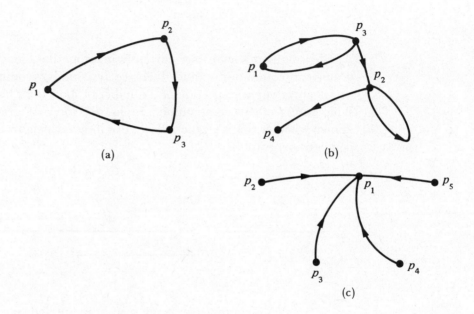

(a)   (b)

(c)

Sketch the digraphs that have the following matrix representations.

2. $\begin{pmatrix} 1 & 0 & 0 & 1 \\ 0 & 1 & 0 & 0 \\ 0 & 0 & 0 & 1 \\ 1 & 0 & 0 & 0 \end{pmatrix}$ **3.** $\begin{pmatrix} 0 & 1 & 1 \\ 1 & 1 & 0 \\ 0 & 1 & 0 \end{pmatrix}$ **4.** $\begin{pmatrix} 1 & 1 & 0 & 0 \\ 0 & 1 & 1 & 0 \\ 0 & 0 & 1 & 1 \\ 1 & 0 & 0 & 1 \end{pmatrix}$

5. Prove that the matrix representation of a digraph is necessarily square.

6. What would be the characteristics of a digraph whose matrix representation contained only zeros as diagonal elements?

7. A digraph is said to be *connected* if there is a path from any one vertex to any other. Give an example of
a) a connected graph   **b)** a graph that is not connected

8. The following matrix defines a communication network.

$$\begin{pmatrix} 0 & 1 & 0 & 1 \\ 1 & 0 & 0 & 0 \\ 0 & 0 & 0 & 1 \\ 1 & 0 & 1 & 0 \end{pmatrix}$$

Sketch the network.

How many links are needed for communication between

a) stations P_1 and P_3 b) stations P_2 and P_3

The following matrices describe communication links. Sketch the networks. Determine the chains of minimum length between every pair of stations in Exercises 9 and 10.

9. $\begin{pmatrix} 0 & 0 & 1 & 1 & 0 \\ 0 & 0 & 0 & 1 & 0 \\ 1 & 0 & 0 & 0 & 0 \\ 1 & 1 & 0 & 0 & 1 \\ 0 & 0 & 0 & 1 & 0 \end{pmatrix}$ 10. $\begin{pmatrix} 0 & 1 & 1 & 0 & 0 \\ 1 & 0 & 0 & 0 & 1 \\ 0 & 0 & 0 & 1 & 1 \\ 0 & 0 & 1 & 0 & 1 \\ 0 & 1 & 0 & 1 & 0 \end{pmatrix}$

11. $\begin{pmatrix} 0 & 1 & 1 & 0 & 0 & 0 \\ 1 & 0 & 0 & 0 & 1 & 1 \\ 1 & 0 & 0 & 1 & 0 & 0 \\ 0 & 0 & 1 & 0 & 1 & 0 \\ 0 & 1 & 0 & 1 & 0 & 0 \\ 0 & 1 & 0 & 0 & 0 & 0 \end{pmatrix}$

12. Represent the information that you obtained in Exercises 9 and 10 in the form of matrices. We call each such matrix the shortest-chain matrix of the network. It represents the minimum number of linkages required for communication between any two stations.

The following tables represent information obtained from questionnaires given to groups of people. In each case construct a graph that describes the leadership structure within the group and the matrix that represents this information. Determine the person(s) who exert most influence within each group.

13.

Group member	Person whose opinion he values most
M_1	M_2
M_2	M_4
M_3	M_2
M_4	Own

14.

Group member	Person whose opinion he values most
M_1	M_4
M_2	Own
M_3	M_5
M_4	M_3
M_5	Own

15.

Group member	Person whose opinion he values most
M_1	M_2
M_2	M_3
M_3	M_1
M_4	M_2
M_5	M_3

The following matrices describe the relationship "friendship" between people of a group. $a_{ij} = 1$ if M_i is a friend of M_j; $a_{ij} = 0$ if he is not. Analyze the relationship for indirect friendships. Note that in this case the matrices that describe the relationship are symmetric. What is the significance of this? Can such a relationship be defined by a nonsymmetric matrix? What would the implication then be?

16.
$$\begin{pmatrix} 0 & 1 & 0 & 0 & 0 \\ 1 & 0 & 0 & 0 & 1 \\ 0 & 0 & 0 & 1 & 0 \\ 0 & 0 & 1 & 0 & 0 \\ 0 & 1 & 0 & 0 & 0 \end{pmatrix}$$

17.
$$\begin{pmatrix} 0 & 0 & 0 & 0 & 0 & 1 \\ 0 & 0 & 1 & 0 & 0 & 0 \\ 0 & 1 & 0 & 0 & 1 & 0 \\ 0 & 0 & 0 & 0 & 0 & 1 \\ 0 & 0 & 1 & 0 & 0 & 0 \\ 1 & 0 & 0 & 1 & 0 & 0 \end{pmatrix}$$

18. What would be the characteristics of a digraph that had a symmetric matrix representation? Give an example (other than the previous one) of a real situation that would be described by such a graph.

19. A structure in a digraph that is of interest, particularly to social scientists, is that of a clique. A *clique* is defined to be a maximal (largest) subset of a digraph consisting of three or more vertices, each pair of which is mutually related. The application of this concept to the relationship "friendship" is immediate: three or more people form a clique if they are all friends and if they have no mutual friendship with anyone outside that set. Give an example of a digraph that contains a clique.

20. The following maps of Cuba and Hawaii give their major highways. Compare their connectivities using the shortest-chain matrix.

CUBA

HAWAII

21. The following are maps of Sri Lanka (formerly Ceylon) and Malaya showing their major railway networks. Compare their connectivities on the basis of these networks using the shortest-chain matrix.

SRI LANKA   MALAYA

In Exercises 22 to 25 draw the interval graphs corresponding to each of the collections of intervals.

22. $p_1 = [-3, 1]$, $p_2 = [-2, 0]$, $p_3 = [0, 4]$, $p_4 = [3, 5]$.

23. $p_1 = [-5, -2]$, $p_2 = [-3, 0]$, $p_3 = [-1, 4]$, $p_4 = [-6, 7]$, $p_5 = [6, 9]$.

24. $p_1 = [-6, -4]$, $p_2 = [-3, -1]$, $p_3 = [-4, 0]$, $p_4 = [-2, 2]$.

25. An examination of graves leads to the following admissable chronological ordering of five pieces of pottery.

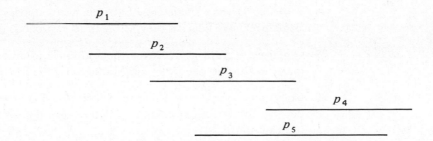

Construct the interval graph corresponding to this ordering.

26. Give four possible chronological orderings represented by the following interval graph.

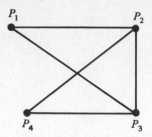

27. Consider the following food web. Give its adjacency matrix and competition graph. Prove that the competition graph is an interval graph by exhibiting an appropriate collection of intervals.

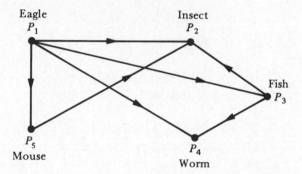

28. Construct a graph to illustrate the relationships among the countries that form NATO, the Warsaw Pact, and SEATO. (Let the countries be the vertices. Draw a single edge between two countries if they are both in a single pact, two edges if they are both in two pacts, etc.)

29. Construct a graph to illustrate the relationships among the countries that make up Europe, the Common Market, and the British Commonwealth.

30. Construct a graph describing the hierarchy of courses in the mathematics department at your university. The relationship of interest will be "is a prerequisite to."

31. Construct a graph that describes the locations of the countries of Europe, the relationship being "have common boundaries." Use your graph to determine the minimum number of countries that must be traversed in traveling from France to Greece.

32. Projects: construct a graph that describes
    a) the hierarchy within the White House
    b) the committee structure within the U.S. Senate

c) your local city government

d) the rank structure within the U.S. Army

e) the shortest distance along interstate highways between the following major cities: New York, Washington, Atlanta, New Orleans, St. Louis, Chicago, Denver, Los Angeles, and San Francisco.

33. Write a computer program for adding the elements in each row of a matrix—a program suitable for analyzing the shortest-chain matrix for a region.

$$2$$

Systems of Linear Equations — A Quantitative Discussion

Many problems in mathematics, engineering, economics, biology, and other sciences reduce to solving systems of linear equations. Historically, linear algebra developed from analyzing methods for solving such equations. This chapter covers numerical techniques for solving systems of linear equations and examples of problems that reduce to solving such equations.

2–1. SYSTEMS OF TWO LINEAR EQUATIONS IN TWO VARIABLES

In this section, we analyze the solving of a system of two linear equations in two variables. This leads to a method for solving larger systems.

Consider the system

$$x - 2y = -1$$
$$3x - y = 2$$

We are asked to determine values of x and the corresponding values of y that satisfy both equations simultaneously. Each such pair (x, y) is called a solution. As it stands, this is an algebraic problem. However, to gain further insight into the problem, we can present it geometrically.

Each equation represents a straight line. A point (x, y) must lie on both lines in order to satisfy both equations simultaneously. Their intersection will represent the only solution (Figure 2-1).

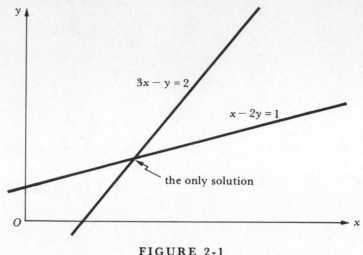

FIGURE 2-1

Let us determine the solution using algebraic techniques. Keep the first equation unchanged but modify the second by adding -3 times the first equation to it. The system becomes

$$x - 2y = -1$$
$$5y = 5$$

This system has the same solution as the original system, but is a simpler system in that the value of y can immediately be seen as 1. Our aim in modifying the second equation in this manner was to eliminate x from it. Thus we have that $y = 1$ at the point of intersection. Substituting back for y into the first equation gives $x = 1$.

Hence the unique solution is $x = 1$, $y = 1$. (Substitute these values back into the original system of equations to see that they are indeed satisfied.)

This discussion raises two questions: Will there always be a solution to such a system? If so, will it always be unique? Let us discuss the existence question first. In the previous example we interpreted the equations geometrically as straight lines, and the solution was the point of intersection of those lines. Equations of a pair of intersecting straight lines have a solution. However, if the equations represent parallel lines, there are no points in common and a solution does not exist. We would expect that a system such as

$$-2x + y = 3$$
$$-4x + 2y = 5$$

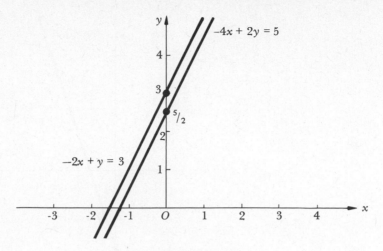

FIGURE 2-2

would not have a solution, for each equation represents a straight line of slope 2 (Figure 2-2).

Let us see what happens when we perform the previous type of algebraic manipulation. To eliminate x from the second equation multiply the first equation by -2 and add it to the second. The system becomes

$$-2x + y = 3$$
$$0y = -1.$$

A solution does not exist.

We now turn to the possibility of having more than one solution to a system. We have seen that solutions correspond geometrically to points of intersection of a pair of straight lines. At first sight it might appear that two straight lines can intersect in, at most, one point; hence there can be, at most, one solution. However, the case where both equations represent the same geometrical straight line should not be excluded. There would then be an infinite number of solutions; each solution would be represented geometrically by a point on the line. For example,

$$6x - 2y = 4$$
$$9x - 3y = 6$$

would be such a system.

Both equations in Figure 2-3 represent a straight line of slope 3 with y intercept -2. Any point on the straight line is a solution. Isolating y in the first equation, one gets $y = 3x - 2$. Hence solutions would be of the form $(x, 3x - 2)$, where x could take on any value.

101

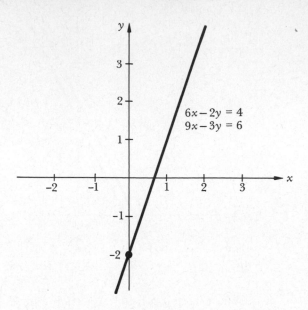

$$6x - 2y = 4$$
$$9x - 3y = 6$$

FIGURE 2-3

In this section we have analyzed solutions to systems of two linear equations in two variables completely. We have found that there may be a unique solution, a solution may not exist, or there may be many solutions.

Our aim is to analyze such systems in many variables. A system of three equations in three variables would be of the form

$$ax + by + cz = p$$
$$dx + ey + fz = q$$
$$gx + hy + iz = r,$$

$a, b, ..., i, p, q, r$ being constants. A solution will be a set of values of x, y, and z that satisfies all the equations simultaneously. Each of the above three equations describes a plane in three dimensional space. Solutions will correspond to points on all three planes. For such a system we can *expect* a solution to be unique, nonexistent, or multiple. We illustrate some of the various possibilities on page 103.

As the number of variables increases, the geometrical interpretation of such a system becomes increasingly complicated and the geometrical discussion used in this section to analyze the question of uniqueness and existence of solutions cannot be employed. One must resort to algebraic methods. Geometrically each equation will represent a space embedded in a larger space. The solutions will correspond to points that

Unique solution :

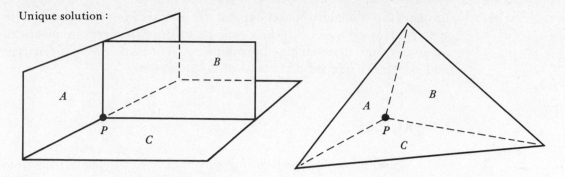

Three planes A, B, C, intersect at a single point P, P corresponds to unique solution

No solutions :

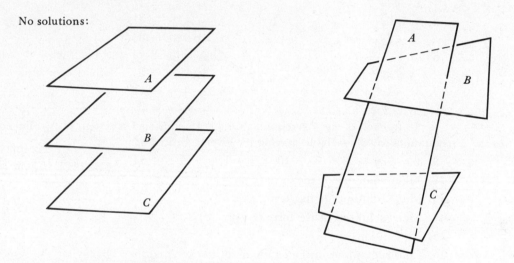

Planes A, B, C have no point of common intersection, no solution

Many solutions :

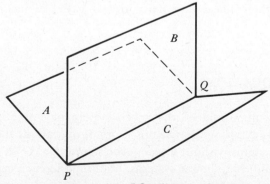

Any point on line PQ will be a solution

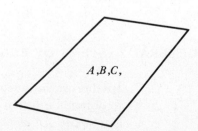

Three equations represent same plane, any point on the plane will be a solution

lie on all the embedded spaces—that is, the points of intersections of these spaces. However, our aim here will be to develop algebraic means of analyzing and determining the solutions. We shall carefully analyze and extend the method that was used in this section.

EXERCISES

Solve, if possible, the following systems of linear equations. Illustrate your answers.

1. $x + y = 1$
$x - y = 2$

2. $3x + 2y = 0$
$x - y = 2$

3. $3x - 6y = 1$
$x - 2y = 2$

4. $x + 2y = 4$
$2x + 4y = 8$

5. $x + y = 7$
$x + 2y = 3$

6. $3x + 6y = 3$
$2x + 4y = 2$

7. $3x - 4y = 2$
$x + y = 1$

8. $2x - y = 1$
$6x - 3y = 2$

In each of the Exercises 9 through 15 construct a system of two linear equations in two variables having the given solution(s).

9. $x = 1$, $y = 2$.   **10.** $x = -1$, $y = 3$.   **11.** $x = 4$, $y = 1$.

12. Many solutions of the form $(x, 4x - 1)$

13. Many solutions of the form $(x, 3x + 4)$

14. Many solutions of the form $(x, -2x + 1)$

15. No solutions.

16. Construct three distinct systems of linear equations all having the same solution, $x = 2$, $y = 3$. Discuss the geometrical significance of this.

17. Construct a system of three linear equations in two variables having solution $x = 1$, $y = 3$. What is the geometrical interpretation of this?

18. Construct a system of three linear equations in two variables having solution $x = -1$, $y = 1$.

2–2. GENERAL SYSTEMS OF LINEAR EQUATIONS

In the previous section we manipulated the given system of equations to obtain a solution. Let us examine the various types of operations that can be performed on such systems without altering the solution. We shall then apply these same techniques to larger systems, obtaining in each case a simpler system that leads to the solution.

Consider the system

$$x - 2y = -1$$
$$3x - y = 2$$

Obviously the order of the equations is immaterial; they can be interchanged.

Secondly, any equation of the system can be multiplied by a nonzero constant; the resulting system will be equivalent to the original one in that it will have the same solution. For example, multiplying the first equation in the system by 3 and leaving the second equation unchanged gives the system

$$3x - 6y = -3$$
$$3x - y = 2$$

which has the same solution as the original system. This is because the values of x and y that satisfy $x - 2y = -1$ also satisfy $3x - 6y = -3$ and vice-versa. Geometrically, these are two equations for the same straight line.

Finally, any multiple of an equation in the system can be added to another equation without affecting the solution of the system. For example, add -3 times the first equation to the second equation to get the system

$$x - 2y = -1$$
$$5y = 5$$

This system has the solution $x = 1$, $y = 1$, which is the solution to the original system.

We now extend these concepts to larger systems of equations. It can be shown (Exercise 20, page 120) that the following operations can be performed on a system of linear equations without altering the solution. We call these operations *elementary transformations*, and systems that are related through elementary transformations are called *equivalent systems*.

1. Interchanging two equations.
2. Multiplying an equation throughout by a nonzero constant.
3. Adding a multiple of one equation to another equation.

We use the notation $\cong$ for equivalent systems.

Using elementary transformations a system of linear equations can be transformed into an equivalent simpler system that leads to the solution. The following example illustrates this approach. We use x_1, x_2, x_3 for the variables here—this is a notation that can be generalized for n variables.

Example 1

Solve, if possible, the system

$$x_1 + x_2 + x_3 = 3$$
$$2x_1 + 3x_2 + x_3 = 5$$
$$x_1 - x_2 - 2x_3 = -5$$

We first use elementary transformations to eliminate x_1 from the second and third equations, using the x_1 of the first equation.

$$x_1 + x_2 + x_3 = 3$$
$$2x_1 + 3x_2 + x_3 = 5 \qquad \underset{\text{Eq. 2}-(2)\text{ Eq. 1}}{\cong} \qquad \begin{aligned} x_1 + x_2 + x_3 &= 3 \\ x_2 - x_3 &= -1 \\ x_1 - x_2 - 2x_3 &= -5 \end{aligned}$$
$$x_1 - x_2 - 2x_3 = -5$$

$$\underset{\text{Eq. 3}-\text{Eq. 1}}{\cong} \qquad \begin{aligned} x_1 + x_2 + x_3 &= 3 \\ x_2 - x_3 &= -1 \\ -2x_2 - 3x_3 &= -8 \end{aligned}$$

x_2 is now eliminated from this last equation by adding twice the second equation to the last equation.

$$\underset{\text{Eq. 3}+(2)\text{ Eq. 2}}{\cong} \qquad \begin{aligned} x_1 + x_2 + x_3 &= 3 \\ x_2 - x_3 &= -1 \\ -5x_3 &= -10 \end{aligned}$$

Multipying the last equation by $-\frac{1}{5}$ leads to a system which can be used to give the solution, through *back substitution*.

$$\underset{(-\frac{1}{5})\text{ Eq. 3}}{\cong} \qquad \begin{aligned} x_1 + x_2 + x_3 &= 3 \\ x_2 - x_3 &= -1 \\ x_3 &= 2. \end{aligned}$$

Substituting for x_3 from the last equation into the second leads to $x_2 = 1$. Substituting $x_3 = 2$ and $x_2 = 1$ into the first equation gives $x_1 = 0$. The solution to this system of equations, and thus also the solution to the original system is $x_1 = 0$, $x_2 = 1$, $x_3 = 2$.

Let us now look at a general system of m linear equations in n variables. Let us label the variables, $x_1, \ldots, x_n$, and the coefficients in the following manner (we shall be discussing the system in terms of matrices):

$$a_{11}x_1 + \cdots + a_{1n}x_n = b_1$$
$$\vdots$$
$$a_{m1}x_1 + \cdots + a_{mn}x_n = b_m.$$

A solution to this system is a set of values for $x_1, \cdots, x_n$ that satisfies all equations simultaneously.

There are two matrices associated with this system.

The matrix $\begin{pmatrix} a_{11} & \cdots & a_{1n} \\ & \vdots & \\ a_{m1} & \cdots & a_{mn} \end{pmatrix}$ is called the *matrix of coefficients* of the

system. The matrix $\begin{pmatrix} a_{11} & \cdots & a_{1n}\, b_1 \\ & \vdots & \\ a_{m1} & \cdots & a_{mn}\, b_m \end{pmatrix}$ that completely characterizes

the system is called the *augmented matrix*.

With every system of m equations in n variables we can associate an augmented matrix having m rows and $n+1$ columns. Conversely, any matrix can be interpreted to be the augmented matrix that defines a system of linear equations. In solving a system of linear equations each equivalent system in the sequence that leads to a simplified system can be represented by an augmented matrix. Thus, it becomes unnecessary to write down $x_1, \ldots, x_n$, since we can work in terms of augmented matrices. Instead of performing elementary transformations on the systems of linear equations, we can perform equivalent transformations, called *elementary matrix transformations*, on the augmented matrices.

Let us analyze elementary matrix transformations. Consider the system

$$
\begin{aligned}
a_{11} x_1 + \cdots + a_{1n} x_n &= b_1 \\
a_{21} x_1 + \cdots + a_{2n} x_n &= b_2 \\
&\vdots \\
a_{m1} x_1 + \cdots + a_{mn} x_n &= b_m
\end{aligned}
$$

having augmented matrix

$$
\begin{pmatrix}
a_{11} & \cdots & b_1 \\
a_{21} & \cdots & b_2 \\
& \vdots & \\
a_{m1} & \cdots & b_m
\end{pmatrix}
$$

Interchanging equations, say the first and second equations, leads to the equivalent system

$$
\begin{aligned}
a_{21} x_1 + \cdots + a_{2n} x_n &= b_2 \\
a_{11} x_1 + \cdots + a_{1n} x_n &= b_1 \\
&\vdots \\
a_{m1} x_1 + \cdots + a_{mn} x_n &= b_m
\end{aligned}
$$

with augmented matrix

$$\begin{pmatrix} a_{21} & \cdots & b_2 \\ a_{11} & \cdots & b_1 \\ & \vdots & \\ a_{m1} & \cdots & b_m \end{pmatrix}$$

Hence the corresponding elementary matrix transformation is to interchange rows in the augmented matrix.

Next consider the elementary transformation that involves multiplying one row, say the first, by a nonzero constant c. The new system is

$$ca_{11}x_1 + \cdots + ca_{1n}x_n = cb_1$$
$$\vdots$$
$$a_{m1}x_1 + \cdots + a_{mn}x_n = b_m$$

with augmented matrix

$$\begin{pmatrix} ca_{11} & \cdots & cb_1 \\ & \vdots & \\ a_{m1} & \cdots & b_m \end{pmatrix}$$

The corresponding elementary matrix transformation is to multiply a row of the augmented matrix by a nonzero constant.

Finally, the matrix transformation that corresponds to adding a multiple of one equation to another equation is that of adding a multiple of one row to another row of the augmented matrix.

In summary, *elementary matrix transformations* are of the following kinds:

1. Interchanging two rows of the matrix.
2. Multiplying a row by a nonzero constant.
3. Adding a multiple of one row of the matrix to another row.

We say that two matrices are *equivalent* if one is obtained from the other using elementary matrix transformations.

We now illustrate this technique of using elementary matrix transformations. Consider the system of Example 1. On the left below is the sequence of equivalent systems of equations that led to the solution, and on the right is the corresponding sequence of equivalent augmented matrices, illustrating the approach we shall be taking from now on.

Example 2

Solve the system of equations

$$x_1 + x_2 + x_3 = 3$$
$$2x_1 + 3x_2 + x_3 = 5$$
$$x_1 - x_2 - 2x_3 = -5$$

Equivalent Systems of Equations	Equivalent Augmented Matrices

Original system
$$x_1 + x_2 + x_3 = 3$$
$$2x_1 + 3x_2 + x_3 = 5$$
$$x_1 - x_2 - 2x_3 = -5$$

Original augmented matrix
$$\begin{pmatrix} 1 & 1 & 1 & 3 \\ 2 & 3 & 1 & 5 \\ 1 & -1 & -2 & -5 \end{pmatrix}$$

$\cong$
Eq. 2 − (2) Eq. 1
$$x_1 + x_2 + x_3 = 3$$
$$x_2 - x_3 = -1$$
$$x_1 - x_2 - 2x_3 = -5$$

$\cong$
Row 2 − (2) Row 1
$$\begin{pmatrix} 1 & 1 & 1 & 3 \\ 0 & 1 & -1 & -1 \\ 1 & -1 & -2 & -5 \end{pmatrix}$$

$\cong$
Eq. 3 − Eq. 1
$$x_1 + x_2 + x_3 = 3$$
$$x_2 - x_3 = -1$$
$$-2x_2 - 3x_3 = -8$$

$\cong$
Row 3 − Row 1
$$\begin{pmatrix} 1 & 1 & 1 & 3 \\ 0 & 1 & -1 & -1 \\ 0 & -2 & -3 & -8 \end{pmatrix}$$

$\cong$
Eq. 3 + (2) Eq. 2
$$x_1 + x_3 + x_3 = 3$$
$$x_2 - x_3 = -1$$
$$-5x_3 = -10$$

$\cong$
Row 3 + (2) Row 2
$$\begin{pmatrix} 1 & 1 & 1 & 3 \\ 0 & 1 & -1 & -1 \\ 0 & 0 & -5 & -10 \end{pmatrix}$$

$\cong$
$(-\tfrac{1}{5})$ Eq. 3
$$x_1 + x_2 + x_3 = 3$$
$$x_2 - x_3 = -1$$
$$x_3 = 2$$

$\cong$
$(-\tfrac{1}{5})$ Row 3
$$\begin{pmatrix} 1 & 1 & 1 & 3 \\ 0 & 1 & -1 & -1 \\ 0 & 0 & 1 & 2 \end{pmatrix}$$

By back substitution we get that the solution is $x_1 = 0, x_2 = 1, x_3 = 2$.

This is the augmented matrix for the system

$$x_1 + x_2 + x_3 = 3$$
$$x_2 - x_3 = -1$$
$$x_3 = 2$$

By back substitution we get that the solution is $x_1 = 0, x_2 = 1, x_3 = 2$.

Observe that this method, using matrices, of reducing a system of equations to an equivalent simpler system, involves creating 1s and 0s in certain locations of the matrices. These numbers are created in a systematic manner. The method is called the *Method of Gaussian Elimination*†.

Method of Gaussian Elimination

1. Write down the augmented matrix of the original system.
2. Use elementary matrix transformations to arrive at a final matrix in which
   (a) any rows consisting entirely of zeros are grouped at the bottom of the matrix, and
   (b) the first nonzero element of each other row is 1, that 1 being positioned to the right of the first nonzero element of the preceding row.

   Such a matrix is said to be in *echelon form*.
3. Write down the system of equations corresponding to this echelon form. Solve this system by back substitution. This is also the solution to the initial system of equations.

† Readers who also desire a discussion of Gauss-Jordan elimination should complete Exercise 19, page 118–119.

There are usually many sequences of elementary transformations that can be used to lead to an echelon form. In these examples we follow a standard pattern—creating zeros in a column, moving on to the next column, constructing the 1 and using it to create zeros in that column, etc. Once the reader has mastered this standard approach he is encouraged to look for his own short cuts.†

Example 3

The following matrices are all in echelon form.

$$\begin{pmatrix} 1 & 2 & 3 & 4 \\ 0 & 1 & 0 & 2 \\ 0 & 0 & 1 & 3 \\ 0 & 0 & 0 & 1 \end{pmatrix}, \begin{pmatrix} 1 & 2 & -1 & 3 \\ 0 & 0 & 1 & 2 \\ 0 & 0 & 0 & 1 \\ 0 & 0 & 0 & 0 \end{pmatrix}, \begin{pmatrix} 1 & -1 & 2 & 1 \\ 0 & 1 & 2 & 3 \\ 0 & 0 & 1 & 3 \end{pmatrix}, \begin{pmatrix} 1 & 2 & 3 & 4 & -1 \\ 0 & 1 & 2 & 3 & 1 \\ 0 & 0 & 1 & 2 & 3 \end{pmatrix},$$

$$\begin{pmatrix} 1 & 2 & 3 & 1 \\ 0 & 1 & 2 & 2 \\ 0 & 0 & 0 & 0 \\ 0 & 0 & 0 & 0 \end{pmatrix}, \begin{pmatrix} 1 & -1 & 2 & 1 & 3 \\ 0 & 0 & 1 & 2 & 2 \\ 0 & 0 & 0 & 0 & 1 \end{pmatrix}$$

The following matrices are not in echelon form.

$$\begin{pmatrix} 1 & 2 & 3 & 1 \\ 0 & 0 & 0 & 0 \\ 0 & 1 & 2 & 1 \\ 0 & 0 & 1 & 3 \end{pmatrix}, \qquad \begin{pmatrix} 1 & 2 & 3 \\ 0 & 1 & 1 \\ 0 & 1 & 2 \end{pmatrix}, \qquad \begin{pmatrix} 1 & -1 & 2 \\ 0 & 2 & 1 \\ 0 & 0 & 1 \end{pmatrix}$$

There is a row consisting of zeros that is not at the bottom of the matrix. *The first non zero element in the third row is not positioned to the right of the first non zero element of the second row.* *The first non zero element in the second row is not 1.*

We now illustrate techniques of using elementary matrix transformations to arrive at echelon forms, and the interpretations of the echelon forms.

Example 4

Solve, if possible, the system of equations

$$x_1 - 2x_2 + 3x_3 = 1$$

$$3x_1 - 4x_2 + 5x_3 = 3$$

$$2x_1 - 3x_2 + 4x_3 = 2$$

We shall find that this system has many solutions.

---

† This pattern should be mastered first! We shall use it later in computing inverses of matrices.

The augmented matrix is

$$\begin{pmatrix} 1 & -2 & 3 & 1 \\ 3 & -4 & 5 & 3 \\ 2 & -3 & 4 & 2 \end{pmatrix}$$

Using elementary matrix transformations, we create zeros below the first element in the first row:

$$\begin{pmatrix} 1 & -2 & 3 & 1 \\ 3 & -4 & 5 & 3 \\ 2 & -3 & 4 & 2 \end{pmatrix} \begin{array}{c} \\ \text{Row 2} - (3)\,\text{Row 1} \\ \text{Row 3} - (2)\,\text{Row 1} \end{array} \cong \begin{pmatrix} 1 & -2 & 3 & 1 \\ 0 & 2 & -4 & 0 \\ 0 & 1 & -2 & 0 \end{pmatrix}$$

Divide the second row by 2, (an elementary matrix operation) to make the first nonzero element in this row 1:

$$\begin{array}{c} \cong \\ (\tfrac{1}{2})\,\text{Row 2} \end{array} \begin{pmatrix} 1 & -2 & 3 & 1 \\ 0 & 1 & -2 & 0 \\ 0 & 1 & -2 & 0 \end{pmatrix}.$$

Create a zero in the $(3,2)$ location:

$$\begin{array}{c} \cong \\ \text{Row 3} - \text{Row 2} \end{array} \begin{pmatrix} 1 & -2 & 3 & 1 \\ 0 & 1 & -2 & 0 \\ 0 & 0 & 0 & 0 \end{pmatrix}.$$

This is an echelon form of the original augmented matrix. This matrix is the augmented matrix of the system

$$x_1 - 2x_2 + 3x_3 = 1$$

$$x_2 - 2x_3 = 0.$$

The last equation gives $x_2 = 2x_3$. Substituting for x_2 into the first equation gives $x_1 = x_3 + 1$. We express the result

$$x_1 = x_3 + 1,$$

$$x_2 - 2x_3.$$

There are many solutions; x_3 can take on any value. For example when $x_3 = 1$, we get that $x_1 = 2$, $x_2 = 2$, $x_3 = 1$ is a solution. When $x_3 = 2$, we see that $x_1 = 3$, $x_2 = 4$, $x_3 = 2$ is a solution.

In general, when there are many solutions, these solutions are given as here by expressing some of the variables in terms of others. It is customary to express leading variables in terms of others.

The solutions can also be expressed

$$(x_3 + 1, \quad 2x_3, \quad x_3),$$

x_3 taking on any value. We shall find this way of expressing the solutions useful in a later geometrical discussion of the set of solutions.

Example 5

This example further illustrates the interpretation of the final system of equations in the case of many solutions.

Solve, if possible, the system

$$x_1 - x_2 + x_3 + 2x_4 - 2x_5 = 1$$

$$2x_1 - x_2 - x_3 + 3x_4 - x_5 = 3$$

$$-x_1 - x_2 + 5x_3 \qquad - 4x_5 = -3.$$

$$\begin{pmatrix} 1 & -1 & 1 & 2 & -2 & 1 \\ 2 & -1 & -1 & 3 & -1 & 3 \\ -1 & -1 & 5 & 0 & -4 & -3 \end{pmatrix}$$

$$\begin{matrix} \cong \\ \text{Row } 2-(2)\text{ Row } 1 \\ \text{Row } 3+\text{Row } 1 \end{matrix} \begin{pmatrix} 1 & -1 & 1 & 2 & -2 & 1 \\ 0 & 1 & -3 & -1 & 3 & 1 \\ 0 & -2 & 6 & 2 & -6 & -2 \end{pmatrix}$$

$$\begin{matrix} \cong \\ \text{Row } 3+(2)\text{ Row } 2 \end{matrix} \begin{pmatrix} 1 & -1 & 1 & 2 & -2 & 1 \\ 0 & 1 & -3 & -1 & 3 & 1 \\ 0 & 0 & 0 & 0 & 0 & 0 \end{pmatrix}$$

This is the augmented matrix of the system

$$x_1 - x_2 + x_3 + 2x_4 - 2x_5 = 1$$

$$x_2 - 3x_3 - x_4 + 3x_5 = 1.$$

The last equation gives $x_2 = 3x_3 + x_4 - 3x_5 + 1$. Substituting for x_2 into the first equation we get

$$x_1 - (3x_3 + x_4 - 3x_5 + 1) + x_3 + 2x_4 - 2x_5 = 1$$

giving

$$x_1 = 2x_3 - x_4 - x_5 + 2.$$

The solutions are

$$x_1 = 2x_3 - x_4 - x_5 + 2$$
$$x_2 = 3x_3 + x_4 - 3x_5 + 1,$$

where x_3, x_4 and x_5 can take on any values. For example, when $x_3 = 1$, $x_4 = -1$ and $x_5 = 1$, we get that $x_1 = 4$, $x_2 = 0$, $x_3 = 1$, $x_4 = -1$, $x_5 = 1$ is a solution.

We can also express the solutions as

$$(2x_3 - x_4 - x_5 + 2, 3x_3 + x_4 - 3x_5 + 1, x_3, x_4, x_5).$$

Since there are three variables that are free to take on any values in this solution we say that the set of solutions has *three degrees of freedom*.

Example 6

Solve, if possible, the system

$$x_1 - x_2 + 2x_3 = 3$$
$$2x_1 - 2x_2 + 5x_3 = 4$$
$$x_1 + 2x_2 - x_3 = -3$$
$$2x_2 + 2x_3 = 1$$

We shall see that this system has no solutions.
Use elementary matrix transformations to create zeros in the first column.

$$\begin{pmatrix} 1 & -1 & 2 & 3 \\ 2 & -2 & 5 & 4 \\ 1 & 2 & -1 & -3 \\ 0 & 2 & 2 & 1 \end{pmatrix}$$

$$\begin{matrix} \cong \\ \text{Row } 2 - (2) \text{ Row } 1 \\ \text{Row } 3 - \text{Row } 1 \end{matrix} \begin{pmatrix} 1 & -1 & 2 & 3 \\ 0 & 0 & 1 & -2 \\ 0 & 3 & -3 & -6 \\ 0 & 2 & 2 & 1 \end{pmatrix}$$

At this stage we want a 1 in the $(2, 2)$ location. We can obtain it by interchanging the second and third rows (an elementary matrix transformation).

$$\begin{matrix} \cong \\ \text{interchange} \\ \text{Rows 2 and 3} \end{matrix} \begin{pmatrix} 1 & -1 & 2 & 3 \\ 0 & 3 & -3 & -6 \\ 0 & 0 & 1 & -2 \\ 0 & 2 & 2 & 1 \end{pmatrix}$$

$$\begin{matrix} \cong \\ (\tfrac{1}{3}) \text{ Row } 2 \end{matrix} \begin{pmatrix} 1 & -1 & 2 & 3 \\ 0 & 1 & -1 & -2 \\ 0 & 0 & 1 & -2 \\ 0 & 2 & 2 & 1 \end{pmatrix}$$

We now create zeros in the second column using the 1 in the $(2, 2)$ location.

$$\begin{matrix} \cong \\ \text{Row } 4 - (2) \text{ Row } 2 \end{matrix} \begin{pmatrix} 1 & -1 & 2 & 3 \\ 0 & 1 & -1 & -2 \\ 0 & 0 & 1 & -2 \\ 0 & 0 & 4 & 5 \end{pmatrix}$$

Create zero in the $(4, 3)$ location using the third row.

$$\begin{matrix} \cong \\ \text{Row } 4 - (4) \text{ Row } 3 \end{matrix} \begin{pmatrix} 1 & -1 & 2 & 3 \\ 0 & 1 & -1 & -2 \\ 0 & 0 & 1 & -2 \\ 0 & 0 & 0 & 13 \end{pmatrix}$$

$$\begin{array}{c} \cong \\ \left(\frac{1}{13}\right) \text{ Row } 4 \end{array} \begin{pmatrix} 1 & -1 & 2 & 3 \\ 0 & 1 & -1 & -2 \\ 0 & 0 & 1 & -2 \\ 0 & 0 & 0 & 1 \end{pmatrix}$$

This last matrix is in echelon form.   It corresponds to the system

$$x_1 - x_2 + 2x_3 = \quad 3$$
$$x_2 - \quad x_3 = -2$$
$$x_3 = -2$$
$$0 = \quad 1$$

This system cannot be satisfied for any values of x_1, x_2, x_3; thus the original system has no solutions.

Example 7

This example illustrates the procedure to follow if on performing a transformation an early row becomes zero.

Reduce the following matrix to echelon form.

$$\begin{pmatrix} 1 & -1 & 1 & 2 \\ -2 & 2 & -2 & -4 \\ 2 & 1 & 2 & 2 \\ 1 & 1 & 1 & 0 \end{pmatrix}$$

We find that

$$\begin{pmatrix} 1 & -1 & 1 & 2 \\ -2 & 2 & -2 & -4 \\ 2 & 1 & 2 & 2 \\ 1 & 1 & 1 & 0 \end{pmatrix}$$

$$\begin{array}{c} \cong \\ \text{Row } 2 + (2) \text{ Row } 1 \\ \text{Row } 3 - (2) \text{ Row } 1 \\ \text{Row } 4 - \text{Row } 1 \end{array} \begin{pmatrix} 1 & -1 & 1 & 2 \\ 0 & 0 & 0 & 0 \\ 0 & 3 & 0 & -2 \\ 0 & 2 & 0 & -2 \end{pmatrix}$$

At this stage interchange the row containing the zeros, with the last row.

$$\begin{array}{c} \cong \\ \text{Interchange} \\ \text{Row } 2 \text{ and Row } 4 \end{array} \begin{pmatrix} 1 & -1 & 1 & 2 \\ 0 & 2 & 0 & -2 \\ 0 & 3 & 0 & -2 \\ 0 & 0 & 0 & 0 \end{pmatrix}$$

$$\begin{array}{c} \cong \\ \left(\frac{1}{2}\right) \text{ Row } 2 \end{array} \begin{pmatrix} 1 & -1 & 1 & 2 \\ 0 & 1 & 0 & -1 \\ 0 & 3 & 0 & -2 \\ 0 & 0 & 0 & 0 \end{pmatrix}$$

$$\begin{array}{c} \cong \\ \text{Row } 3 - (3) \text{ Row } 2 \end{array} \begin{pmatrix} 1 & -1 & 1 & 2 \\ 0 & 1 & 0 & -1 \\ 0 & 0 & 0 & 1 \\ 0 & 0 & 0 & 0 \end{pmatrix}$$

This matrix is in echelon form.

In this section we have seen examples of systems of equations that have unique solutions, many solutions, and no solutions at all. In general, without further analysis, one cannot look at a system and tell which of these possibilities applies. However, when the number of variables is greater than the number of equations in a homogeneous system, one does know in advance that there will be many solutions:

THEOREM 2-1   *A homogeneous system of linear equations which has more variables than equations always has many solutions.*

Proof: Consider the system

$$a_{11}x_1 + \cdots + a_{1n}x_n = 0$$
$$\vdots \qquad\qquad \vdots$$
$$a_{m1}x_1 + \cdots + a_{mn}x_n = 0$$

where $n > m$.

Reduce the augmented matrix to echelon form and write down the corresponding system of equations. This system will be, in general, of the form

$$\begin{aligned}
x_1 + p_{12}x_2 + p_{13}x_3 + \cdots &= 0 \\
x_2 + p_{23}x_3 + \cdots &= 0 \\
x_3 + \cdots &= 0 \\
& \qquad\qquad\qquad\qquad\qquad (1) \\
x_m + p_{m(m+1)}x_{m+1} + \cdots &= 0
\end{aligned}$$

Isolate x_1 in the first equation, x_2 in the second, etc. to get the system

$$x_1 = -\sum(\ \ )$$
$$x_2 = -\sum(\ \ )$$
$$\vdots$$
$$x_m = -\sum(\ \ )$$

where $\sum(\ \ )$ denotes the remaining terms in each equation. The $\sum(\ \ )$ in the last equation involves the variables x_{m+1} to x_n. These

115

variables can be given any values, leading to many solutions for the system.

For some systems the echelon form does not lead to the general system (1). A certain variable may be missing from the lead position. We illustrate this situation with x_2 missing from the lead position in the second equation. Suppose the simplified system is

$$x_1 + p_{12}x_2 + p_{13}x_3 + p_{14}x_4 + \cdots = 0$$
$$x_3 + p_{24}x_4 + \cdots = 0$$
$$x_4 + \cdots = 0$$
$$\vdots \qquad \vdots$$

(No x_2 term in lead position here)

For such a system, the variable x_2 can take on any value, leading to many solutions. In general, for this type of situation the variables missing from lead positions can take on any values; such systems have many solutions.

EXERCISES

1. Determine the matrix of coefficients and the augmented matrix for each of the following systems of linear equations.

a) $x_1 - x_2 = 1$
    $2x_1 + x_2 = 3$

b) $x_1 - x_2 + x_3 = 4$
    $x_1 - x_2 + 2x_3 = 5$
    $2x_1 + 2x_2 - x_3 = -2$

c) $2x_1 - x_2 + x_3 = -1$
    $x_1 + x_2 \qquad = 4$
    $-x_1 + x_2 - 2x_3 = 5$
    $x_1 + 2x_2 - x_3 = 2$

2. Any matrix can be interpreted as an augmented matrix defining a system of linear equations. Write out the systems defined by the matrices.

a) $\begin{pmatrix} 1 & 2 & 1 \\ -1 & 0 & 3 \end{pmatrix}$

b) $\begin{pmatrix} -1 & 2 & 3 & -1 \\ 4 & 1 & 0 & 1 \\ 2 & 3 & 1 & 0 \end{pmatrix}$

c) $\begin{pmatrix} 2 & 1 & 1 & 3 \\ -1 & 1 & 4 & 1 \\ -2 & 1 & 1 & 0 \\ -1 & 2 & 3 & 1 \end{pmatrix}$

3. The following matrices are all in echelon form. Interpret each one as the augmented matrix of a system of linear equations. Solve each system.

a) $\begin{pmatrix} 1 & -1 & 0 \\ 0 & 1 & 1 \\ 0 & 0 & 0 \end{pmatrix}$
b) $\begin{pmatrix} 1 & 2 & 3 & -1 \\ 0 & 1 & 2 & 1 \\ 0 & 0 & 1 & 2 \end{pmatrix}$

c) $\begin{pmatrix} 1 & -1 & 0 & 1 \\ 0 & 1 & 2 & 2 \\ 0 & 0 & 0 & 1 \end{pmatrix}$
d) $\begin{pmatrix} 1 & 1 & 0 & -1 & 0 \\ 0 & 0 & 1 & 2 & 0 \\ 0 & 0 & 0 & 1 & 0 \end{pmatrix}$

e) $\begin{pmatrix} 1 & 2 \\ 0 & 1 \end{pmatrix}$
f) $\begin{pmatrix} 1 & 0 & 2 & 3 \\ 0 & 1 & 2 & 2 \end{pmatrix}$

g) $\begin{pmatrix} 1 & 0 & 0 & 0 & 0 & 3 \\ 0 & 1 & -1 & 1 & 0 & 2 \\ 0 & 0 & 1 & 1 & -1 & 1 \end{pmatrix}$

4. Give examples (other than the ones in Exercise 3) of an echelon matrix that corresponds to a system of equations

a) with a unique solution
b) with many solutions involving one parameter
c) with many solutions involving two parameters
d) with many solutions involving five parameters
e) with no solutions

5. State whether or not the following matrices are in echelon form. If a matrix is not in echelon form, determine an echelon form.

a) $\begin{pmatrix} 1 & 2 & 1 \\ 0 & 1 & 3 \\ 0 & 0 & 0 \end{pmatrix}$
b) $\begin{pmatrix} 1 & 0 & 1 \\ -1 & 1 & -1 \\ 0 & 2 & 2 \end{pmatrix}$
c) $\begin{pmatrix} 1 & 2 & 3 \\ 0 & 0 & 1 \\ 0 & 0 & 0 \end{pmatrix}$

d) $\begin{pmatrix} 1 & 2 & -1 & 0 \\ 2 & 4 & 1 & 2 \\ 0 & 1 & 2 & 3 \end{pmatrix}$
e) $\begin{pmatrix} 1 & 2 & -1 & 1 \\ 2 & 4 & -2 & 2 \\ 0 & 0 & -1 & 1 \\ 1 & -1 & 0 & 2 \end{pmatrix}$

f) $\begin{pmatrix} 1 & 2 & -1 & 1 \\ 0 & 0 & 0 & 1 \\ 0 & 1 & 1 & 2 \\ 0 & -1 & 0 & 1 \end{pmatrix}$
g) $\begin{pmatrix} 1 & 2 & -1 & 1 \\ 0 & 0 & 0 & 0 \\ 2 & 1 & -1 & 1 \\ 1 & -1 & 1 & 3 \end{pmatrix}$

h) $\begin{pmatrix} 1 & -1 & 1 & 2 & 1 \\ 2 & -1 & 0 & 3 & 0 \\ -1 & 1 & 1 & 1 & -1 \\ 0 & 1 & 0 & 1 & 1 \end{pmatrix}$
i) $\begin{pmatrix} 1 & 1 & 1 & -2 \\ 2 & 2 & 0 & 1 \\ 2 & 2 & 2 & 3 \\ -1 & 1 & 1 & -1 \end{pmatrix}$

Solve, if possible, the following systems of equations.

6.
$$x_1 + x_2 + x_3 = 6$$
$$x_1 - x_2 + x_3 = 2$$
$$x_1 + 2x_2 + 3x_3 = 14$$

7.
$$2x_1 + x_2 + x_3 = 0$$
$$x_1 - x_2 + x_3 = 0$$
$$3x_1 + 2x_2 + 2x_3 = 0$$

8.
$$2x_1 - x_2 - x_3 = 2$$
$$x_1 - x_2 + x_3 = 2$$
$$3x_1 - 2x_2 + x_3 = 5$$

9.
$$x_1 - x_2 = 0$$
$$x_2 + x_3 = 1$$
$$2x_1 + 3x_2 + 5x_3 = 4$$

10.
$$2x_1 + x_2 + x_3 = 3$$
$$x_1 - x_2 + 2x_3 = 0$$
$$3x_1 + x_2 + x_3 = 2$$

11.
$$x_1 - x_2 + x_3 = 1$$
$$2x_1 - 2x_2 + 3x_3 = 3$$
$$x_1 - x_2 - x_3 = -1$$

12.
$$x_1 + 2x_2 - x_3 + x_4 = 4$$
$$2x_1 + 3x_2 + x_3 - 2x_4 = 6$$
$$x_1 + x_2 + 3x_3 + 2x_4 = 5$$
$$3x_1 + x_2 - x_3 - 3x_4 = 4$$

13.
$$x_1 + x_2 + x_3 + x_4 = 3$$
$$x_1 - x_2 - 2x_3 + 3x_4 = 4$$
$$2x_1 + x_2 + x_3 + 3x_4 = 7$$
$$x_1 + 2x_2 - x_3 - 2x_4 = -5$$

14.
$$x_1 - x_2 + 2x_3 = 1$$
$$2x_1 - 2x_2 + 2x_3 + 3x_4 = 2$$
$$x_1 + 2x_2 - 3x_4 = 1$$
$$3x_1 + x_2 - 4x_3 + 3x_4 = 3$$

15.
$$x_1 - x_2 - x_3 + 2x_4 = 4$$
$$2x_1 - x_2 + 3x_3 - x_4 = 2$$

16.
$$x_1 + 2x_2 - x_3 + x_4 + 2x_5 = 0$$
$$x_2 + 2x_3 - 3x_4 - 3x_5 = 0$$
$$x_3 + 5x_4 + 3x_5 = 0$$

17.
$$x_1 + 2x_2 + 3x_3 = -1$$
$$-2x_1 + x_2 = 3$$
$$3x_1 + x_2 + 3x_3 = 2$$

18.
$$x_1 + 2x_2 + x_3 + 3x_4 = 2$$
$$-x_1 - x_2 + 3x_3 - x_4 = -3$$
$$2x_1 + 5x_2 + 5x_3 + 8x_4 = 3$$

19. It is sometimes convenient to modify the Gaussian elimination procedure by creating zeros above as well as below the 1 that is the first nonzero element in each row. In this manner one arrives at the *reduced echelon form.* This method is called the *Gauss-Jordan* method. We illustrate the method with the following example.

$$x_1 + x_2 + x_3 = 3$$
$$x_1 - x_2 - 2x_3 = -5$$
$$2x_1 + 3x_2 + x_3 = 5$$

Using Gauss-Jordan elimination,

$$\begin{pmatrix} 1 & 1 & 1 & 3 \\ 1 & -1 & -2 & -5 \\ 2 & 3 & 1 & 5 \end{pmatrix}$$

$$\begin{array}{c} \cong \\ \text{Row } 2 - \text{Row } 1 \\ \text{Row } 3 - (2)\,\text{Row } 1 \end{array} \begin{pmatrix} 1 & 1 & 1 & 3 \\ 0 & -2 & -3 & -8 \\ 0 & 1 & -1 & -1 \end{pmatrix}$$

$$\begin{array}{c} \cong \\ (-\tfrac{1}{2})\,\text{Row } 2 \end{array} \begin{pmatrix} 1 & 1 & 1 & 3 \\ 0 & 1 & \tfrac{3}{2} & 4 \\ 0 & 1 & -1 & -1 \end{pmatrix}$$

At this stage we create zeros above as well as below the $(2, 2)$ element:

$$\begin{array}{c} \cong \\ \text{Row } 1 - \text{Row } 2 \\ \text{Row } 3 - \text{Row } 2 \end{array} \begin{pmatrix} 1 & 0 & -\frac{1}{2} & -1 \\ 0 & 1 & \frac{3}{2} & 4 \\ 0 & 0 & -\frac{5}{2} & -5 \end{pmatrix}$$

$$\begin{array}{c} \cong \\ (-\frac{2}{5}) \text{ Row } 3 \end{array} \begin{pmatrix} 1 & 0 & -\frac{1}{2} & -1 \\ 0 & 1 & \frac{3}{2} & 4 \\ 0 & 0 & 1 & 2 \end{pmatrix}$$

Now create zeros above the $(3, 3)$ element.

$$\begin{array}{c} \cong \\ \text{Row } 1 + (\frac{1}{2}) \text{ Row } 3 \\ \text{Row } 2 - (\frac{3}{2}) \text{ Row } 3 \end{array} \begin{pmatrix} 1 & 0 & 0 & 0 \\ 0 & 1 & 0 & 1 \\ 0 & 0 & 1 & 2 \end{pmatrix} \leftarrow \textit{Reduced echelon form}$$

This matrix defines a system of equations having the same solutions as the original system. It was obtained from the original augmented matrix using elementary matrix transformations.
This reduced echelon form gives

$$\begin{aligned} x_1 & = 0 \\ x_2 & = 1 \\ x_3 & = 2, \end{aligned}$$

The Gauss-Jordan method, in effect, incorporates the back substitution. In practice, for larger systems, the method of Gaussian elimination is more efficient and accurate than the Gauss-Jordan method in that it involves fewer operations of addition and multiplication. The Gauss-Jordan method is of theoretical importance and is suitable for small systems of equations.

The following are examples of matrices in reduced echelon form.

$$\begin{pmatrix} 1 & 0 & 0 & 2 \\ 0 & 1 & 0 & 4 \\ 0 & 0 & 1 & 6 \end{pmatrix}, \quad \begin{pmatrix} 1 & 0 & 3 & 0 & 2 \\ 0 & 1 & 2 & 0 & 1 \\ 0 & 0 & 0 & 1 & 4 \end{pmatrix},$$

$$\begin{pmatrix} 1 & 2 & 0 & 3 & 0 & 2 & 3 \\ 0 & 0 & 1 & -1 & 0 & 4 & 2 \\ 0 & 0 & 0 & 0 & 1 & -1 & 4 \\ 0 & 0 & 0 & 0 & 0 & 0 & 0 \end{pmatrix}.$$

Solve the following systems of equations using the Gauss-Jordan elimination method.

a)
$$\begin{aligned} x_1 + x_2 - x_3 & = 2 \\ x_1 + x_3 & = 3 \\ -x_1 + x_2 + 2x_3 & = 6 \end{aligned}$$

b)
$$\begin{aligned} x_1 + x_2 + x_3 & = 2 \\ x_1 + 2x_2 + 2x_3 & = 3 \\ 2x_1 + 4x_2 + 5x_3 & = 5 \end{aligned}$$

c)
$$\begin{aligned} x_1 + x_2 + x_3 & = 4 \\ -x_1 + x_2 + 3x_3 & = -2 \\ -2x_1 + 3x_2 + 8x_3 + x_4 & = -4 \end{aligned}$$

d)
$$\begin{aligned} x_1 + x_2 + x_3 + 2x_4 & = 5 \\ x_1 + 2x_2 + x_3 + x_4 & = 3 \\ 2x_1 + 4x_2 + 3x_3 + 3x_4 & = 9 \end{aligned}$$

20. Prove that elementary transformations leave the solution to a system of linear equations unchanged.

2–3.* MODELS INVOLVING SYSTEMS OF LINEAR EQUATIONS

Many models in such diverse fields as electrical engineering, economics, and traffic analysis involve solving systems of linear equations. We now present some of these models.

Electrical Network Analysis   Systems of linear equations are involved in analyzing currents through various circuits in a network. In the construction of electrical networks containing resistances, it is necessary to predict the current load through the various circuits. Two laws that form the foundation of this mathematical analysis are *Kirchhoff's laws*.

Law 1. All the current flowing into a junction must flow out of it.
Law 2. The sum of the *IR* terms (*I* denotes current, *R* resistance) in any direction around a closed path is equal to the total voltage in the path in that direction.

These laws are based on experimental verification in the laboratory.

Example 1

Consider the network in Figure 2-4. Let us determine the current through each branch.

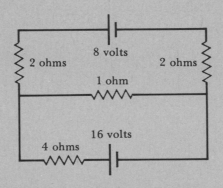

FIGURE 2-4

The batteries (denoted ┤├) are 8 volts and 16 volts. The following convention is used to indicate out of which terminal of the battery the current flows: ⟳ .   The resistances (denoted -⋁⋁⋁-) are one 1-ohm, one 4-ohm, and two 2-ohm.   The current entering each battery will be the same as that leaving it.

In Figure 2-5 we label the currents I_1, I_2, and I_3, and the batteries and junctions A, B, C, and D.

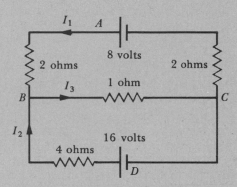

FIGURE 2-5

Applying Law 1 to each junction,

Junction B, $I_1 + I_2 = I_3$

Junction C, $I_1 + I_2 = I_3$

giving $I_1 + I_2 - I_3 = 0$.   Applying Law 2 to various paths,

Path $ABCA$, $2I_1 + 1I_3 + 2I_1 = 8$

Path $DBCD$, $4I_2 + 1I_3 \qquad = 16$

It is not necessary to look further at path $ABDCA$. We now have a system of three equations in three unknowns—I_1, I_2, I_3. Path $ABDCA$ leads to an equation that is a combination of the last two; there is no new information.

The problem thus reduces to solving the system of linear equations

$$I_1 + \; I_2 - I_3 = 0$$

$$4I_1 \qquad + I_3 = 8$$

$$4I_2 + I_3 = 16$$

We find, using the method of Gaussian elimination, that

$$\begin{pmatrix} 1 & 1 & -1 & 0 \\ 4 & 0 & 1 & 8 \\ 0 & 4 & 1 & 16 \end{pmatrix} \cong \begin{pmatrix} 1 & 1 & -1 & 0 \\ 0 & -4 & 5 & 8 \\ 0 & 4 & 1 & 16 \end{pmatrix}$$

$$\cong \begin{pmatrix} 1 & 1 & -1 & 0 \\ 0 & -4 & 5 & 8 \\ 0 & 0 & 6 & 24 \end{pmatrix} \cong \begin{pmatrix} 1 & 1 & -1 & 0 \\ 0 & 1 & -\frac{5}{4} & -2 \\ 0 & 0 & 1 & 4 \end{pmatrix}$$

leading to the system

$$I_1 + I_2 - I_3 = 0$$
$$I_2 - \tfrac{5}{4}I_3 = -2$$
$$I_3 = 4$$

The equations give the currents to be $I_3 = 4$, $I_2 = 3$, $I_1 = 1$. The units are amps.

The solution is unique, as is to be expected in this physical situation.

Example 2

Determine the currents through the various branches of the network in Figure 2-6.

Law 1 gives

Junction B, $I_1 + I_3 = I_2$

Junction C, $I_1 + I_3 = I_2$

giving $I_1 - I_2 + I_3 = 0$.

From Law 2,

Path $ABCA$, $1I_1 - 1I_3 = 2 - 4$

$$I_1 - I_3 = -2$$

Path $ABDCA$, $1I_1 + 1I_2 = 2$

$$I_1 + I_2 = 2$$

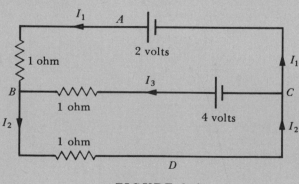

FIGURE 2-6

It is not necessary to look further at path *BCD*.   We now have three equations in three unknowns, I_1, I_2, and I_3.   Path *BCD* would not give any new information.

The mathematical model is thus the system

$$I_1 - I_2 + I_3 = 0$$

$$I_1 \qquad - I_3 = -2$$

$$I_1 + I_2 \qquad = 2$$

Using Gaussian elimination, we get $I_1 = 0, I_2 = 2, I_3 = 2$; the units are amps.   There would thus be no current passing out of battery *A*.

In practice, electrical networks can involve many resistances and circuits; the problem involves solving a large system of equations on the computer.

Traffic Flow   Network analysis, as we saw in the previous example, plays an important role in electrical engineering.   In recent years the concepts and tools of network analysis have been found to be very useful in many other fields, such as information theory and the study of transportation systems.   The following analysis of traffic flow through a road network during a peak period illustrates how a system of linear equations with many solutions can arise in practice.

Consider the typical road network in Figure 2-7.   It represents an area of downtown Jacksonville, Florida.   The flow in and out of the

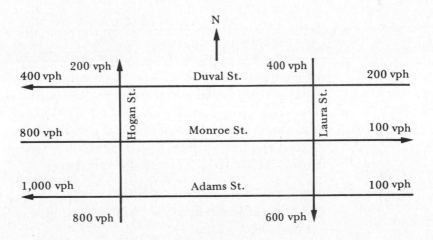

FIGURE 2-7

network is measured in vehicles per hour (vph). Figures are based on midweek peak traffic hours, 7 A.M. to 9 A.M. and 4 P.M. to 6 P.M. An increase of two percent in the overall flow should be allowed for during the Friday evening traffic flow. Let us construct a mathematical model that can be used to analyze this network.

Suppose it becomes necessary to perform road work on the stretch of Adams Street between Laura and Hogan. It is desirable to have as small a flow of traffic as possible along this stretch of road. The flows can be controlled along various branches by means of traffic lights at the junctions. What is the minimum flow possible along Adams that would not lead to traffic congestion? What are the flows along the other branches when this is attained? Our model will enable us to answer these questions.

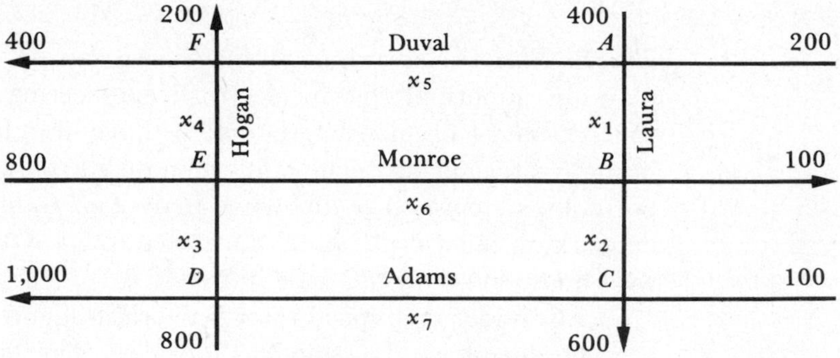

FIGURE 2-8

Let the traffic flows along the various branches be as represented in Figure 2-8. It is reasonable to assume that all traffic entering a junction must leave that junction if there is to be no congestion. This conservation of flow constraint (compare it with the first of Kirchhoff's laws for electrical networks) leads to a system of linear equations.

Consider Junction A. The traffic flowing in is $400 + 200$, and that flowing out is $x_1 + x_5$. This leads to the equation

$$x_1 + x_5 = 600$$

If we look at Junction B, the traffic flowing in is $x_1 + x_6$, and that flowing out is $x_2 + 100$. Thus

$$x_1 + x_6 = x_2 + 100$$

Rearranging this equation we get

$$x_1 - x_2 + x_6 = 100$$

Carrying out this procedure for each junction and writing the resulting equations in convenient form, we get the following system.

Junction A x_1 $+ x_5$ $=$ 600

Junction B $x_1 - x_2$ $+ x_6$ $=$ 100

Junction C x_2 $- x_7 =$ 500

Junction D $- x_3$ $+ x_7 =$ 200

Junction E $x_3 - x_4$ $- x_6$ $= -800$

Junction F $x_4 + x_5$ $=$ 600

We use the method of Gaussian elimination to analyze the system.

$$\begin{pmatrix} 1 & 0 & 0 & 0 & 1 & 0 & 0 & 600 \\ 1 & -1 & 0 & 0 & 0 & 1 & 0 & 100 \\ 0 & 1 & 0 & 0 & 0 & 0 & -1 & 500 \\ 0 & 0 & -1 & 0 & 0 & 0 & 1 & 200 \\ 0 & 0 & 1 & -1 & 0 & -1 & 0 & -800 \\ 0 & 0 & 0 & 1 & 1 & 0 & 0 & 600 \end{pmatrix}$$

$$\cong \begin{pmatrix} 1 & 0 & 0 & 0 & 1 & 0 & 0 & 600 \\ 0 & 1 & 0 & 0 & 1 & -1 & 0 & 500 \\ 0 & 1 & 0 & 0 & 0 & 0 & -1 & 500 \\ 0 & 0 & -1 & 0 & 0 & 0 & 1 & 200 \\ 0 & 0 & 1 & -1 & 0 & -1 & 0 & -800 \\ 0 & 0 & 0 & 1 & 1 & 0 & 0 & 600 \end{pmatrix}$$

$$\cong \begin{pmatrix} 1 & 0 & 0 & 0 & 1 & 0 & 0 & 600 \\ 0 & 1 & 0 & 0 & 1 & -1 & 0 & 500 \\ 0 & 0 & 0 & 0 & 1 & -1 & 1 & 0 \\ 0 & 0 & -1 & 0 & 0 & 0 & 1 & 200 \\ 0 & 0 & 1 & -1 & 0 & -1 & 0 & -800 \\ 0 & 0 & 0 & 1 & 1 & 0 & 0 & 600 \end{pmatrix}$$

$$\cong \begin{pmatrix} 1 & 0 & 0 & 0 & 1 & 0 & 0 & 600 \\ 0 & 1 & 0 & 0 & 1 & -1 & 0 & 500 \\ 0 & 0 & 1 & -1 & 0 & -1 & 0 & -800 \\ 0 & 0 & -1 & 0 & 0 & 0 & 1 & 200 \\ 0 & 0 & 0 & 0 & 1 & -1 & 1 & 0 \\ 0 & 0 & 0 & 1 & 1 & 0 & 0 & 600 \end{pmatrix}$$

$$\cong \begin{pmatrix} 1 & 0 & 0 & 0 & 1 & 0 & 0 & 600 \\ 0 & 1 & 0 & 0 & 1 & -1 & 0 & 500 \\ 0 & 0 & 1 & -1 & 0 & -1 & 0 & -800 \\ 0 & 0 & 0 & -1 & 0 & -1 & 1 & -600 \\ 0 & 0 & 0 & 0 & 1 & -1 & 1 & 0 \\ 0 & 0 & 0 & 1 & 1 & 0 & 0 & 600 \end{pmatrix}$$

$$\cong \begin{pmatrix} 1 & 0 & 0 & 0 & 1 & 0 & 0 & 600 \\ 0 & 1 & 0 & 0 & 1 & -1 & 0 & 500 \\ 0 & 0 & 1 & -1 & 0 & -1 & 0 & -800 \\ 0 & 0 & 0 & 1 & 0 & 1 & -1 & 600 \\ 0 & 0 & 0 & 0 & 1 & -1 & 1 & 0 \\ 0 & 0 & 0 & 0 & 1 & -1 & 1 & 0 \end{pmatrix}$$

$$\cong \begin{pmatrix} 1 & 0 & 0 & 0 & 1 & 0 & 0 & 600 \\ 0 & 1 & 0 & 0 & 1 & -1 & 0 & 500 \\ 0 & 0 & 1 & -1 & 0 & -1 & 0 & -800 \\ 0 & 0 & 0 & 1 & 0 & 1 & -1 & 600 \\ 0 & 0 & 0 & 0 & 1 & -1 & 1 & 0 \\ 0 & 0 & 0 & 0 & 0 & 0 & 0 & 0 \end{pmatrix}$$

This is an echelon form of the augmented matrix. The simplified system is

$$x_1 \qquad\qquad + x_5 \qquad\qquad = 600$$
$$\qquad x_2 \qquad + x_5 - x_6 \qquad = 500$$
$$\qquad\qquad x_3 - x_4 \qquad - x_6 \qquad = -800$$
$$\qquad\qquad\qquad x_4 \qquad + x_6 - x_7 = 600$$
$$\qquad\qquad\qquad\qquad x_5 - x_6 + x_7 = 0$$

Using substitution, we can express all the variables in terms of x_6 and x_7 to get

$$x_1 = -x_6 + x_7 + 600$$
$$x_2 = x_7 + 500$$
$$x_3 = x_7 - 200$$
$$x_4 = -x_6 + x_7 + 600$$
$$x_5 = x_6 - x_7$$

As was perhaps to be expected, the system has no unique solution—the traffic flow in each branch is not uniquely defined.

We are interested in minimizing the flow x_7. Since all the flows must be greater than or equal to zero, the third equation implies that the minimum flow for x_7 is 200, for otherwise x_3 could become negative. (A negative flow would be interpreted as traffic moving in the direction opposite to the one permitted on the one-way street.) Thus road work must allow for a flow of at least 200 cars per hour on the branch CD in the peak period.

Let us now examine what the flows in the other branches will be when this is attained. $x_7 = 200$ gives

$$x_1 = -x_6 + 800$$

$$x_2 = 700$$

$$x_3 = 0$$

$$x_4 = -x_6 + 800$$

$$x_5 = x_6 - 200$$

Since $x_7 = 200$ implies that $x_3 = 0$ and vice versa, we see that the minimum flow in the branch x_7 can be attained by making x_3 zero i.e., by closing DE to traffic. The traffic x_2 in BC is then determined uniquely as 700 cars per hour. There is still some freedom in the flows x_1, x_4, x_5, and x_6 along AB, EF, AF, and EB, respectively, since these variables are still not uniquely determined. For example, one possible set of flows in these branches is $x_1 = 500, x_4 = 500, x_5 = 100, x_6 = 300$.

However, since $x_1 \geq 0, x_6 \leq 800$; and since $x_5 \geq 0, x_6 \geq 200$. Thus $200 \leq x_6 \leq 800$. The flow of traffic along Monroe between Hogan and Laura will be between 200 and 800 vph.

Supply and Demand Models   In this section we construct and discuss a mathematical model that is used in economics for analyzing supplies and demands for commodities in terms of prices.

First consider a model for a single commodity. Let D be the quantity in demand, S the quantity in supply, and P the price of the commodity. Demand and supply will both depend upon the price. The following mathematical model, called the *Equilibrium Model*, leads to a situation when demand and supply are equal. The model consists of the following equations describing relationships between demand D, supply S, and price P.

$$D = S$$

$$D = -bP + a$$

$$S = dP + c$$

a, b, c, and d are constants, b and d positive, called the *parameters* of the model; they define the model. They are arrived at empirically.

In practice a great deal of attention must be paid to evaluating them. The reliability of the model depends upon them. The interpretation of the first equation is that demand is equal to supply. The second equation implies that demand is a linear function of price; the graph of the function has a negative slope, for as the price goes up the demand does down. The third equation implies that supply is also a linear function of price; the graph has a positive slope, for as the price increases so does the supply. The values of D, S, and P that satisfy this system form the *equilibrium solution* of the model. The following figure gives a geometric representation of the system.

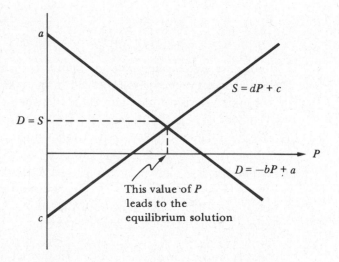

We will now illustrate the model by looking at a specific one-commodity situation.

Example 1

The demand and supply functions of a one-commodity market have been arrived at empirically. The resulting model is

$$D = S$$
$$D = -2P + 30$$
$$S = 3P + 20$$

Rewrite these equations

$$D - S \qquad = 0$$
$$D \qquad + 2P = 30$$
$$S - 3P = 20$$

Solve, using the method of Gaussian elimination,

$$\begin{pmatrix} 1 & -1 & 0 & 0 \\ 1 & 0 & 2 & 30 \\ 0 & 1 & -3 & 20 \end{pmatrix} \cong \begin{pmatrix} 1 & -1 & 0 & 0 \\ 0 & 1 & 2 & 30 \\ 0 & 1 & -3 & 20 \end{pmatrix}$$

$$\cong \begin{pmatrix} 1 & -1 & 0 & 0 \\ 0 & 1 & 2 & 30 \\ 0 & 0 & -5 & -10 \end{pmatrix} \cong \begin{pmatrix} 1 & -1 & 0 & 0 \\ 0 & 1 & 2 & 30 \\ 0 & 0 & 1 & 2 \end{pmatrix}$$

Thus

$$D - S \quad = \quad 0$$

$$S + 2P = 30$$

$$P = \quad 2$$

On back substitution we get $D = S = 26$, $P = 2$.

Thus the equilibrium solution of this model is $D = S = 26$, $P = 2$.

A marketing situation may suddenly change for some reason, the demand suddenly outstripping the supply available or vice-versa. The model can then be used to give some idea of the outcomes. We will now use such an approach in analyzing the effect of raising the price of gasoline.

Example 2

In November 1973 an extreme gasoline shortage in the United States was caused by the Arab oil boycott following the Middle East war. The Nixon administration was faced with two alternatives to combat this shortage, rationing or price increase. The latter approach was adopted. In this example we use the equilibrium model to examine the effect of increasing the price of gasoline.

An oil company sold 600 million gallons of regular gasoline per month through its gas stations in a certain marketing region, at a price of 40 cents per gallon, before the crisis. What would be the effect of raising the price to 50 cents per gallon? A marketing research group estimated that supply and demand in units of a million gallons in the region were related to price P per gallon as follows:

$$D = -2P + 680$$

$$S = \quad P + 560$$

Observe that when $D = S$, $P = 40$ cents per gallon. The research group constructed a model based on the pre-crisis situation being the equilibrium solution, using the available statistics from the effects of past fluctuations in prices to determine the constants.

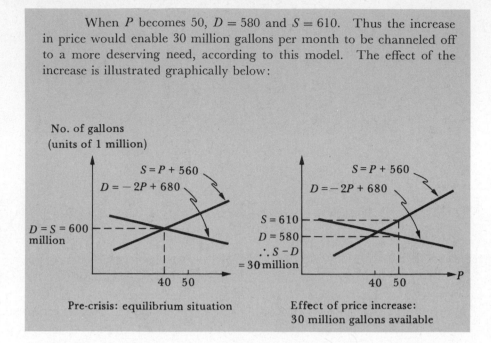

When P becomes 50, $D = 580$ and $S = 610$. Thus the increase in price would enable 30 million gallons per month to be channeled off to a more deserving need, according to this model. The effect of the increase is illustrated graphically below:

No. of gallons
(units of 1 million)

$S = P + 560$

$D = -2P + 680$

$D = S = 600$
million

40 50

Pre-crisis: equilibrium situation

$S = P + 560$

$D = -2P + 680$

$S = 610$

$D = 580$

$\therefore S - D$
$= 30$ million

40 50 P

Effect of price increase:
30 million gallons available

We now proceed to discuss a model for a two-commodity market, leading up to the n-commodity market. Consider two commodities, the price of each affecting the supply and demand of the other. Let the demands, supplies, and prices of the commodities be D_1, S_1, P_1, and D_2, S_2, P_2. Then a model would be of the type

$$D_1 = S_1$$
$$D_1 = a_1 P_1 + a_2 P_2 + a_3$$
$$S_1 = b_1 P_1 + b_2 P_2 + b_3$$
$$D_2 = S_2$$
$$D_2 = c_1 P_1 + c_2 P_2 + c_3$$
$$S_2 = d_1 P_1 + d_2 P_2 + d_3$$

The model involves a system of six equations in the six variables D_1, S_1, P_1 and D_2, S_2, P_2. The selection of the parameters is again crucial to the reliability of the model.

An n-commodity model would involve a system of $3n$ equations in the $3n$ variables $D_1, S_1, P_1, ..., D_n, S_n, P_n$.

We complete this discussion by looking at a specific two-commodity model.

Example 3

The demand and supply functions of a market have been arrived at empirically. The model is

$$D_1 = S_1$$

$$D_1 = -P_1 + P_2 + 8$$

$$S_1 = P_1 + 4P_2 - 2$$

$$D_2 = S_2$$

$$D_2 = P_1 + 3P_2 + 4$$

$$S_2 = 3P_1 + P_2 - 1$$

What will the equilibrium solution be?

The system of equations is rewritten

$$D_1 \qquad - S_1 \qquad\qquad\qquad\qquad = 0$$

$$D_1 \qquad\qquad\qquad + P_1 - P_2 = 8$$

$$\qquad S_1 \qquad - P_1 - 4P_2 = -2$$

$$D_2 \qquad - S_2 \qquad\qquad\qquad = 0$$

$$D_2 \qquad\qquad - P_1 - 3P_2 = 4$$

$$\qquad S_2 - 3P_1 - P_2 = -1$$

The solution is found using the method of Gaussian elimination to be $D_1 = S_1 = 5.5$, $P_1 = 3.5$, $D_2 = S_2 = 10.5$, $P_2 = 1$.

There is a significant difference between these economic models and the electrical network model. In the latter, the systems of equations fitted the real situation. In models of economic relationships, the equations may only approximately describe the real situations.

In a model of the type,

$$D = S$$

$$D = -bP + a$$

$$S = dP + c$$

we attempt to select the parameters a, b, c, and d so that $D = -bP + a$ and $S = dP + c$ resemble as closely as possible real dependences. In reality, these relationships may not be linear. It is not possible to predict the exact relationship. Let us symbolize any actual relationship between D and P by $D = f(P)$ in Figure 2-9. The graph of $D = -bP + a$ should be "close" to that of $D = f(P)$ over the region of interest. The supply and

131

demand models discussed here are *linear models* that can be used to approximate given situations.

Sometimes one imposes a system of linear approximations on a system of nonlinear equations that exactly describes a situation, particularly when the nonlinear system is too cumbersome mathematically to give much information. In such a case, one might be willing to sacrifice a certain amount of realism in order to obtain a solution or perhaps a more complete overall analysis of the situation.

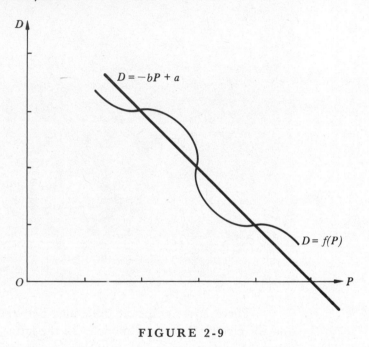

FIGURE 2-9

For further discussions of linear economic models, see *Economic Models*, E. F. Beach, John Wiley & Sons., Inc., New York, 1957. For a linear economic model for the United States, involving six linear equations in eleven variables, fitted to quarterly data over a span of twenty-one years, see "A System of Equations Explaining the United States Trade Cycles, 1921–41", Colin Clark, *Econometrica*, **17**: 93–124.

EXERCISES

1. Determine the currents through the various branches of the electrical networks in Figure 2-10.

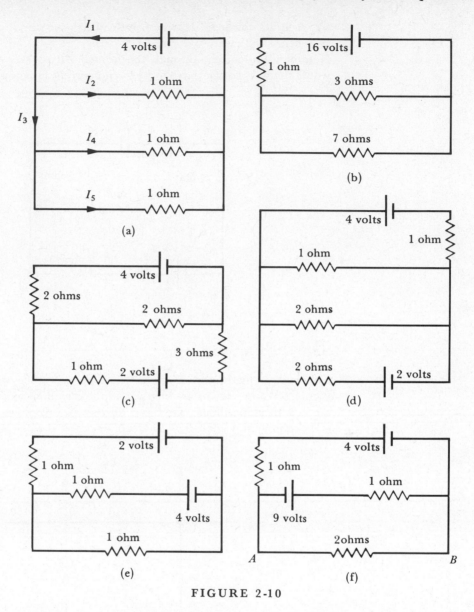

FIGURE 2-10

(*Hint*: In (f) it is difficult to decide on the direction of the current along *AB*. Make a guess. A negative result for that current means that your guess was the wrong one—the current is in the opposite direction. However the magnitude will still be correct. There is no need to rework the problem.)

2. Determine the currents through the various branches of the electrical network in Figure 2-11.

a) when the voltage of battery C is 9 volts

b) when it is 23 volts

Note how the current through the branch AB is reversed in (b). What would the voltage of C have to be for no current to flow through AB?

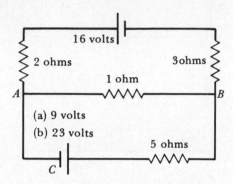

FIGURE 2-11

3. Construct a mathematical model that describes the traffic network in Figure 2-12. All streets are one-way in the directions indicated. The units are in vehicles per hour. Give two distinct possible flows of traffic. What is the minimum possible flow that can be expected along AB?

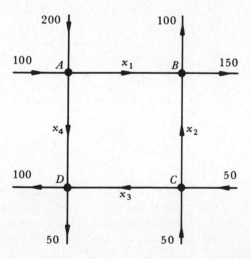

FIGURE 2-12

4. Figure 2-13 represents the traffic entering and leaving a "roundabout" road junction. Such junctions are very common in Europe. Construct a mathematical model that describes the flow of traffic along various branches. What is the minimum flow theoretically possible along branch BC? Is this ever likely to be realized in practice?

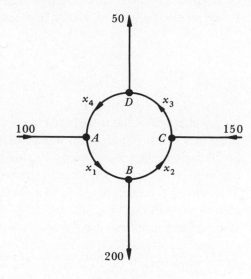

FIGURE 2-13

5. Determine the equilibrium solutions to the following one- or two-commodity markets.

a) $D = S$
   $D = -3P + 15$
   $S = 2P - 5$

b) $D_1 = S_1$
   $D_1 = -P_1 - 2P_2 + 8$
   $S_1 = 3P_1 - 4$
   $D_2 = S_2$
   $D_2 = -2P_1 + P_2 + 3$
   $S_2 = 2P_1 - P_2 - 1$

c) $D_1 = S_1$
   $D_1 = 2P_1 - 5P_2 + 10$
   $S_1 = 6P_1 - P_2 - 2$
   $D_2 = S_2$
   $D_2 = 3P_2 + 3$
   $S_2 = 3P_1 + 2P_2 + 2$

6. In a stable situation, gasoline is sold at a certain station in quantities of 135 gallons of regular per hour at a price of 45 cents per gallon. Construct a mathematical model to analyze the effect of raising the price to 65 cents per gallon.

135

7. Write a general system of equations that could be used to describe a three-commodity market.

8. Note that the augmented matrix of the system of equations involved in a commodity market model contains many zero elements arising from the equilibrium conditions $D_i = S_i$. Matrices involving large numbers of zero elements are called *sparse*. Can you develop special techniques that reduce the work involved in solving the special systems of equations of these models?

2–4.* USE OF PIVOTS

In this chapter we have used the Gaussian elimination method to determine the solution to a system of linear equations, and we have seen how practical problems can be solved by using these tools. The examples considered involved systems of linear equations with uncomplicated coefficients. The aim of the examples was to enable the reader to understand various techniques without being distracted by arithmetic. However, in practice, the coefficients often arise from measurements taken and can involve numbers to many and varying decimal places. When calculations are performed, only a finite number of digits can be carried—numbers are rounded off. Thus round-off errors can occur. In many cases these errors cannot be eliminated entirely, but they can be minimized by selection of *pivots*.

The method of Gaussian elimination involves using the first equation to eliminate the first variable from other equations, and so on. In general, at the rth stage, the rth equation is used to eliminate the rth variable from the remaining equations. However, we are not restricted to using the rth equation to eliminate the rth variable. At any stage any given equation can be used to eliminate one variable from the other equations; the coefficient of that variable in the given equation is called a *pivot*. The selection of pivots is very important in obtaining accurate numerical solutions.

Example 1

Solve the system

$$10^{-3}x_1 - x_2 = 1$$

$$2x_1 + x_2 = 0,$$

assuming that you are working to three significant figures.

Let us solve this system to three significant figures in two ways, using two distinct selections of pivots.   The augmented matrix is:

$$\begin{pmatrix} 10^{-3} & -1 & 1 \\ 2 & 1 & 0 \end{pmatrix}$$

1. Let us select the $(1,1)$ element as pivot:

$$\begin{matrix} pivot \\ \downarrow \end{matrix}$$

$$\begin{pmatrix} \boxed{10^{-3}} & -1 & 1 \\ 2 & 1 & 0 \end{pmatrix} \quad \underset{(10^3)\ \text{Row 1}}{\cong} \quad \overset{A}{\begin{pmatrix} \boxed{1} & -1000 & 1000 \\ 2 & 1 & 0 \end{pmatrix}}$$

$$\underset{\text{Row } 2-(2)\ \text{Row 1}}{\cong} \quad \begin{pmatrix} 1 & -1000 & 1000 \\ 0 & 2000 & -2000 \end{pmatrix}$$

$$\uparrow$$

(*This element should have been* 2001, *but has been rounded to three significant figures.*)

$$\underset{(1/2000)\ \text{Row 2}}{\cong} \quad \begin{pmatrix} 1 & -1000 & 1000 \\ 0 & 1 & -1 \end{pmatrix}$$

The final system is

$$x_1 - 1000x_2 = 1000$$

$$x_2 = -1$$

Substituting for x_2 into the first equation gives $x_1 = 0$, $x_2 = -1$.

2. Let us now solve the system by pivoting on the $(2,1)$ element:

$$Pivot \rightarrow \begin{pmatrix} 10^{-3} & -1 & 1 \\ \boxed{2} & 1 & 0 \end{pmatrix}$$

$$\underset{\substack{\text{Interchange} \\ \text{Rows 1 and 2}}}{\cong} \quad \begin{pmatrix} \boxed{2} & 1 & 0 \\ 10^{-3} & -1 & 1 \end{pmatrix}$$

(Note that once the pivot has been selected we interchange rows, if necessary, to bring it into the customary location for creating zeros in that particular column.)

$$\underset{(\frac{1}{2})\ \text{Row 1}}{\cong} \quad \begin{pmatrix} 1 & \frac{1}{2} & 0 \\ 10^{-3} & -1 & 1 \end{pmatrix}$$

$$\underset{\text{Row } 2-(10^{-3})\ \text{Row 1}}{\cong} \quad \begin{pmatrix} 1 & \frac{1}{2} & 0 \\ 0 & -1 & 1 \end{pmatrix}$$

$$\underset{(-1)\ \text{Row 2}}{\cong} \quad \begin{pmatrix} 1 & \frac{1}{2} & 0 \\ 0 & 1 & -1 \end{pmatrix}$$

The final system is

$$x_1 + \tfrac{1}{2}x_2 = 0$$
$$x_2 = -1,$$

leading to $x_1 = \tfrac{1}{2}$, $x_2 = -1$.

The exact solution to the system is in fact $x_1 = \frac{1000}{2001}$, $x_2 = -\frac{2000}{2001}$. The second solution, obtained by pivoting on the $(2, 1)$ element is a reasonably good approximation; the first solution, obtained by pivoting on the $(1, 1)$ element is not.

Having seen that the choice of pivots is important, let us now use the above example to decide how pivots can be selected to improve accuracy.

Why was the selection of the element $(2, 1)$ a better choice for pivot than the $(1, 1)$ element?

In the first method note that the choice of the $(1, 1)$ element as pivot caused the introduction of large numbers, -1000 and 1000, into the matrix A. In the final system,

$$x_1 - 1000x_2 = 1000$$
$$x_2 = -1$$

it is these large numbers that cause the error in x_1. When substituting $x_2 = -1$ into the first equation, the x_1 term (which should have been approximately $\tfrac{1}{2}$) is completely overpowered and lost in comparison with the $-1000x_2$ term, working to three significant figures.

In the second method, involving the use of the $(2, 1)$ element as pivot, no such large numbers are introduced. The solution obtained is more accurate.

This discussion illustrates the fact that the introduction of large numbers during the elimination process can cause round-off errors. The following procedures minimize this cause of error. It should be stressed that these procedures work most of the time and are the ones most widely adopted. There are, however, exceptions which we shall not go into here.

Procedures used for minimizing round-off errors First of all, we properly scale the equations. By scaling a set of equations we mean:

1. Multiplying any equation by a nonzero constant.
2. Replacing any unknown by a new unknown that is a multiple of the old one.

By properly scaling we mean scaling so that the largest element in each row and column of the augmented matrix is of order unity. There is no automatic method for scaling.

We then select pivots, eliminating the unknowns in the natural order $x_1, x_2, \ldots$ At the rth stage the pivot is generally taken to be the coefficient of x_r, in the remaining $n - r + 1$ equations, that has the largest absolute value. In terms of elementary matrix transformations, we scan that column of the last $n - r + 1$ rows, choose the row that has the number with the largest absolute value, and interchange this row with the rth row. The pivot is then employed to create a 1 in its own location (that row is divided by the pivot), and zeros elsewhere in the column, by adding suitable multiples of the pivot row to other rows.

The scaling procedure is an attempt to make the coefficients comparable in magnitude. The choice of pivot ensures that division in eliminating a variable is by the largest number available. (Division by a small number introduces a large number.)

We now give further examples to illustrate these techniques.

Example 2

Solve the system of equations

$$2x_1 - x_2 + 3x_3 + 3x_4 = 1$$
$$x_1 - x_2 + x_3 + 2x_4 = 1$$
$$-x_1 + x_2 - x_3 - 2x_4 = -1$$
$$2x_1 + x_2 + 7x_3 + 2x_4 = 2$$

The augmented matrix is

$$\begin{pmatrix} 2 & -1 & 3 & 3 & 1 \\ 1 & -1 & 1 & 2 & 1 \\ -1 & 1 & -1 & -2 & -1 \\ 2 & 1 & 7 & 2 & 2 \end{pmatrix}$$

The system is such that the largest element in each row and column of the augmented matrix is of order unity. Thus it is already properly scaled.

Scanning the first column of the augmented matrix, we find that the absolute value of the element in the $(1, 1)$ location, namely 2, is as large as the absolute value of any other element. Hence there is no need to interchange rows.

Make the leading element in the first row 1 by dividing the first row by 2. This new first row is then used to create zeros in the first column.

$$\begin{pmatrix} 2 & -1 & 3 & 3 & 1 \\ 1 & -1 & 1 & 2 & 1 \\ -1 & 1 & -1 & -2 & -1 \\ 2 & 1 & 7 & 2 & 2 \end{pmatrix} \cong \begin{pmatrix} 1 & -\frac{1}{2} & \frac{3}{2} & \frac{3}{2} & \frac{1}{2} \\ 1 & -1 & 1 & 2 & 1 \\ -1 & 1 & -1 & -2 & -1 \\ 2 & 1 & 7 & 2 & 2 \end{pmatrix}$$

$$\cong \begin{pmatrix} 1 & -\frac{1}{2} & \frac{3}{2} & \frac{3}{2} & \frac{1}{2} \\ 0 & -\frac{1}{2} & -\frac{1}{2} & \frac{1}{2} & \frac{1}{2} \\ 0 & \frac{1}{2} & \frac{1}{2} & -\frac{1}{2} & -\frac{1}{2} \\ 0 & 2 & 4 & -1 & 1 \end{pmatrix} \cong \begin{pmatrix} 1 & -\frac{1}{2} & \frac{3}{2} & \frac{3}{2} & \frac{1}{2} \\ 0 & 2 & 4 & -1 & 1 \\ 0 & \frac{1}{2} & \frac{1}{2} & -\frac{1}{2} & -\frac{1}{2} \\ 0 & -\frac{1}{2} & -\frac{1}{2} & \frac{1}{2} & \frac{1}{2} \end{pmatrix}$$

Scan these elements to determine the one with the largest absolute value. It is 2. Interchange second and fourth rows.

Make the leading nonzero element 1 by dividing the second row by 2.

$$\cong \begin{pmatrix} 1 & -\frac{1}{2} & \frac{3}{2} & \frac{3}{2} & \frac{1}{2} \\ 0 & 1 & 2 & -\frac{1}{2} & \frac{1}{2} \\ 0 & \frac{1}{2} & \frac{1}{2} & -\frac{1}{2} & -\frac{1}{2} \\ 0 & -\frac{1}{2} & -\frac{1}{2} & \frac{1}{2} & \frac{1}{2} \end{pmatrix} \cong \begin{pmatrix} 1 & -\frac{1}{2} & \frac{3}{2} & \frac{3}{2} & \frac{1}{2} \\ 0 & 1 & 2 & -\frac{1}{2} & \frac{1}{2} \\ 0 & 0 & -\frac{1}{2} & -\frac{1}{4} & -\frac{3}{4} \\ 0 & 0 & \frac{1}{2} & \frac{1}{4} & \frac{3}{4} \end{pmatrix}$$

Scanning these elements, we find they have the same absolute value. There is no need to interchange rows. Multiply row 3 by −2 to make leading nonzero element 1.

$$\cong \begin{pmatrix} 1 & -\frac{1}{2} & \frac{3}{2} & \frac{3}{2} & \frac{1}{2} \\ 0 & 1 & 2 & -\frac{1}{2} & \frac{1}{2} \\ 0 & 0 & 1 & \frac{1}{2} & \frac{3}{2} \\ 0 & 0 & \frac{1}{2} & \frac{1}{4} & \frac{3}{4} \end{pmatrix} \cong \begin{pmatrix} 1 & -\frac{1}{2} & \frac{3}{2} & \frac{3}{2} & \frac{1}{2} \\ 0 & 1 & 2 & -\frac{1}{2} & \frac{1}{2} \\ 0 & 0 & 1 & \frac{1}{2} & \frac{3}{2} \\ 0 & 0 & 0 & 0 & 0 \end{pmatrix}$$

The system of equations reduces to the simpler system

$$x_1 - \tfrac{1}{2}x_2 + \tfrac{3}{2}x_3 + \tfrac{3}{2}x_4 = \tfrac{1}{2}$$

$$x_2 + 2x_3 - \tfrac{1}{2}x_4 = \tfrac{1}{2}$$

$$x_3 + \tfrac{1}{2}x_4 = \tfrac{3}{2}$$

The solution is not unique. The complete solution is

$$x_1 = -3$$

$$x_2 = -\tfrac{5}{2} + \tfrac{3}{2}x_4$$

$$x_3 = \tfrac{3}{2} - \tfrac{1}{2}x_4$$

where x_4 can take on any value.

Example 3   Solve the system of equations

$$0.002x_1 + 4x_2 - 2x_3 = 1$$

$$0.001x_1 + 2.0001x_2 + x_3 = 2$$

$$0.001x_1 + 3x_2 + 3x_3 = -1$$

Assume that calculations can be carried out to five significant figures.

First of all, we scale the system by introducing the new variable $y_1 = 0.001x_1$. All the coefficients in the first column of the augmented matrix will then be of order unity. For consistency of notation, let $y_2 = x_2$ and $y_3 = x_3$.

We now have the properly scaled system

$$2y_1 + \qquad 4y_2 - 2y_3 = \quad 1$$

$$y_1 + 2.0001y_2 + \quad y_3 = \quad 2$$

$$y_1 + \qquad 3y_2 + 3y_3 = -1$$

Proceed using the pivoting sequence of transformations.

$$\begin{pmatrix} 2 & 4 & -2 & 1 \\ 1 & 2.0001 & 1 & 2 \\ 1 & 3 & 3 & -1 \end{pmatrix}$$

$$\cong \begin{pmatrix} 1 & 2 & -1 & 0.5 \\ 1 & 2.0001 & 1 & 2 \\ 1 & 3 & 3 & -1 \end{pmatrix}$$

$$\cong \begin{pmatrix} 1 & 2 & -1 & 0.5 \\ 0 & 0.0001 & 2 & 1.5 \\ 0 & 1 & 4 & -1.5 \end{pmatrix}$$

$$\cong \begin{pmatrix} 1 & 2 & -1 & 0.5 \\ 0 & 1 & 4 & -1.5 \\ 0 & 0.0001 & 2 & 1.5 \end{pmatrix}$$

$$\cong \begin{pmatrix} 1 & 2 & -1 & 0.5 \\ 0 & 1 & 4 & -1.5 \\ 0 & 0 & 1.9996 & \boxed{1.5002} \end{pmatrix}$$

This element should have been 1.50015, but it has been rounded off to five significant figures.

Thus the system reduces to

$$y_1 + 2y_2 - \qquad y_3 = 0.5$$

$$y_2 + \qquad 4y_3 = -1.5$$

$$1.9996y_3 = 1.5002$$

The solution to five significant figures is $y_1 = 10.25$, $y_2 = -4.5001$, $y_3 = 0.75003$. In terms of the original variables, the solution is

$$x_1 = 10{,}250, \quad x_2 = -4.5001, \quad x_3 = 0.75003$$

It is interesting to compare this solution with the solution that would have been obtained using the Gaussian elimination method, which does not incorporate the pivoting and scaling refinements. The solution to five significant figures would have been $x_1 = 3,250.1$, $x_2 = -1$, $x_3 = 0.75003$. Substituting into the original equations, it can be seen that this solution is very unsatisfactory, whereas the solution obtained using the refinements is very accurate.

EXERCISES

Solve the following systems of equations using the pivoting procedure and the scaling technique when appropriate.

1.
$$- x_2 + 2x_3 = 4$$
$$x_1 + 2x_2 - x_3 = 1$$
$$x_1 + 2x_2 + 2x_3 = 4$$

2.
$$-x_1 + x_2 + 2x_3 = 8$$
$$2x_1 + 4x_2 - x_3 = 10$$
$$-x_1 + 2x_2 + 2x_3 = 2$$

3.
$$-x_1 + 0.002x_2 = 0$$
$$x_1 + 2x_3 = -2$$
$$0.001x_2 + x_3 = 1$$

4.
$$2x_1 + x_3 = 1$$
$$0.0001x_1 + 0.0002x_2 + 0.0004x_3 = 0.0004$$
$$x_1 - 2x_2 - 3x_3 = -3$$

5.
$$x_2 - 0.1x_3 = 2$$
$$0.02x_1 + 0.01x_2 = 0.01$$
$$x_1 - 4x_2 + 0.1x_3 = 2$$

6.
$$- 0.4x_2 + 0.002x_3 = 0.2$$
$$100x_1 - x_2 - 0.01x_3 = 1$$
$$200x_1 + 2x_2 - 0.02x_3 = 1$$

7.
$$-x_1 + 2x_3 = 1$$
$$0.1x_1 + 0.001x_2 = 0.1$$
$$0.1x_1 + 0.002x_2 = 0.2$$

8.
$$-x_2 + 0.001x_3 = 6$$
$$0.1x_1 + 0.0002x_3 = -0.2$$
$$0.01x_1 + 0.01x_2 + 0.00003x_3 = 0.02$$

9.
$$3x_1 + 2x_2 - x_3 = 1$$
$$x_1 + x_2 - x_3 = 2$$
$$-x_1 + 3x_2 - x_3 = -1$$

10.
$$x_1 - 2x_2 - x_3 = 1$$
$$x_1 - 0.001x_2 - x_3 = 2$$
$$x_1 + 3x_2 - 0.002x_3 = -1$$

11.
$$x_1 - 2x_2 + x_3 = 2$$
$$0.01x_1 - 0.03x_2 + 0.07x_3 = 0.01$$
$$x_1 + x_2 - x_3 = 4$$

12.
$$x_1 - x_2 + 3x_3 + x_4 - x_5 = 2$$
$$2x_1 + 3x_2 - x_3 + x_4 + 2x_5 = 0$$
$$x_1 - 3x_2 + x_3 + 2x_4 - x_5 = -1$$
$$x_1 + 2x_2 - x_3 + 4x_4 + x_5 = 2$$
$$3x_1 - x_2 + 2x_3 - x_4 + 2x_5 = 0$$

13. Write a computer program for determining an echelon form that incorporates the pivoting technique. Check your program on the matrix.

$$\begin{pmatrix} 2.1231 & 3.0144 & 2.011 & 3.4325 \\ 6.4868 & 8.0484 & 7.4286 & 2.6132 \\ 3.7528 & 4.3294 & 4.6591 & 6.7255 \end{pmatrix}$$

Echelon form is

$$\begin{pmatrix} 1 & 1.24074 & 1.14519 & 0.402849 \\ 0 & 1 & -1.10561 & 6.77866 \\ 0 & 0 & 1 & 80517.2 \end{pmatrix}$$

Check your answers to Exercises 1–6 using your program.

14. Modify your program in Exercise 13 to get a printout after each transformation, with a statement describing the transformation.

2–5. INVERSES OF MATRICES

In this section we introduce the concept of matrix inverse. We shall see how an inverse can be used to solve certain systems of equations.

   Consider the set of real numbers. All numbers except zero have a multiplicative inverse. The inverse of 4 is $\frac{1}{4}$ (also written 4^{-1}), the inverse of $\frac{3}{5}$ is $\frac{5}{3}$, etc. For the real numbers we say that the number b is the inverse of a if and only if

$$a \times b = 1$$

and

$$b \times a = 1.$$

This is the concept that we extend to matrices.

   An $n \times n$ matrix A is said to be *invertible* if and only if there exists a matrix B such that

$$AB = BA = I_n.$$

Here I_n is the unit $n \times n$ matrix. B is called a *multiplicative inverse* of A and is denoted A^{-1}.

   We now show that, as in the case of real numbers, an inverse of a matrix, if it exists, is unique.

   THEOREM 2-2 *Let A be an invertible matrix. Its multiplicative inverse is unique.*

   Proof: Let B and C be multiplicative inverses of A.
Thus

$$AB = I_n$$

143

Multiplying both sides by C,

$$C\,(AB) = CI_n$$

$$(CA)B = C.$$

$$I_n B = C,$$

$$B = C.$$

Thus an invertible matrix A has only one inverse.

Example 1

Prove that the matrix A, $\begin{pmatrix} 1 & 2 \\ 3 & 4 \end{pmatrix}$ has inverse B, $\begin{pmatrix} -2 & 1 \\ \frac{3}{2} & -\frac{1}{2} \end{pmatrix}$.

We have that

$$AB = \begin{pmatrix} 1 & 2 \\ 3 & 4 \end{pmatrix}\begin{pmatrix} -2 & 1 \\ \frac{3}{2} & -\frac{1}{2} \end{pmatrix} = \begin{pmatrix} 1 & 0 \\ 0 & 1 \end{pmatrix}$$

and

$$BA = \begin{pmatrix} -2 & 1 \\ \frac{3}{2} & -\frac{1}{2} \end{pmatrix}\begin{pmatrix} 1 & 2 \\ 3 & 4 \end{pmatrix} = \begin{pmatrix} 1 & 0 \\ 0 & 1 \end{pmatrix}$$

Thus the inverse of A is the matrix B.

Note that the definition of inverse is limited to square matrices. For both the products AB and BA to give I_n, A and B must both be $n \times n$ matrices. Not every square matrix has an inverse. The reader will see an example of such a matrix in example 4 on pps. 147–148.

There are a number of methods that have been developed for determining the inverse of a matrix. Here we will see the *Gauss-Jordan elimination method*. Historically it is derived from the Gaussian elimination method. The method is based on associating with a matrix A (whose inverse is required) a system of equations.

Consider the system of equations

$$a_{11}x_1 + \cdots + a_{1n}x_n = y_1$$
$$\vdots$$
$$a_{m1}x_1 + \cdots + a_{mn}x_n = y_m.$$

This system can be written as a single matrix equation

$$\begin{pmatrix} a_{11} & \cdots & a_{1n} \\ & \vdots & \\ a_{m1} & \cdots & a_{mn} \end{pmatrix}\begin{pmatrix} x_1 \\ \vdots \\ x_n \end{pmatrix} = \begin{pmatrix} y_1 \\ \vdots \\ y_m \end{pmatrix}$$

for, on multiplying the matrices

$$\begin{pmatrix} a_{11}x_1 + \cdots + a_{1n}x_n \\ \vdots \\ a_{m1}x_1 + \cdots + a_{mn}x_n \end{pmatrix} = \begin{pmatrix} y_1 \\ \vdots \\ y_m \end{pmatrix}.$$

The elements of the matrix on the left of this equation are equal to the corresponding elements of the matrix on the right, giving the system of equations.

Writing the matrix of coefficients as A and the column vectors as X and Y, this matrix form of writing the system of equations becomes

$$AX = Y.$$

Example 2

Consider the system of equations

$$2x_1 + 3x_2 - x_3 = 4$$

$$x_1 - x_2 + x_3 = 1$$

$$3x_1 + x_2 - x_3 = 2$$

The matrix of coefficients is

$$\begin{pmatrix} 2 & 3 & -1 \\ 1 & -1 & 1 \\ 3 & 1 & -1 \end{pmatrix}$$

In terms of matrices, the system can be written

$$\begin{pmatrix} 2 & 3 & -1 \\ 1 & -1 & 1 \\ 3 & 1 & -1 \end{pmatrix} \begin{pmatrix} x_1 \\ x_2 \\ x_3 \end{pmatrix} = \begin{pmatrix} 4 \\ 1 \\ 2 \end{pmatrix}.$$

Conversely, any matrix A can be interpreted to be the matrix of coefficients for a system of linear equations. Our method of determining the inverse of A is based on interpreting A in this manner.

Let A be an $n \times n$ matrix, and Y an arbitrary $n \times 1$ matrix. Consider the system of equations.

$$AX = Y.$$

This system can also be written

$$AX = I_n Y \qquad\qquad (1)$$

Multiplying both sides of this equation by A^{-1} gives

$$A^{-1}AX = A^{-1}I_n Y$$

leading to

$$I_n X = A^{-1} Y. \tag{2}$$

The solutions to system (1), as we have seen, remain unchanged under elementary transformations. The method of determining the inverse of A involves transforming (1) into (2); A^{-1} can then be read off. In practice one aims at transforming the A in (1) into the I_n of (2) using elementary transformations. These same transformations are carried out on the right hand sides of the equations involved so that I_n in (1) is simultaneously transformed into the A^{-1} of (2). In practice X and Y can be omitted; we perform the transformations on the matrices A and I_n.

The following examples illustrate the method.

Example 3

Determine the inverse of the matrix

$$A = \begin{pmatrix} 1 & 2 & 0 \\ 2 & 1 & -1 \\ 3 & 1 & 1 \end{pmatrix}$$

We set up the matrices A and I_3 as follows.

$$A | I_3 = \begin{pmatrix} 1 & 1 & 0 \\ 2 & 1 & -1 \\ 3 & 1 & 1 \end{pmatrix} \middle| \begin{pmatrix} 1 & 0 & 0 \\ 0 & 1 & 0 \\ 0 & 0 & 1 \end{pmatrix}$$

$\downarrow$ $\downarrow$

Aim is to transform   *Perform same transformations*
this matrix into I_3   *here. This matrix will be*
using elementary   *transformed into A^{-1}.*
transformations.

We have

$$A | I_3 = \begin{pmatrix} 1 & 2 & 0 \\ 2 & 1 & -1 \\ 3 & 1 & 1 \end{pmatrix} \middle| \begin{pmatrix} 1 & 0 & 0 \\ 0 & 1 & 0 \\ 0 & 0 & 1 \end{pmatrix}$$

$$\begin{matrix} \cong \\ \text{Row } 2 - (2) \text{ Row } 1 \\ \text{Row } 3 - (3) \text{ Row } 1 \end{matrix} \begin{pmatrix} 1 & 2 & 0 \\ 0 & -3 & -1 \\ 0 & -5 & 1 \end{pmatrix} \middle| \begin{pmatrix} 1 & 0 & 0 \\ -2 & 1 & 0 \\ -3 & 0 & 1 \end{pmatrix}$$

$$\begin{matrix} \cong \\ -(\frac{1}{3}) \text{ Row } 2 \end{matrix} \begin{pmatrix} 1 & 2 & 0 \\ 0 & 1 & \frac{1}{3} \\ 0 & -5 & 1 \end{pmatrix} \middle| \begin{pmatrix} 1 & 0 & 0 \\ \frac{2}{3} & -\frac{1}{3} & 0 \\ -3 & 0 & 1 \end{pmatrix}$$

So far we have used the transformations in the same pattern as in Gaussian elimination. At this stage, in Gauss-Jordan elimination we use the 1 in the $(2,2)$ location to create a zero above as well as below. In general, the pivot element is used to create zeros in all other locations in that column.

$$\begin{array}{c} \cong \\ \text{Row } 1-(2)\,\text{Row } 2 \\ \text{Row } 3+(5)\,\text{Row } 2 \end{array}\left(\begin{array}{ccc} 1 & 0 & -\frac{2}{3} \\ 0 & 1 & \frac{1}{3} \\ 0 & 0 & \frac{8}{3} \end{array}\right)\left|\left(\begin{array}{ccc} -\frac{1}{3} & \frac{2}{3} & 0 \\ \frac{2}{3} & -\frac{1}{3} & 0 \\ \frac{1}{3} & -\frac{5}{3} & 1 \end{array}\right)\right.$$

$$\begin{array}{c} \cong \\ (\frac{3}{8})\,\text{Row } 3 \end{array}\left(\begin{array}{ccc} 1 & 0 & -\frac{2}{3} \\ 0 & 1 & \frac{1}{3} \\ 0 & 0 & 1 \end{array}\right)\left|\left(\begin{array}{ccc} -\frac{1}{3} & \frac{2}{3} & 0 \\ \frac{2}{3} & -\frac{1}{3} & 0 \\ \frac{1}{8} & -\frac{5}{8} & \frac{3}{8} \end{array}\right)\right.$$

We now use the 1 in the $(3, 3)$ location to create zeros above it, so obtaining the identity matrix on the left side. The matrix on the right will become A^{-1}.

$$\begin{array}{c} \cong \\ \text{Row } 1+(\frac{2}{3})\,\text{Row } 3 \\ \text{Row } 2-(\frac{1}{3})\,\text{Row } 3 \end{array}\left(\begin{array}{ccc} 1 & 0 & 0 \\ 0 & 1 & 0 \\ 0 & 0 & 1 \end{array}\right)\left|\left(\begin{array}{ccc} -\frac{1}{4} & \frac{1}{4} & \frac{1}{4} \\ \frac{5}{8} & -\frac{1}{8} & -\frac{1}{8} \\ \frac{1}{8} & -\frac{5}{8} & \frac{3}{8} \end{array}\right)\right.$$

Thus $A^{-1} = \left(\begin{array}{ccc} -\frac{1}{4} & \frac{1}{4} & \frac{1}{4} \\ \frac{5}{8} & -\frac{1}{8} & -\frac{1}{8} \\ \frac{1}{8} & -\frac{5}{8} & \frac{3}{8} \end{array}\right).$

As has previously been mentioned, certain square matrices do not have inverses. We now illustrate how this fact shows up in this method.

Example 4

Determine the inverse, if it exists, of the matrix

$$A = \left(\begin{array}{cccc} 1 & -1 & 0 & 2 \\ 3 & 1 & 5 & -1 \\ -1 & 1 & 0 & 0 \\ 2 & -2 & 0 & 1 \end{array}\right).$$

We have that

$$A|I_4 = \left(\begin{array}{cccc} 1 & -1 & 0 & 2 \\ 3 & 1 & 5 & -1 \\ -1 & 1 & 0 & 0 \\ 2 & -2 & 0 & 1 \end{array}\right)\left|\left(\begin{array}{cccc} 1 & 0 & 0 & 0 \\ 0 & 1 & 0 & 0 \\ 0 & 0 & 1 & 0 \\ 0 & 0 & 0 & 1 \end{array}\right)\right.$$

$$\begin{array}{c} \cong \\ \text{Row } 2-(3)\,\text{Row } 1 \\ \text{Row } 3+\quad\text{Row } 1 \\ \text{Row } 4-(2)\,\text{Row } 1 \end{array}\left(\begin{array}{cccc} 1 & -1 & 0 & 2 \\ 0 & 4 & 5 & -7 \\ 0 & 0 & 0 & 2 \\ 0 & 0 & 0 & -3 \end{array}\right)\left|\left(\begin{array}{cccc} 1 & 0 & 0 & 0 \\ -3 & 1 & 0 & 0 \\ 1 & 0 & 1 & 0 \\ -2 & 0 & 0 & 1 \end{array}\right)\right.$$

$$\underset{(\frac{1}{4})\ \text{Row 2}}{\cong}\quad \begin{pmatrix} 1 & -1 & 0 & 2 \\ 0 & 1 & \frac{5}{4} & -\frac{7}{4} \\ 0 & 0 & 0 & 2 \\ 0 & 0 & 0 & -3 \end{pmatrix} \begin{pmatrix} 1 & 0 & 0 & 0 \\ -\frac{3}{4} & \frac{1}{4} & 0 & 0 \\ 1 & 0 & 1 & 0 \\ -2 & 0 & 0 & 1 \end{pmatrix}$$

$$\underset{\text{Row 1} + \text{Row 2}}{\cong}\quad \begin{pmatrix} 1 & 0 & \frac{5}{4} & \frac{1}{4} \\ 0 & 1 & \frac{5}{4} & -\frac{7}{4} \\ 0 & 0 & 0 & 2 \\ 0 & 0 & 0 & -3 \end{pmatrix} \begin{pmatrix} \frac{1}{4} & \frac{1}{4} & 0 & 0 \\ -\frac{3}{4} & \frac{1}{4} & 0 & 0 \\ 1 & 0 & 1 & 0 \\ -2 & 0 & 0 & 1 \end{pmatrix}$$

At this stage we should use this element to create zeros above and below it in the third column. We need a nonzero diagonal element to do this. If possible, interchange this row with one of the following row(s) to obtain a nonzero element in this location. When this is not possible, as here, because all elements below the diagonal element are zero, the inverse does not exist. In such a case it is not possible to transform A into a unit matrix using elementary transformations.

Thus the matrix $\begin{pmatrix} 1 & -1 & 0 & 2 \\ 3 & 1 & 5 & -1 \\ -1 & 1 & 0 & 0 \\ 2 & -2 & 0 & 1 \end{pmatrix}$ is not invertible.

We now see a role of invertible matrices in the theory of systems of linear equations.

THEOREM 2-2 *Let $AX = Y$ be a system of n equations in n variables having an invertible matrix of coefficients. Then a solution exists, is unique and is given by $X = A^{-1}Y$.*

Proof: Let us first prove that a solution exists. Consider $X = A^{-1}Y$. $AX = A(A^{-1})Y = I_nY = Y$. This value of X thus satisfies the system; it is a solution. We now prove that the solution is unique. Let X_1 and X_2 be solutions to $AX = Y$. Thus

$$AX_1 = Y \quad \text{and} \quad AX_2 = Y.$$

Equating,

$$AX_1 = AX_2.$$

Multiplying both sides by A^{-1},

$$A^{-1}AX_1 = A^{-1}AX_2,$$

$$I_nX_1 = I_nX_2.$$

Thus $X_1 = X_2$; the solution is unique.

Example 5

Solve the system of equations

$$x_1 + 2x_2 \qquad = \quad 4$$

$$2x_1 + \ x_2 - x_3 = \quad 2$$

$$3x_1 + \ x_2 + x_3 = -2$$

This system can be written in matrix form

$$\begin{pmatrix} 1 & 2 & 0 \\ 2 & 1 & -1 \\ 3 & 1 & 1 \end{pmatrix} \begin{pmatrix} x_1 \\ x_2 \\ x_3 \end{pmatrix} = \begin{pmatrix} 4 \\ 2 \\ -2 \end{pmatrix}$$

Hence

$$\begin{pmatrix} x_1 \\ x_2 \\ x_3 \end{pmatrix} = \begin{pmatrix} 1 & 2 & 0 \\ 2 & 1 & -1 \\ 3 & 1 & 1 \end{pmatrix}^{-1} \begin{pmatrix} 4 \\ 2 \\ -2 \end{pmatrix}$$

The inverse was found in Example 3.

$$\begin{pmatrix} x_1 \\ x_2 \\ x_3 \end{pmatrix} = \begin{pmatrix} -\frac{1}{4} & \frac{1}{4} & \frac{1}{4} \\ \frac{5}{8} & -\frac{1}{8} & -\frac{1}{8} \\ \frac{1}{8} & -\frac{5}{8} & \frac{3}{8} \end{pmatrix} \begin{pmatrix} 4 \\ 2 \\ -2 \end{pmatrix} = \begin{pmatrix} -1 \\ \frac{5}{2} \\ -\frac{3}{2} \end{pmatrix}$$

Thus the unique solution is $x_1 = -1$, $x_2 = \frac{5}{2}$, $x_3 = -\frac{3}{2}$.

Exercise 16 on p. 150 illustrates a situation that arises in a number of models in which one is interested in solving a number of systems of equations, all having a unique solution, all having the same matrix of coefficients. The inverse matrix method can be more efficient than Gaussian elimination for solving such problems; it can involve fewer operations (addition and multiplication).

The following section introduces the reader to a model in economics that uses this approach.

EXERCISES

Determine the inverses (if they exist) of the following matrices using the Gauss-Jordan elimination method.

1. $\begin{pmatrix} 1 & 0 \\ 2 & 1 \end{pmatrix}$ **2.** $\begin{pmatrix} 2 & 1 \\ 4 & 3 \end{pmatrix}$ **3.** $\begin{pmatrix} 0 & 2 \\ -\frac{1}{3} & \frac{1}{3} \end{pmatrix}$

4. $\begin{pmatrix} 1 & 2 & 3 \\ 0 & 1 & 2 \\ 4 & 5 & 3 \end{pmatrix}$ **5.** $\begin{pmatrix} 2 & 0 & 4 \\ -1 & 3 & 1 \\ 0 & 1 & 2 \end{pmatrix}$ **6.** $\begin{pmatrix} 0 & 3 & 3 \\ 1 & 2 & 3 \\ 1 & 4 & 6 \end{pmatrix}$

7. $\begin{pmatrix} 1 & 2 & -1 \\ 3 & -1 & 0 \\ 2 & -3 & 1 \end{pmatrix}$ **8.** $\begin{pmatrix} 1 & 2 & 3 \\ 2 & -1 & 4 \\ 0 & -1 & 1 \end{pmatrix}$ **9.** $\begin{pmatrix} 1 & 2 & -1 \\ 2 & 4 & -3 \\ 1 & -2 & 0 \end{pmatrix}$

10. $\begin{pmatrix} -3 & -1 & 1 & -2 \\ -1 & 3 & 2 & 1 \\ 1 & 2 & 3 & -1 \\ -2 & 1 & -1 & -3 \end{pmatrix}$ **11.** $\begin{pmatrix} 1 & 1 & 0 & 0 \\ 0 & 1 & 1 & 0 \\ 1 & 0 & 0 & 1 \\ 0 & 0 & 1 & 1 \end{pmatrix}$

Solve the following systems of equations by determining the inverse of the matrix of coefficients and then using matrix multiplication.

12. $\begin{aligned} x_1 + 3x_2 &= 5 \\ 2x_1 + x_2 &= 10 \end{aligned}$

13. $\begin{aligned} x_1 + 2x_2 - x_3 &= 2 \\ x_1 + x_2 + 2x_3 &= 0 \\ x_1 - x_2 - x_3 &= 1 \end{aligned}$

14. $\begin{aligned} x_1 - x_2 &= 1 \\ x_1 + x_2 + 2x_3 &= 2 \\ x_1 + 2x_2 + x_3 &= 0 \end{aligned}$

15. $\begin{aligned} x_1 + x_2 + 2x_3 + x_4 &= 5 \\ 2x_1 + 2x_2 + x_4 &= 6 \\ x_2 + 3x_3 - x_4 &= 1 \\ 3x_1 + 2x_2 + 2x_4 &= 7 \end{aligned}$

16. Solve the following system of equations for three distinct values of $\begin{pmatrix} b_1 \\ b_2 \\ b_3 \end{pmatrix}$:

$$\begin{pmatrix} 1 \\ 2 \\ 3 \end{pmatrix}, \begin{pmatrix} 0 \\ 1 \\ 4 \end{pmatrix}, \text{ and } \begin{pmatrix} 5 \\ 2 \\ 3 \end{pmatrix}.$$

$$\begin{aligned} x + 2y - z &= b_1 \\ x + y + 2z &= b_2 \\ x - y - z &= b_3 \end{aligned}$$

Note that you are required to solve three distinct systems of equations, all with the same matrix of coefficients. Thus the inverse matrix only has to be computed once. This makes the inverse matrix method more efficient than the Gaussian elimination method for this type of problem.

17. Prove that a diagonal matrix is invertible if and only if all its diagonal elements are nonzero. Find a rule that can be used for determining the inverse of an invertible diagonal matrix.

18. An *upper triangular matrix* is a square matrix of the type

$$\begin{pmatrix} a_{11} & a_{12} & \cdots & a_{1n} \\ & a_{22} & & \\ 0 & & \ddots & \\ & & & a_{nn} \end{pmatrix}$$

Prove that an upper triangular matrix is invertible if and only if all its diagonal elements are nonzero.   Prove that the inverse of an invertible upper triangular matrix is itself an upper triangular matrix.

19. Let A be an invertible matrix and c a nonzero scalar.   Prove that

a) A^{-1} is invertible with $(A^{-1})^{-1} = A$.

b) $(cA)^{-1} = \dfrac{1}{c} A^{-1}$

c) A^n is invertible with $(A^n)^{-1} = (A^{-1})^n$, for $n = 1, 2, 3, \ldots$

d) If B and C are matrices such that $AB = AC$, then $B = C$.

20. Let A be an invertible matrix having inverse $\begin{pmatrix} 2 & 1 \\ 4 & 3 \end{pmatrix}$.   Determine the matrix A.   [*Hint:* Use Exercise 19(a).]

21. Consider the matrix $A = \begin{pmatrix} 1 & 2 \\ 3 & 4 \end{pmatrix}$ having inverse $\begin{pmatrix} -2 & 1 \\ 1.5 & -0.5 \end{pmatrix}$. Determine **a)** $(3A)^{-1}$;   **b)** $(A^3)^{-1}$ [*Hint:* Use Exercises 19(b) and (c).]

22. Check your answers to Exercises 1–11 using the computer (Appendix J).

23. Write a program for solving a system of linear equations using the inverse of the matrix of coefficients.   Use your program to solve the system

$$
\begin{aligned}
x_1 + 2x_2 &= 4 \\
2x_1 + x_2 - x_3 &= 2 \\
3x_1 + x_2 + x_3 &= -2
\end{aligned}
$$

The answer is $x_1 = -1$, $x_2 = 2.5$, $x_3 = -1.5$.   Use your program to check your answers to Exercises 12–15.

24. Write a program that can be used for Exercise 16.   Check your answers.

2–6.*   LEONTIEF INPUT-OUTPUT MODELS IN ECONOMICS

In this section we discuss Leontief's input-output analysis of an economic situation.   Consider an economic situation involving n interdependent industries.   The output of any one industry is needed as input by other industries, and even by the industry itself.   We shall see how a mathematical model involving a system of linear equations can be constructed to analyze such a situation.   Let us assume, for the sake of simplicity, that each industry produces one commodity.

Let a_{ij} denote the amount of input of a certain commodity i to produce unit output of commodity j.   The first subscript refers to input, the second to output.   In our model let the amounts of input and output be in dollars.   Thus $a_{34} = 0.45$ means that 45 cents' worth of commodity 3 is required to produce one dollar's worth of commodity 4.

The elements a_{ij}, called *input coefficients*, define a matrix $A = (a_{ij})$, which gives the interdependence of the industries involved. Note that each column of A specifies the input requirements for the production of one unit of the output of a particular industry. For example, if the system involves three industries, and

$$A = \begin{pmatrix} a_{11} & a_{12} & a_{13} \\ a_{21} & a_{22} & a_{23} \\ a_{31} & a_{32} & a_{33} \end{pmatrix} = \begin{pmatrix} 0.25 & 0.40 & 0.50 \\ 0.35 & 0.10 & 0.20 \\ 0.20 & 0.30 & 0.10 \end{pmatrix}$$

then the elements in the second column give the inputs required from each sector to produce one dollar's worth of the commodity produced by the second industry.

Let us make two observations about such a matrix.

1. The sum of the elements in each column corresponds to the total input cost of producing one dollar's worth of output. Let us assume that such a matrix describes an economically feasible situation; therefore, the sum of the elements in each column will be less than unity.

$$\sum_{i=1}^{n} a_{ij} < 1 \qquad j = 1, 2, ..., n$$

2. It follows that each element of the matrix is less than unity, and by the interpretation given to these elements, each one must be greater than or equal to zero.

$$0 \le a_{ij} < 1 \qquad i, j = 1, 2, ..., n$$

Example 1

The following matrix of input coefficients is for the U.S. for 1947. It is based on computations made by the Bureau of Labor Statistics and involves grouping U.S. industries into 10 interdependent classes. These classes are:

1. Agriculture and food
2. Coal and power
3. Building, building materials and timber
4. Chemicals and rubber
5. Textiles and clothing
6. Paper, printing and miscellaneous
7. Metal making
8. Engineering
9. Metal goods
10. Services

	1	2	3	4	5	6	7	8	9	10
1	0	0	0.008	0.087	0.145	0.005	0.001	0	0.001	0.036
2	0.006	0	0.008	0.036	0.009	0.013	0.030	0.008	0.007	0.036
3	0.019	0.043	0	0.020	0.007	0.035	0.034	0.032	0.032	0.049
4	0.063	0.066	0.046	0	0.065	0.040	0.092	0.037	0.019	0.021
5	0.007	0.001	0.010	0.023	0	0.013	0	0.011	0.003	0.001
6	0.012	0.001	0.012	0.023	0.024	0	0.001	0.010	0.027	0.030
7	0	0.006	0.060	0.010	0	0.015	0	0.150	0.186	0.002
8	0.004	0.008	0.041	0	0.003	0.005	0.011	0	0.053	0.014
9	0.014	0.001	0.028	0.011	0.001	0.003	0.002	0.064	0	0.003
10	0.171	0.046	0.178	0.106	0.079	0.082	0.111	0.052	0.055	0

This example showing industrial interdependence in the U.S. for 1947 was adapted from *Mathematical Economics*, R. G. D. Allen, Macmillan and Co., Ltd., 1956. The figures are based on computations given in "The Inter-Industry Relations Study for 1947", W. D. Evans and M. Hoffenberg, *Review of Economics and Statistics*, **34**, 97–142.

So far we have only considered the case where n industries are interdependent. Let us now extend the situation by assuming that this model contains an *open sector*, where labor service, profit, etc. enter in the following way. Let x_i be the total output of industry i required to meet the demand of the open sector and all n industries (in dollars). Then

$$x_i = a_{i1} x_1 + a_{i2} x_2 + \cdots + a_{in} x_n + d_i$$

where d_i denotes the demand of the open sector from the ith industry. Here $a_{ij} x_j$ represents the input requirement of the jth industry from the ith.

Thus the output levels required of the entire set of n industries in order to meet these demands are given by the system of n linear equations

$$x_1 = a_{11} x_1 + a_{12} x_2 + \cdots + a_{1n} x_n + d_1$$
$$x_2 = a_{21} x_1 + a_{22} x_2 + \cdots + a_{2n} x_n + d_2$$
$$\vdots$$
$$x_n = a_{n1} x_1 + a_{n2} x_2 + \cdots + a_{nn} x_n + d_n$$

Rewriting, this system becomes

$$(1 - a_{11}) x_1 - \qquad a_{12} x_2 - \cdots - \qquad a_{1n} x_n = d_1$$
$$-a_{21} x_1 + (1 - a_{22}) x_2 - \cdots - \qquad a_{2n} x_n = d_2$$
$$\vdots$$
$$-a_{n1} x_1 - \qquad a_{n2} x_2 - \cdots + (1 - a_{nn}) x_n = d_n$$

153

The necessary output levels become solutions to the matrix equation

$$\begin{pmatrix} (1-a_{11}) & -a_{12} & \cdots & -a_{1n} \\ -a_{21} & (1-a_{22}) & \cdots & -a_{2n} \\ & & \vdots & \\ -a_{n1} & -a_{n2} & \cdots & (1-a_{nn}) \end{pmatrix} \begin{pmatrix} x_1 \\ x_2 \\ \vdots \\ x_n \end{pmatrix} = \begin{pmatrix} d_1 \\ d_2 \\ \vdots \\ d_n \end{pmatrix}$$

or

$$(I-A)\mathbf{x} = \mathbf{d}$$

At this stage, the inverse of the matrix of coefficients is usually used, as in the previous section, to determine the solutions. In practice, the analysis might involve feeding in various $\mathbf{d}$ vectors to determine what the corresponding output levels would have to be; thus this method is more appropriate than the method of Gaussian elimination. From the nature of the situation it is to be expected that the solution will be unique and that $(I-A)^{-1}$ will exist. Thus

$$\mathbf{x} = (I-A)^{-1}\mathbf{d}$$

Example 2

Consider an economy consisting of three industries, where the input coefficient matrix A is $\begin{pmatrix} 0.20 & 0.20 & 0.30 \\ 0.50 & 0.50 & 0 \\ 0 & 0 & 0.20 \end{pmatrix}$ and the vector $\mathbf{d}$, which gives the demand of the open sector, is $\begin{pmatrix} 9 \\ 12 \\ 16 \end{pmatrix}$ in millions of dollars.

Determine the output levels required of each industry to meet these demands. It can be shown that

$$(I-A)^{-1} = \begin{pmatrix} \frac{5}{3} & \frac{2}{3} & \frac{5}{8} \\ \frac{5}{3} & \frac{8}{3} & \frac{5}{8} \\ 0 & 0 & \frac{5}{4} \end{pmatrix}$$

Therefore,

$$\mathbf{x} = \begin{pmatrix} \frac{5}{3} & \frac{2}{3} & \frac{5}{8} \\ \frac{5}{3} & \frac{8}{3} & \frac{5}{8} \\ 0 & 0 & \frac{5}{4} \end{pmatrix} \begin{pmatrix} 9 \\ 12 \\ 16 \end{pmatrix} = \begin{pmatrix} 33 \\ 57 \\ 20 \end{pmatrix}$$

The output levels of the three industries would have to be 33, 57, and 20 million dollars.

A matrix A in such a model as this will, as we have seen, satisfy two conditions: All its elements will be nonnegative, and each column will add up to less than 1. A method has been developed for determining $(I-A)^{-1}$ for such a matrix. The method is most appropriate if A is large, for then the computation involved is less than the computation involved in the method of elimination.

Consider the following matrix multiplication for any positive integer m.

$$(I-A)(I+A+A^2+\cdots+A^m)$$
$$= I(I+A+A^2+\cdots+A^m)$$
$$\quad - A(I+A+A^2+\cdots+A^m)$$
$$= (I+A+A^2+\cdots+A^m)$$
$$\quad - (A+A^2+A^3+\cdots+A^{m+1})$$
$$= I - A^{m+1}$$

In this case, the elements of successive powers of matrix A become smaller; as m becomes larger, A^{m+1} approaches the zero matrix. Thus for a large enough m (depending on the accuracy to which one is working)

$$(I-A)(I+A+A^2+\cdots+A^m) = I$$

The inverse of $(I-A)$ is $I+A+A^2+\cdots+A^m$. It can be found, using this expression, on a computer.

Readers who desire a further knowledge of these models should consult *Fundamental Methods of Mathematical Economics*, by Alpha C. Chiang, McGraw-Hill Book Company, 1967. This text is also a good source for additional references.

EXERCISES

1. Consider the following input-output matrix that defines the interdependency of the given five industries.

	1	2	3	4	5
1. Auto	0.15	0.10	0.05	0.05	0.10
2. Steel	0.40	0.20	0.10	0.10	0.10
3. Electricity	0.10	0.25	0.20	0.10	0.20
4. Coal	0.10	0.20	0.30	0.15	0.10
5. Chemical	0.05	0.10	0.05	0.02	0.05

Determine:

a) the amount of electricity consumed in producing $1 worth of steel.

b) the amount of steel consumed in producing $1 worth in the auto industry.

c) the largest consumer of coal.

d) the largest consumer of electricity.

e) On which industry is the auto industry most dependent?

Consider the following economies of either two or three industries. Determine the output levels required of each industry in each situation to meet the demands of the other industries and of open sector. The units are millions of dollars.

2. $A = \begin{pmatrix} 0.20 & 0.60 \\ 0.40 & 0.10 \end{pmatrix}$, $D = \begin{pmatrix} 24 \\ 12 \end{pmatrix}$, $\begin{pmatrix} 8 \\ 6 \end{pmatrix}$, $\begin{pmatrix} 0 \\ 12 \end{pmatrix}$ in turn.

3. $A = \begin{pmatrix} 0.10 & 0.40 \\ 0.30 & 0.20 \end{pmatrix}$, $D = \begin{pmatrix} 6 \\ 12 \end{pmatrix}$, $\begin{pmatrix} 18 \\ 6 \end{pmatrix}$, $\begin{pmatrix} 24 \\ 12 \end{pmatrix}$ in turn.

4. $A = \begin{pmatrix} 0.30 & 0.60 \\ 0.35 & 0.10 \end{pmatrix}$, $D = \begin{pmatrix} 42 \\ 84 \end{pmatrix}$, $\begin{pmatrix} 0 \\ 10 \end{pmatrix}$, $\begin{pmatrix} 14 \\ 7 \end{pmatrix}$, $\begin{pmatrix} 42 \\ 42 \end{pmatrix}$ in turn.

5. $A = \begin{pmatrix} 0.20 & 0.20 & 0.10 \\ 0 & 0.40 & 0.20 \\ 0 & 0.20 & 0.60 \end{pmatrix}$, $D = \begin{pmatrix} 4 \\ 8 \\ 8 \end{pmatrix}$, $\begin{pmatrix} 0 \\ 8 \\ 16 \end{pmatrix}$, $\begin{pmatrix} 8 \\ 24 \\ 8 \end{pmatrix}$ in turn.

6. $A = \begin{pmatrix} 0.20 & 0.20 & 0 \\ 0.40 & 0.40 & 0.60 \\ 0.40 & 0.10 & 0.40 \end{pmatrix}$, $D = \begin{pmatrix} 36 \\ 72 \\ 36 \end{pmatrix}$, $\begin{pmatrix} 36 \\ 0 \\ 18 \end{pmatrix}$, $\begin{pmatrix} 36 \\ 0 \\ 0 \end{pmatrix}$, $\begin{pmatrix} 0 \\ 18 \\ 18 \end{pmatrix}$ in turn.

Consider the following economies of either two or three industries. The output level of each industry is given. Determine the amounts available for the open sector from each industry.

7. $A = \begin{pmatrix} 0.20 & 0.40 \\ 0.50 & 0.10 \end{pmatrix}$, $X = \begin{pmatrix} 8 \\ 10 \end{pmatrix}$

8. $A = \begin{pmatrix} 0.10 & 0.20 & 0.30 \\ 0 & 0.10 & 0.40 \\ 0.50 & 0.40 & 0.20 \end{pmatrix}$, $X = \begin{pmatrix} 10 \\ 10 \\ 20 \end{pmatrix}$

9. $A = \begin{pmatrix} 0.10 & 0.10 & 0.20 \\ 0.20 & 0.10 & 0.30 \\ 0.40 & 0.30 & 0.15 \end{pmatrix}$, $X = \begin{pmatrix} 6 \\ 4 \\ 5 \end{pmatrix}$

10. Write a computer program that can be used to analyze the situations of Exercises 2 through 6. Let the program use the built in matrix inverse function MAT B = INV(C) of BASIC to compute $(I-A)^{-1}$. Check your answers to Exercises 2 through 6.

11. Let A be a matrix that describes the interdependence of industries; all its elements will be nonnegative and each column will add up to less than 1. Write a computer program for determining output levels that incorporates the method $(I-A)^{-1} = I+A+A^2+\ldots+A^m$ for determining the inverse of $I-A$.

2–7.*   ITERATIVE METHODS FOR LINEAR SYSTEMS

We have already discussed the Gaussian elimination method for solving systems of equations. We shall now discuss iterative methods. At the close of the section we shall compare the merits of the Gaussian elimination method and the Gauss-Seidel iterative method.

We introduce an iterative method by means of an example.

Example 1

Solve the system of equations

$$6x + 2y - z = 4$$
$$x + 5y + z = 3 \tag{1}$$
$$2x + y + 4z = 27$$

Rewriting the system in the form

$$x = \frac{4 - 2y + z}{6}$$
$$y = \frac{3 - x - z}{5} \tag{2}$$
$$z = \frac{27 - 2x - y}{4}$$

we have isolated x in the first equation, y in the second, and z in the third.

Now make a guess at the solution, say $x = 1, y = 1, z = 1$. The accuracy of the guess affects only the speed with which we get a good approximation. Let us call these values $x^{(0)} = 1, y^{(0)} = 1, z^{(0)} = 1$. Substitute these values into the right side of system (2) to get values of x, y, and z that we denote $x^{(1)}$, $y^{(1)}$, and $z^{(1)}$.

$$x^{(1)} = \tfrac{1}{2}, \qquad y^{(1)} = \tfrac{1}{5}, \qquad z^{(1)} = 6$$

These values are then substituted again into system (2) to get

$$x^{(2)} = 1.6, \qquad y^{(2)} = -0.7, \qquad z^{(2)} = 6.45$$

The process is repeated to get $x^{(3)}$, $y^{(3)}$, and $z^{(3)}$, etc. Repeating the iteration will give us a better approximation of the exact solution each time. For this simple system, the solution is easily seen to be $x = 2$, $y = -1$, $z = 6$, so that after the second iteration one has quite a long way to go.

The following theorem gives one set of conditions under which this iterative method can be used.

THEOREM 2-3 *Let A be the matrix of coefficients of a system of n equations in n variables. If*

$$|a_{ii}| > \sum_{\substack{j=1 \\ j \neq i}}^{n} |a_{ij}|, \qquad where\ i = 1, 2, ..., n$$

$\uparrow$ $\qquad\qquad$ $\uparrow$

$\begin{pmatrix} diagonal \\ element \\ of\ ith\ row \end{pmatrix}$ $\quad$ $\begin{pmatrix} sum\ of\ absolute\ values\ of \\ all\ other\ elements\ in\ ith\ row \end{pmatrix}$

then the system of equations has a unique solution and the iterative method converges to this unique solution no matter what values are selected for the initial guess.

Thus this method can be used when the diagonal elements dominate the rows. We see that the system in our example satisfies these conditions.

In some systems it may be necessary to rearrange the equations before the above condition is satisfied and this method becomes applicable.

The better the initial guess the sooner one gets a result to within a required degree of accuracy. Note that this method has an advantage; if an error is made at any stage, you merely make a new initial guess at that stage.

There are many theorems similar to the one above that guarantee convergence under varying conditions. Investigations have made available various swiftly convergent methods for various special systems of equations.

The Gauss-Seidel method, a refinement of the above, leads to a more rapid convergence. The latest value of each variable is substituted at each stage.

This method also works if the diagonal elements dominate the rows. We illustrate this method for the previous system of equations.

As before, let $x^{(0)} = 1, y^{(0)} = 1, z^{(0)} = 1$ be the initial guess. Substituting the latest value of each variable into (2) every time,

$$x^{(1)} = \frac{4 - 2y^{(0)} + z^{(0)}}{6} = \frac{1}{2} = 0.5$$

$$y^{(1)} = \frac{3 - x^{(1)} - z^{(0)}}{5} = \frac{3}{10} = 0.3$$

$$z^{(1)} = \frac{27 - 2x^{(1)} - y^{(1)}}{4} = \frac{25.7}{4} = 6.4250$$

Thus after one iteration, $x^{(1)} = 0.5, y^{(1)} = 0.3, z^{(1)} = 6.4250$. Notice that we have used $x^{(1)}$, the most up-to-date value of x, to get $y^{(1)}$, and we have used $x^{(1)}$ and $y^{(1)}$ to get $z^{(1)}$.

Continuing,

$$x^{(2)} = \frac{4 - 2y^{(1)} + z^{(1)}}{6} = 1.6375$$

$$y^{(2)} = \frac{3 - x^{(2)} - z^{(1)}}{5} = -1.0125$$

$$z^{(2)} = \frac{27 - 2x^{(2)} - y^{(2)}}{4} = 6.1844$$

Note how, after only two iterations, this set is much closer than the previous set $x^{(2)} = 1.6$, $y^{(2)} = -0.7$, $z^{(2)} = 6.45$ to the exact solution of $x = 2$, $y = -1$, $z = 6$. Tables 2-1 and 2-2 give the results obtained for this particular system by both methods. They illustrate the Gauss-Seidel method's more rapid convergence to the exact solution.

Table 2-1. *First Method*

Iteration	x	y	z
Initial Guess	1	1	1
1	0.5	0.2	6
2	1.6	−0.7	6.45
3	1.975	−1.01	6.125
4	2.024167	−1.02	6.015
5	2.009167	−1.007833	5.992917
6	2.001431	−1.000417	5.997375

Table 2-2. *Gauss-Seidel Method*

Iteration	x	y	z
Initial Guess	1	1	1
1	0.5	0.3	6.425
2	1.6375	−1.0125	6.184375
3	2.034896	−1.043854	5.993516
4	2.013537	−1.001411	5.993584
5	1.999401	−0.998597	5.999949
6	1.999524	−0.9998945	6.000212

Table 2-3

	x	y	z
First Method	0.001431	0.000417	0.002625
Gauss-Seidel Method	0.000476	0.0001055	0.000212

Table 2-3 gives the differences between the solutions obtained in the two methods and the actual solution after six iterations. The Gauss-Seidel method converges much more rapidly.

Let us now compare the Gaussian elimination method with the Gauss-Seidel iterative method.

The Gaussian elimination method is finite and leads to a solution for any system of linear equations. The Gauss-Seidel method converges only for special systems of equations; thus it can only be used for such systems.

A second factor of comparison must be the efficiency of the two methods, a function of the number of arithmetic operations (addition, subtraction, multiplication, and division) involved in each method. For a system of n equations in n variables where the solution is unique, Gaussian elimination involves $(4n^3 + 9n^2 - 7n)/6$ arithmetic operations. The Gauss-Seidel method requires $2n^2 - n$ arithmetic operations per iteration. For large values of n, the number of arithmetic operations required by each method is, respectively, approximately $2n^3/3$ and $2n^2$ per iteration. Therefore, if the number of iterations is less than or equal to $n/3$, then the iterative method requires fewer arithmetic operations. As a specific example, consider a system of 300 equations in 300 variables. Elimination requires 18,000,000 operations, whereas iteration requires 180,000 operations per iteration. For 100 or fewer iterations the Gauss-Seidel method involves less arithmetic; it is more efficient. It should be stated that the Gaussian elimination method involves movement of data; for example, several rows may need to be interchanged. This is time-consuming and costly on computers. Iterative processes suffer much less from this factor. Thus, even if the number of iterations is more than $n/3$, iteration may require less computer time.

A final factor in the comparison of the two methods is the accuracy of the methods. Round-off errors are minimized in the Gaussian elimination method by using the pivoting technique. However they can still be sizeable. The errors in the Gauss-Seidel method, on the other hand, are only the round-off errors committed in the final iteration—the result of the next-to-last iteration can be interpreted as a very good initial guess! Thus, in general, when the Gauss-Seidel method is applicable, it is more accurate than the Gaussian elimination method. This fact often justifies the use of the Gauss-Seidel method over the Gaussian elimination method even when the total amount of computation time involved is greater.

For more in-depth discussions of iterative methods the reader is referred to *Numerical Methods with Fortran IV Case Studies*, by William S. Dorn and Daniel D. McCracken, John Wiley & Sons, Chap. 4. Interesting and important surveys of this field are to be found in "Solving Linear Equations Can Be Interesting," by George R. Forsythe, *Bulletin of the American Mathematical Society*, Vol. **59**, 1953, pp. 299–329, and "On the Solution of Linear Systems by Iteration," by David Young, *Proceedings of Symposia in Applied Mathematics*, Vol. **6**, McGraw-Hill Book Company, 1956.

EXERCISES

Determine approximate solutions to the following systems using the Gauss-Seidel iterative method. Work to two decimal places if you are performing the computations by hand. Use the given initial value. [Initial values are given solely that answers may be checked.]

1.
$$4x + y - z = 8$$
$$5y + 2z = 6$$
$$x - y + 4z = 10$$
Let $x^{(0)} = 1, y^{(0)} = 2, z^{(0)} = 3$.

2.
$$4x - y = 6$$
$$2x + 4y - z = 4$$
$$x - y + 5z = 10$$
Let $x^{(0)} = 1, y^{(0)} = 2, z^{(0)} = 3$.

3.
$$5x - y + z = 20$$
$$2x + 4y = 30$$
$$x - y + 4z = 10$$
Let $x^{(0)} = 0, y^{(0)} = 0, z^{(0)} = 0$.

4.
$$6x + 2y - z = 30$$
$$-x + 8y + 2z = 20$$
$$2x - y + 10z = 40$$
Let $x^{(0)} = 5, y^{(0)} = 6, z^{(0)} = 7$.

5.
$$5x - y + 2z = 40$$
$$2x + 4y - z = 10$$
$$-2x + 2y + 10z = 8$$
Let $x^{(0)} = 20, y^{(0)} = 30, z^{(0)} = -40$.

6.
$$6x - y + z + w = 20$$
$$x + 8y + 2z = 30$$
$$-x + y + 6z + 2w = 40$$
$$2x - 3z + 10w = 10$$
Let $x^{(0)} = y^{(0)} = z^{(0)} = w^{(0)} = 0$.

7. Apply the Gauss-Seidel method to the following system to observe that the values of $x^{(n)}, y^{(n)}, z^{(n)}$ do not converge. The method does not lead to the solution here.
$$5x + 4y - z = 8$$
$$x - y + z = 4$$
$$2x + y + 2z = 1.$$
Let $x^{(0)} = y^{(0)} = z^{(0)} = 0$.

8. Write a computer program for the Gauss-Seidel method.   Check your answers to the exercises above.

[*Hint*: Suppose you use 10 iterations.   List values of **x** as components of a vector $(x_1, \ldots, x_{10})$.   Use following dimension statement for this vector, DIM X (10).   **x** is a one-dimensional array—do not need two subscripts to handle it.]

3

Vector Spaces

In the previous two chapters the reader has been introduced to vectors, matrices, and systems of linear equations and has seen some of the many applications of these ideas. In this chapter and Chapter 4 we further develop these mathematical concepts. This development will be interesting from a mathematical viewpoint and will broaden the scope and power of these tools in applications. It will also give us a deeper understanding of these tools. We will, for example, obtain a clearer understanding of the behavior of systems of linear equations, leading to a better understanding of any situation they describe.

3–1. SUBSPACES

In Chapter 1 we developed the structure of the vector space $\mathbf{R}^n$. This was a set of elements called vectors on which two operations were defined, addition and scalar multiplication. Observe that on adding two elements of $\mathbf{R}^n$ we get an element of $\mathbf{R}^n$. $\mathbf{R}^n$ is said to be *closed under addition*. $\mathbf{R}^n$ is also *closed under scalar multiplication*; the scalar multiple of an element of $\mathbf{R}^n$ is again in $\mathbf{R}^n$. These two closure properties give vector spaces a certain completeness under the operations. In this section we see that certain subsets of $\mathbf{R}^n$ have these characteristics and can be interpreted as vector spaces in their own right.

To illustrate the concepts involved let us look below at the geometrical interpretation of $\mathbf{R}^3$, three-space. The xy plane is embedded in this three-space. We have seen how a plane can be interpreted as a vector space in itself, using $\mathbf{R}^2$ to characterize its points. It is natural that

163

the *xy* plane can be interpreted as a vector space also when considered as a subset of three-space, using certain elements of **R³** to describe it.

The *xy* plane is described by the elements of **R³** of the type $(a, b, 0)$, having zero as a third component.

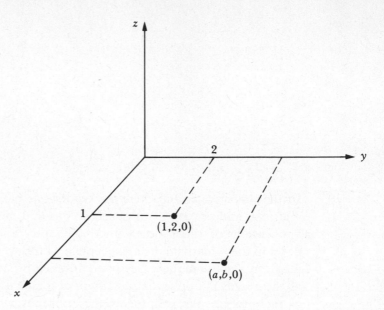

We make the observation that if we add two elements that lie in the *xy* plane such as $(a, b, 0)$ and $(c, d, 0)$ we get $(a+c, b+d, 0)$ that also lies in the plane. Further, if we multiply an element that lies in the plane by a scalar we get another element in the plane; $p(a, b, 0)$ gives $(pa, pa, 0)$ which again lies in the plane. The *xy* plane, considered as a subset of **R³** has operations of addition and scalar multiplication defined on it and it is closed under these operations. The *xy* plane has all the characteristics of a vector space when considered as a subset of **R³**. We call such an embedded vector space a subspace of the larger space.

We now formally define subspace.

DEFINITION 3-1 A nonempty subset of a vector space forms a *subspace* of that vector space if it is closed under addition and under scalar multiplication.†

---

† Note that this definition does imply that a vector space is a subspace of itself. The subspaces which are proper subsets can be distinguished if desired by calling them *proper* subspaces.

Example 1

Consider the vector space $\mathbf{R}^3$. All vectors of the form $(a, 0, 0)$ (with zero as second and third components) form a subset U of $\mathbf{R}^3$. Let us show that this subset is a subspace of $\mathbf{R}^3$.

Consider two arbitrary elements of U, $(a, 0, 0)$ and $(b, 0, 0)$. Their sum is $(a+b, 0, 0)$, an element of U. U is closed under addition.

Let p be an arbitrary scalar. Then $p(a, 0, 0) = (pa, 0, 0)$, an element of U. U is closed under scalar multiplication. Hence U is a subspace of $\mathbf{R}^3$.

Let us look at the geometrical interpretation of U. Let $\mathbf{R}^3$ be the set of all points in three-space. U is the x-axis.

Example 2

Prove that the subset V of $\mathbf{R}^3$ consisting of all vectors of the form (a, a, b) is a subspace of $\mathbf{R}^3$. Interpret this result geometrically.

First of all, we see that the subset of interest is the subset of $\mathbf{R}^3$ consisting of vectors that have a common number as the first two components. For example, $(1, 1, 2)$ and $(-\frac{1}{2}, -\frac{1}{2}, 3)$ would be elements of the subset; $(1, 2, 3)$ would not. We have to show that V is closed under addition and under scalar multiplication.

Let (a, a, b) and (c, c, d) be arbitrary elements of V. Then $(a, a, b) + (c, c, d) = (a+c, a+c, b+d)$. The first two components are equal; hence this is an element of V.

Let p be an arbitrary scalar. $p(a, a, b) = (pa, pa, pb)$ is an element of V. V is closed under addition and scalar multiplication; it is a subspace of $\mathbf{R}^3$.

We now interpret V geometrically. If $\mathbf{R}^3$ is three-space then the points (a, a, b), having equal x and y components, make up the plane in Figure 3-1.

Example 3

Consider the subset of $\mathbf{R}^3$ consisting of vectors of the form (a, b, a^2). We show that this subset of $\mathbf{R}^3$ is not a subspace, illustrating the fact that not all subsets of $\mathbf{R}^3$ are subspaces.

The subset consists of all elements whose third components are the squares of the first. For example, $(2, 5, 4)$ is in the subset, whereas the vector $(2, 5, 3)$ is not.

Let (a, b, a^2) and (c, d, c^2) be arbitrary elements of the subset. $(a, b, a^2) + (c, d, c^2) = (a+c, b+d, a^2+c^2)$.

In general, the third component here is not the square of the first, $a^2 + c^2 \neq (a+c)^2$. Thus the subset is not closed under addition; it is not a subspace of $\mathbf{R}^3$.

The following theorem gives us an important characteristic of all subspaces:

THEOREM 3-1   *Let V be a subspace of the vector space $\mathbf{R}^n$. Then V contains the zero vector of $\mathbf{R}^n$.*

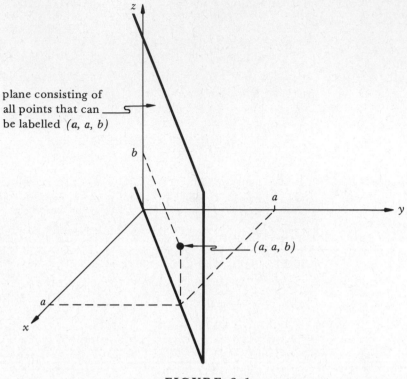

plane consisting of
all points that can
be labelled *(a, a, b)*

FIGURE 3-1

Proof: Let v be an arbitrary element of V; v will have n components. Let 0 be the zero scalar.

Since V is a subspace then $0v$ will be in V. But $0v = (0, \ldots, 0)$, the zero vector, proving the theorem.

If we interpret $\mathbf{R}^3$ as being three-dimensional space, then this theorem tells us that all subspaces of $\mathbf{R}^3$ contain the origin (that is, they pass through the origin). We do indeed see that this is true for the subspaces we have discussed in this section.

In the following sections we develop tools that will enable us to understand subspaces further. We shall find that all the significant subspaces of $\mathbf{R}^3$ are lines or planes through the origin, and that those of $\mathbf{R}^2$ are lines through the origin.

EXERCISES

1. Prove that the subset of $\mathbf{R}^3$ consisting of all vectors of the form $(a, 0, b)$ forms a subspace of $\mathbf{R}^3$. Interpret this result geometrically.

2. Prove that the subset of $\mathbf{R}^3$ of all vectors of the form $(0, b, 0)$ forms a subspace of $\mathbf{R}^3$.   Interpret this result geometrically.

3. Prove that the subset of $\mathbf{R}^3$ of all vectors of the form $(a, 2a, b)$ forms a subspace of $\mathbf{R}^3$.   Interpret this result geometrically.

4. Prove that the subset of $\mathbf{R}^4$ of all vectors of the form $(a, 2a, b, 3b)$ forms a subspace of $\mathbf{R}^4$.

In Exercises 5–11 determine which subsets form subspaces of $\mathbf{R}^3$.

5. The subset of vectors of the form $(a, b, a+2)$.

6. The subset of vectors of the form $(a, 2a, 3a)$.

7. The subset of vectors of the form $(a, b, 3b)$.

8. The subset of vectors of the form $(a, b, 2)$.

9. The subset of vectors of the form (a, b, c) satisfying the condition
   a) $a+b+c = 1$.   **b)** $a+b+c = 0$.

10. The subset of vectors of the form $(b+c, b, c)$.

11. The subset consisting of vectors of the form $p(1, 2, 3)$ and vectors of the form $q(-1, 0, 2)$.

12. Give an example of a subset of $\mathbf{R}^2$ that is
    a) closed under addition but not under scalar multiplication.
    b) closed under scalar multiplication but not under addition.
    Such examples illustrate the independence of these two conditions.

3–2.   LINEAR COMBINATIONS OF VECTORS

In Example 2 of the previous section we discussed the subspace of $\mathbf{R}^2$ consisting of all vectors of the form (a, a, b).   Observe that an arbitrary vector in this space can be written

$$(a, a, b) = a(1, 1, 0) + b(0, 0, 1)$$

All vectors in the subspace can be expressed in terms of $(1, 1, 0)$ and $(0, 0, 1)$—these vectors do, in some sense, characterize the space.   In this section and the following ones we pursue this approach to understanding vector spaces.   We develop mathematical techniques for analyzing vector spaces in terms of certain vectors that lie in those spaces.

DEFINITION 3-2   Consider m vectors $\mathbf{v}_1, ..., \mathbf{v}_m$, of a vector space V. We say that a vector $\mathbf{v}$ of V is a *linear combination* of these vectors if there exist m scalars, $a_1, ..., a_m$, such that

$$\mathbf{v} = a_1\mathbf{v}_1 + \cdots + a_m\mathbf{v}_m$$

167

Example 1

In $\mathbf{R^3}$, the vector $(1, 2, 9)$ is a linear combination of the three vectors $(2, 1, 0)$, $(1, 3, 0)$, and $(0, 0, 3)$, since it can be expressed

$$(1, 2, 9) = \tfrac{1}{5}(2, 1, 0) + \tfrac{3}{5}(1, 3, 0) + 3(0, 0, 3)$$

The problem of determining whether or not a vector is a linear combination of given vectors becomes that of solving a system of linear equations:

Example 2

Determine whether or not the vector $(-1, 1, 5)$ is a linear combination of the vectors $(1, 2, 3)$, $(0, 1, 4)$, and $(2, 3, 6)$.

We examine the identity

$$a_1(1, 2, 3) + a_2(0, 1, 4) + a_3(2, 3, 6) = (-1, 1, 5)$$

If scalars a_1, a_2, and a_3 can be found that satisfy this identity, $(-1, 1, 5)$ is a linear combination of the given vectors.

The identity becomes

$$(a_1, 2a_1, 3a_1) + (0, a_2, 4a_2) + (2a_3, 3a_3, 6a_3) = (-1, 1, 5)$$

$$(a_1 + 2a_3, 2a_1 + a_2 + 3a_3, 3a_1 + 4a_2 + 6a_3) = (-1, 1, 5)$$

Thus

$$a_1 + \qquad 2a_3 = -1$$

$$2a_1 + a_2 + 3a_3 = 1$$

$$3a_1 + 4a_2 + 6a_3 = 5$$

There exists a unique solution to this system; $a_1 = 1$, $a_2 = 2$, $a_3 = -1$. Thus the vector $(-1, 1, 5)$ is a linear combination of $(1, 2, 3)$, $(0, 1, 4)$, and $(2, 3, 6)$:

$$(-1, 1, 5) = 1(1, 2, 3) + 2(0, 1, 4) - 1(2, 3, 6)$$

Since the solution to the system of equations is unique, it is possible to express $(-1, 1, 5)$ in only one way as a linear combination of the other three vectors.

The following example illustrates that certain vectors can be represented in more than one way as linear combinations of others.

Example 3

Discuss the representation of the vector $(4, 5, 5)$ as a linear combination of $(1, 2, 3)$ $(-1, 1, 4)$ and $(3, 3, 2)$.

The identity

$$a_1(1, 2, 3) + a_2(-1, 1, 4) + a_3(3, 3, 2) = (4, 5, 5)$$

becomes

$$(a_1 - a_2 + 3a_3, 2a_1 + a_2 + 3a_3, 3a_1 + 4a_2 + 2a_3) = (4, 5, 5)$$

Thus

$$a_1 - a_2 + 3a_3 = 4$$
$$2a_1 + a_2 + 3a_3 = 5$$
$$3a_1 + 4a_2 + 2a_3 = 5$$

This system of equations has many solutions; $a_1 = -2a_3 + 3$, $a_2 = a_3 - 1$. Thus the linear combination is not unique,

$$(4, 5, 5) = (-2a_3 + 3)(1, 2, 3) + (a_3 - 1)(-1, 1, 4) + a_3(3, 3, 2),$$

for any value of a_3.

That it may not be possible to express a given vector as a linear combination of others is illustrated by the following example.

Example 4

Prove that the vector $(0, 0, 2)$ is not a linear combination of the vectors $(2, 1, 0)$ and $(-1, 2, 1)$.

We examine the identity

$$a_1(2, 1, 0) + a_2(-1, 2, 1) = (0, 0, 2)$$

It becomes

$$(2a_1 - a_2, a_1 + 2a_2, a_2) = (0, 0, 2)$$

giving

$$2a_1 - a_2 = 0$$
$$a_1 + 2a_2 = 0$$
$$a_2 = 2$$

This system has no solution.   Thus $(0, 0, 2)$ is not a linear combination of $(2, 1, 0)$ and $(-1, 2, 1)$.

DEFINITION 3-3   The vectors $\mathbf{v}_1, \ldots, \mathbf{v}_m$ are said to *span* a vector space if every vector in the space can be expressed as a linear combination of these vectors.

Such a set of vectors in a sense defines the vector space since every vector in the space can be obtained from this set.

Example 5

Consider the subspace of $\mathbf{R}^3$ consisting of vectors of the form (a, a, b). An arbitrary vector in the subspace can be written

$$(a, a, b) = a(1, 1, 0) + b(0, 0, 1)$$

Thus the vectors $(1, 1, 0)$ and $(0, 0, 1)$ span the subspace.

169

Example 6

Prove that the vectors $(1,2,0)$, $(0,1,-1)$, and $(1,1,2)$ span $\mathbf{R}^3$. Let (x,y,z) represent an arbitrary element of $\mathbf{R}^3$. We have to show that (x, y, z) can be expressed as a linear combination of the vectors. We prove that there exist scalars a_1, a_2, a_3 such that

$$(x,y,z) = a_1(1,2,0) + a_2(0,1,-1) + a_3(1,1,2)$$

This identity may be written

$$(a_1+a_3, 2a_1+a_2+a_3, -a_2+2a_3) = (x,y,z)$$

Thus

$$a_1 \qquad + a_3 = x$$
$$2a_1 + a_2 + a_3 = y$$
$$- a_2 + 2a_3 = z$$

This system has solution $a_1 = 3x-y-z$, $a_2 = -4x+2y+z$, $a_3 = -2x+y+z$. Thus the arbitrary vector (x,y,z) of $\mathbf{R}^3$ can be expressed as a linear combination of the vectors $(1,2,0)$ $(0,1,-1)$, and $(1,1,2)$.

$$(x,y,z) = (3x-y-z)(1,2,0) + (-4x+2y+z)(0,1,-1)$$
$$+ (-2x+y+z)(1,1,2)$$

These three vectors span $\mathbf{R}^3$.

We have developed a way of looking at vector spaces in terms of certain vectors in the space, spanning sets of vectors. It is also useful to be able to do the converse—to use a given set of vectors to construct a vector space. Subspaces often arise in discussions of vector spaces in this manner. We now pursue this approach.

Let $\mathbf{v_1}, ..., \mathbf{v_m}$ be vectors in a vector space V. Let U be the set consisting of all possible linear combinations of $\mathbf{v_1}, ..., \mathbf{v_m}$. We say that U is *generated* by $\mathbf{v_1}, ..., \mathbf{v_m}$. The following theorem tells us that such a set U is a subspace of V.

THEOREM 3-2   *Let $\mathbf{v_1}, ..., \mathbf{v_m}$ be a given set of vectors in a vector space V. The set generated by $\mathbf{v_1}, ..., \mathbf{v_m}$ is a subspace of V. The vectors $\mathbf{v_1}, ..., \mathbf{v_m}$ span this subspace.*

Proof: Let U be the set generated by $\mathbf{v_1}, ..., \mathbf{v_m}$.

Let $\mathbf{v} = a_1\mathbf{v_1} + \cdots + a_m\mathbf{v_m}$ and $\mathbf{v}' = b_1\mathbf{v_1} + \cdots + b_m\mathbf{v_m}$ be arbitrary elements of U and p an arbitrary scalar. Then

$$\mathbf{v} + \mathbf{v}' = (a_1+b_1)\mathbf{v_1} + \cdots + (a_m+b_m)\mathbf{v_m}$$

This is an element of U since it is expressed as a linear combination of $\mathbf{v_1}, ..., \mathbf{v_m}$.

Further,

$$p\mathbf{v} = pa_1\mathbf{v_1} + \cdots + pa_m\mathbf{v_m}$$

$p\mathbf{v}$ is an element of U.

Thus U is a subspace.

The vectors $\mathbf{v_1}, ..., \mathbf{v_m}$ span U since an arbitrary element of U can be expressed as a linear combination of these vectors.

Example 7

Discuss the subspace of $\mathbf{R}^3$ generated by the vectors $(1, 1, 1)$ and $(2, 1, 3)$.

The subspace will consist of all possible linear combinations of these vectors. An arbitrary element of the subspace will be

$$\mathbf{v} = a_1(1, 1, 1) + a_2(2, 1, 3)$$

The vector $(5, 3, 7)$ will, for example, be in the subspace since

$$(5, 3, 7) = (1, 1, 1) + 2(2, 1, 3)$$

The vector (x, y, z) is in the subspace if and only if there exist scalars a_1 and a_2 such that

$$(x, y, z) = a_1(1, 1, 1) + a_2(2, 1, 3)$$

This condition leads to

$$a_1 + 2a_2 = x$$

$$a_1 + a_2 = y$$

$$a_1 + 3a_2 = z$$

Solving, we get

$$\begin{pmatrix} 1 & 2 & x \\ 1 & 1 & y \\ 1 & 3 & z \end{pmatrix} \cong \begin{pmatrix} 1 & 2 & x \\ 0 & -1 & y-x \\ 0 & 1 & z-x \end{pmatrix}$$

$$\cong \begin{pmatrix} 1 & 2 & x \\ 0 & 1 & x-y \\ 0 & 0 & -2x+y+z \end{pmatrix}$$

giving $a_1 = -x+2y, a_2 = x-y, -2x+y+z = 0$.

The interpretation of this result is that the vector (x, y, z) is in the subspace if and only if it satisfies the condition $-2x+y+z = 0$. If it is in the subspace it can then be represented as a linear combination of $(1, 1, 1)$ and $(2, 1, 3)$:

$$(x, y, z) = (-x+2y)(1, 1, 1) + (x-y)(2, 1, 3)$$

All the vectors in $\mathbf{R}^3$ that satisfy $-2x + y + z = 0$ lie in a plane through the origin—this is the geometrical representation of the subspace generated by $(1, 1, 1)$ and $(-2, 1, 3)$. The subspace is the plane defined by the vectors $(1, 1, 1)$ and $(-2, 1, 3)$.

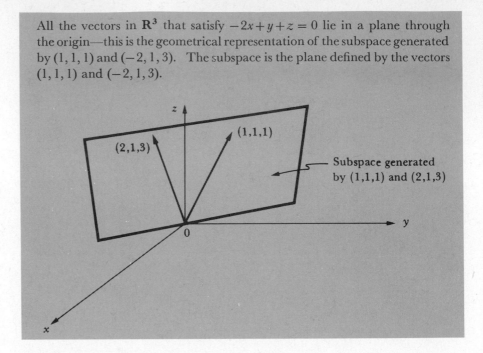

This result can be generalized:

Consider a subspace generated by two non-colinear vectors $\mathbf{u}$ and $\mathbf{v}$, (that is, they are not in the same or opposite directions). An arbitrary vector $\mathbf{w}$ in this subspace can be expressed

$$\mathbf{w} = a\mathbf{u} + b\mathbf{v}.$$

The geometrical interpretation of this subspace is that it is the plane defined by $\mathbf{u}$ and $\mathbf{v}$:

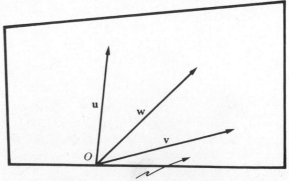

Subspace generated by $\mathbf{u}$ and $\mathbf{v}$
is plane defined by $\mathbf{u}$ and $\mathbf{v}$

EXERCISES

In Exercises 1–10 determine whether or not the first vector is a linear combination of the following vectors.   Determine the combination(s) if it is.

1. $(-1, 7)$; $(1, -1)$, $(2, 4)$
2. $(4, 0)$; $(-1, 2)$, $(3, 2)$, $(6, 4)$
3. $(6, 22)$; $(2, 3)$, $(-1, 5)$
4. $(-3, 3, 7)$; $(1, -1, 2)$, $(2, 1, 0)$, $(-1, 2, 1)$
5. $(2, 7, 13)$; $(1, 2, 3)$, $(-1, 2, 4)$, $(1, 6, 10)$
6. $(0, 10, 8)$; $(-1, 2, 3)$, $(1, 3, 1)$, $(1, 8, 5)$
7. $(2, 2, -2)$; $(1, 1, -1)$, $(2, 1, 3)$, $(4, 3, 1)$
8. $(4, 3, 8)$; $(-1, 0, 1)$, $(2, 1, 3)$, $(0, 1, 5)$
9. $(-2, -4, -3)$; $(1, 1, -1)$, $(0, 1, 2)$, $(2, 3, 0)$
10. $(2, 0, 12, 3)$; $(1, 2, 3, 0)$, $(1, -1, 1, 2)$, $(3, \ 4, -1, 1)$
11. Prove that the vectors $(1, 2)$ and $(3, 1)$ span $\mathbf{R}^2$.
12. Prove that the vectors $(2, 1, 0)$, $(-1, 3, 1)$, and $(4, 5, 0)$ span $\mathbf{R}^3$.
13. Do the vectors $(3, -1)$, $(2, 3)$, and $(4, 0)$ span $\mathbf{R}^2$?
14. Do the vectors $(4, 0, 1)$, $(0, 1, 0)$, and $(0, 0, 1)$ span $\mathbf{R}^3$?
15. Do the vectors $(1, 2, 1)$, $(-1, 3, 0)$, and $(0, 5, 1)$ span $\mathbf{R}^3$?
16. Do the vectors $(1, 1, -1)$, $(1, 3, 2)$, and $(3, 5, 0)$ span $\mathbf{R}^3$?
17. Do the vectors $(1, -1, -1)$, $(0, 1, 2)$, and $(1, 2, 1)$ span $\mathbf{R}^3$?
18. Give three vectors in the subspace of $\mathbf{R}^3$ generated by $(1, 2, 3)$ and $(1, 2, 0)$. (Other than the given vectors.)   Sketch the subspace.
19. Give three vectors in the subspace of $\mathbf{R}^3$ generated by $(1, 2, 1)$ and $(2, 1, 4)$. Sketch the subspace.
20. Give three vectors in the subspace of $\mathbf{R}^3$ generated by the vector $(1, 2, 3)$. Sketch the subspace.
21. Give three vectors in the subspace of $\mathbf{R}^2$ generated by the vector $(1, 2)$. Sketch the subspace.
22. Give three vectors in the subspace of $\mathbf{R}^4$ generated by the vectors $(1, 2, -1, 3)$ and $(1, 2, 1, 1)$.

3–3. LINEAR DEPENDENCE OF VECTORS

In this section we continue the development of vector space structure. We focus on developing concepts of dependence and independence of vectors.   These will be useful tools in constructing "efficient" spanning sets of vectors—spanning sets in which there are no redundant vectors. Such a set is called a *basis* of a vector space.

Let us motivate a concept of dependence of vectors by looking at a specific example. Observe that the vector $(-2, 6, 1)$ is a linear combination of $(1, 2, 3)$, $(0, 1, 4)$, and $(-1, 2, 2)$ for it can be expressed

$$(-2, 6, 1) = (1, 2, 3) - 2(0, 1, 4) + 3(-1, 2, 2) \tag{1}$$

The above equation could be rewritten in a number of ways. Each of the vectors could in turn be represented in terms of the others:

$$(1, 2, 3) = (-2, 6, 1) + 2(0, 1, 4) - 3(-1, 2, 2)$$
$$(0, 1, 4) = \tfrac{1}{2}(1, 2, 3) - \tfrac{1}{2}(-2, 6, 1) + \tfrac{3}{2}(-1, 2, 2)$$
$$(-1, 2, 2) = \tfrac{1}{3}(-2, 6, 1) - \tfrac{1}{3}(1, 2, 3) + \tfrac{2}{3}(0, 1, 4)$$

Each of the four vectors is in fact dependent upon the other three. We express this fact by writing equation (1)

$$(-2, 6, 1) - (1, 2, 3) + 2(0, 1, 4) - 3(-1, 2, 2) = \mathbf{0}$$

This concept of dependence of vectors is made precise with the following definition.

DEFINITION 3-4   *m* vectors $\mathbf{v_1}, \ldots, \mathbf{v_m}$ in a vector space V are said to be *linearly dependent* if there exist scalars $a_1, \ldots, a_m$, not all zero, such that

$$a_1\mathbf{v_1} + \cdots + a_m\mathbf{v_m} = \mathbf{0}$$

If the vectors are not linearly dependent, they are said to be *linearly independent*.

Note that in the above definition the $\mathbf{0}$ on the right of the equation denotes the zero vector of the vector space $(0, 0, \ldots, 0)$.

Example 1

Prove that the vectors $(1, 2, 3)$, $(-2, 1, 1)$, $(8, 6, 10)$ are linearly dependent vectors in $\mathbf{R}^3$.

To arrive at this conclusion we examine the identity

$$a_1(1, 2, 3) + a_2(-2, 1, 1) + a_3(8, 6, 10) = \mathbf{0}$$

This becomes

$$(a_1 - 2a_2 + 8a_3, 2a_1 + a_2 + 6a_3, 3a_1 + a_2 + 10a_3) = \mathbf{0}$$

giving

$$a_1 - 2a_2 + 8a_3 = 0$$
$$2a_1 + a_2 + 6a_3 = 0$$
$$3a_1 + a_2 + 10a_3 = 0$$

This system has $a_1 = 4$, $a_2 = -2$, $a_3 = -1$ as one of its solutions, a set of a's, not all zero. Thus the vectors are linearly dependent.

Example 2

Determine whether the vectors $(1, -1, 1, 2, 1)$, $(4, -1, 6, 6, 2)$, $(-2, -1, 1, -2, -2)$ are linearly dependent or independent in $\mathbf{R}^5$.

We examine the identity

$$a_1(1, -1, 1, 2, 1) + a_2(4, -1, 6, 6, 2) + a_3(-2, -1, 1, -2, -2) = \mathbf{0}$$

This gives

$$(a_1 + 4a_2 - 2a_3, -a_1 - a_2 - a_3, a_1 + 6a_2 + a_3,$$
$$2a_1 + 6a_2 - 2a_3, a_1 + 2a_2 - 2a_3) = \mathbf{0}$$

$$a_1 + 4a_2 - 2a_3 = 0$$

$$-a_1 - a_2 - a_3 = 0$$

$$a_1 + 6a_2 + a_3 = 0$$

$$2a_1 + 6a_2 - 2a_3 = 0$$

$$a_1 + 2a_2 - 2a_3 = 0$$

This system can be shown to have the unique solution $a_1 = a_2 = a_3 = 0$. The vectors are thus linearly independent.

We now present two important general results in the analysis of vector spaces.

THEOREM 3-4 *In a vector space V any collection of vectors that contains the zero vector is linearly dependent.*

Proof: Consider the collection $\mathbf{0}, \mathbf{v_2}, \ldots, \mathbf{v_m}$. Then

$$a_1\mathbf{0} + 0\mathbf{v_2} + \cdots + 0\mathbf{v_m} = \mathbf{0}$$

for any real number a_1. Hence this set is linearly dependent.

THEOREM 3-5 *Let $\mathbf{v_1}, \ldots, \mathbf{v_m}$ be linearly dependent vectors in a vector space V. Any set of vectors that contains these vectors will be a set of linearly dependent vectors.*

Proof: The vectors $\mathbf{v_1}, \ldots, \mathbf{v_m}$ are linearly dependent, hence there exist scalars $a_1, \ldots, a_m$ that are not all zero such that

$$a_1\mathbf{v_1} + \cdots + a_m\mathbf{v_m} = \mathbf{0}$$

Consider the set $\mathbf{v_1}, \ldots, \mathbf{v_m}, \mathbf{u_1}, \ldots, \mathbf{u_k}$, which contains the vectors $\mathbf{v_1}, \ldots, \mathbf{v_m}$. There are scalars $a_1, \ldots, a_m, 0, \ldots, 0$ that are not all zero such that

$$a_1\mathbf{v_1} + \cdots + a_m\mathbf{v_m} + 0\mathbf{u_1} + \cdots + 0\mathbf{u_m} = \mathbf{0}$$

Therefore this set is linearly dependent.

175

EXERCISES

1. Prove that the vectors $(1, 0)$ and $(0, 1)$ are linearly independent in $\mathbf{R}^2$.

2. Prove that the vectors $(2, 1)$ and $(1, 2)$ are linearly independent in $\mathbf{R}^2$.

3. Prove that the vectors $(1, -2)$ and $(3, -6)$ are linearly dependent in $\mathbf{R}^2$.

4. Are the vectors $(2, 0)$, $(0, 1)$, and $(3, 4)$ linearly dependent in $\mathbf{R}^2$?

5. Prove that the vectors $(-1, 3, 1)$, $(2, 1, 0)$, and $(1, 4, 1)$ are linearly dependent in $\mathbf{R}^3$.

6. Prove that the vectors $(2, 3, 0)$, $(-1, 4, 0)$, and $(0, 0, 2)$ are linearly independent in $\mathbf{R}^3$.

7. Prove that the vectors $(1, 1, 1)$, $(2, 1, 0)$, $(3, 1, 4)$, and $(1, 2, -2)$ are linearly dependent in $\mathbf{R}^3$.

8. **a)** Prove that the vectors $(1, 1)$ and $(0, 2)$ are linearly independent in $\mathbf{R}^2$.
   b) Prove that the vectors $(1, 1, 2)$, $(0, -1, 3)$, and $(0, 0, -1)$ are linearly independent in $\mathbf{R}^3$.
   c) Prove that the vectors $(-1, 1, 2, 3)$, $(0, 1, 2, 3)$, $(0, 0, -1, 2)$, and $(0, 0, 0, -3)$ are linearly independent in $\mathbf{R}^4$.
   d) Discuss the pattern of zero components in the vectors of (a), (b), and (c). How can you use this pattern to construct any number of sets of five linearly independent vectors in $\mathbf{R}^5$? Any number of sets of n linearly independent vectors in $\mathbf{R}^n$?
   e) Give a geometrical interpretation of the concepts involved in (a) and (b).

9. Prove that the set of vectors $(-1, 3, 1)$, $(2, 1, 0)$, $(1, 4, 1)$, $(5, 7, 9)$, and $(8, -1, 0)$ is linearly dependent. (*Hint*: Use Exercise 5 above and Theorem 3-5.)

10. Give an expression that defines a general vector in the space spanned by the vectors $(2, 1, 0)$ and $(-1, 2, 0)$.

11. Prove that two vectors are linearly dependent if and only if it is possible to express one vector as a scalar multiple of the other.

12. Prove that the vectors $\mathbf{v}_1, \dots, \mathbf{v}_m$ of a vector space V are linearly dependent if and only if it is possible to express one of the vectors as a linear combination of the others.

13. Let $\mathbf{v}_1$, $\mathbf{v}_2$, and $\mathbf{v}_3$ be linearly independent vectors in V and let c be a non-zero scalar. Prove that $\mathbf{v}_1$, $c\mathbf{v}_2$, and $\mathbf{v}_3$ are also linearly independent. Further prove that $\mathbf{v}_1 + c\mathbf{v}_2$, $\mathbf{v}_2$, and $\mathbf{v}_3$ are linearly independent.

14. Extend the results of Exercise 13 to m linearly independent vectors in V.

15. If $\mathbf{v}_1$, $\mathbf{v}_2$, and $\mathbf{v}_3$ are linearly dependent vectors in V and c a nonzero scalar, prove that both the sets $\mathbf{v}_1$, $c\mathbf{v}_2$, $\mathbf{v}_3$ and $\mathbf{v}_1 + c\mathbf{v}_2$, $\mathbf{v}_2$, $\mathbf{v}_3$ are linearly dependent.

16. Extend the results of Exercise 15 to m linearly dependent vectors in V.

3–4. BASES

We now bring together the concepts of spanning set and linear independence.

DEFINITION 3-5   A finite set of vectors $\mathbf{v}_1, ..., \mathbf{v}_m$ in a vector space V forms a *basis* for the space if it spans the space and is linearly independent.

In this section we give examples of bases for various vector spaces and develop results concerning bases. We shall see that a basis is more appropriate for discussing a vector space than is a general spanning set.

Example 1

The set of vectors $(1, 0, ..., 0)$, $(0, 1, 0, ..., 0)$, ..., $(0, ..., 0, 1)$ forms a basis for $\mathbf{R}^n$ called the *canonical basis*.

We have to show that this set spans $\mathbf{R}^n$ and that the vectors are linearly independent.

Let $(x_1, ..., x_n)$ be an arbitrary element of $\mathbf{R}^n$. We can express this vector as

$$(x_1, ..., x_n) = x_1(1, 0, ..., 0) + \cdots + x_n(0, ..., 0, 1)$$

Hence this set of vectors spans $\mathbf{R}^n$.

Further, consider the identity

$$a_1(1, 0, ..., 0) + \cdots + a_n(0, ..., 0, 1) = \mathbf{0}$$

This implies that

$$(a_1, 0, ..., 0) + \cdots + (0, ..., 0, a_n) = \mathbf{0}$$

$$(a_1, ..., a_n) = \mathbf{0}$$

giving $a_1 = a_2 = \cdots = a_n = 0$.

The vectors are thus linearly independent and the set is a basis for $\mathbf{R}^n$.

For any given vector space there can be many bases. The following theorems leads up to a key result concerning all bases for a given vector space—all bases consist of the same number of vectors.

THEOREM 3-6   *Let* $\mathbf{v}_1, ..., \mathbf{v}_n$ *be a basis for a given vector space. If* $\mathbf{w}_1, ..., \mathbf{w}_m$ *is a set of more than n vectors in that space, then this set is linearly dependent.*

Proof: We examine the identity

$$a_1\mathbf{w}_1 + \cdots + a_m\mathbf{w}_m = \mathbf{0} \tag{1}$$

We shall prove that values of $a_1, ..., a_m$, not all zero, exist, such that this identity holds, proving that the vectors are linearly dependent.

$\mathbf{v}_1, ..., \mathbf{v}_n$ is a basis. Thus each of $\mathbf{w}_1, ..., \mathbf{w}_m$ can be expressed in terms of these vectors. Let

$$\mathbf{w}_1 = b_{11}\mathbf{v}_1 + b_{12}\mathbf{v}_2 + \cdots + b_{1n}\mathbf{v}_n$$
$$\vdots$$
$$\mathbf{w}_m = b_{m1}\mathbf{v}_1 + b_{m2}\mathbf{v}_2 + \cdots + b_{mn}\mathbf{v}_n$$

Substituting for $\mathbf{w}_1, ..., \mathbf{w}_m$ into (1),

$$a_1(b_{11}\mathbf{v}_1 + \cdots + b_{1n}\mathbf{v}_n) + \cdots + a_m(b_{m1}\mathbf{v}_1 + \cdots + b_{mn}\mathbf{v}_n) = \mathbf{0}$$

Rearranging we get

$$(a_1 b_{11} + a_2 b_{21} + \cdots + a_m b_{m1})\mathbf{v}_1 + \cdots$$
$$+ (a_1 b_{1n} + a_2 b_{2n} + \cdots + a_m b_{mn})\mathbf{v}_m = \mathbf{0}$$

Since $\mathbf{v}_1, ..., \mathbf{v}_m$ are linearly independent this identity can only be satisfied if the coefficients are all zero. Thus

$$b_{11}a_1 + b_{21}a_2 + \cdots + b_{m1}a_m = 0$$
$$\vdots$$
$$b_{1n}a_1 + b_{2n}a_2 + \cdots + b_{mn}a_m = 0$$

Thus finding a's that satisfy (1) reduces to finding a's that are solutions to this system of n equations in m variables. Since n, the number of equations, is less than m, the number of variables, we know that many solutions exist; there must be nonzero a's that satisfy (1). Thus the vectors $\mathbf{w}_1, ..., \mathbf{w}_m$ are linearly dependent.

THEOREM 3-7 *Any two bases for a given vector space V have the same number of vectors.*

Proof: Let $\mathbf{v}_1, ..., \mathbf{v}_n$ and $\mathbf{w}_1, ..., \mathbf{w}_m$ be two bases for V.

Interpreting $\mathbf{v}_1, ..., \mathbf{v}_n$ as a basis for V and $\mathbf{w}_1, ..., \mathbf{w}_m$ as a set of m linearly independent vectors in V, the previous theorem implies that $m \leq n$. Conversely, interpreting $\mathbf{w}_1, ..., \mathbf{w}_m$ as a basis for V and $\mathbf{v}_1, ..., \mathbf{v}_n$ as a set of n vectors in V, the theorem tells us that $n \leq m$. Thus $n = m$.

We call the number of vectors that make up a basis for a vector space the *dimension* of that space. Thus $\mathbf{R}^n$ is of dimension n.

Note that we have defined a basis in terms of a *finite* set of vectors that spans the space and is linearly independent. Such a set does not exist for all vector spaces. When a basis, as defined thus, exists for a vector space, the space is said to be *finite dimensional*. If such a set of vectors does not exist, the space is said to be *infinite dimensional*. Some of the function spaces of Section 3-7 are infinite dimensional vector spaces. We are primarily interested in finite dimensional vector spaces in this text.

Example 2

In Example 6 of Section 3-2 we saw that the vectors $(1, 2, 0)$, $(0, 1, -1)$, and $(1, 1, 2)$ span $\mathbf{R}^3$. It can be shown that these vectors are also linearly independent. They thus form a basis for $\mathbf{R}^3$.

Example 3

In Example 2 of previous section 3-3 we saw that the vectors $(1, -1, 1, 2, 1)$, $(4, -1, 6, 6, 2)$, $(-2, -1, 1, -2, -2)$ are linearly independent in $\mathbf{R}^5$. They can be used to generate a subspace V of $\mathbf{R}^5$. These vectors will span that subspace. They form a basis for the subspace; its dimension will be 3.

An arbitrary vector $\mathbf{v}$ in V can be expressed

$$\mathbf{v} = a(1, -1, 1, 2, 1) + b(4, -1, 6, 6, 2) + c(-2, -1, 1, -2, -2)$$

Example 4

Consider the subspace V of $\mathbf{R}^3$ consisting of vectors of the form (a, a, b). V is a plane through the origin. [Example 2, Section 3.1] We can express (a, a, b)

$$(a, a, b) = a(1, 1, 0) + b(0, 0, 1)$$

The vectors $(1, 1, 0)$ and $(0, 0, 1)$ span the space. They are also linearly independent. They form a basis for the space; its dimension is 2.

In this last example we saw that a certain plane through the origin was a two-dimensional subspace of $\mathbf{R}^3$. The following theorem tells us more about the geometrical interpretations of subspaces of $\mathbf{R}^3$.

THEOREM 3-8
a. *The origin is a subspace of* $\mathbf{R}^3$. *This subspace is defined to be of dimension zero.*
b. *The one dimensional subspaces of* $\mathbf{R}^3$ *are lines through the origin.*
c. *The two dimensional subspaces of* $\mathbf{R}^3$ *are planes through the origin.*

Proof: (a) The origin of $\mathbf{R}^3$ is the vector $(0, 0, 0)$, the zero vector. Consider the subset of $\mathbf{R}^3$ consisting of this single vector only. It is closed under addition and under scalar multiplication:

$$(0, 0, 0) + (0, 0, 0) = (0, 0, 0),$$

and

$$p(0, 0, 0) = (0, 0, 0) \text{ for arbitrary scalar } p.$$

Thus the subset consisting of the zero vector is a subspace of $\mathbf{R}^3$. We define the dimension of this subspace to be zero.

(b) Let us consider the one-dimensional subspaces of $\mathbf{R}^3$.

Let **v** be a basis for such a one-dimensional subspace. Every vector in the space can be expressed in the form a**v** for some scalar a. These vectors make up the line through the origin defined by the vector **v**. Thus the arbitrary one-dimensional subspace is a line through the origin.

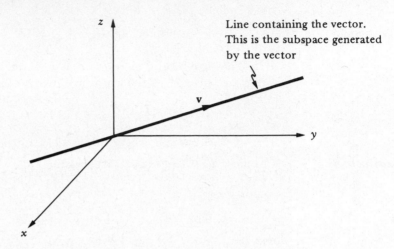

Line containing the vector. This is the subspace generated by the vector

(c) Finally we look at the two-dimensional subspaces of $\mathbf{R}^3$.

Let **u** and **v** be a basis for such a subspace. If **w** is an arbitrary vector in the subspace, then it can be expressed

$$\mathbf{w} = a\mathbf{u} + b\mathbf{v},$$

for scalars a and b. By the geometrical interpretation of vector addition, **w** lies in the plane of a**u** and b**v**—it is in fact the diagonal of the parallelogram defined by a**u** and b**v**.

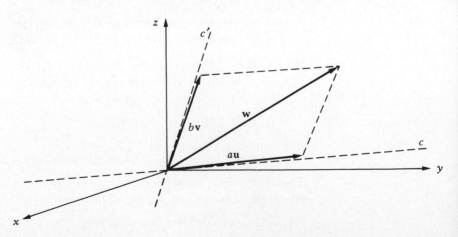

As a and b vary, the vectors $a\mathbf{u}$ and $b\mathbf{v}$ will always lie on the lines c and c'. $\mathbf{w}$ will always lie in the plane defined by these two lines. Furthermore, the plane is made up of vectors that can be expressed in the form $a\mathbf{u}+b\mathbf{v}$. Thus the arbitrary two-dimensional subspace is a plane through the origin.

Example 5*

In this example†, a color matching experiment is analyzed using algebraic techniques. In this model, the operations of vector addition and scalar multiplication are seen to have direct physical significance, and the concept of linear dependence is seen to have physical interpretation.

Consider an experiment in which we are given two screens. On screen 1 a light of a certain color is flashed. On screen 2 we attempt to duplicate both in color and intensity the light shown on screen 1 by projecting combinations of various lights of given colors. Experimentation in color perception has shown that any color can be duplicated by using a proper additive mixture of the primary colors: red (denoted s_1), green (s_2), and blue (s_3). Assume that we can project combinations of these colors onto screen 2.

Let $(1,0,0)$, $(0,1,0)$, and $(0,0,1)$ denote, respectively, the projection of one foot-candle of red, green, and blue onto screen 2. (A foot-candle is a unit of illumination equal to the amount of direct light thrown by one standard candle on a surface one foot away.)

s_1 foot-candles of red light would be represented by the vector $s_1(1,0,0)$, that is, $(s_1,0,0)$; s_2 foot-candles of green light by the vector $s_2(0,1,0)$, that is $(0,s_2,0)$; and s_3 foot-candles of blue light by $s_3(0,0,1)$, that is $(0,0,s_3)$.

A combination of s_1, s_2, and s_3 foot-candles of red, green, and blue would be represented mathematically by the linear combination

$$s_1(1,0,0) + s_2(0,1,0) + s_3(0,0,1)$$

or the vector

$$(s_1,s_2,s_3)$$

Let (s_1,s_2,s_3) and (p_1,p_2,p_3) be two colors. If both are flashed onto the screen simultaneously, the resulting color will be $(s_1,s_2,s_3)+(p_1,p_2,p_3)$, that is, $(s_1+p_1,s_2+p_2,s_3+p_3)$.

Thus we have represented unit brightness of each of the primary colors by the base vectors of $\mathbf{R}^3$, $(1,0,0)$, $(0,1,0)$, and $(0,0,1)$. All other colors can be represented by elements of $\mathbf{R}^3$ that are linear combinations of these vectors.

---

† This example is adapted from "A Psychophysical Matching Experiment," by Richard F. Baum in *Some Mathematical Models in Biology*, NTIS (952.85), U.S. Dept. of Commerce, Springfield, Virginia 22151, 1969.

The color (s_1, s_2, s_3) can be varied in brightness using scalar multiplication. $a(s_1, s_2, s_3)$, where $a > 0$, would represent the same color with a change in brightness.

The operation of vector addition on $\mathbf{R}^3$ corresponds to mixing colors; the operation of scalar multiplication by a positive scalar corresponds to varying the brightness.   Note that the model is a subset of $\mathbf{R}^3$— namely, all the nonzero vectors (x_1, x_2, x_3) with nonnegative components. The other elements of $\mathbf{R}^3$ do not have physical interpretation.

Using this model, an analysis and classification of colors can be carried out.   The color purple, for example, is represented by the vector $a(1, 0, 1)$, $a > 0$; it is a combination of red and blue in equal intensities.

We have discussed the fact that a set of vectors that spans a space characterizes the space since every vector can be expressed as a linear combination of the spanning set.   In general, there may be more than one way of expressing a vector as such a linear combination.   For example, if we consider the space spanned by the vectors $(1, 2, 3)$, $(-1, 1, 4)$, and $(3, 3, 2)$, we see that we can express the vector $(4, 5, 5)$, which lies in this space, in more than one way in terms of these vectors:

$$(4, 5, 5) = -(1, 2, 3) + (-1, 1, 4) + 2(3, 3, 2)$$

and

$$(4, 5, 5) = 5(1, 2, 3) - 2(-1, 1, 4) - (3, 3, 2)$$

There are many other representations of $(4, 5, 5)$ in terms of the vectors $(1, 2, 3)$, $(-1, 1, 4)$, and $(3, 3, 2)$.   (See Example 3, Section 3-2).   However, if the spanning set is a basis, the following theorem tells us that each such linear combination is unique.   A basis thus characterizes the space more exactly than a general spanning set does.

THEOREM 3-9   *Let $\mathbf{v}_1, \ldots, \mathbf{v}_m$ be a basis for a vector space V.   Then each vector in V can be expressed uniquely as a linear combination of these vectors.*

If $\mathbf{v}_1, \ldots, \mathbf{v}_m$ is a spanning set which is not a basis then each vector in V can be expressed in more than one way as a linear combination of these vectors.

Proof: Assume that $\mathbf{v}_1, \ldots, \mathbf{v}_m$ is a basis for V.

By the definition of a basis we know that each vector in V can be expressed as a linear combination of $\mathbf{v}_1, \ldots, \mathbf{v}_m$.   Hence we need only prove the uniqueness of the combination.

Suppose the linear combination is not unique for some vector $\mathbf{v}$; suppose there exist two distinct sets of scalars, $a_1, \ldots, a_m$ and $b_1, \ldots, b_m$, such that

$$\mathbf{v} = a_1 \mathbf{v}_1 + \cdots + a_m \mathbf{v}_m$$

and

$$\mathbf{v} = b_1 \mathbf{v_1} + \cdots + b_m \mathbf{v_m}$$

Hence

$$a_1 \mathbf{v_1} + \cdots + \qquad a_m \mathbf{v_m} = b_1 \mathbf{v_1} + \cdots + b_m \mathbf{v_m}$$

$$(a_1 - b_1) \mathbf{v_1} + \cdots + (a_m - b_m) \mathbf{v_m} = \mathbf{0}$$

Since the vectors $\mathbf{v_1}, \ldots, \mathbf{v_m}$ form a basis, they are linearly independent and each coefficient must be zero.

Thus $a_1 = b_1, \ldots, a_m = b_m$. This contradicts the assumption that the two linear combinations are distinct. Therefore each linear combination is unique.

To prove the second part of the theorem, let $\mathbf{v_1}, \ldots, \mathbf{v_m}$ be a spanning set which is not a basis, that is, a spanning set of linearly dependent vectors. Let $\sum_{i=1}^{m} c_i \mathbf{v_i} = \mathbf{0}$, where at least one $c_i \neq 0$. Let $\mathbf{v}$ be an arbitrary vector of V and let it be expressed $\mathbf{v} = \sum_{i=1}^{m} d_i \mathbf{v_i}$ in terms of the spanning set. Then

$$\mathbf{v} = \sum_{i=1}^{m} d_i \mathbf{v_i} + \mathbf{0} = \sum_{i=1}^{m} d_i \mathbf{v_i} + \sum_{i=1}^{m} c_i \mathbf{v_i} = \sum_{i=1}^{m} (d_i + c_i) \mathbf{v_i}$$

giving another distinct expression for $\mathbf{v}$ in terms of $\mathbf{v_1}, \ldots, \mathbf{v_m}$.

This theorem will be important to the theoretical discussion of the existence and uniqueness of solutions to systems of linear equations (see Section 6-2).

We complete this discussion of bases with two further results. The following theorem gives a sufficient condition for a given set of vectors to be a basis.

THEOREM 3-10   *In a vector space of dimension n, any set of n linearly independent vectors forms a basis.*

Proof: Let $\mathbf{v_1}, \ldots, \mathbf{v_n}$ be a set of n linearly independent vectors in a vector space V of dimension n. Let $\mathbf{w}$ be another vector in the space. The set $\mathbf{v_1}, \ldots, \mathbf{v_n}, \mathbf{w}$, being a set of more than n vectors in V, is linearly dependent by Theorem 3-6. Thus there exist scalars $a_1, \ldots, a_n, a$, not all zero, such that

$$a_1 \mathbf{v_1} + \cdots + a_n \mathbf{v_n} + a\mathbf{w} = \mathbf{0}.$$

a cannot be zero for otherwise $\mathbf{v_1}, \ldots, \mathbf{v_n}$ would be linearly dependent. Thus $\mathbf{w}$ can be expressed

$$\mathbf{w} = \left(-\frac{a_1}{a} \right) \mathbf{v_1} + \cdots + \left(-\frac{a_n}{a} \right) \mathbf{v_n}$$

proving that $\mathbf{v}_1, ..., \mathbf{v}_n$ span V. They are given to be linearly independent; they form a basis for V.

The final theorem of this section discusses the construction of bases using given linearly independent vectors.

THEOREM 3-11 *Let V be a vector space of dimension n, and let $\mathbf{v}_1, ..., \mathbf{v}_r$, with $r < n$, be a set of linearly independent vectors in V. Then there exist vectors $\mathbf{v}_{r+1}, ..., \mathbf{v}_n$ in V such that the set $\mathbf{v}_1, ..., \mathbf{v}_r, \mathbf{v}_{r+1}, ..., \mathbf{v}_n$ forms a basis for V. That is, a given set of linearly independent vectors can always be extended to form a basis for the space.*

Proof: The set $\mathbf{v}_1, ..., \mathbf{v}_r$ cannot span V, for otherwise the dimension would be r and not n. Hence there exists a vector $\mathbf{v}_{r+1}$ in V which is not a linear combination of $\mathbf{v}_1, ..., \mathbf{v}_r$. If $r + 1 = n$, the set $\mathbf{v}_1, ..., \mathbf{v}_{r+1}$ is a set of n linearly independent vectors in V and hence forms a basis for V. If $n > r + 1$, then a vector $\mathbf{v}_{r+2}$ can be added to the set such that the set $\mathbf{v}_1, ..., \mathbf{v}_{r+2}$ is linearly independent. Continuing in this way, i vectors can be added until $r + i = n$. Then the set $\mathbf{v}_1, ..., \mathbf{v}_r,$ $\mathbf{v}_{r+1}, ..., \mathbf{v}_n$ forms a basis for V.

In later sections we shall be interested in constructing bases, starting off from given sets of linearly independent vectors. This theorem will be useful then.

EXERCISES

1. Prove that the vectors $(-1, 3)$ and $(2, 2)$ form a basis for $\mathbf{R}^2$.

2. Prove that the vectors $(-1, 2, 1)$, $(3, -1, 0)$, and $(2, 2, -2)$ form a basis for $\mathbf{R}^3$.

3. Give examples of three distinct bases for $\mathbf{R}^2$.

4. Give examples of three distinct bases for $\mathbf{R}^3$.

5. Prove that the set of vectors $(1, 2)$, $(-2, 3)$, and $(1, 1)$ spans $\mathbf{R}^2$ but is not linearly independent. Thus it is not a basis for $\mathbf{R}^2$. Prove that the vector $(3, 4)$, an element of $\mathbf{R}^2$, can be expressed in more than one way as a linear combination of these vectors.

6. Prove that the vectors $(1, 0, 0)$, $(-1, 1, 0)$, $(1, 1, 2)$, and $(1, 0, 1)$ span $\mathbf{R}^3$, are not linearly independent, and thus do not form a basis for $\mathbf{R}^3$. Illustrate the fact that the vector $(1, 1, 1)$ can be expressed in more than one way as a linear combination of these vectors. Select a subset of this set of four vectors that does form a basis. Show that $(1, 1, 1)$ is expressed uniquely as a linear combination of these basis vectors.

7. The vector $(1, 2, 0)$ can be expressed as a linear combination of the vectors $(1, 0, 0)$ and $(0, 1, 0)$ in $\mathbf{R}^3$, for $(1, 2, 0) = (1, 0, 0) + 2(0, 1, 0)$. What is

the geometrical interpretation of this equation? What is the geometrical characteristic of all vectors in $\mathbf{R}^3$ that can be expressed as linear combinations of $(1, 0, 0)$ and $(0, 1, 0)$?

8. Use Exercise 8 of the previous section to demonstrate how any number of distinct bases can be constructed for

 a) $\mathbf{R}^2$ b) $\mathbf{R}^3$ c) $\mathbf{R}^n$

9. $(1, 2, 0)$ and $(2, 3, 0)$ are vectors in $\mathbf{R}^3$. Consider the subspace of $\mathbf{R}^3$ generated by these vectors. What is the dimension of this subspace? Illustrate this subspace geometrically.

10. $(-1, 2, 1)$, $(3, 0, 1)$, and $(1, 4, 3)$ are vectors in $\mathbf{R}^3$. Prove that the set of vectors generated by these three vectors is a two-dimensional subspace of $\mathbf{R}^3$.

11. $(1, -1, 2)$ is a vector in $\mathbf{R}^3$. Use this vector to generate a one-dimensional subspace of $\mathbf{R}^3$. Illustrate this subspace geometrically.

12. Let $\mathbf{v}_1, \ldots, \mathbf{v}_m$ be m vectors in $\mathbf{R}^n$, where $m < n$. Use these vectors to generate a subspace of $\mathbf{R}^n$. Discuss the dimension of this subspace.

13. Prove that the vector $(1, 2, -1)$ lies in the two-dimensional subspace of $\mathbf{R}^3$ generated by the vectors $(1, 3, 1)$ and $(1, 4, 3)$.

14. Prove that the vector $(2, 1, 4)$ lies in the two-dimensional subspace of $\mathbf{R}^3$ generated by the vectors $(1, 0, 2)$ and $(1, 1, 2)$.

15. Prove that the vector $(1, 2, -1, 3)$ lies in the three-dimensional subspace of $\mathbf{R}^4$ generated by the vectors $(1, 2, 1, 0)$, $(1, -2, 3, 1)$, and $(-2, 4, -8, 1)$.

16. Does the vector $(1, 2, -1)$ lie in the subspace of $\mathbf{R}^3$ generated by the vectors $(1, -1, 0)$ and $(3, -1, 2)$?

17. Determine a basis for the subspace of $\mathbf{R}^3$ consisting of vectors of the form $(a, -b, 3a)$.

18. Determine a basis for the subspace of $\mathbf{R}^3$ consisting of vectors of the form $(a+b, 2a, a-b)$. Give the dimension of the subspace.

19. Determine a basis for the subspace of $\mathbf{R}^4$ consisting of vectors of the form $(a, -a+b, 3a+2b+c, a-c)$. Give the dimension of the subspace.

20. Consider the vector space $\mathbf{R}^2$. Prove that
 a) the zero vector forms a subspace. The dimension of this subspace is defined to be zero; and
 b) the one dimensional subspaces are lines through the origin.

21. Prove that if $\{\mathbf{v}_1, \ldots, \mathbf{v}_n\}$ is a basis for a vector space V, it is a maximal set of linearly independent vectors in V. That is, any set of vectors in V containing this set as a proper subset is a set of linearly dependent vectors.

22. In Example 5, p. 181 the primary colors are represented by linearly independent vectors—the vectors $(1, 0, 0)$, $(0, 1, 0)$, and $(0, 0, 1)$. What is the physical significance of linear independence here? Can colors other than the primary colors be used to generate all the other colors? (Filters not allowed.)

3–5. INNER PRODUCTS, NORMS, ANGLES AND DISTANCES

In this section we add additional structure to the vector space $\mathbf{R}^n$. This additional structure leads to concepts of magnitudes of vectors and angles between vectors.

DEFINITION 3-6   If $(x_1, ..., x_n)$ and $(y_1, ..., y_n)$ are vectors in $\mathbf{R}^n$, their *inner product* (or dot product) is

$$(x_1, ..., x_n) \cdot (y_1, ..., y_n) = x_1 y_1 + \cdots + x_n y_n$$

The inner product assigns to each pair of vectors a real number.

Example 1

$(1, 2)$ and $(3, -1)$ are elements of $\mathbf{R}^2$. Their inner product is

$$(1, 2) \cdot (3, -1) = (1 \times 3) + (2 \times -1) = 3 - 2 = 1$$

$(0, -1, 4)$ and $(5, -2, -1)$ are elements of $\mathbf{R}^3$. Their inner product is

$$(0, -1, 4) \cdot (5, -2, -1) = (0 \times 5) + (-1 \times -2) + (4 \times -1)$$

$$= 0 + 2 - 4 = -2$$

The inner product has the following three properties (Theorems 3-12, 3-13, 3-14). These results are of theoretical value in algebraic manipulations involving the inner product. The reader will, for example, use these results in discussing angles between vectors of $\mathbf{R}^2$ in this section and in handling projections in the following section. In more advanced work they become axioms for generalizing this inner product.

THEOREM 3-12   *If* $\mathbf{u}$ *and* $\mathbf{v}$ *are vectors in* $\mathbf{R}^n$ *then* $\mathbf{u} \cdot \mathbf{v} = \mathbf{v} \cdot \mathbf{u}$

Proof: Let $\mathbf{u} = (x_1, ..., x_n)$ and $\mathbf{v} = (y_1, ..., y_n)$
Then

$$\mathbf{u} \cdot \mathbf{v} = (x_1, ..., x_n) \cdot (y_1, ..., y_n)$$

$$= x_1 y_1 + \cdots + x_n y_n$$

Using the commutative property of real numbers under multiplication, this can be written

$$\mathbf{u} \cdot \mathbf{v} = y_1 x_1 + \cdots + y_n x_n$$

$$= \mathbf{v} \cdot \mathbf{u}$$

THEOREM 3-13   *If* $\mathbf{u}$, $\mathbf{v}$, *and* $\mathbf{w}$ *are vectors in* $\mathbf{R}^n$, *then*

$$(\mathbf{u} + \mathbf{v}) \cdot \mathbf{w} = \mathbf{u} \cdot \mathbf{w} + \mathbf{v} \cdot \mathbf{w}$$

and

$$\mathbf{w} \cdot (\mathbf{u} + \mathbf{v}) = \mathbf{w} \cdot \mathbf{u} + \mathbf{w} \cdot \mathbf{v}$$

THEOREM 3-14   *If a is an arbitrary scalar and* $\mathbf{v}$, $\mathbf{u}$ *are elements of* $\mathbf{R}^n$, *then*

$$a\mathbf{v} \cdot \mathbf{u} = \mathbf{v} \cdot a\mathbf{u}$$

Theorems 3-13 and 3-14 are proved using similar techniques to those adopted in proving Theorem 3-12. The vectors are expressed as having *n* components and the definition of the inner product is used.

Example 2

In working with inner products, one finds the need to simplify algebraic expressions. In such manipulations it is important to keep track of scalars and vectors.

For example, each of the following expressions will result in a scalar: $\mathbf{u} \cdot \mathbf{v}$, $(\mathbf{u} \cdot \mathbf{v})(\mathbf{u} \cdot \mathbf{w}) + 4$, $(3\mathbf{u} \cdot \mathbf{v} - 4\mathbf{w} \cdot \mathbf{v})(\mathbf{u} \cdot \mathbf{w} - \mathbf{x} \cdot \mathbf{y})$.

The following will all result in vectors. $3(\mathbf{u} \cdot \mathbf{v})\mathbf{w}$, $(\mathbf{u} \cdot \mathbf{v})\mathbf{w} + \mathbf{u}$, $(\mathbf{u} \cdot \mathbf{v} - \mathbf{v} \cdot \mathbf{w})\mathbf{w}$.

We now illustrate the use of the above theorems in simplifying algebraic expressions. Let us simplify

$$\left((\mathbf{u} \cdot \mathbf{u})\mathbf{v} - \frac{\mathbf{u} \cdot \mathbf{v}}{3} \frac{\mathbf{u}}{2} \right) \cdot 3\mathbf{u} = (\mathbf{u} \cdot \mathbf{u})\mathbf{v} \cdot 3\mathbf{u} - \frac{\mathbf{u} \cdot \mathbf{v}}{3} \frac{\mathbf{u}}{2} \cdot 3\mathbf{u}$$

$$= 3(\mathbf{u} \cdot \mathbf{u})(\mathbf{v} \cdot \mathbf{u}) - \frac{3}{3 \times 2}(\mathbf{u} \cdot \mathbf{v})(\mathbf{u} \cdot \mathbf{u})$$

$$= 3(\mathbf{u} \cdot \mathbf{u})(\mathbf{v} \cdot \mathbf{u}) - \tfrac{1}{2}(\mathbf{u} \cdot \mathbf{u})(\mathbf{v} \cdot \mathbf{u})$$

$$= \tfrac{5}{2}(\mathbf{u} \cdot \mathbf{u})(\mathbf{v} \cdot \mathbf{u})$$

Norm of a Vector   We now use the concept of inner product to define the *norm* of a vector. The norm is also called the *magnitude* or *length* of the vector.

DEFINITION 3-7   The norm of the arbitrary vector $(x_1, \ldots, x_n)$ of $\mathbf{R}^n$ is denoted by $\|(x_1, \ldots, x_n)\|$ and given by

$$\|(x_1, \ldots, x_n)\| = \sqrt{(x_1, \ldots, x_n) \cdot (x_1, \ldots, x_n)}$$
$$= \sqrt{x_1{}^2 + \cdots + x_n{}^2}$$

This definition is compatible with the geometric lengths of position vectors in $\mathbf{R}^2$ and $\mathbf{R}^3$:

If $\mathbf{u}$ is the position vector (a, b) in $\mathbf{R}^2$ its length is indeed $\sqrt{a^2 + b^2}$:

187

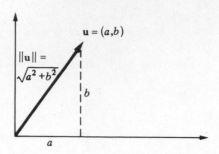

If $\mathbf{v}$ is the position vector (a, b, c) in $\mathbf{R}^3$ its length is $\sqrt{a^2 + b^2 + c^2}$:

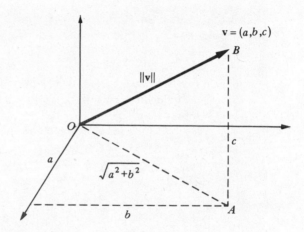

$$OA = \sqrt{a^2 + b^2}$$

$$\|\mathbf{v}\| = OB = \sqrt{OA^2 + AB^2} = \sqrt{a^2 + b^2 + c^2}$$

Example 3*

We have seen how forces can be represented by vectors. The magnitude of the force will be the norm of the vector. Let the vector $(4, 3)$ represent a force acting on a particle. The magnitude of this force is

$$\|(4, 3)\| = \sqrt{(4, 3) \cdot (4, 3)} = \sqrt{16 + 9} = 5$$

A vector whose norm is 1 is said to be a *unit vector*. The vector $(1, 0)$ in $\mathbf{R}^2$ is a unit vector, for

$$\|(1, 0)\| = \sqrt{(1, 0) \cdot (1, 0)} = \sqrt{(1 \times 1) + (0 \times 0)} = 1$$

Similarly, the vector $(0, 1)$ of $\mathbf{R}^2$ is a unit vector, and so are the vectors $(1, 0, 0)$, $(0, 1, 0)$, and $(0, 0, 1)$ of $\mathbf{R}^3$.

THEOREM 3-15   *If* $\mathbf{v}$ *is any nonzero vector in* $\mathbf{R}^n$, *the vector* $(1/\|\mathbf{v}\|)\mathbf{v}$
derived from $\mathbf{v}$ *is a unit vector. This procedure of constructing a unit vector
from a given nonzero vector is called normalizing the vector.*

Proof: We set out to prove that $(1/\|\mathbf{v}\|)\mathbf{v}$ is a unit vector.
Let $\mathbf{v}$ be the vector $(x_1, \ldots, x_n)$.   Then

$$\|\mathbf{v}\| = \sqrt{x_1^2 + \cdots + x_n^2}$$

$(1/\|\mathbf{v}\|)\mathbf{v}$ is thus the vector $(1/\sqrt{x_1^2 + \cdots + x_n^2})(x_1, \ldots, x_n)$.

That is, it becomes the vector $(x_1/\sqrt{}, \ldots, x_n/\sqrt{})$, where
$\sqrt{} = \sqrt{x_1^2 + \cdots + x_n^2}$ for convenience of notation.   Hence

$$\left\| \frac{1}{\|\mathbf{v}\|}\mathbf{v} \right\| = \sqrt{\frac{x_1^2}{(\sqrt{})^2} + \cdots + \frac{x_n^2}{(\sqrt{})^2}}$$

$$= \sqrt{\frac{x_1^2 + \cdots + x_n^2}{(\sqrt{})^2}}$$

$$= \sqrt{\frac{x_1^2 + \cdots + x_n^2}{x_1^2 + \cdots + x_n^2}} = 1$$

The unit vector obtained on normalizing a given vector will be a
unit vector in the same direction as the original vector; in Section 1-2 we
discussed that multiplication of a vector by a positive scalar leaves it
unchanged in direction.

Example 4

$(1, -2, 3)$ is a nonzero vector in $\mathbf{R}^3$.

$$\|(1, -2, 3)\| = \sqrt{1^2 + 2^2 + 3^2}$$

$$= \sqrt{14}$$

Thus, according to Theorem 3-15, the vector $(1/\sqrt{14})(1, -2, 3)$ is a unit
vector.   Let us verify this.

$$\left\| \frac{1}{\sqrt{14}}(1, -2, 3) \right\| = \left\| \left(\frac{1}{\sqrt{14}}, \frac{-2}{\sqrt{14}}, \frac{3}{\sqrt{14}} \right) \right\|$$

$$= \sqrt{\left(\frac{1}{\sqrt{14}} \right)^2 + \left(\frac{-2}{\sqrt{14}} \right)^2 + \left(\frac{3}{\sqrt{14}} \right)^2}$$

$$= \sqrt{\frac{1}{14} + \frac{4}{14} + \frac{9}{14}}$$

$$= 1$$

Angles between vectors

DEFINITION 3-8   If $\mathbf{u}$ and $\mathbf{v}$ are vectors in $\mathbf{R}^n$, the cosine of the angle between them is given by

$$\cos\theta = \frac{1}{\|\mathbf{u}\|}\,\mathbf{u}\cdot\frac{1}{\|\mathbf{v}\|}\,\mathbf{v}$$

If one interprets $\mathbf{u}$ and $\mathbf{v}$ as position vectors in $\mathbf{R}^2$ or $\mathbf{R}^3$, this definition leads to the expected angle between the vectors. We prove this result for $\mathbf{R}^2$.

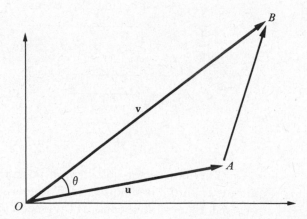

Let $\mathbf{u} = (a, b)$, $\mathbf{v} = (c, d)$ in the above diagram.   The vector $AB = \mathbf{v} - \mathbf{u} = (c - a, d - b)$. Thus the lengths of the sides of the triangle OAB are

$$OA = \|\mathbf{u}\| = \sqrt{a^2 + b^2},$$

$$OB = \|\mathbf{v}\| = \sqrt{c^2 + d^2},$$

$$AB = \|\mathbf{v} - \mathbf{u}\| = \sqrt{(c - a)^2 + (d - b)^2}.$$

The law of cosines gives

$$AB^2 = OA^2 + OB^2 - 2(OA)(OB)\cos\theta$$

Thus

$$\cos\theta = \frac{OA^2 + OB^2 - AB^2}{2(OA)(OB)}$$

$$= \frac{a^2 + b^2 + c^2 + d^2 - (c - a)^2 - (d - b)^2}{2(OA)(OB)}$$

$$= \frac{2ac + 2db}{2(OA)(OB)} = \frac{ac + db}{(OA)(OB)}$$

$$= \frac{\mathbf{u} \cdot \mathbf{v}}{\|\mathbf{u}\| \, \|\mathbf{v}\|}$$

Using the result of Theorem 3-14 this can be expressed in the more meaningful form

$$\cos \theta = \frac{1}{\|\mathbf{u}\|} \mathbf{u} \cdot \frac{1}{\|\mathbf{v}\|} \mathbf{v}$$

To obtain the cosine of the angle between two vectors one normalizes each vector and then takes their inner product.

Example 5

Determine the angle between the vectors $(1, 0)$ and $(1, 1)$ in $\mathbf{R}^2$. Using the definition, we find that

$$\cos \theta = \frac{1}{\|(1, 0)\|} (1, 0) \cdot \frac{1}{\|(1, 1)\|} (1, 1)$$

$$= (1, 0) \cdot \frac{1}{\sqrt{2}} (1, 1)$$

$$= (1, 0) \cdot \left(\frac{1}{\sqrt{2}}, \frac{1}{\sqrt{2}} \right) = \frac{1}{\sqrt{2}}$$

Hence $\theta = 45°$.

If we interpret $(1, 0)$ and $(1, 1)$ as position vectors in $\mathbf{R}^2$, we see that the angle between them is indeed $45°$ (Figure 3-2).

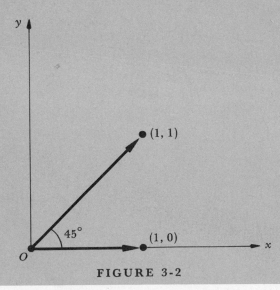

FIGURE 3-2

Example 6

Determine the angle between the vectors $(2, 0)$ and $(0, 3)$ in $\mathbf{R}^2$. Using the definition,

$$\cos \theta = \frac{1}{\|(2,0)\|} (2,0) \cdot \frac{1}{\|(0,3)\|} (0,3)$$

$$= \frac{1}{\|(2,0)\| \, \|(0,3)\|} (2,0) \cdot (0,3)$$

$$= 0$$

Hence $\theta = 90°$; the vectors are at right angles.

We say that two nonzero vectors are *orthogonal* if they are at right angles.

THEOREM 3-16   *Two vectors are orthogonal if and only if their inner product is zero.*

Proof: Let $\mathbf{v}_1$ and $\mathbf{v}_2$ be orthogonal vectors. Hence θ, the angle between them, is $90°$ and $\cos \theta = 0$.

$$\frac{1}{\|\mathbf{v}_1\|} \mathbf{v}_1 \cdot \frac{1}{\|\mathbf{v}_2\|} \mathbf{v}_2 = 0$$

implying that

$$\frac{1}{\|\mathbf{v}_1\| \, \|\mathbf{v}_2\|} \mathbf{v}_1 \cdot \mathbf{v}_2 = 0$$

and hence that $\mathbf{v}_1 \cdot \mathbf{v}_2 = 0$.

Conversely, if $\mathbf{v}_1 \cdot \mathbf{v}_2 = 0$, $\cos \theta = 0$, and θ is $90°$.

Example 7

The vectors $(1, 0)$ and $(0, 1)$ in $\mathbf{R}^2$ are orthogonal, since

$$(1,0) \cdot (0,1) = 0 + 0 = 0$$

Example 8

The vectors $(1, 2, -1)$ and $(2, 4, 10)$ in $\mathbf{R}^3$ are orthogonal, since

$$(1,2,-1) \cdot (2,4,10) = (1 \times 2) + (2 \times 4) + (-1 \times 10)$$

$$= 2 + 8 - 10 = 0$$

A basis consisting of unit vectors, any two of which are orthogonal, is called an *orthonormal basis*.

THEOREM 3-17   *The canonical basis for* $\mathbf{R}^n$, *namely*

$$(1, 0, \ldots, 0), \ldots, (0, \ldots, 0, 1)$$

is an orthonormal basis.

Proof: Consider the vector $(0, \ldots, 1, \ldots, 0)$.   Its magnitude is

$$\sqrt{(0, \ldots, 1, \ldots, 0) \cdot (0, \ldots, 1, \ldots, 0)} = \sqrt{1} = 1$$

Thus it is a unit vector.

Consider two distinct vectors: $\mathbf{v}_1 = (0, \ldots, 1, \ldots, 0, \ldots, 0)$ with the 1 in the ith slot, and $\mathbf{v}_2 = (0, \ldots, 0, \ldots, 1, \ldots, 0)$ with the 1 in the jth slot.   Since they are both of unit magnitude,

$$\cos \theta = \mathbf{v}_1 \cdot \mathbf{v}_2$$
$$= \sqrt{(0, \ldots, 1, \ldots, 0, \ldots, 0) \cdot (0, \ldots, 0, \ldots, 1, \ldots, 0)}$$
$$= 0$$

Hence they are orthogonal.

Thus the basis $\mathbf{i}, \mathbf{j}$ is an orthonormal basis for $\mathbf{R}^2$ and $\mathbf{i}, \mathbf{j}, \mathbf{k}$ is an orthonormal basis for $\mathbf{R}^3$ (Figure 3-3).   In $\mathbf{R}^2$, $\mathbf{i} \cdot \mathbf{j} = 0$ and $\|\mathbf{i}\| = \|\mathbf{j}\| = 1$. In $\mathbf{R}^3$, $\mathbf{i} \cdot \mathbf{j} = \mathbf{j} \cdot \mathbf{k} = \mathbf{i} \cdot \mathbf{k} = 0$ and $\|\mathbf{i}\| = \|\mathbf{j}\| = \|\mathbf{k}\| = 1$.

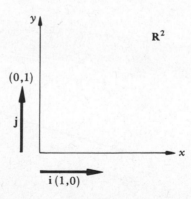

FIGURE 3-3

Two vectors are said to be *parallel* if the angle between them is either 0 or π; that is, if they are in the same or opposite directions.

193

Example 9

The vectors $(1, 2)$ and $(3, 6)$ are parallel, since

$$\cos \theta = \frac{(1, 2)}{\|(1, 2)\|} \cdot \frac{(3, 6)}{\|(3, 6)\|} = \frac{15}{\sqrt{5}\sqrt{45}} = 1$$

$\theta = 0$ and the vectors are parallel; they are, in fact, in the same direction. The vectors $(1, -2, 3)$ and $(-2, 4, -6)$ are parallel, since

$$\cos \theta = \frac{(1, -2, 3)}{\|(1, -2, 3)\|} \cdot \frac{(-2, 4, -6)}{\|(-2, 4, -6)\|}$$

$$= \frac{-2 - 8 - 18}{\sqrt{1^2 + 2^2 + 3^2}\sqrt{2^2 + 4^2 + 6^2}} = -1$$

$\theta = \pi$, and the vectors are in opposite directions.

Distance between Points   We now use the concept of norm to define distance between points in $\mathbf{R}^n$. Once again our definition will be compatible with known results for two and three-dimensional spaces.

Let X and Y be arbitrary points in a vector space. These points define position vectors. Let us denote the position vectors $\mathbf{x}$ and $\mathbf{y}$ (below). Construct the vector $-\mathbf{y}$. Observe that the sum of the vectors $\mathbf{x}$ and $-\mathbf{y}$ has magnitude equal to the distance between the points X and Y.

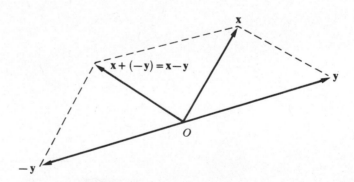

Denote the distance between X and Y, $d(X, Y)$, and let

$$d(X, Y) = \|\mathbf{x} - \mathbf{y}\|.$$

Let us express $d(X, Y)$ in terms of the components of $\mathbf{x}$ and $\mathbf{y}$. Suppose $\mathbf{x} = (x_1, \ldots, x_n)$ and $\mathbf{y} = (y_1, \ldots, y_n)$. Then

$$\mathbf{x} - \mathbf{y} = (x_1 - y_1, \ldots, x_n - y_n).$$

and

$$d(X, Y) = \sqrt{(x_1 - y_1)^2 + \cdots + (x_n - y_n)^2}$$

Note that this reduces to the usual expression for distance when the vector space is $\mathbf{R}^2$. Then $\mathbf{x} = (x_1, x_2)$ and $\mathbf{y} = (y_1, y_2)$ and

$$d(X, Y) = \sqrt{(x_1 - y_1)^2 + (x_2 - y_2)^2}.$$

Example 10

Determine the distance between the points $(2, 4, 5)$ and $(1, 2, 3)$ of $\mathbf{R}^3$. We have that

$$d(X, Y) = \sqrt{(x_1 - y_1)^2 + (x_2 - y_2)^2 + (x_3 - y_3)^2}$$
$$= \sqrt{(2-1)^2 + (4-2)^2 + (5-3)^2}$$
$$= \sqrt{1^2 + 2^2 + 2^2} = \sqrt{9} = 3$$

Example 11

Determine the distance between the points $X(18, 4, 3, 3, -2)$ and $Y(14, 1, 0, 2, -3)$ of $\mathbf{R}^5$.

$$d(X, Y) = \sqrt{(x_1 - y_1)^2 + (x_2 - y_2)^2 + (x_3 - y_3)^2 + (x_4 - y_4)^2 + (x_5 - y_5)^2}$$
$$= \sqrt{(18 - 14)^2 + (4 - 1)^2 + (3 - 0)^2 + (3 - 2)^2 + (-2 + 3)^2}$$
$$= \sqrt{4^2 + 3^2 + 3^2 + 1^2 + 1^2} = \sqrt{36} = 6$$

EXERCISES

1. Determine the inner product of the following elements of $\mathbf{R}^2$: $(2, 1)$ and $(3, 0)$.

2. Determine the inner product of the vectors $(-1, 3, 0)$ and $(2, 1, 5)$ in $\mathbf{R}^3$.

3. Determine the norms of the vectors $(1, 2)$ and $(3, 4)$ in $\mathbf{R}^2$.

4. Determine the norms of the vectors $(2, 0, 0)$, $(0, 4, 0)$, and $(1, 2, 3)$ in $\mathbf{R}^3$.

5. Normalize the vectors $(2, 3)$ and $(-1, 2)$ in $\mathbf{R}^2$.

6. Normalize the vectors $(1, 2, 3)$, $(-1, 2, 0)$, and $(0, 0, 1)$ in $\mathbf{R}^3$.

7. Using the definition of the inner product, prove that $2(1, 2, 3) \cdot (-1, 0, 4) = (1, 2, 3) \cdot 2(-1, 0, 4)$ (Illustrates Theorem 3-14.)

8. Prove that the norm of a vector is always greater than or equal to zero. When is the norm equal to zero?

9. Determine the angle between the vectors $(-1, 1)$ and $(0, 1)$ in $\mathbf{R}^2$.

10. Determine the angle between the vectors $(2, 0)$ and $(1, \sqrt{3})$ in $\mathbf{R}^2$.

11. Determine the angle between the vectors $(1, 2, 0)$ and $(0, 1, 1)$ in $\mathbf{R}^3$.

12. Prove that the vectors $(1, 2)$ and $(2, -1)$ are orthogonal in $\mathbf{R}^2$.

13. Prove that the vectors $(4, -2, 2)$ and $(2, 3, -1)$ are orthogonal in $\mathbf{R}^3$.

14. Determine a vector orthogonal to $(2, 4)$ in $\mathbf{R}^2$.

15. Determine a vector orthogonal to $(-1, 2, 3)$ in $\mathbf{R}^3$.

16. Determine a vector orthogonal to $(\frac{1}{2}, 2, -1)$ in $\mathbf{R}^3$.

17. Prove that the set of vectors orthogonal to the vector $(1, 0, 0)$ in $\mathbf{R}^3$ forms a subspace of $\mathbf{R}^3$.

18. Prove that the set of vectors orthogonal to the vector $(1, 2, 1)$ in $\mathbf{R}^3$ forms a subspace of $\mathbf{R}^3$.

19. a) Prove that the vectors $(1, 2, -1)$ and $(-1, -2, 1)$ are parallel.
    b) Prove that the vectors $(2, 3, 1)$ and $(4, 6, 2)$ are parallel.

20. In each of the following compute the distance between X and Y.
    a) $X(6, 5)$, $Y(2, 2)$
    b) $X(3, 1, 1)$, $Y(1, -1, 0)$
    c) $X(6, 6, 2)$, $Y(-2, 4, 1)$
    d) $X(3, 2, 3, 1)$, $Y(-2, -2, 1, -1)$
    e) $X(3, -2, 4)$, $Y(-1, 2, 0)$
    f) $X(4, -1, 3, 2, 4)$, $Y(4, 5, 2, 1, 1)$

21. In Example 8 we proved that the vectors $(1, 2)$ and $(3, 6)$ are in the same direction. Note that $(3, 6) = 3(1, 2)$. Prove that, in general, two vectors $\mathbf{u}$ and $\mathbf{v}$ are in the same direction if and only if there exists a nonzero positive scalar c such that $\mathbf{u} = c\mathbf{v}$.

    Further, in Example 8 we saw that $(1, -2, 3)$ and $(-2, 4, -6)$ were in opposite directions. Note that $(-2, 4, -6) = -2(1, -2, 3)$. Prove that two vectors $\mathbf{u}$ and $\mathbf{v}$ are in opposite directions if and only if there exists a nonzero negative scalar c such that $\mathbf{u} = c\mathbf{v}$.

22. Let $\mathbf{u}$, $\mathbf{v}$, and $\mathbf{w}$ be vectors in $\mathbf{R}^n$; c and d, be non zero scalars, and α an angle. Each of the following expressions will be a vector, scalar, or make no sense. Identify which:

    a) $(\mathbf{u} \cdot \mathbf{v})\mathbf{w}$ b) $(\mathbf{u} \cdot \mathbf{v}) \cdot \mathbf{w}$ c) $\mathbf{u} \cdot \mathbf{v} + c\mathbf{w}$

    d) $\mathbf{u} \cdot \mathbf{w} + c$ e) $c\mathbf{u} \cdot d\mathbf{w} + \|\mathbf{w}\|\mathbf{v}$ f) $\|\mathbf{u} \cdot \mathbf{v}\|$

    g) $c(\mathbf{u} \cdot \mathbf{v}) + d\mathbf{w}$ h) $\dfrac{\mathbf{w} + \mathbf{u}}{\mathbf{w} \cdot \mathbf{u}}$ i) $\|\mathbf{u} + c\mathbf{v}\| + d$

    j) $\|\mathbf{v}\| \cos\alpha \dfrac{\mathbf{u}}{\|\mathbf{u}\|}$ k) $\left(\mathbf{v} \cdot \dfrac{\mathbf{u}}{\|\mathbf{u}\|}\right)\dfrac{\mathbf{u}}{\|\mathbf{u}\|}$ l) $\mathbf{u} - \left(\dfrac{\mathbf{u}}{\|\mathbf{u}\|} \cdot \mathbf{v}\right)\dfrac{\mathbf{v}}{\|\mathbf{v}\|}$

    [We shall see the significance of (j), (k), and (l) in the following section.]

23. Prove that if $\mathbf{u}$ is orthogonal to $\mathbf{v}$ and to $\mathbf{w}$ then $\mathbf{u}$ is orthogonal to every vector in the subspace generated by $\mathbf{v}$ and $\mathbf{w}$.

24. Let $\mathbf{u}$ be orthogonal to each of the vectors $\mathbf{v}_1, \ldots, \mathbf{v}_m$. Prove that $\mathbf{u}$ is orthogonal to each vector in the subspace generated by $\mathbf{v}_1, \ldots, \mathbf{v}_m$. We say that $\mathbf{u}$ is orthogonal to this subspace.

25. Let **v** be a vector in a space V of dimension n. Show that all the vectors orthogonal to **v** form a subspace of V.

26. Let U and V be two subspaces of $\mathbf{R}^n$. U is said to be orthogonal to V if and only if every vector in U is orthogonal to every vector in V. Give an example of two orthogonal subspaces of $\mathbf{R}^3$.

27. Let $\mathbf{u}, \mathbf{v}_1, ..., \mathbf{v}_m$ be vectors in a given vector space and $a_1, ..., a_m$ be scalars. Prove that

$$\mathbf{u} \cdot (a_1 \mathbf{v}_1 + a_2 \mathbf{v}_2 + \cdots + a_m \mathbf{v}_m) = a_1 \mathbf{u} \cdot \mathbf{v}_1 + a_2 \mathbf{u} \cdot \mathbf{v}_2 + \cdots a_m \mathbf{u} \cdot \mathbf{v}_m$$

28. Let $\mathbf{v}_1, ..., \mathbf{v}_m$ be a set of m mutually orthogonal vectors. (Any pair of these vectors are orthogonal.) Prove that $\mathbf{v}_1, ..., \mathbf{v}_m$ are linearly independent. [*Hint*: Use the result of Exercise 27.]

29. **a)** If **u** and **v** are vectors in $\mathbf{R}^n$ then $|\mathbf{u} \cdot \mathbf{v}| \leq \|\mathbf{u}\| \, \|\mathbf{v}\|$. [The bars on the left of this equation mean absolute value]. This identity is called the *Schwarz inequality*. Prove that the Schwarz inequality holds for $\mathbf{R}^2$. [*Hint*: Let $\mathbf{u} = (a, b)$ and $\mathbf{v} = (c, d)$.]

    b) Use the Schwarz inequality to prove that

    $$-1 \leq \frac{1}{\|\mathbf{u}\|} \mathbf{u} \cdot \frac{1}{\|\mathbf{v}\|} \mathbf{v} \leq 1$$

    This is a necessary condition for our definition of angle between vectors,

    $$\cos \theta = \frac{1}{\|\mathbf{u}\|} \mathbf{u} \cdot \frac{1}{\|\mathbf{v}\|} \mathbf{v}$$

    to be a valid one.

    The reader may be interested in reading an article "Pythagoras and the Cauchy-Schwarz Inequality", by Ladnor Geissinger in *The American Mathematical Monthly*, January 1976. In this paper the author uses geometric considerations to motivate the concept of inner product, leading to a definition of angle.

30. Write a program for computing the inner product of two vectors. Test your program for the vectors $(1, -1, 2, 3)$ and $(0, 1, 4, 2)$. Their inner product is 13.

31. Write a program for computing norms of vectors. Test your program on the vector $(1, 2, 3, 4)$. Its norm is 5.4772255.

32. Write a program for normalizing a vector. Test your program on the vector $(1, 0, 3, -4, 5)$. The normalized vector is

    $$(0.140028 \quad 0 \quad 0.420084 \quad -0.560112 \quad 0.70014).$$

33. Write a program to determine the cosine of the angle between two given vectors. Test your program on the vectors $(1, 2, 3, 0, -1)$ and $(-1, 0, 1, 4, 1)$. The cosine of the angle is 5.923489 E -2.

3–6. PROJECTIONS AND THE GRAM-SCHMIDT ORTHOGONALIZATION PROCESS

Let **u** and **v** be vectors with angle α between them.   The *scalar projection* (or *component*) of **v** in the direction of **u** is defined to be $\|\mathbf{v}\| \cos \alpha$.   The geometrical interpretation of this projection for $\mathbf{R}^2$ and $\mathbf{R}^3$ is given below as the length OA.

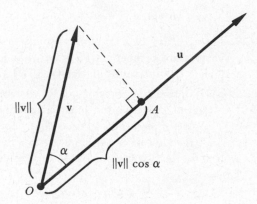

Note that if $90° < \alpha \leq 180°$, then $\cos \alpha$ is negative and the scalar projection will be negative (figure below).

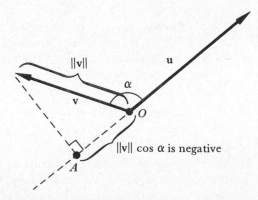

In certain contexts we are interested in the *vector projection* of **v** onto **u**.   The vector projection is defined to be

$$\|\mathbf{v}\| \cos \alpha \frac{\mathbf{u}}{\|\mathbf{u}\|}$$

This is the vector OA in the above figures.

To compute scalar and vector projections it is easier to rewrite the definitions:

Scalar projection of **v** onto **u**

$$= \|\mathbf{v}\| \cos \alpha = \|\mathbf{v}\| \left(\frac{\mathbf{v}}{\|\mathbf{v}\|} \cdot \frac{\mathbf{u}}{\|\mathbf{u}\|} \right) = \mathbf{v} \cdot \frac{\mathbf{u}}{\|\mathbf{u}\|}$$

Note that this is the inner product of **v** with the unit vector in the direction of **u**.

Vector projection of **v** onto **u**

$$= \|\mathbf{v}\| \cos \alpha \, \frac{\mathbf{u}}{\|\mathbf{u}\|} = \left(\mathbf{v} \cdot \frac{\mathbf{u}}{\|\mathbf{u}\|} \right) \frac{\mathbf{u}}{\|\mathbf{u}\|}$$

This is the scalar projection times the unit vector in the direction of **u**.

Example 1

Determine the scalar and vector projections of $(1, 2, 3)$ onto $(4, -1, 3)$.

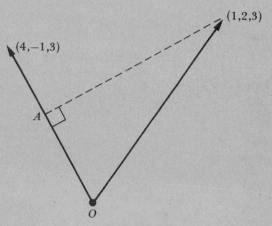

Scalar projection of $(1, 2, 3)$ onto $(4, -1, 3)$

$$= (1, 2, 3) \cdot \frac{(4, -1, 3)}{\|(4, -1, 3)\|} = \frac{11}{\sqrt{26}}$$

$$= 2.157 - \text{the length } OA \text{ in the above diagram.}$$

Vector projection of $(1, 2, 3)$ onto $(4, -1, 3)$

$$= \left(\frac{11}{\sqrt{26}} \right) \frac{(4, -1, 3)}{\|(4, -1, 3)\|} = \frac{11}{26} (4, -1, 3)$$

$$= 0.423 \, (4, -1, 3) - \text{the vector } OA \text{ above.}$$

The next example paves the way for a method of constructing an orthogonal set of vectors from a given set.

Example 2

Let $\mathbf{u}$ and $\mathbf{v}$ be elements of $\mathbf{R}^n$. Prove that when one subtracts from $\mathbf{v}$ its vector projection onto $\mathbf{u}$, the resulting vector is orthogonal to $\mathbf{u}$.

By subtracting the vector projection we get the vector

$$\mathbf{v} - \left(\mathbf{v} \cdot \frac{\mathbf{u}}{\|\mathbf{u}\|}\right)\frac{\mathbf{u}}{\|\mathbf{u}\|}$$

We are interested in proving that

$$\left[\mathbf{v} - \left(\mathbf{v} \cdot \frac{\mathbf{u}}{\|\mathbf{u}\|}\right)\frac{\mathbf{u}}{\|\mathbf{u}\|}\right] \cdot \mathbf{u} = 0$$

The left side becomes

$$\mathbf{v} \cdot \mathbf{u} - \left(\mathbf{v} \cdot \frac{\mathbf{u}}{\|\mathbf{u}\|}\right)\frac{\mathbf{u}}{\|\mathbf{u}\|} \cdot \mathbf{u}$$

$$= \mathbf{v} \cdot \mathbf{u} - (\mathbf{v} \cdot \mathbf{u})\frac{\mathbf{u}}{\|\mathbf{u}\|} \cdot \frac{\mathbf{u}}{\|\mathbf{u}\|}$$

$$= \mathbf{v} \cdot \mathbf{u} - \mathbf{v} \cdot \mathbf{u} = 0$$

We now use the results of this example to construct a set of orthogonal vectors from any given set of linearly independent vectors in $\mathbf{R}^n$. The procedure is called the *Gram-Schmidt orthogonalization procedure.*

Let $\mathbf{v}_1, \ldots, \mathbf{v}_m$ be a set of linearly independent vectors in $\mathbf{R}^n$. We shall construct a set of orthogonal vectors $\mathbf{u}_1, \ldots, \mathbf{u}_m$ from this set.

Let $\mathbf{u}_1 = \mathbf{v}_1$. Next let $\mathbf{u}_2 = \mathbf{v}_2 - [\mathbf{v}_2 \cdot (\mathbf{u}_1/\|\mathbf{u}_1\|)](\mathbf{u}_1/\|\mathbf{u}_1\|)$. We constructed $\mathbf{u}_2$ by subtracting from $\mathbf{v}_2$ the vector projection of $\mathbf{v}_2$ onto $\mathbf{u}_1$. Example 2 tells us that $\mathbf{u}_2$ is orthogonal to $\mathbf{u}_1$.

We continue in the same way.

$$\mathbf{u}_3 = \mathbf{v}_3 - \left(\mathbf{v}_3 \cdot \frac{\mathbf{u}_1}{\|\mathbf{u}_1\|}\right)\frac{\mathbf{u}_1}{\|\mathbf{u}_1\|} - \left(\mathbf{v}_3 \cdot \frac{\mathbf{u}_2}{\|\mathbf{u}_2\|}\right)\frac{\mathbf{u}_2}{\|\mathbf{u}_2\|}$$

$\mathbf{u}_3$ is obtained by subtracting from $\mathbf{v}_3$ the vector projections of $\mathbf{v}_3$ onto both $\mathbf{u}_1$ and $\mathbf{u}_2$. $\mathbf{u}_3$ will be orthogonal to both $\mathbf{u}_1$ and $\mathbf{u}_2$.

The general vector $\mathbf{u}_i$ is

$$\mathbf{u}_i = \mathbf{v}_i - \left(\mathbf{v}_i \cdot \frac{\mathbf{u}_1}{\|\mathbf{u}_1\|}\right)\frac{\mathbf{u}_1}{\|\mathbf{u}_1\|} - \cdots - \left(\mathbf{v}_i \cdot \frac{\mathbf{u}_{i-1}}{\|\mathbf{u}_{i-1}\|}\right)\frac{\mathbf{u}_{i-1}}{\|\mathbf{u}_{i-1}\|}$$

The vectors $\mathbf{u}_1, \ldots, \mathbf{u}_m$ then form an orthogonal set. In order to obtain a set of unit orthogonal vectors all one has to do is normalize each of these vectors.

This procedure can be used to construct an orthonormal basis from a given basis that defines a subspace. If $v_1, ..., v_m$ is a given basis for a subspace of R^n, then by using this procedure one can construct an orthonormal basis for this subspace. The following example illustrates the method.

Example 3

The set $(3, 0, 0, 0)$, $(0, 1, 2, 1)$, $(0, -1, 3, 2)$ is a set of three linearly independent vectors in R^4. The set defines a three-dimensional subspace of R^4. Using the Gram-Schmidt orthogonalization procedure, construct an orthonormal basis for the space.

Using the notation of the above theory, let $v_1 = (3, 0, 0, 0)$, $v_2 = (0, 1, 2, 1)$, and $v_3 = (0, -1, 3, 2)$. Then

$$u_1 = v_1 = (3, 0, 0, 0)$$

$$u_2 = v_2 - \left(v_2 \cdot \frac{u_1}{\|u_1\|}\right) \frac{u_1}{\|u_1\|}$$

$$= (0, 1, 2, 1) - \left[(0, 1, 2, 1) \cdot \frac{(3, 0, 0, 0)}{3}\right] \frac{(3, 0, 0, 0)}{3}$$

$$= (0, 1, 2, 1)$$

$$u_3 = v_3 - \left(v_3 \cdot \frac{u_1}{\|u_1\|}\right) \frac{u_1}{\|u_1\|} - \left(v_3 \cdot \frac{u_2}{\|u_2\|}\right) \frac{u_2}{\|u_2\|}$$

$$= (0, -1, 3, 2) - \left[(0, -1, 3, 2) \cdot \frac{(3, 0, 0, 0)}{3}\right] \frac{(3, 0, 0, 0)}{3}$$

$$- \left[(0, -1, 3, 2) \cdot \frac{(0, 1, 2, 1)}{\sqrt{1^2 + 2^2 + 1^2}}\right] \frac{(0, 1, 2, 1)}{\sqrt{1^2 + 2^2 + 1^2}}$$

$$= (0, -1, 3, 2) - (0, 0, 0, 0) - \tfrac{7}{6}(0, 1, 2, 1)$$

$$= (0, -\tfrac{13}{6}, \tfrac{2}{3}, \tfrac{5}{6})$$

Thus the vectors $(3, 0, 0, 0)$, $(0, 1, 2, 1)$, and $(0, -\tfrac{13}{6}, \tfrac{2}{3}, \tfrac{5}{6})$ form an orthogonal set. The first two vectors are obviously in the subspace of interest; the third also lies in this subspace because it is a linear combination of the first two and v_3. This set of vectors is therefore an orthogonal basis for the subspace. On normalizing each vector we get the orthonormal basis $(1, 0, 0, 0)$, $(1/\sqrt{6})(0, 1, 2, 1)$, $(6/\sqrt{210})(0, -13, 4, 5)$.

The following examples illustrate the use of the concept of scalar projection in the analyses of forces. The term *component* is more commonly used in this context.

Example 4*

The vector $(1, 2, 3)$ represents a force acting on a body. What is the component of the force in the direction of the vector $(4, -1, 3)$?

Its component will be the scalar projection of $(1, 2, 3)$ in the direction $(4, -1, 3)$. This is $\|(1, 2, 3)\| \cos \alpha$, which has been calculated to be 2.157 (Example 1). Physically, this will be the effect of the force in this direction.

Example 5*

This example of stress analysis in frameworks illustrates how a problem in engineering is approached using components of forces. The problem reduces to solving a system of linear equations.

In designing a structure such as a bridge, it is necessary to know the forces that will come into play in different parts of the structure when it is subject to various loads. In many cases the structure is composed of bars connected together to form a rigid framework. The weights of the bars are usually insignificant compared to those of the forces they carry, and they are neglected in analyzing the distribution of forces over the bars. The bars are usually pin-jointed at their ends.

Each bar is in equilibrium under the action of forces at its two ends. Hence these two forces acting along the bar are equal and opposite. The bar itself exerts equal and opposite forces at its end joints. They may both be outward or both be inward. The force in the bar is called a *stress*.

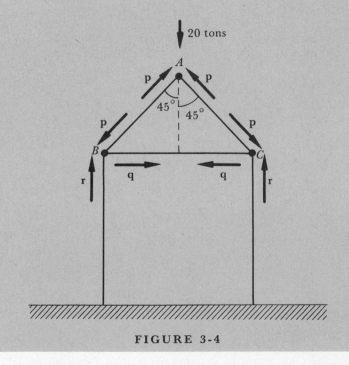

FIGURE 3-4

Consider the framework in Figure 3-4, which carries a load of 20 tons at its upper joint. Let us determine the stresses in the bars.

Consider the joint A. Intuitively, one can see that the forces exerted by the two bars at this joint must be upward in order to balance the 20 tons if the framework is going to remain in equilibrium. By the symmetry of the situation, the forces exerted by the bars are equal. Label these forces p. The forces at B and C are labelled p, q, and r. The student should be able to see that these bars either push or pull at the joints. To determine p, q, and r we analyze the situation at various joints. Each joint is in equilibrium under the forces acting on it. This is usually described mathematically by stating that the sum of the components of the forces at a joint in any two directions is zero. (The resultant vector at the joint is then zero.) We choose the directions as convenient perpendicular directions.

Let us examine the forces at joint A. Select directions $\uparrow$ and $\rightarrow$. We have that

$$\uparrow -20 + p \cos 45° + p \cos 45° = 0, \text{ giving } p = 10\sqrt{2}$$

$$\rightarrow p \cos 45° - p \cos 45° = 0, \text{ no information}$$

At joint B, using the same pair of directions,

$$\uparrow -p \cos 45° + r = 0, \text{ giving } \frac{-p}{\sqrt{2}} + r = 0$$

$$\rightarrow -p \cos 45° + q = 0, \text{ giving } \frac{-p}{\sqrt{2}} + q = 0$$

The analysis at joint C leads to the same equations as joint B.

Thus the problem reduces to solving the system of linear equations

$$p \qquad\qquad = 10\sqrt{2}$$

$$-\frac{p}{\sqrt{2}} + r \quad = 0$$

$$-\frac{p}{\sqrt{2}} \quad + q = 0$$

This system has the solution $p = 10\sqrt{2}, r = 10, q = 10$. (A negative number for any stress implies that the stress is actually in the direction opposite to the one assumed. Hence the correct initial choice of directions is not crucial to the final analysis.)

The actual stresses in the bars themselves are necessarily equal in magnitude and in a direction opposite to the forces exerted by the bars on the joints. The stresses in the bars are represented in Figure 3-5.

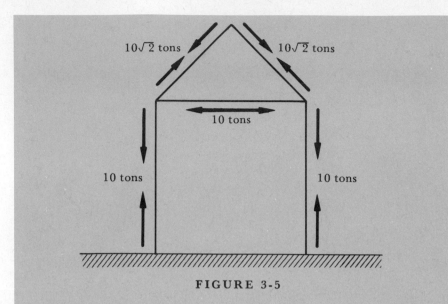

FIGURE 3-5

In practice, these structures can involve many bars and the mathematical analysis reduces to solving a system involving many equations in many variables. From the nature of the problem, we can expect the solution to be unique.

EXERCISES

In Exercises 1–6, determine the scalar and vector projections of **u** onto **v**. Illustrate these concepts geometrically when possible.

1. $\mathbf{u} = (1, 2)$, $\mathbf{v} = (2, 5)$.
2. $\mathbf{u} = (-1, 3)$, $\mathbf{v} = (2, 4)$.
3. $\mathbf{u} = (1, 2, 3)$, $\mathbf{v} = (1, 2, 0)$.
4. $\mathbf{u} = (2, 1, 4)$, $\mathbf{v} = (-1, -3, 2)$.
5. $\mathbf{u} = (1, -1, 0, 1)$, $\mathbf{v} = (2, 3, -2, 1)$.
6. $\mathbf{u} = (2, -1, 3, 1)$, $\mathbf{v} = (-1, 2, 1, 3)$.
7. Let $\mathbf{v} = (a_1, ..., a_n)$ be an arbitrary element of $\mathbf{R}^n$. Prove that the components of **v** onto the vectors $(1, 0, ..., 0)$, ..., $(0, ..., 0, 1)$ are $a_1, ..., a_n$, respectively. This illustrates the compatability of the meaning of the term *components* introduced in Section 1-1 (the scalars that make up the vector) and our use of the term in this section.
8. Determine a vector whose vector projection onto the vector $(1, 0)$ is $(1/\sqrt{2})(1, 0)$. Then give an arbitrary vector whose vector projection onto the vector $(1, 0)$ is $(1/\sqrt{2})(1, 0)$. Illustrate your answer geometrically.
9. Construct an orthonormal basis for the subspace of $\mathbf{R}^3$ defined by the vectors

   a) $(1, 0, 2), (-1, 0, 1)$
   b) $(1, -1, 1), (1, 2, -1)$
10. Construct the following bases for the subspaces of $\mathbf{R}^4$ defined by the vectors
    a) $(1, 2, 3, 4), (-1, 1, 0, 1)$—orthonormal basis
    b) $(1, 0, -1, 2), (1, 1, 2, 0), (1, 2, -3, -1)$—orthogonal basis
11. Construct a vector in $\mathbf{R}^4$ orthogonal to the vector $(1, 2, -1, 1)$. $\left(+/ +/ \ z \ / \ \right)$
12. Construct a vector in $\mathbf{R}^5$ orthogonal to the vector $(2, 0, 0, 1, 1)$. $(\not{1} 7 0 0 2)$
13. If $\mathbf{v}_1, \mathbf{v}_2$, and $\mathbf{v}_3$ are linearly independent in $\mathbf{R}^3$, $\mathbf{u}_1 = \mathbf{v}_1$, and

$$\mathbf{u}_2 = \mathbf{v}_2 - \frac{\mathbf{v}_2 \cdot \mathbf{u}_1}{\|\mathbf{u}_1\|} \frac{\mathbf{u}_1}{\|\mathbf{u}_1\|}$$

    prove that the vector $\mathbf{u}_3$ defined by

$$\mathbf{u}_3 = \mathbf{v}_3 - \frac{\mathbf{v}_3 \cdot \mathbf{u}_1}{\|\mathbf{u}_1\|} \frac{\mathbf{u}_1}{\|\mathbf{u}_1\|} - \frac{\mathbf{v}_3 \cdot \mathbf{u}_2}{\|\mathbf{u}_2\|} \frac{\mathbf{u}_2}{\|\mathbf{u}_2\|}$$

    is orthogonal to both $\mathbf{u}_1$ and $\mathbf{u}_2$.
14. Discuss the significance of the fact that $\mathbf{u}_2$ is actually equal to $\mathbf{v}_2$ in Example 3.
15. $(1, 2)$ represents a force acting on a body.   Find the component of the force
    a) in the direction $(1, 1)$
    b) in the **i** direction
    c) in the **j** direction
16. $(1, -1, 3)$ represents a force acting on a body.   Find the component of the force
    a) in the direction $(1, 1, 1)$
    b) in the direction $(-1, 1, -1)$
    c) in each of the directions **i**, **j**, and **k**
17. Determine the stresses in the bars of the structures in Figure 3-6.

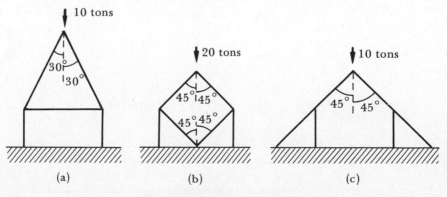

(a)   (b)   (c)

FIGURE 3-6

18. a) Determine the tensions T_1 and T_2 in the strings in Figure 3-7.

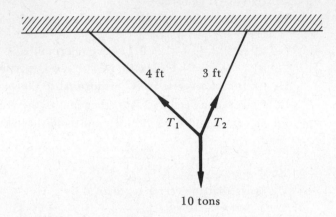

4 ft 3 ft

T_1 T_2

10 tons

FIGURE 3-7

b) Figure 3-8 is a lamp structure attached to a wall. Determine the stresses in the bar AB and in the wire BC as functions of W and θ. Note that the tension in BC is greater than W! As θ decreases, this tension increases. This type of device can be used to break wires.

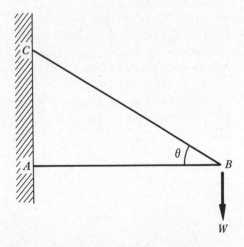

FIGURE 3-8

19. Write a computer program to determine scalar and vector projections. Test your program with the vectors of Example 1: The scalar and vector projections of $(1, 2, 3)$ onto $(4, -1, 3)$ are 2.1572776 and (1.6923076, -0.4230769, 1.2692307), respectively.

20. Write a computer program that uses the Gram-Schmidt procedure to construct an orthonormal set of vectors from two given vectors. Test your program for the vectors $(1, 2, -3, 4)$ and $(1, 2, 4, -3)$. An orthonormal set, constructed by normalizing the vector $(1, 2, -3, 4)$ to get the first unit vector, is

$$(0.1825742, 0.3651484, -0.547726, 0.7302967),$$

$$(0.3853373, 0.7706746, 0.4954337, -0.1100964)$$

3–7.* FUNCTION SPACES

There are vector spaces other than the $\mathbf{R}^n$ vector spaces and their subspaces that we have already considered. All such spaces have similar mathematical structures. Two operations, addition and scalar multiplication are defined, and the space is closed under these operations. Furthermore, these operations have to satisfy certain axioms. Rather than enter into a discussion of these axioms, we will focus our attention on another important type of vector space, a *function space*.† The reader should aim at understanding the structure of a function space by comparing the analysis of the space with that of $\mathbf{R}^n$.

Example 1

Let V be the set of real-valued functions defined on the interval $[-1, 1]$. We visualize V as in Figure 3-9. Every element of V maps every number between -1 and 1 into the real line. Addition and scalar multiplication have been defined on $\mathbf{R}^n$; we now define such operations on V.

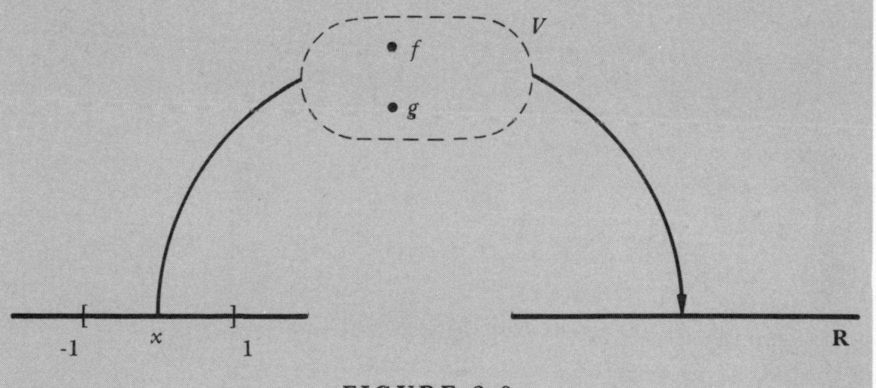

FIGURE 3-9

---

† To develop an understanding of the axiomatic definition of a real vector space, complete Exercise 7 at the end of this section. For further reading consult the author's *A Course in Linear Algebra*, Gordon and Breach, Science Publishers, 1972.

Let f and g be two elements of V. We define their sum $f+g$ to be a function such that

$$(f+g)(x) = f(x) + g(x)$$

This defines $f+g$ as a function on $[-1,1]$, since it tells us how it operates on every element of $[-1,1]$. We see that $f+g$ is a function of $[-1,1]$ into **R** and is thus an element of V.

If c is an arbitrary scalar, then the scalar multiple of f is defined as

$$cf(x) = c[f(x)]$$

cf is a function of $[-1,1]$ into **R** and, therefore, an element of V.

Under these operations V is a vector space, and the functions are called *vectors* in this context. There will be many such function vector spaces. For example, the set of real-valued functions on the interval $[-\pi,\pi]$ with the operations of addition and scalar multiplication defined as above would be another such space.

Example 2

Let V be the vector space of real-valued functions defined on the interval $[-1,1]$. Then $f(x) = x$ and $g(x) = x^2$, where $-1 \leq x \leq 1$, are elements of V. Let us analyze the functions $f+g$ and $3f$ to develop an understanding of the concepts of addition and scalar multiplication.

Using our definitions of addition and scalar multiplication, $f+g$ is defined by

$$(f+g)(x) = f(x) + g(x) = x^2 + x$$

and $3f$ by

$$(3f)(x) = 3[f(x)] = 3x$$

Thus

$$(f+g)(x) = x^2 + x \quad \text{and} \quad (3f)(x) = 3x$$

To find the value of $f+g$ at a point x, we add the value of f at x to the value of g at x (Figure 3-10). This operation is called *pointwise addition*. To find cf at a point x, we multiply the value of f at x by c (Figure 3-11). This operation is called *pointwise scalar multiplication*.

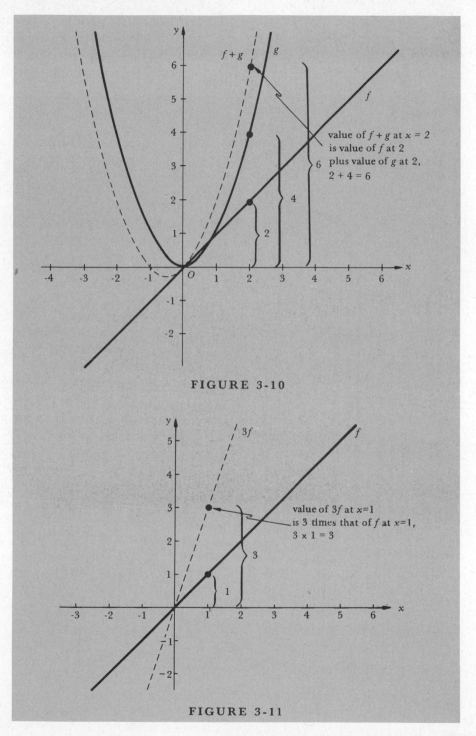

FIGURE 3-10

FIGURE 3-11

The concepts of a spanning set, linear dependence and independence of vectors, basis, and dimension are defined for certain function spaces in the same way as they were for $\mathbf{R}^n$. The following example illustrates these concepts for a function space.

Example 3

Let V be the vector space generated by the functions x, x^2 defined on the interval $[-1, 1]$.

We shall show that V is a two-dimensional vector space by proving that these two functions, interpreted as vectors, are linearly independent.

Consider the identity

$$ax + bx^2 = 0$$

The 0 here is the zero function on the interval $[-1, 1]$; this function takes every element of this interval into the number 0. It is the zero vector of the function space.

This identity must hold for all x in the interval $[-1, 1]$. $x = -1$ leads to $-a + b = 0$ and $x = 1$ leads to $a + b = 0$. Thus the identity can only hold for all x in the interval $[-1, 1]$ if and only if $a = b = 0$. The vectors are thus linearly independent.

The vectors span V since they generate the space. The dimension of V is therefore 2. The vectors x and x^2 are a basis for V. An arbitrary element $\mathbf{v}$ of V can be represented as a linear combination of these vectors $\mathbf{v} = px + qx^2$, where p and q are real numbers.

Inner products and norms can be defined on function spaces. The most useful inner products and norms on such spaces are defined in terms of integrals. The following example illustrates these concepts on the function space of the above example.

Example 4

Let V be the two-dimensional vector space generated by the functions x and x^2 defined on the interval $[-1, 1]$.

We define an inner product on V. Let $f(x)$ and $g(x)$ be two arbitrary vectors. The inner product of $f(x)$ and $g(x)$ on this space is usually denoted $< f(x), g(x) >$ rather than $f(x) \cdot g(x)$.

$$< f(x), g(x) > = \int_{-1}^{1} f(x) g(x) \, dx$$

Two vectors are said to be orthogonal if their inner product is zero. Thus x and x^2 are orthogonal vectors in this space for

$$< x, x^2 > = \int_{-1}^{1} xx^2 \, dx = \int_{-1}^{1} x^3 \, dx = \left. \frac{x^4}{4} \right|_{1}^{-1} = 0.$$

x and x^2 form an orthogonal basis for this space.

The inner product induces a norm on this space, as in the case of the vector space $\mathbf{R}^n$:

$$\|f(x)\| = \sqrt{\langle f(x), f(x) \rangle} = \sqrt{\int_{-1}^{1} [f(x)]^2 \, dx}$$

We compute the norms of the vectors x and x^2:

$$\|x\| = \sqrt{\int_{-1}^{1} x^2 \, dx} = \sqrt{x^3/3 |_{-1}^{1}} = \sqrt{2/3}$$

$$\|x^2\| = \sqrt{\int_{-1}^{1} x^4 \, dx} = \sqrt{x^5/5 |_{-1}^{1}} = \sqrt{2/5}$$

Thus the vector x is of magnitude $\sqrt{2/3}$ and x^2 is of magnitude $\sqrt{2/5}$.

We can normalize these vectors to get the vectors $\sqrt{3/2}\, x$ and $\sqrt{5/2}\, x^2$. These vectors form an orthonormal basis for this space.

We have already mentioned one major development of twentieth-century physics, relativity theory. The other major development was quantum mechanics, the mathematical theory that attempts to describe the behavior of atomic-sized systems. This theory was initially formulated by such people as Erwin Schrödinger and Werner Heisenberg in the early 1920's. Schrödinger's work revolved about a certain differential equation, whereas Heisenberg's was phrased in terms of matrices. John von Neumann gave a definitive mathematical treatment of the subject. The appropriate mathematical setting turned out to be a function vector space with an inner product and norm defined in terms of integrals (as in the last example). The space is called a *Hilbert space*. Certain functions represent *states*, and the inner product is used to represent values of physical quantities and probabilities of finding systems in certain states.

For an introduction to quantum mechanics the reader is referred to *Introduction to Modern Physics*, by F. K. Richtmyer, E. H. Lauritsen, and T. Lauritsen, McGraw-Hill Book Company, 1969.

EXERCISES

1. Let V be the real vector space generated by the functions $1, x, x^2$ defined on the interval $[0, 1]$. Prove that these vectors are linearly independent, and hence that the vector space is three-dimensional. What does an arbitrary vector of this space look like? Give two distinct sets of bases for this space.

2. Prove that e^x and e^{2x} are linearly independent vectors in the vector space of real-valued functions on the interval $[0, 1]$. Define a subspace of the vector space generated by e^x and e^{2x}. Give an alternative basis for this space.

3. Prove that two functions f and g in the same vector space are linearly dependent if and only if there exists a constant c such that one of the functions, say f, can be expressed in terms of the other $f = cg$. Knowledge of this result makes it apparent whether two functions are linearly dependent or not. For example, all functions linearly dependent on e^x must be of the form ce^x.

4. Let V be the vector space of real-valued continuous functions on the interval $[-\pi/2, \pi/2]$ with the inner product defined by $\langle f,g \rangle = \int_{-\pi/2}^{\pi/2} f(x)g(x)\,dx$. Prove that the vectors $\sin x$ and $\cos x$ are orthogonal.

5. Let V be the vector space of real-valued continuous functions on the interval $[-1, 1]$ with the inner product defined by $\langle f,g \rangle = \int_{-1}^{1} f(x)g(x)\,dx$.

    a) Find the induced norm on this space and determine the magnitudes of the vectors $x, x+1, x^2$, and x^4.

    b) Prove that the vectors $1, x, x^2 - \frac{1}{3}$, and $x^3 - \frac{3}{5}x$ are an orthogonal set of vectors (any pair is orthogonal).

6. a) Consider the set of all polynomials of degrees less than or equal to three. Prove that this set forms a vector space. Give a basis for this space.

    b) Consider the set of all polynomials of degrees less than or equal to n, where n is an arbitrary nonnegative integer. Prove that this set forms a vector space and give a basis for this space.

7. The axiomatic definition of a real vector space is as follows: A set V is a real vector space if it has two operations, addition and scalar multiplication, that satisfy the following conditions.

    I. If $\mathbf{u}$ and $\mathbf{v}$ are arbitrary elements of V, then $\mathbf{u}+\mathbf{v}$ exists and is an element of V.

    II. If c is an arbitrary real number, then $c\mathbf{u}$ exists and is an element of V.

    III. $(\mathbf{u}+\mathbf{v})+\mathbf{w} = \mathbf{u}+(\mathbf{v}+\mathbf{w})$ for $\mathbf{u}$, $\mathbf{v}$, and $\mathbf{w}$, arbitrary elements of V.

    IV. There is an element of V called the zero vector, $\mathbf{0}$, such that

    $$\mathbf{u} + \mathbf{0} = \mathbf{0} + \mathbf{u} = \mathbf{0}$$

    V. For every element $\mathbf{u}$ of V, there exists another element denoted $-\mathbf{u}$ such that

    $$\mathbf{u} + (-\mathbf{u}) = \mathbf{0}$$

    VI. $\mathbf{u}+\mathbf{v} = \mathbf{v}+\mathbf{u}$ for all elements $\mathbf{u}$ and $\mathbf{v}$ of V.

    VII. $c(\mathbf{u}+\mathbf{v}) = c\mathbf{u}+c\mathbf{v}$ for all real numbers c.

    VIII. $(c+d)\mathbf{u} = c\mathbf{u}+d\mathbf{u}$ for all real numbers c and d.

    IX. $(cd)\mathbf{u} = c(d\mathbf{u})$.

    X. $1\mathbf{u} = \mathbf{u}$.

    a) Prove that the set $\mathbf{R}^n$ with the operations of addition and scalar multiplication as we have defined them satisfies the above conditions.

    b) Prove that the set of all real-valued continuous functions defined on the interval $[0, 1]$ with operations of pointwise addition and pointwise scalar multiplication forms a vector space according to the above definition.

c) Consider the set of all real-valued functions (not necessarily continuous) on the interval $[0, 1]$. Is this set a vector space under operations of pointwise addition and pointwise multiplication? If so, is the set of continuous real-valued functions of (b) a subspace of this space?

d) Is the set of all discontinuous real-valued functions on the interval $[0, 1]$ a vector space under the same two operations?

e) Prove that the set of $m \times n$ matrices is a vector space. Give a basis for the vector space of 2×3 matrices.

3–8.*   THE INNER PRODUCT OF SPECIAL RELATIVITY

In Section 3-5 we defined an inner product on $\mathbf{R}^n$. This assigned to two elements, $(x_1, \ldots, x_n)$ and $(y_1, \ldots, y_n)$, the number $x_1 y_1 + \cdots + x_n y_n$. We found that the inner product could be used to give magnitudes to the elements of $\mathbf{R}^n$ and also to define angles. These concepts fitted in with our usual interpretations of magnitudes of vectors and angles between vectors in $\mathbf{R}^2$ and $\mathbf{R}^3$, where these sets represent two- and three-dimensional space. We saw that when a force is represented mathematically by a vector in $\mathbf{R}^2$, the magnitude of that vector derived from the dot product gives the magnitude of the force.

Mathematicians develop models to describe various situations. The aim of the mathematician is to develop the mathematical model that most accurately describes the situation. This model can then be used to further our understanding of the situation. The dot product is not the only inner product that can be defined on $\mathbf{R}^n$; in this section we develop an alternative inner product that is used on $\mathbf{R}^4$ when $\mathbf{R}^4$ represents spacetime. There are many other inner products. In general, all inner products follow certain rules to associate a number with every pair of vectors. In more advanced linear algebra we axiomize the concept of inner product; that is, we draw up a list of conditions that all inner products must satisfy.

Special relativity was developed by Albert Einstein in an attempt to describe the physical world that we live in. At the time Newtonian mechanics was the theory used to describe the motions of bodies under forces. However, experiments had led scientists to believe that the large-scale motions of bodies, such as planetary motions, were not accurately described by Newtonian mechanics. One of Einstein's main contributions to science was the development of more accurate mathematical models. First he developed special relativity, which did not incorporate gravitation; later on he incorporated gravitation in his theory of general relativity. Here we introduce the student to the mathematical model of special relativity; we derive one of its predictions by describing the prediction and then showing how it arises out of the special theory.

The nearest star to earth other than the sun is Alpha Centauri; it is about four light-years away. (A light-year is the distance light travels in one year, 5.88×10^{12} miles.) Consider a pair of twins who are separated immediately after birth. Twin 1 remains on earth and Twin 2 is flown off to Alpha Centauri in a rocket at 0.8 the speed of light. On arriving at Alpha Centauri, he immediately returns to earth at the same high speed. On reaching earth, the mathematical theory of special relativity predicts that he will find that his twin who remained on earth is ten years old (in every sense of the word), while he himself will be only six years old. (These times would vary according to the speed of Twin 2.)

There is experimental verification of this phenomenon. Physicists have found that certain radioactive particles moving in the atmosphere decay more slowly than ones at rest on earth. The effect is not likely to be realized on this scale by humans, since the energies required to produce such high speeds in a macroscopic body are prohibitive. This is probably fortunate from the sociological viewpoint. Theoretically a man could go off on such a trip, return to earth, and be of an age to marry his great granddaughter! This phenomenon was experienced to a much lesser degree by astronauts who went to the moon—on arrival back on earth they were fractionally younger than they would have been if they had remained on earth.

Let us now see how this phenomenon arises out of Einstein's model. The model is one of space-time; hence it involves four coordinates —three space coordinates, x_1, x_2, x_3, and a time coordinate, x_4. We use $\mathbf{R}^4$ to represent space-time; we call each element of $\mathbf{R}^4$ an *event*. Each event has a location in space given by x_1, x_2, and x_3 and occurs at a certain time x_4. Further, there is additional mathematical structure on $\mathbf{R}^4$, an inner product. For two arbitrary elements (x_1, x_2, x_3, x_4) and (y_1, y_2, y_3, y_4), it is defined as follows:

$$(x_1, x_2, x_3, x_4) \cdot (y_1, y_2, y_3, y_4) = -x_1 y_1 - x_2 y_2 - x_3 y_3 + x_4 y_4$$

Using this inner product, we can find the norm† of a vector (x_1, x_2, x_3, x_4):

$$\|(x_1, x_2, x_3, x_4)\| = \sqrt{|(x_1, x_2, x_3, x_4) \cdot (x_1, x_2, x_3, x_4)|}$$
$$= \sqrt{|-x_1{}^2 - x_2{}^2 - x_3{}^2 + x_4{}^2|}$$

---

† It is necessary to introduce absolute values in the norm of this model. The norm is actually a *pseudo norm*.

214

$\mathbf{R}^4$ with these structures is called Minkowski space. The sweeping innovations in special relativity were the introduction of this inner product and norm on a four-dimensional space, implying that space and time are not completely independent, as was assumed in the earlier Newtonian model. We shall see now how the physical interpretation of the norm leads to prediction of an age difference between the twins.

We draw a space-time diagram. For convenience, assume that Alpha Centauri lies in the direction of the x_1 axis from earth. The twin on earth advances in time, x_4, while the twin in the rocket advances in time and also moves in the direction of x_1. The space-time diagram is shown in Figure 3-12.

The path of Twin 1 is PQ; he is advancing only in time, x_4. The path of Twin 2 to Alpha Centauri is PR, advancing in time and in the direction of increasing x_1. His return path to earth is RQ; he rejoins his earth twin at Q. There is no motion in either the x_2 or the x_3 coordinate, hence we supress these dimensions in the diagram. We can use the inner product of special relativity to find certain norms for vectors; that is, we can calculate lengths of various vectors such as PR. The theory gives a physical interpretation to such lengths. They are the actual times recorded by observers moving along these paths. For example, the length of PQ is the time recorded by Twin 1 in traveling between P and Q. The length of PR is the time recorded by Twin 2 in traveling between P and R, that is, from earth to Alpha Centauri; $PR + RQ$ is his age at Q.

Let us first look at Twin 1, who stays on earth. The rocket travels

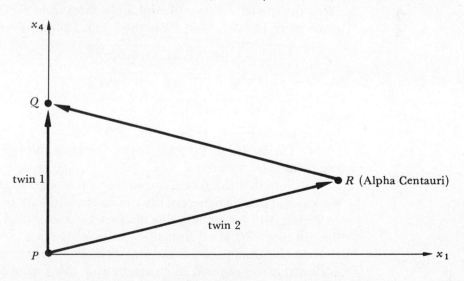

FIGURE 3-12

with 0.8 the speed of light relative to earth and covers a distance of 8 light years (4 there and 4 back). Thus the round trip, from the point of view of earth takes 8/.8 years, that is 10 years. (Time = distance/velocity.) The age of Twin 1 at Q is 10 years.

   We now examine the situation for Twin 2. Let P be the origin in $\mathbf{R}^4$, $(0, 0, 0, 0)$ (Figure 3-13). Q is the point $(0, 0, 0, 10)$ and S is $(0, 0, 0, 5)$.

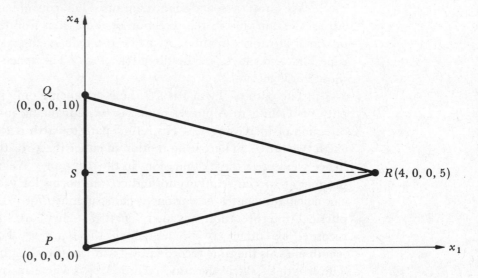

FIGURE 3-13

SR is the spatial distance of Alpha Centauri from earth—four. Thus R is the point $(4, 0, 0, 5)$ and PR is the vector $(4, 0, 0, 5)$.

$$\|PR\| = \sqrt{|(4, 0, 0, 5) \cdot (4, 0, 0, 5)|}$$
$$\sqrt{|-(4)^2 - 0 - 0 + (5)^2|}$$
$$\sqrt{|-16 + 25|} = 3$$

By the symmetry of the situation, $\|RQ\| = 3$.

   The length of $PR + RQ$ is six; therefore the age of Twin 2 when the twins meet is six!

   Note that this model introduces a new kind of geometry in which the straight line is not necessarily the shortest distance between two points. In Figure 3-13 the straight line distance between P and Q is ten, whereas the distance PRQ is six, a smaller distance. In fact, it turns out that the straight line distance PQ is the longest distance between P and Q! Thus not only is the physical interpretation of this model fascinating, but it opens up a new trend in geometrical thinking.

In the general theory of relativity, gravity is taken into account and is represented mathematically by an inner product. The space becomes curved in nature; curves lead to extreme distances between points instead of straight lines. Those who are interested in looking further into special relativity will find an interesting, very readable article by Alfred Schild entitled "The Clock Paradox in Relativity Theory" in the *American Mathematical Monthly*, Vol. **66**, No. 1, January, 1959. For those whose geometrical curiosity has been aroused, this branch of geometry (more general than Euclidean geometry) is called *Riemannian geometry*. Erwin Schrödinger relates Riemannian geometry to space-time in *Space-Time Structure*, Cambridge University Press, 1950. For a more mathematical development of the subject, look at a book such as J. J. Stoker's *Differential Geometry*, John Wiley & Sons, 1969. A knowledge of calculus is required to read both of these books. Two books that make very enjoyable light reading on geometries and dimensions are *Flatland*, by Edwin Abbott, Barnes & Noble, 1963, a romance of many dimensions; and *Sphereland*, by Dionys Burger, Thomas Y. Crowell, 1965, a fantasy about curved spaces and an expanding universe.

EXERCISES

1. The star Sirius is eight light-years from earth. Sirius is the nearest star other than the sun that is visible to the naked eye from most parts of the United States. It is the brightest appearing of all stars. Light reaches us from the sun in eight minutes and from Sirius in eight years. Suppose a rocket ship leaves for Sirius and returns to earth 20 years later. What is the duration of the voyage for a person on the space ship? What is the speed of the ship?

2. A rocket ship makes a round-trip flight to the bright star Capella, which is 45 light-years away from earth. If the time lapse on earth is 120 years, what is the length of the voyage from the traveler's viewpoint?

3. The star cluster Pleiades in the constellation Taurus is 410 light-years from earth. A traveller to the cluster ages 40 years on a round trip from earth. By the time he returns, how many centuries will have passed on earth since he started the voyage?

4. Write a program to determine inner products of vectors using the inner product of special relativity. Use your program to determine the inner products of the following vectors.
   a) $(1, 2, -1, 3)$ and $(2, 1, 0, 1)$
   b) $(2, 0, 1, 4)$ and $(0, 1, 3, -2)$

5. Write a program to determine norms using the norm of special relativity. Use your program to determine the norms of the following vectors.

a) $(0, 1, 2, 4)$

b) $(-1, 2, 3, 7)$

6. Write a program to solve Exercise 3 above. Use your program to determine the number of years that have passed on earth during the following space voyages.

   a) to the star cluster Praesepe in the constellation Cancer, 515 light-years from earth. Duration for the traveler is 50 years.

   b) to the cluster Hyades in Taurus, 130 light-years from earth. Duration for the traveler is 20 years.

7. Prove that the inner product of special relativity satisfies the following axoims (as does that of Section 3-5):

   I. $\mathbf{u}_1 \cdot \mathbf{u}_2 = \mathbf{u}_2 \cdot \mathbf{u}_1$ for any pair of vectors $\mathbf{u}_1$ and $\mathbf{u}_2$. This property is called the *symmetry property of the inner product.*

   II. $\mathbf{u}_1 \cdot (\mathbf{u}_2 + \mathbf{u}_3) = \mathbf{u}_1 \cdot \mathbf{u}_2 + \mathbf{u}_1 \cdot \mathbf{u}_3$ for any triple of vectors $\mathbf{u}_1$, $\mathbf{u}_2$, and $\mathbf{u}_3$.

   III. If c is a scalar, then $c\mathbf{u}_1 \cdot \mathbf{u}_2 = c(\mathbf{u}_1 \cdot \mathbf{u}_2) = \mathbf{u}_1 \cdot (c\mathbf{u}_2)$ for any pair of vectors $\mathbf{u}_1$ and $\mathbf{u}_2$.

   In general, any rule that associates with every pair of vectors $\mathbf{u}_1$ and $\mathbf{u}_2$ a scalar $\mathbf{u}_1 \cdot \mathbf{u}_2$ that satisfies these axioms is called an *inner product.*
   A *norm* can then be generated from the inner product: $\|\mathbf{u}\| = \sqrt{|\mathbf{u} \cdot \mathbf{u}|}$.

8. For the model in this section we needed a norm other than the usual norm on $\mathbf{R}^4$ to describe the physical situation. In Example 5 of Section 3-4 a model is constructed for the analysis of colors. All colors can be represented by a subset of $\mathbf{R}^3$, and vector addition and scalar multiplication have interpretation. (s_1, s_2, s_3) corresponds to a color. However, the usual norm on $\mathbf{R}^3$ does not correspond to the intensity of brightness of the color. Define a norm on this subset of $\mathbf{R}^3$ that would give a measure of brightness.

4

Mappings

In previous sections we have discussed vector spaces and have seen applications of the theory developed. It is often of interest to relate elements of a vector space to other elements within the same space and also to elements within other spaces. In this chapter we develop the methods for doing this. The theory developed also gives us greater insight into the properties of systems of linear equations.

4–1. MAPPINGS DEFINED BY SQUARE MATRICES

We have seen how the elements of $\mathbf{R}^2$ can be used to represent the locations of points in a plane. This is what we are actually doing when we introduce a coordinate system. In this section we shall see, by means of examples, how a 2×2 matrix can be used to move points around in the plane.

Example 1

> Let the coordinate system in Figure 4-1 define the relative location of points in a plane. We shall use the elements of $\mathbf{R}^2$ in the form of columns rather than rows, as this permits us to perform the desired matrix multiplications.
>
> Consider the matrix $\begin{pmatrix} 0 & -1 \\ 1 & 0 \end{pmatrix}$. We can multiply any element of $\mathbf{R}^2$ by this matrix to get another element of $\mathbf{R}^2$. For example, $\begin{pmatrix} 0 & -1 \\ 1 & 0 \end{pmatrix}\begin{pmatrix} 2 \\ 1 \end{pmatrix} = \begin{pmatrix} -1 \\ 2 \end{pmatrix}$. Interpreting $\begin{pmatrix} 2 \\ 1 \end{pmatrix}$ as a point in the plane, this matrix multiplication "takes it into" the point $\begin{pmatrix} -1 \\ 2 \end{pmatrix}$. We say that

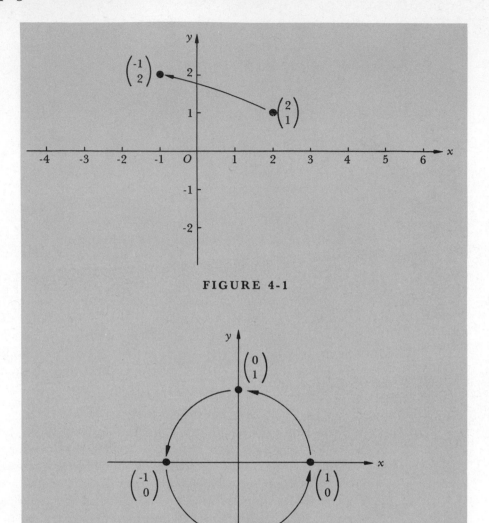

FIGURE 4-1

FIGURE 4-2

the matrix *maps* the point $\begin{pmatrix} 2 \\ 1 \end{pmatrix}$ into the point $\begin{pmatrix} -1 \\ 2 \end{pmatrix}$. In this manner the matrix can be used to map a point in the plane into another point.   Let us consider the points in Figure 4-2.   The point $\begin{pmatrix} 1 \\ 0 \end{pmatrix}$ is mapped into the point $\begin{pmatrix} 0 \\ 1 \end{pmatrix}$, since $\begin{pmatrix} 0 & -1 \\ 1 & 0 \end{pmatrix}\begin{pmatrix} 1 \\ 0 \end{pmatrix} = \begin{pmatrix} 0 \\ 1 \end{pmatrix}$.   We write $\begin{pmatrix} 1 \\ 0 \end{pmatrix} \mapsto \begin{pmatrix} 0 \\ 1 \end{pmatrix}$.

Further, $\begin{pmatrix} 0 \\ 1 \end{pmatrix} \mapsto \begin{pmatrix} -1 \\ 0 \end{pmatrix}$, $\begin{pmatrix} -1 \\ 0 \end{pmatrix} \mapsto \begin{pmatrix} 0 \\ -1 \end{pmatrix}$, and $\begin{pmatrix} 0 \\ -1 \end{pmatrix} \mapsto \begin{pmatrix} 1 \\ 0 \end{pmatrix}$.

(Verify these.)

At this stage, you may conjecture that the effect of the matrix multiplication is to rotate the points that make up the plane through an angle of $\pi/2$ in a counterclockwise direction. We shall prove that this is the case in the following example.

Example 2

Determine the matrix that can be used to rotate the points that make up a plane through an angle θ about the origin in a counterclockwise direction.

Such a rotation about O would take the point A into the point B (Figure 4-3). The distance OA is equal to OB for this rotation about O; let this distance be r. Let the angle AOC be α. Then we know that

$$x' = OC = r\cos(\alpha+\theta) = r\cos\alpha\cos\theta - r\sin\alpha\sin\theta$$

$$= x\cos\theta - y\sin\theta$$

$$y' = BC = r\sin(\alpha+\theta) = r\sin\alpha\cos\theta + r\cos\alpha\sin\theta$$

$$= y\cos\theta + x\sin\theta$$

$$= x\sin\theta + y\cos\theta$$

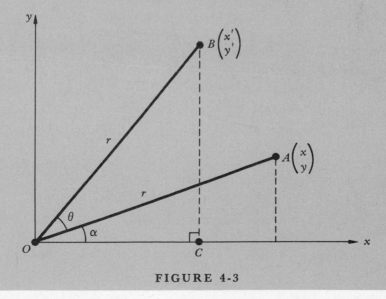

FIGURE 4-3

counter clockwise

We can write these two equations in matrix form.

$$\begin{pmatrix} x' \\ y' \end{pmatrix} = \begin{pmatrix} \cos\theta & -\sin\theta \\ \sin\theta & \cos\theta \end{pmatrix} \begin{pmatrix} x \\ y \end{pmatrix}$$

Thus the matrix that maps A into B is $\begin{pmatrix} \cos\theta & -\sin\theta \\ \sin\theta & \cos\theta \end{pmatrix}$. Multiplying any point in the plane by this matrix will rotate it through an angle θ about the origin. Multiplying every point in the plane by this matrix rotates the plane through an angle θ.

To rotate the points through $\pi/2$, we let $\theta = \pi/2$. The matrix is $\begin{pmatrix} 0 & -1 \\ 1 & 0 \end{pmatrix}$, verifying our conjecture from the previous example.

DEFINITION 4-1   A *mapping* of a set V into a set U is a rule that associates with each element of V a single element of U.

We have seen how the matrix $\begin{pmatrix} \cos\theta & -\sin\theta \\ \sin\theta & \cos\theta \end{pmatrix}$ defines a mapping of $\mathbf{R}^2$ into $\mathbf{R}^2$. We now continue with a discussion of the use of matrices as mappings.

Example 3

Determine the geometrical effect of multiplying an element of $\mathbf{R}^2$ by the matrix $\begin{pmatrix} 3 & 0 \\ 0 & 3 \end{pmatrix}$.

Let us examine the effect of multiplying $\begin{pmatrix} 1 \\ 2 \end{pmatrix}$ by the matrix (Figure 4-4). We find that $\begin{pmatrix} 3 & 0 \\ 0 & 3 \end{pmatrix}\begin{pmatrix} 1 \\ 2 \end{pmatrix} = \begin{pmatrix} 3 \\ 6 \end{pmatrix}$. The element $\begin{pmatrix} 3 \\ 6 \end{pmatrix}$ represents a point B in the plane that is in the same direction as A from the origin but at a distance three times the distance of OA from the origin. $\begin{pmatrix} 1 \\ 2 \end{pmatrix} \mapsto \begin{pmatrix} 3 \\ 6 \end{pmatrix}$. For an arbitrary point $\begin{pmatrix} x \\ y \end{pmatrix}$, $\begin{pmatrix} 3 & 0 \\ 0 & 3 \end{pmatrix}\begin{pmatrix} x \\ y \end{pmatrix} = \begin{pmatrix} 3x \\ 3y \end{pmatrix} = 3\begin{pmatrix} x \\ y \end{pmatrix}$. This point is in the same direction from the origin as $\begin{pmatrix} x \\ y \end{pmatrix}$, but it is three times as far away. Hence the effect of this matrix is to move points out from the origin to a location three times their original distance from the origin (Figure 4-5). The matrix can be thought of as expanding the plane. Such a mapping is called a *dilatation*.

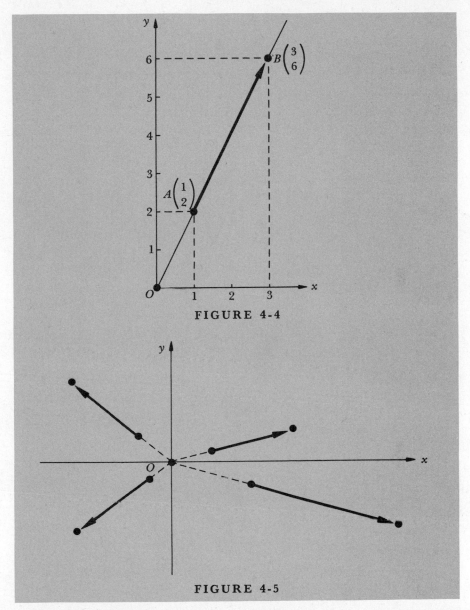

FIGURE 4-4

FIGURE 4-5

Let us now turn to another example to analyze the geometrical effect of multiplying an element of $\mathbf{R^2}$ by a general matrix.

Example 4   Determine the geometrical effect of multiplying an element of $\mathbf{R^2}$ by the matrix $\begin{pmatrix} 4 & 2 \\ 2 & 3 \end{pmatrix}$. Consider the vertices of a square, points $A\begin{pmatrix} 1 \\ 0 \end{pmatrix}$,

223

$B\begin{pmatrix} 1 \\ 1 \end{pmatrix}$, $C\begin{pmatrix} 0 \\ 1 \end{pmatrix}$, and $O\begin{pmatrix} 0 \\ 0 \end{pmatrix}$.

We have that

$$A, \begin{pmatrix} 1 \\ 0 \end{pmatrix} \mapsto \begin{pmatrix} 4 & 2 \\ 2 & 3 \end{pmatrix} \begin{pmatrix} 1 \\ 0 \end{pmatrix} = \begin{pmatrix} 4 \\ 2 \end{pmatrix}, P$$

$$B, \begin{pmatrix} 1 \\ 1 \end{pmatrix} \mapsto \begin{pmatrix} 4 & 2 \\ 2 & 3 \end{pmatrix} \begin{pmatrix} 1 \\ 1 \end{pmatrix} = \begin{pmatrix} 6 \\ 5 \end{pmatrix}, Q$$

$$C, \begin{pmatrix} 0 \\ 1 \end{pmatrix} \mapsto \begin{pmatrix} 4 & 2 \\ 2 & 3 \end{pmatrix} \begin{pmatrix} 0 \\ 1 \end{pmatrix} = \begin{pmatrix} 2 \\ 3 \end{pmatrix}, R$$

$$O, \begin{pmatrix} 0 \\ 0 \end{pmatrix} \mapsto \begin{pmatrix} 4 & 2 \\ 2 & 3 \end{pmatrix} \begin{pmatrix} 0 \\ 0 \end{pmatrix} = \begin{pmatrix} 0 \\ 0 \end{pmatrix}, O$$

This is represented geometrically in Figure 4-6.

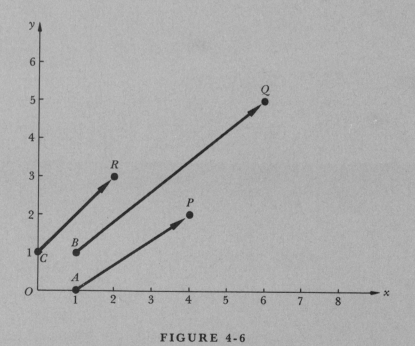

FIGURE 4-6

We now demonstrate that this mapping does in fact take the line OA into OP, $AB \to PQ$, $BC \to QR$, and $OC \to OR$, deforming the square $OABC$ into the figure $OPQR$.

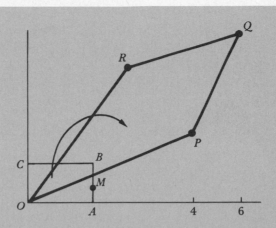

We show that AB is mapped into PQ; the proofs for the other sides are similar. A general point on the line AB is M, $\begin{pmatrix} 1 \\ y \end{pmatrix}$ with $0 \le y \le 1$. Let $\begin{pmatrix} 1 \\ y \end{pmatrix} \mapsto \begin{pmatrix} x' \\ y' \end{pmatrix}$, N. Then

$$\begin{pmatrix} x' \\ y' \end{pmatrix} = \begin{pmatrix} 4 & 2 \\ 2 & 3 \end{pmatrix} \begin{pmatrix} 1 \\ y \end{pmatrix} = \begin{pmatrix} 4+2y \\ 2+3y \end{pmatrix}.$$

Thus $x' = 4 + 2y$ and $y' = 2 + 3y$ with $0 \le y \le 1$. We get, on eliminating y between these identities, $3x' - 2y' = 8$ with $4 \le x' \le 6$. Thus N lies on the line $3x - 2y = 8$ with values of x, $4 \le x \le 6$. When $x = 4$, $y = 2$, giving the point P; when $x = 6$, $y = 5$, the point Q. Points on the line $3x - 2y = 8$ with values of x between 4 and 6 will lie on PQ between P and Q. Thus AB is mapped onto PQ.

The general geometrical effect of multiplying points in a plane by a 2×2 matrix is to map points into points such that straight lines go into straight lines or points, while the origin remains fixed (Exercise 9 p. 231). Such a mapping can be illustrated by considering a geometrical figure (the square $ABCO$ in this case) and the figure into which it is transformed (called its image).

Mappings are of practical as well as of theoretical value. The square could represent a physical body deformed into the shape $PQRO$. Solid bodies can be represented geometrically and deformations analyzed in this manner. The fields of science that investigate such problems are *elasticity* and *plasticity*. When loads are applied to bodies, changes in shape, called *deformations*, occur. If the body returns to its original shape when the loads are removed, the property of elasticity has been displayed. In many cases deformations can be represented by matrix mappings.

Here we have seen how 2×2 matrices can be used to define mappings of $\mathbf{R}^2$ into itself. Similarly, any 3×3 matrix defines a mapping of $\mathbf{R}^3$ into itself and any $n \times n$ matrix a mapping of $\mathbf{R}^n$ into itself.

Let A be an $n \times n$ matrix and $\mathbf{x}$ an element of $\mathbf{R}^n$. $A\mathbf{x}$ will be an element of R^n. We say that A maps $\mathbf{x}$ into $A\mathbf{x}$ and indicate this

$$A: \quad \mathbf{x} \mapsto A\mathbf{x}$$

$A\mathbf{x}$ is called the *image* of $\mathbf{x}$ under the mapping.

Example 5

Interpret the matrix $A = \begin{pmatrix} 1 & 2 & 0 & -1 \\ 3 & 0 & 3 & 6 \\ 2 & 4 & 2 & 1 \\ 1 & 1 & -1 & 2 \end{pmatrix}$ as a mapping

of $\mathbf{R}^4$ into itself. Determine the images of the following two vectors $\mathbf{x}$ and $\mathbf{y}$ under this mapping.

$$\mathbf{x} = \begin{pmatrix} 1 \\ 2 \\ -1 \\ 0 \end{pmatrix} \quad \mathbf{y} = \begin{pmatrix} 3 \\ -1 \\ 2 \\ 1 \end{pmatrix}$$

We have that

$$A: \begin{pmatrix} 1 \\ 2 \\ -1 \\ 0 \end{pmatrix} \mapsto \begin{pmatrix} 1 & 2 & 0 & -1 \\ 3 & 0 & 3 & 6 \\ 2 & 4 & 2 & 1 \\ 1 & 1 & -1 & 2 \end{pmatrix} \begin{pmatrix} 1 \\ 2 \\ -1 \\ 0 \end{pmatrix}$$

which equals $\begin{pmatrix} 5 \\ 0 \\ 8 \\ 4 \end{pmatrix}$

and

$$A: \begin{pmatrix} 3 \\ -1 \\ 2 \\ 1 \end{pmatrix} \mapsto \begin{pmatrix} 1 & 2 & 0 & -1 \\ 3 & 0 & 3 & 6 \\ 2 & 4 & 2 & 1 \\ 1 & 1 & -1 & 2 \end{pmatrix} \begin{pmatrix} 3 \\ -1 \\ 2 \\ 1 \end{pmatrix}$$

which equals $\begin{pmatrix} 0 \\ 21 \\ 7 \\ 2 \end{pmatrix}$

Thus the image of $\begin{pmatrix} 1 \\ 2 \\ -1 \\ 0 \end{pmatrix}$ is $\begin{pmatrix} 5 \\ 0 \\ 8 \\ 4 \end{pmatrix}$ and the image of $\begin{pmatrix} 3 \\ -1 \\ 2 \\ 1 \end{pmatrix}$ is $\begin{pmatrix} 0 \\ 21 \\ 7 \\ 2 \end{pmatrix}$

under the mapping defined by A.

Let us look again at the matrix $\begin{pmatrix} \cos\theta & -\sin\theta \\ \sin\theta & \cos\theta \end{pmatrix}$ that defines rotation in a plane. Its columns can be viewed as vectors, $\begin{pmatrix} \cos\theta \\ \sin\theta \end{pmatrix}$ and $\begin{pmatrix} -\sin\theta \\ \cos\theta \end{pmatrix}$. These are unit orthogonal vectors. A square matrix whose columns form a set of unit, mutually orthogonal vectors is called an *orthogonal matrix*. Such matrices play a fundamental role in the application of matrix theory because of their properties. Their most significant property is given in the following theorem.

THEOREM 4-1 *Let A be an $n \times n$ orthogonal matrix. A can be interpreted as defining a mapping of $\mathbf{R}^n$ into itself. This mapping preserves magnitudes and angles. Thus, if $\mathbf{x}$ and $\mathbf{y}$ are arbitrary elements of $\mathbf{R}^n$ interpreted as column vectors, then $\|\mathbf{x}\| = \|A\mathbf{x}\|$ (the magnitude of $\mathbf{x}$ is equal to the magnitude of its image, $A\mathbf{x}$) and*

$$\frac{\mathbf{x}}{\|\mathbf{x}\|} \cdot \frac{\mathbf{y}}{\|\mathbf{y}\|} = \frac{A\mathbf{x}}{\|A\mathbf{x}\|} \cdot \frac{A\mathbf{y}}{\|A\mathbf{y}\|}$$

(the angle between $\mathbf{x}$ and $\mathbf{y}$ is equal to the angle between $A\mathbf{x}$ and $A\mathbf{y}$).

Proof: Let $A = (a_{ij})$, $\mathbf{x} = \begin{pmatrix} x_1 \\ \vdots \\ x_n \end{pmatrix}$, and $\mathbf{y} = \begin{pmatrix} y_1 \\ \vdots \\ y_n \end{pmatrix}$. Then

$$A\mathbf{x} = \begin{pmatrix} a_{11} & \cdots & a_{1n} \\ & \vdots & \\ a_{n1} & \cdots & a_{nn} \end{pmatrix} \begin{pmatrix} x_1 \\ \vdots \\ x_n \end{pmatrix} = \begin{pmatrix} \sum a_{1i} x_i \\ \vdots \\ \sum a_{ni} x_i \end{pmatrix}$$

$$A\mathbf{y} = \begin{pmatrix} \sum a_{1j} y_j \\ \vdots \\ \sum a_{nj} y_j \end{pmatrix}$$

$$\|\mathbf{x}\|^2 = \begin{pmatrix} x_1 \\ \vdots \\ x_n \end{pmatrix} \cdot \begin{pmatrix} x_1 \\ \vdots \\ x_n \end{pmatrix} = x_1^2 + \cdots + x_n^2$$

$$\|A\mathbf{x}\|^2 = \begin{pmatrix} \sum a_{1i} x_i \\ \vdots \\ \sum a_{ni} x_i \end{pmatrix} \cdot \begin{pmatrix} \sum a_{1j} x_j \\ \vdots \\ \sum a_{nj} x_j \end{pmatrix}$$

$$= \sum_k \left(\sum_i a_{ki} x_i \right) \left(\sum_j a_{kj} x_j \right)$$

$$= \sum_i \sum_j \left(\sum_k a_{ki} a_{kj} \right) x_i x_j$$

However, if A is orthogonal, $\sum_k a_{ki} a_{kj}$ is the inner product of the ith column and the jth column of A. Hence $\sum_k a_{ki} a_{kj} = \delta_{ij}$. $\|A\mathbf{x}\|^2 = \sum_i \sum_j \delta_{ij} x_i x_j = x_1{}^2 + \cdots + x_n{}^2 = \|\mathbf{x}\|^2$, proving that the mapping preserves magnitudes of vectors.

The proof that the mapping preserves angles is similar. It is left as an exercise.

The following theorem gives another important property of orthogonal matrices.

THEOREM 4-2 *If A is an orthogonal matrix then $A^{-1} = A^t$.*

Thus the inverse of an orthogonal matrix always exists. It can be found by taking the transpose. We do not have the necessary tools for proving this theorem at this time. The reader will be given the proof in Section 6-4.

Example 6

Consider the matrix $A = \begin{pmatrix} 0 & 0 & -1 \\ 1 & 0 & 0 \\ 0 & -1 & 0 \end{pmatrix}$. Its columns are the

vectors $\begin{pmatrix} 0 \\ 1 \\ 0 \end{pmatrix}$, $\begin{pmatrix} 0 \\ 0 \\ -1 \end{pmatrix}$, and $\begin{pmatrix} -1 \\ 0 \\ 0 \end{pmatrix}$.

These vectors are unit, mutually orthogonal vectors. Thus the matrix is an orthogonal matrix. Let us show that when this matrix is used as a mapping, it preserves magnitudes and angles. Let $\mathbf{x} = \begin{pmatrix} a \\ b \\ c \end{pmatrix}$

and $\mathbf{y} = \begin{pmatrix} p \\ q \\ r \end{pmatrix}$ be arbitrary elements of $\mathbf{R}^3$. Then $A\mathbf{x} = \begin{pmatrix} -c \\ a \\ -b \end{pmatrix}$ and

$A\mathbf{y} = \begin{pmatrix} -r \\ p \\ -q \end{pmatrix}$. Since $\|\mathbf{x}\| = \sqrt{a^2 + b^2 + c^2} = \|A\mathbf{x}\|$, the mapping

preserves magnitudes.

Further,

$$\frac{\mathbf{x}}{\|\mathbf{x}\|} \cdot \frac{\mathbf{y}}{\|\mathbf{y}\|} = \frac{\begin{pmatrix} a \\ b \\ c \end{pmatrix}}{\sqrt{a^2 + b^2 + c^2}} \cdot \frac{\begin{pmatrix} p \\ q \\ r \end{pmatrix}}{\sqrt{p^2 + q^2 + r^2}}$$

$$= \frac{ap + bq + cr}{\sqrt{a^2 + b^2 + c^2} \sqrt{p^2 + q^2 + r^2}}$$

and

$$\frac{A\mathbf{x}}{\|A\mathbf{x}\|} \cdot \frac{A\mathbf{y}}{\|A\mathbf{y}\|} = \frac{\begin{pmatrix} -c \\ a \\ -b \end{pmatrix}}{\sqrt{c^2 + a^2 + b^2}} \cdot \frac{\begin{pmatrix} -r \\ p \\ -q \end{pmatrix}}{\sqrt{r^2 + p^2 + q^2}}$$

$$= \frac{cr + ap + bq}{\sqrt{a^2 + b^2 + c^2}\sqrt{p^2 + q^2 + r^2}} = \frac{\mathbf{x}}{\|\mathbf{x}\|} \cdot \frac{\mathbf{y}}{\|\mathbf{y}\|}$$

Thus the mapping preserves angles.

The inverse of A is its transpose; $A^{-1} = A^t = \begin{pmatrix} 0 & 1 & 0 \\ 0 & 0 & -1 \\ -1 & 0 & 0 \end{pmatrix}$

We have seen how certain square matrices have inverses. Let us now focus on such matrices when interpreted as mappings. Let A be an $n \times n$ matrix having inverse A^{-1} (A not necessarily orthogonal). A defines a mapping of $\mathbf{R}^n$ into $\mathbf{R}^n$. Let $\mathbf{x}$ be an element of $\mathbf{R}^n$ that is mapped into the element $\mathbf{y}$,

$$A\mathbf{x} = \mathbf{y}$$

Multiply both sides of this equation by A^{-1},

$$A^{-1} A\mathbf{x} = A^{-1}\mathbf{y}$$

giving $\qquad I_n \mathbf{x} = A^{-1}\mathbf{y}$

$$\mathbf{x} = A^{-1}\mathbf{y}$$

Thus A^{-1} maps $\mathbf{y}$ into $\mathbf{x}$. A^{-1} interpreted as a mapping of $\mathbf{R}^n$ into $\mathbf{R}^n$ is the *inverse mapping* of the mapping defined by A; it brings the vector back.

$$A \text{ maps } \mathbf{x} \mapsto \mathbf{y}$$

$$A^{-1} \text{ maps } \mathbf{y} \mapsto \mathbf{x}$$

If given the image vector $\mathbf{y}$, A^{-1} can be used to determine the original vector $\mathbf{x}$.

Example 7

Interpret the matrix $\begin{pmatrix} 1 & 2 \\ 3 & 4 \end{pmatrix}$ as a mapping of $\mathbf{R}^2$ into $\mathbf{R}^2$. Determine the vector that is mapped into $\begin{pmatrix} 3 \\ -1 \end{pmatrix}$ by this matrix.

We require the vector $\mathbf{x}$ such that

$$\begin{pmatrix} 1 & 2 \\ 3 & 4 \end{pmatrix} \mathbf{x} = \begin{pmatrix} 3 \\ -1 \end{pmatrix}$$

In Section 2-5 we found that the matrix $\begin{pmatrix} 1 & 2 \\ 3 & 4 \end{pmatrix}$ had inverse $\begin{pmatrix} -2 & 1 \\ \frac{3}{2} & -\frac{1}{2} \end{pmatrix}$.

Thus

$$\mathbf{x} = \begin{pmatrix} -2 & 1 \\ \frac{3}{2} & -\frac{1}{2} \end{pmatrix} \begin{pmatrix} 3 \\ -1 \end{pmatrix}$$

giving that

$$\mathbf{x} = \begin{pmatrix} -7 \\ 5 \end{pmatrix}$$

This discussion is in fact another way of looking at the matrix inverse method of determining a solution to a system of linear equations (Section 2-5).   Let $AX = Y$ be a system of n equations in n variables, where A^{-1} exists.   Here Y is given and we are to find X. X is found by computing $A^{-1}Y$.   Mappings enable us to discuss systems of linear equations in a very elegant and powerful manner.   We shall pursue this approach for more general linear systems in following sections.

EXERCISES

1. Determine the matrix that could be used to rotate the points that make up a plane through each of the following angles about the origin.
   a) $\pi/4$ in a clockwise direction   **b)** $\pi/2$ in a clockwise direction
   c) π in a counterclockwise direction
   Determine the point that $\begin{pmatrix} 2 \\ 1 \end{pmatrix}$ is mapped into in each case.   This point is the *image* of $\begin{pmatrix} 2 \\ 1 \end{pmatrix}$ under the mapping.

2. Prove that the matrix $\begin{pmatrix} \cos\theta & \sin\theta \\ \sin\theta & -\cos\theta \end{pmatrix}$ is an orthogonal matrix.

3. Determine the matrix that could be used to map each point of a plane into its mirror image in the x axis (the matrix that maps $\begin{pmatrix} 1 \\ 1 \end{pmatrix} \mapsto \begin{pmatrix} 1 \\ -1 \end{pmatrix}$, $\begin{pmatrix} 1 \\ 0 \end{pmatrix} \mapsto \begin{pmatrix} 1 \\ 0 \end{pmatrix}$, $\begin{pmatrix} -2 \\ -2 \end{pmatrix} \mapsto \begin{pmatrix} -2 \\ 2 \end{pmatrix}$, etc.).

4. Determine the matrix that could be used to map each point into its mirror image in the y axis.

5. Determine the matrix that expands the plane outward from the origin so that each point moves to a point
   a) 2 times   **b)** c times
   as far from the origin.

What would be the characteristics of a matrix that could be used to shrink the plane in a similar manner (a matrix that would map the points into points closer to the origin)? Such a mapping is called a *contraction*.

6. Determine the matrix that rotates the points in the plane through an angle of $\pi/2$, counterclockwise, about the origin, and at the same time expands the points to twice the distance from the origin. (*Hint*: Determine the two matrices and multiply.) Is this equivalent to a rotation followed by an expansion? Is this equivalent to an expansion followed by a rotation? Discuss this from the geometrical and algebraic viewpoints.

7. Determine the general matrix that could be used to define a rotation and expansion about the origin in a plane. Show geometrically and algebraically that this is equivalent both to a rotation followed by an expansion and to an expansion followed by a rotation.

8. From a geometrical viewpoint discuss the following matrices as mappings by considering the deformations of a square with vertices $\begin{pmatrix} 0 \\ 0 \end{pmatrix}$, $\begin{pmatrix} 1 \\ 0 \end{pmatrix}$, $\begin{pmatrix} 1 \\ 1 \end{pmatrix}$, and $\begin{pmatrix} 0 \\ 1 \end{pmatrix}$.

a) $\begin{pmatrix} 0 & -1 \\ 1 & 0 \end{pmatrix}$ b) $\begin{pmatrix} 2 & 0 \\ 0 & 2 \end{pmatrix}$ c) $\begin{pmatrix} 3 & 0 \\ 1 & 4 \end{pmatrix}$

d) $\begin{pmatrix} 4 & -1 \\ 1 & 5 \end{pmatrix}$ e) $\begin{pmatrix} -2 & -3 \\ 0 & 4 \end{pmatrix}$ f) $\begin{pmatrix} -2 & -4 \\ -4 & -1 \end{pmatrix}$

9. Prove that a 2×2 matrix always maps straight lines into straight lines or points. Prove that the origin is mapped into the origin.

10. Determine the matrix that could be used to rotate the points that make up a three-dimensional space through an angle of $\pi/2$ about the z axis. (You may consider either direction.)

11. Determine the matrix that expands three-dimensional space outward from the origin so that each point moves to a point three times as far away.

12. Prove that each of the following matrices is an orthogonal matrix. Determine the inverse of each matrix.

a) $\begin{pmatrix} 1 & 0 \\ 0 & 1 \end{pmatrix}$ b) $\begin{pmatrix} 0 & -1 \\ 1 & 0 \end{pmatrix}$ c) $\begin{pmatrix} \dfrac{1}{\sqrt{2}} & \dfrac{1}{\sqrt{2}} \\ -\dfrac{1}{\sqrt{2}} & \dfrac{1}{\sqrt{2}} \end{pmatrix}$

d) $\begin{pmatrix} 1 & 0 & 0 \\ 0 & 0 & -1 \\ 0 & 1 & 0 \end{pmatrix}$ e) $\begin{pmatrix} 0 & \dfrac{1}{\sqrt{2}} & -\dfrac{1}{\sqrt{2}} \\ -\dfrac{2}{\sqrt{6}} & \dfrac{1}{\sqrt{6}} & \dfrac{1}{\sqrt{6}} \\ \dfrac{1}{\sqrt{3}} & \dfrac{1}{\sqrt{3}} & \dfrac{1}{\sqrt{3}} \end{pmatrix}$

13. Prove that when an orthogonal matrix is used as a mapping, it preserves angles.

14. Let A and B be orthogonal matrices of the same kind. Prove that AB is also orthogonal. Interpret this result geometrically from the point of view of preserving magnitudes and angles.

15. The matrix $\begin{pmatrix} 2 & 1 \\ 4 & 3 \end{pmatrix}$ has inverse $\begin{pmatrix} \frac{3}{2} & -\frac{1}{2} \\ -2 & 1 \end{pmatrix}$. Interpret $\begin{pmatrix} 2 & 1 \\ 4 & 3 \end{pmatrix}$ as a mapping of $\mathbf{R^2}$ into $\mathbf{R^2}$. Determine the vectors that are mapped into $\begin{pmatrix} 2 \\ -4 \end{pmatrix}$ and $\begin{pmatrix} 0 \\ 2 \end{pmatrix}$.

16. The matrix $A = \begin{pmatrix} 0 & 3 & 3 \\ 1 & 2 & 3 \\ 1 & 4 & 6 \end{pmatrix}$ has inverse $\begin{pmatrix} 0 & 2 & -1 \\ 1 & 1 & -1 \\ -\frac{2}{3} & -1 & 1 \end{pmatrix}$. Interpret A as a mapping of $\mathbf{R^3}$ into $\mathbf{R^3}$. Determine the vectors that are mapped into $\begin{pmatrix} 1 \\ 0 \\ 0 \end{pmatrix}$ and $\begin{pmatrix} 3 \\ 6 \\ -3 \end{pmatrix}$.

Interpret your results in terms of solutions to systems of linear equations.

17. Write a program that can be used to perform rotations in a plane. The data should include the relevant angle and point. The output should be the image point. Check your answers to Exercise 1 using your program.

4–2. LINEAR MAPPINGS

In the previous section we saw how any $n \times n$ matrix can be used to define a mapping of $\mathbf{R^n}$ into itself. In this section we shall generalize this concept to mappings between vector spaces.

Consider the vector spaces $\mathbf{R^n}$ and $\mathbf{R^m}$ and an $m \times n$ matrix A. (We do not exclude the possibility that $n = m$.) A can be used to define a mapping of $\mathbf{R^n}$ into $\mathbf{R^m}$. Let $\mathbf{x}$ be an arbitrary element of $\mathbf{R^n}$ interpreted as a column vector. Then $A\mathbf{x}$ exists and is a column vector, an element of $\mathbf{R^m}$. We say that A maps the element $\mathbf{x}$ of $\mathbf{R^n}$ into the element $A\mathbf{x}$ of $\mathbf{R^m}$. We indicate this, as in the case of mappings of $\mathbf{R^n}$ into itself,

$$A: \mathbf{x} \mapsto A\mathbf{x}$$

$A\mathbf{x}$ is the *image* of $\mathbf{x}$ under the mapping.

Example 1

Consider the 2×3 matrix $\begin{pmatrix} 1 & -1 & 0 \\ 2 & 1 & 3 \end{pmatrix}$. This matrix can be used to define a mapping of $\mathbf{R^3}$ into $\mathbf{R^2}$ (Figure 4-7).

✳ Let $\begin{pmatrix} x \\ y \\ z \end{pmatrix}$ be an arbitrary element of $\mathbf{R}^3$. We have that

$$\begin{pmatrix} 1 & -1 & 0 \\ 2 & 1 & 3 \end{pmatrix} : \begin{pmatrix} x \\ y \\ z \end{pmatrix} \mapsto \begin{pmatrix} 1 & -1 & 0 \\ 2 & 1 & 3 \end{pmatrix} \begin{pmatrix} x \\ y \\ z \end{pmatrix}$$

$$= \begin{pmatrix} x-y \\ 2x+y+3z \end{pmatrix}$$

The image of the element $\begin{pmatrix} x \\ y \\ z \end{pmatrix}$ of $\mathbf{R}^3$ under this mapping is the element

$\begin{pmatrix} x-y \\ 2x+y+3z \end{pmatrix}$ of $\mathbf{R}^2$. The image of a specific element of $\mathbf{R}^3$, such as

$\begin{pmatrix} 1 \\ -2 \\ 3 \end{pmatrix}$, would be $\begin{pmatrix} 1+2 \\ 2-2+9 \end{pmatrix}$, or $\begin{pmatrix} 3 \\ 9 \end{pmatrix}$.

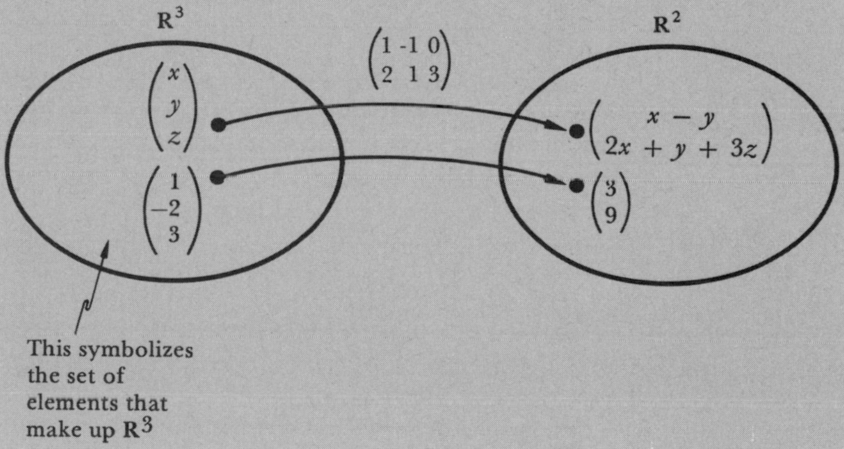

FIGURE 4-7

Let A be an $m \times n$ matrix interpreted as a mapping of $\mathbf{R}^n$ into $\mathbf{R}^m$. If $\mathbf{x}$ and $\mathbf{y}$ are arbitrary elements of $\mathbf{R}^n$ and c an arbitrary scalar, then we know from the properties of matrices that

$$A(\mathbf{x}+\mathbf{y}) = A\mathbf{x} + A\mathbf{y}$$

and

$$A(c\mathbf{x}) = cA\mathbf{x}$$

(Exercise 8.) These results are very significant. The first equation implies that A maps the vector $\mathbf{x}+\mathbf{y}$, an element of $\mathbf{R}^n$ which is the sum of $\mathbf{x}$ and $\mathbf{y}$, into the vector $A\mathbf{x}+A\mathbf{y}$, an element of $\mathbf{R}^m$ which is the sum of $A\mathbf{x}$ and $A\mathbf{y}$. Thus the operation of addition is preserved on these spaces.

The second equation implies that A maps the vector $c\mathbf{x}$, an element of $\mathbf{R}^n$ which is the scalar multiple of $\mathbf{x}$ by c, into the element $cA\mathbf{x}$, which is the scalar multiple of $A\mathbf{x}$ by c. The operation of scalar multiplication is thus preserved. Figure 4-8 illustrates these properties.

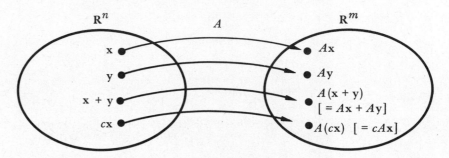

FIGURE 4-8

Mappings between vector spaces that preserve addition and scalar multiplication are of great importance in the analyses and applications of vector spaces. These mappings preserve the mathematical structures of the vector spaces. They are called linear mappings. The formal definition of a linear mapping follows.

DEFINITION 4-2   If V and U are vector spaces, then f is a *linear mapping* if and only if

$$f(\mathbf{v_1}+\mathbf{v_2}) = f(\mathbf{v_1}) + f(\mathbf{v_2})$$
$$f(c\mathbf{v_1}) = cf(\mathbf{v_1})$$

for arbitrary vectors $\mathbf{v_1}$ and $\mathbf{v_2}$ of V and arbitrary scalar c. V is called the *domain* of f.

Thus a matrix interpreted as a mapping is a linear mapping.

There are two further vector spaces that are important in the discussion of linear mappings. The set of all vectors that are mapped onto the zero vector is called the *kernel* of the mapping (Figure 4-9). The *range* of a mapping is the set consisting of all images of vectors in the domain (Figure 4-10). When f is linear, both the kernel and the range are subspaces. We shall prove this in the case of the kernel and leave the proof of the range as an exercise.

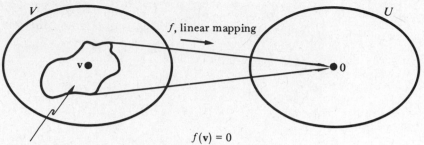

Kernel of *f*. (Kernel is a subspace.)

$f(\mathbf{v}) = 0$

FIGURE 4-9

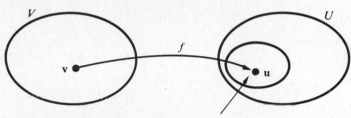

Range of *f*. Every element of this set is
the image of a vector in *V*. The range
may be the whole of *U* or just part of *U*.
(Range is a subspace.)

FIGURE 4-10

THEOREM 4-3   *The kernel of a linear mapping forms a subspace of the
domain.*

Proof: Let $\mathbf{v_1}$ and $\mathbf{v_2}$ be arbitrary elements of the kernel of a linear
mapping *f*, and *c* an arbitrary scalar.

Then $f(\mathbf{v_1}) = \mathbf{0}$ and $f(\mathbf{v_2}) = \mathbf{0}$.   Thus

$$f(\mathbf{v_1}) + f(\mathbf{v_2}) = \mathbf{0}$$

By the linearity of *f*,

$$f(\mathbf{v_1} + \mathbf{v_2}) = \mathbf{0}$$

This implies that $\mathbf{v_1} + \mathbf{v_2}$ is an element of the kernel.

Further, $f(\mathbf{v_1}) = \mathbf{0}$ implies that $cf(\mathbf{v_1}) = \mathbf{0}$.   Since *f* is linear,

$$f(c\mathbf{v_1}) = \mathbf{0}$$

implying that $c\mathbf{v_1}$ is in the kernel.   Thus the kernel is a subspace of
the domain.

Example 2 illustrates these concepts for a specific mapping.

235

Example 2

Consider the mapping f of $\mathbf{R}^3$ into $\mathbf{R}^3$ defined by $f: (x, y, z) \mapsto (x, y, 0)$. We shall prove that this mapping is linear and interpret it geometrically.

Let (x_1, y_1, z_1) and (x_2, y_2, z_2) be arbitrary elements of $\mathbf{R}^3$ and c an arbitrary scalar. Then

$$f\left[(x_1, y_1, z_1) + (x_2, y_2, z_2)\right] = f\left[(x_1 + x_2, y_1 + y_2, z_1 + z_2)\right]$$
$$= (x_1 + x_2, y_1 + y_2, 0)$$
$$= (x_1, y_1, 0) + (x_2, y_2, 0)$$
$$= f\left[(x_1, y_1, z_1)\right] + f\left[(x_2, y_2, z_2)\right]$$

The first requirement of a linear mapping is satisfied—it preserves addition. Further,

$$f\left[c(x_1, y_1, z_1)\right] = f\left[(cx_1, cy_1, cz_1)\right]$$
$$= (cx_1, cy_1, 0)$$
$$= c(x_1, y_1, 0)$$
$$= cf\left[(x_1, y_1, z_1)\right]$$

Thus f preserves the operation of scalar multiplication; it is linear.

Geometrically we have Figure 4-11. f maps the vector (x, y, z) into the vector $(x, y, 0)$. This vector lies in the xy plane. f projects $\mathbf{R}^3$ onto the xy plane; it is called a *projection mapping*.

The range of f consists of all vectors of the type $(x, y, 0)$. Thus the range is the xy plane. It is a subspace of $\mathbf{R}^3$.

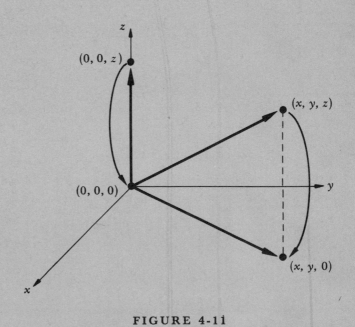

FIGURE 4-11

> The kernel of f is the subset of $\mathbf{R}^3$ that is mapped onto the vector $(0, 0, 0)$. Since $f: (x, y, z) \mapsto (x, y, 0)$, (x, y, z) will be in the kernel if and only if $x = y = 0$. Thus the kernel consists of all vectors of the form $(0, 0, z)$. Geometrically this is the z axis, a one-dimensional subspace of $\mathbf{R}^3$.

We complete this example with a discussion of an application of projection mappings. The world in which we live has three spacial dimensions. When we observe an object, however, we get a two-dimensional impression of that object, the view changing from location to location. Projection mappings can be used to illustrate what three-dimensional objects look like from various locations. Such mappings are used in architecture and the aerospace industry, for example. The outline of the object of interest, relative to a suitable coordinate system, is fed into a computer. The computer program contains an appropriate projection mapping that maps the object onto a plane. The output gives a two-dimensional view of the object; the outline being graphed by the computer. In this manner various mappings can be used to lead to various perspectives of an object. The General Electric Plant at Daytona Beach, Florida uses such a computer graphics system for simulating aircraft. The Grumman Aerospace Corporation at Bethpage, N.Y. uses such a graphics system in the designing of aircraft. We illustrate these concepts below for an aircraft. The projection mapping used is onto the xz plane. The image represents the view an observer at A has of the aircraft; it would be graphed out by the computer.

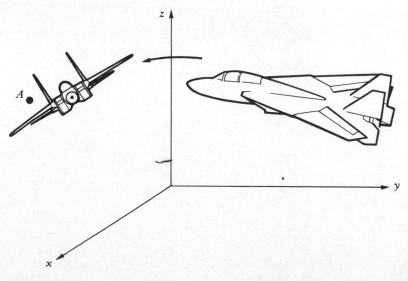

Both the kernel and range of the linear mapping of this last example were easily found on inspection. The following example illustrates a general approach.

Example 3

Interpreting the matrix $\begin{pmatrix} 1 & 2 & 3 \\ 0 & -1 & 1 \\ 1 & 1 & 4 \end{pmatrix}$ as a linear mapping f of $\mathbf{R}^3$ into $\mathbf{R}^3$ let us determine its kernel and range. The kernel will consist of all elements $\begin{pmatrix} x \\ y \\ z \end{pmatrix}$ of $\mathbf{R}^3$ such that

$$\begin{pmatrix} 1 & 2 & 3 \\ 0 & -1 & 1 \\ 1 & 1 & 4 \end{pmatrix} \begin{pmatrix} x \\ y \\ z \end{pmatrix} = \begin{pmatrix} 0 \\ 0 \\ 0 \end{pmatrix}$$

That is, such that x, y, and z satisfy the system of homogeneous equations:

$$x + 2y + 3z = 0$$
$$- y + z = 0$$
$$x + y + 4z = 0$$

This system has many solutions; they can be represented $\begin{pmatrix} -5z \\ z \\ z \end{pmatrix}$, z being a parameter.

Thus the kernel of f consists of all elements of $\mathbf{R}^3$ of the form $\begin{pmatrix} -5z \\ z \\ z \end{pmatrix}$. The kernel is a one-dimensional subspace of $\mathbf{R}^3$ generated by the vector $\begin{pmatrix} -5 \\ 1 \\ 1 \end{pmatrix}$.

Let us now determine the range of f. The range will consist of all elements $\begin{pmatrix} x \\ y \\ z \end{pmatrix}$ of $\mathbf{R}^3$ for which there exist elements $\begin{pmatrix} a \\ b \\ c \end{pmatrix}$ such that

$$\begin{pmatrix} 1 & 2 & 3 \\ 0 & -1 & 1 \\ 1 & 1 & 4 \end{pmatrix} \begin{pmatrix} a \\ b \\ c \end{pmatrix} = \begin{pmatrix} x \\ y \\ z \end{pmatrix}$$

$\begin{pmatrix} x \\ y \\ z \end{pmatrix}$ would be the image of such an $\begin{pmatrix} a \\ b \\ c \end{pmatrix}$.

This condition may be written

$$a + 2b + 3c = x \cdot \quad a + 2b + 3c = x$$
$$-\ b +\ c = y \cong \quad -\ b +\ c = y$$
$$a +\ b + 4c = z \qquad -\ b +\ c = z - x$$
$$a + 2b + 3c = x$$
$$\cong \quad -\ b +\ c = y$$
$$0 = z - x - y$$

Let us now interpret this result.

The last equation implies that $z = x + y$. If an element $\begin{pmatrix} x \\ y \\ z \end{pmatrix}$ is in the range its components satisfy this condition. Let us now turn our attention to the first two equations, $a + 2b + 3c = x$ and $-b + c = y$. Remembering that $\begin{pmatrix} a \\ b \\ c \end{pmatrix}$ is in the domain, a, b, and c can each take on all values, implying that x and y are independent of each other. The range of f will consist of all elements of the form $\begin{pmatrix} x \\ y \\ x+y \end{pmatrix}$. Observe that the range is a two-dimensional subspace of $\mathbf{R}^3$, the plane $z = x + y$, spanned by $\begin{pmatrix} 1 \\ 0 \\ 1 \end{pmatrix}$ and $\begin{pmatrix} 0 \\ 1 \\ 1 \end{pmatrix}$.

The following example illustrates a nonlinear mapping.

Example 4

Examine the mapping $f(x, y, z) = (xy, z)$ of $\mathbf{R}^3$ into $\mathbf{R}^2$ for linearity.

Let (x_1, y_1, z_1) and (x_2, y_2, z_2) be two arbitrary elements of $\mathbf{R}^3$. We have that

$$f[(x_1, y_1, z_1) + (x_2, y_2, z_2)] = f(x_1 + x_2, y_1 + y_2, z_1 + z_2)$$
$$= ((x_1 + x_2)(y_1 + y_2), z_1 + z_2) \text{ by definition of } f.$$
$$= (x_1 y_1 + x_2 y_2 + x_1 y_2 + x_2 y_1, z_1 + z_2)$$

and

$$f(x_1, y_1, z_1) + f(x_2, y_2, z_2) = (x_1 y_1, z_1) + (x_2 y_2, z_2)$$
$$= (x_1 y_1 + x_2 y_2, z_1 + z_2)$$

We see that, in general

$$f[(x_1, y_1, z_1) + (x_2, y_2, z_2)] \neq f(x_1, y_1, z_1) + f(x_2, y_2, z_2)$$

Since the first linearity condition is not satisfied, vector addition is not preserved; f is not linear. In practice one need not check the second condition. We do so here to illustrate the manner in which this condition is violated for this mapping also. Let c be an arbitrary scalar. Then

$$f[c(x_1, y_1, z_1)] = f[(cx_1, cy_1, cz_1)]$$
$$= (cx_1 cy_1, cz_1) \text{ by definition of } f$$
$$= (c^2 x_1 y_1, cz_1).$$

However,

$$cf(x_1, y_1, z_1) = c(x_1 \ y_1, z_1) = (cx_1 y_1, cz_1)$$

Thus in general $f[c(x_1, y_1, z_1)] \neq cf(x_1, y_1, z_1)$; the second condition is also violated.

Example 5*

Section 3-7 is a prerequisite for this example of a linear mapping on the second type of vector space, a function space.

Let V be the vector space of real-valued, differentiable functions of all orders on **R**. (It is a vector space under the operations of pointwise addition and scalar multiplication as defined in Section 3-7). Let D be the operation of taking the derivative. Then D is a linear mapping of V into itself, for

$$D(f + g) = Df + Dg$$

and

$$D(cf) = cDf$$

for arbitrary elements f and g of V and an arbitrary scalar c. These properties of D are known from those of derivatives discussed in calculus.

Further, the kernel of D will be the set consisting of all functions f such that $Df = 0$. These are, of course, the constant functions. Thus the kernel of the linear mapping D is the set of constant functions. They form a subspace of the set of differentiable functions. (Prove this.)

The range of D is the space V itself. (When this is so we say that the mapping is *onto*.) Every element of V is the image of an element under D; f is the image of $\int f$, for $D(\int f) = f$.

We complete this section with a theorem that gives a relationship between the kernel and range of a linear mapping.

THEOREM 4-4   *Let f be a linear mapping of a vector space V into a vector* *space U.   Then*

$$\text{dimension of kernel of } f + \text{dimension of range of } f$$
$$= \text{dimension of } V, \text{ the domain of } f$$

Proof: Let us assume that the kernel consists of more than the zero vector, and that it is not the whole of V (the reader is asked to consider these special cases in Exercise 17, p. 244).

Let $\mathbf{v}_1, \dots, \mathbf{v}_m$ be a basis for the kernel. Complete this set to a basis $\mathbf{v}_1, \dots, \mathbf{v}_n$ for V by adding the vectors $\mathbf{v}_{m+1}, \dots, \mathbf{v}_n$. We shall show that the vectors $f(\mathbf{v}_{m+1}), \dots, f(\mathbf{v}_n)$ form a basis for the range, thus verifying the theorem.

Let $\mathbf{v}$ be an arbitrary vector in V. Then $\mathbf{v}$ can be expressed

$$\mathbf{v} = a_1 \mathbf{v}_1 + \cdots + a_m \mathbf{v}_m + a_{m+1} \mathbf{v}_{m+1} + \cdots + a_n \mathbf{v}_n$$

and

$$f(\mathbf{v}) = f(a_1 \mathbf{v}_1 + \cdots + a_m \mathbf{v}_m + a_{m+1} \mathbf{v}_{m+1} + \cdots + a_n \mathbf{v}_n)$$

The linearity of f gives (Exercise 14)

$$f(\mathbf{v}) = a_1 f(\mathbf{v}_1) + \cdots + a_m f(\mathbf{v}_m) + a_{m+1} f(\mathbf{v}_{m+1}) + \cdots + a_n f(\mathbf{v}_n)$$

Since $\mathbf{v}_1, \dots, \mathbf{v}_m$ are in the kernel of f, this reduces to

$$f(\mathbf{v}) = a_{m+1} f(\mathbf{v}_{m+1}) + \cdots + a_n f(\mathbf{v}_n)$$

$f(\mathbf{v})$ is an arbitrary vector in the range. Thus $f(\mathbf{v}_{m+1}), \dots, f(\mathbf{v}_n)$ span the range.

It remains to prove that these vectors are linearly independent. Consider the identity

$$b_{m+1} f(\mathbf{v}_{m+1}) + \cdots + b_n f(\mathbf{v}_n) = \mathbf{0}, \tag{1}$$

where $b_{m+1}, \dots, b_n$ are arbitrary scalars, labeled thus for convenience. The linearity of f implies that

$$f(b_{m+1} \mathbf{v}_{m+1} + \cdots + b_n \mathbf{v}_n) = \mathbf{0}$$

that is, the vector $b_{m+1} \mathbf{v}_{m+1} + \cdots + b_n \mathbf{v}_n$ is an element of the kernel. Thus it can be expressed as a linear combination of $\mathbf{v}_1, \dots, \mathbf{v}_m$. Let

$$b_{m+1} \mathbf{v}_{m+1} + \cdots + b_n \mathbf{v}_n = c_1 \mathbf{v}_1 + \cdots + c_m \mathbf{v}_m$$

That is

$$c_1 \mathbf{v}_1 + \cdots + c_m \mathbf{v}_m - b_{m+1} \mathbf{v}_{m+1} - \cdots - b_n \mathbf{v}_n = \mathbf{0}$$

Since the vectors $\mathbf{v}_1, \dots, \mathbf{v}_m, \mathbf{v}_{m+1}, \dots, \mathbf{v}_n$ are linearly independent, this can only be satisfied if the coefficients are all zero. In particular $b_{m+1} = \cdots = b_n = 0$. Thus, in (1), $f(\mathbf{v}_{m+1}), \dots, f(\mathbf{v}_n)$ are linearly independent. They are a basis for the range, and the theorem is proved.

Example 6

Let us look further at the linear mappings discussed in Examples 2 and 3 of this section.

Consider the linear mapping of $\mathbf{R}^3$ into $\mathbf{R}^3$ defined by $f : (x, y, z) \rightarrow (x, y, 0)$. The kernel of this mapping was found to be the z axis, a one-dimensional subspace of $\mathbf{R}^3$. The range was found to be the xy plane, a two-dimensional subspace of $\mathbf{R}^3$. We see that the dimension of the kernel + the dimension of the range is 3, the dimension of the domain of f.

Consider the linear mapping of $\mathbf{R}^3$ into $\mathbf{R}^3$ defined by the matrix $\begin{pmatrix} 1 & 2 & 3 \\ 0 & -1 & 1 \\ 1 & 1 & 4 \end{pmatrix}$. We found that the kernel was the one-dimensional

space with basis $\begin{pmatrix} -5 \\ 1 \\ 1 \end{pmatrix}$. The range was a two-dimensional space

with basis $\begin{pmatrix} 1 \\ 0 \\ 1 \end{pmatrix}, \begin{pmatrix} 0 \\ 1 \\ 1 \end{pmatrix}$. The dimension of the domain, $\mathbf{R}^3$, is 3.

The dimension of the kernel plus the dimension of the range is equal to the dimension of the domain.

EXERCISES

1. Interpret the matrix $\begin{pmatrix} 1 & 2 \\ -1 & 3 \\ 1 & 2 \end{pmatrix}$ as a mapping of $\mathbf{R}^2$ into $\mathbf{R}^3$. What

are the images of the following elements under this mapping?

a) $\begin{pmatrix} -1 \\ 1 \end{pmatrix}$ b) $\begin{pmatrix} 2 \\ 3 \end{pmatrix}$ c) $\begin{pmatrix} 1 \\ 4 \end{pmatrix}$ d) $\begin{pmatrix} -3 \\ 3 \end{pmatrix}$

Observe   that $\begin{pmatrix} 1 \\ 4 \end{pmatrix} = \begin{pmatrix} -1 \\ 1 \end{pmatrix} + \begin{pmatrix} 2 \\ 3 \end{pmatrix}$.   Comment   on   a

similar relation between the images of these three vectors.

Observe   that $\begin{pmatrix} -3 \\ 3 \end{pmatrix} = 3\begin{pmatrix} -1 \\ 1 \end{pmatrix}$.   Comment   on   a   similar

relation between the images of $\begin{pmatrix} -3 \\ 3 \end{pmatrix}$ and $\begin{pmatrix} -1 \\ 1 \end{pmatrix}$.

2. Let A be an $m \times n$ matrix and $\mathbf{c}$ a column vector having n components. Prove that the mapping f of $\mathbf{R}^n$ into $\mathbf{R}^m$ defined by $f : \mathbf{x} \mapsto A\mathbf{x} + \mathbf{c}$ is not linear.

3. Prove that the mapping f of $\mathbf{R}^3$ into $\mathbf{R}^3$ defined by $f(x, y, z) = (0, y, 0)$

is linear. Illustrate this mapping geometrically and determine its kernel and range.

4. Prove that the mapping f of $\mathbf{R}^3$ into $\mathbf{R}^2$ defined by $f(x, y, z) = (2x, y + z)$ is linear. Determine the images of the elements $(1, 2, 3)$ and $(-1, 4, 3)$ under f.

5. Prove that the mapping of $\mathbf{R}^2$ into $\mathbf{R}^3$ defined by $f(x, y) = (3x + y, 2y, x - y)$ is linear. Determine the images of the elements $(1, 2)$ and $(2, -1)$ under f.

6. Prove that the mapping f of $\mathbf{R}^2$ into $\mathbf{R}$ defined by $f(x, y) = x + a$, where a is constant, is not linear.

7. Prove that the following mappings of $\mathbf{R}^3$ into $\mathbf{R}^2$ are not linear:
   a) $f(x, y, z) = (x^2, y^2)$, b) $f(x, y, z) = (x + 2, y)$

8. A is an $m \times n$ matrix, X and Y are elements of $\mathbf{R}^n$ in column form, and c is an arbitrary scalar. Prove that
   a) $A(X + Y) = AX + AY$, b) $A(cX) = cAX$
   Thus A interpreted as a mapping of $\mathbf{R}^n \to \mathbf{R}^m$ preserves addition and scalar multiplication.

9. Interpret the following matrices as mappings between appropriate vector spaces. Determine the kernel and range of each mapping.

   a) $\begin{pmatrix} 1 & 2 \\ 3 & 0 \end{pmatrix}$ b) $\begin{pmatrix} 2 & 0 \\ 3 & 0 \end{pmatrix}$ c) $\begin{pmatrix} 2 & 4 \\ 4 & 8 \end{pmatrix}$ d) $\begin{pmatrix} 1 & 2 \\ -1 & 3 \end{pmatrix}$

   e) $\begin{pmatrix} 1 & 2 & 3 \\ 0 & 1 & 2 \end{pmatrix}$ f) $\begin{pmatrix} 1 & 0 & 0 \\ 0 & 2 & 0 \\ 0 & 0 & 3 \end{pmatrix}$ g) $\begin{pmatrix} 0 & 1 & 0 \\ 0 & 2 & 0 \\ 0 & 0 & 4 \end{pmatrix}$

10. Prove that the range of a linear mapping is a subspace.

11. Let V be the vector space of real-valued, differentiable functions of all orders on $\mathbf{R}$; let D^2 be the operation of taking the second derivative. Prove that D^2 may be interpreted as a linear mapping of V into itself. What is the kernel of this mapping? Further prove that D^n, the operation of taking the nth derivative, is linear.

12. Let A and B be $m \times n$ and $p \times m$ matrices, respectively. B may be interpreted as a linear mapping of the range of A into $\mathbf{R}^p$. BA thus defines a mapping of $\mathbf{R}^n$ into $\mathbf{R}^p$. Prove that this *composite* mapping is linear.

13. Let A be an invertible matrix. Show that AB has the same kernel and range as B.

14. Let f be a linear mapping of a vector space V into a vector space U, let $\mathbf{v}_1, ..., \mathbf{v}_m$ be m arbitrary vectors in V, and let $a_1, ..., a_m$ be m arbitrary scalars. Prove that $f(a_1 \mathbf{v}_1 + \cdots + a_m \mathbf{v}_m) = a_1 f(\mathbf{v}_1) + \cdots + a_m f(\mathbf{v}_m)$.

15. Prove that a mapping f between vector spaces is linear if and only if

$$f(a\mathbf{u} + b\mathbf{v}) = af(\mathbf{u}) + bf(\mathbf{v})$$

for arbitrary vectors $\mathbf{u}$, $\mathbf{v}$ and arbitrary scalars a, b. This is an alternative definition for a linear mapping.

16. Determine the kernel and range of each of the following linear mappings:
    a) f mapping $\mathbf{R}^3$ into $\mathbf{R}^3$ defined by $f(x, y, z) = (x, 0, 0)$
    b) f mapping $\mathbf{R}^3$ into $\mathbf{R}^2$ defined by $f(x, y, z) = (x+y, z)$
    c) f mapping $\mathbf{R}^3$ into $\mathbf{R}$ defined by $f(x, y, z) = x+y+z$

17. Let f be a linear mapping of a vector space V into a vector space U. Prove that

$$\text{dimension of kernel of } f + \text{dimension of range of } f = \text{dimension of } V$$

When
    a) The kernel of f consists of only the zero vector.
    b) The kernel of f is the whole of V.

4–3. MAPPINGS AND SYSTEMS OF EQUATIONS

Linear mappings and the concepts of kernel and range play an important role in the analyses of systems of linear equations. They also enable one to "see" geometrically what is happening in a given system. Consider a system of m equations in n variables, $AX = Y$. The matrix of coefficients A is an $m \times n$ matrix. Interpret A as a mapping of $\mathbf{R}^n$ into $\mathbf{R}^m$. The set of solutions is made up of all the vectors in $\mathbf{R}^n$ that are mapped onto the vector Y by A. We now look at systems of linear equations from this viewpoint.

THEOREM 4-5   *The set of solutions to a system of m homogeneous equations in n variables, $AX = 0$, is a subspace of $\mathbf{R}^n$.*

Proof: Interpret A as a mapping of $\mathbf{R}^n$ into $\mathbf{R}^m$. The set of solutions is the kernel of this mapping and is thus a subspace of $\mathbf{R}^n$.

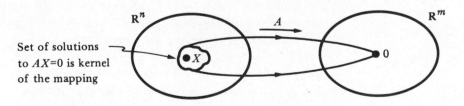

Set of solutions to $AX=0$ is kernel of the mapping

Example 1

Consider the system of linear homogeneous equations

$$x_1 + 2x_2 + 3x_3 = 0$$
$$- x_2 + x_3 = 0$$
$$x_1 + x_2 + 4x_3 = 0$$

Let us solve this system and interpret the set of solutions in this geometrical manner. We get, using the method of Gaussian elimination,

$$\begin{pmatrix} 1 & 2 & 3 & 0 \\ 0 & -1 & 1 & 0 \\ 1 & 1 & 4 & 0 \end{pmatrix} \cong \begin{pmatrix} 1 & 2 & 3 & 0 \\ 0 & -1 & 1 & 0 \\ 0 & -1 & 1 & 0 \end{pmatrix}$$

$$\cong \begin{pmatrix} 1 & 2 & 3 & 0 \\ 0 & 1 & -1 & 0 \\ 0 & 0 & 0 & 0 \end{pmatrix}$$

Thus

$$x_1 + 2x_2 + 3x_3 = 0$$

$$x_2 - x_3 = 0$$

The last equation gives $x_2 = x_3$. Substitution for x_2 into the first equation gives $x_1 = -5x_3$.

The set of solution is $(-5x_3, x_3, x_3)$.

This may also be expressed $x_3(-5, 1, 1)$.

Thus the set of solutions is the one-dimensional subspace of $\mathbf{R}^3$ spanned by the vector $(-5, 1, 1)$. It is the line defined by the vector $(-5, 1, 1)$, below.

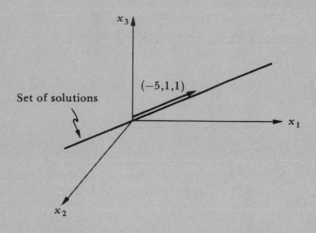

In Example 3 of the previous section we discussed the linear mapping defined by the matrix of coefficients of this system. Observe that the set of solutions is indeed the kernel of this mapping.

Let us now look at nonhomogeneous systems. We find that the set of solutions to such a system, $AX = Y$, is not a subspace.

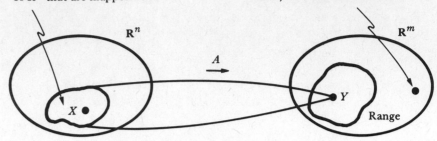

Set of solutions — all elements of $\mathbf{R}^n$ that are mapped onto Y.

If Y is here, not in range of A, solution will not exist.

Let X_1 and X_2 be distinct solutions.   Then

$$AX_1 = Y \text{ and } AX_2 = Y$$

Thus

$$AX_1 + AX_2 = 2Y$$

$$A(X_1 + X_2) = 2Y$$

$X_1 + X_2$ is not a solution.   Thus the set of solutions is not closed under addition; it is not a subspace.

The following is a significant result for nonhomogeneous systems.

THEOREM 4-6   *Consider a system of m nonhomogeneous equations in n variables, $AX = Y$. Let X_1 be a particular solution. Every other solution can be written in the form $X_1 + Z$, where Z is an element of the kernel of the mapping defined by A. The solution is unique if the kernel consists of the zero vector only.*

Proof: X_1 is a solution, thus $AX_1 = Y$.   Let X_2 be another solution, $AX_2 = Y$.
Equating,

$$AX_1 = Y = AX_2$$

$$AX_2 - AX_1 = 0$$

$$A(X_2 - X_1) = 0$$

Thus $X_2 - X_1$ is an element of the kernel of A; call it Z.

$$X_2 - X_1 = Z$$

Thus

$$X_2 = X_1 + Z \quad \text{where } Z \text{ is an element of the kernel}$$

The solution is unique; $X_2 = X_1$ if and only if the only value Z can assume is $\mathbf{0}$. That is, if and only if the kernel is the zero vector.

This result implies that the set of solutions to a system of non-homogeneous equations can be generated from a particular solution and the kernel of the matrix of coefficients. If we take any vector in the kernel and add X_1 to it we get a solution. Geometrically, this means that the set of solutions is that obtained by sliding the kernel in the direction and distance defined by the vector X_1, in $\mathbf{R}^n$.

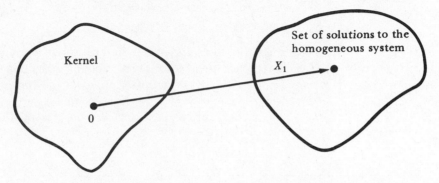

We illustrate this way of looking at such a set of solutions with the following example.

Example 2

Solve the system of equations

$$x_1 + 2x_2 + 3x_3 = 11$$
$$- x_2 + x_3 = -2$$
$$x_1 + x_2 + 4x_3 = 9$$

and discuss the set of solutions.

We get

$$\begin{pmatrix} 1 & 2 & 3 & 11 \\ 0 & -1 & 1 & -2 \\ 1 & 1 & 4 & 9 \end{pmatrix} \cong \begin{pmatrix} 1 & 2 & 3 & 11 \\ 0 & -1 & 1 & -2 \\ 0 & -1 & 1 & -2 \end{pmatrix}$$

$$\cong \begin{pmatrix} 1 & 2 & 3 & 11 \\ 0 & 1 & -1 & 2 \\ 0 & 0 & 0 & 0 \end{pmatrix}$$

Thus

$$x_1 + 2x_2 + 3x_3 = 11$$
$$x_2 - x_3 = 2$$

The last equation gives $x_2 = x_3 + 2$. Substitution for x_2 into the first equation gives $x_1 = -5x_3 + 7$.

The set of solutions is $(-5x_3 + 7, x_3 + 2, x_3)$.

"Pull this solution apart", separating the parameter. The interpretations of the parts are:

$$(7, 2, 0) + (-5x_3, x_3, x_3)$$

↑   ↑

A particular   Kernel of the mapping
solution to the   defined by the matrix
above system.   of coefficients.

One can verify that $(7, 2, 0)$ is indeed a particular solution to this system. $(-5x_3, x_3, x_3)$ is the kernel of the mapping defined by the matrix of coefficients. This is the set of solutions to the corresponding system of homogeneous equations (see previous example).

Geometrically we have the following picture of the set of solutions:

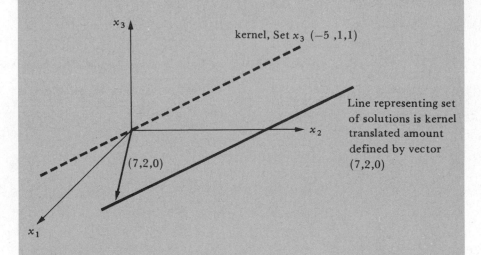

x_3

kernel, Set x_3 $(-5, 1, 1)$

x_2

Line representing set of solutions is kernel translated amount defined by vector $(7, 2, 0)$

$(7, 2, 0)$

x_1

Example 3   Discuss the set of solutions to the system

$$x_1 - 2x_2 + 3x_3 + x_4 = 1$$
$$2x_1 - 3x_2 + 2x_3 - x_4 = 4$$
$$3x_1 - 5x_2 + 5x_3 = 5$$
$$x_1 - x_2 - x_3 - 2x_4 = 3$$

We get

$$\begin{pmatrix} 1 & -2 & 3 & 1 & 1 \\ 2 & -3 & 2 & -1 & 4 \\ 3 & -5 & 5 & 0 & 5 \\ 1 & -1 & -1 & -2 & 3 \end{pmatrix} \cong \begin{pmatrix} 1 & -2 & 3 & 1 & 1 \\ 0 & 1 & -4 & -3 & 2 \\ 0 & 1 & -4 & -3 & 2 \\ 0 & 1 & -4 & -3 & 2 \end{pmatrix}$$

$$\cong \begin{pmatrix} 1 & -2 & 3 & 1 & 1 \\ 0 & 1 & -4 & -3 & 2 \\ 0 & 0 & 0 & 0 & 0 \\ 0 & 0 & 0 & 0 & 0 \end{pmatrix}$$

giving

$$x_1 - 2x_2 + 3x_3 + x_4 = 1$$

$$x_2 - 4x_3 - 3x_4 = 2$$

We express the variables x_1 and x_2 in terms of x_3 and x_4,

$$x_1 = 1 + 5x_3 - 5x_4$$

$$x_2 = 2 + 4x_3 + 3x_4$$

Express these solutions

$$(1 + 5x_3 - 5x_4, 2 + 4x_3 + 3x_4, x_3, x_4)$$

Separating the parameters, the solutions are

$$\underset{\underset{\substack{Particular \\ solution}}{\uparrow}}{(1, 2, 0, 0)} + \underbrace{x_3(5, 4, 1, 0) + x_4(-5, 3, 0, 1)}_{\substack{\textit{The kernel of the mapping defined} \\ \textit{by the matrix of cofficients—this is} \\ \textit{the set of solutions to the} \\ \textit{corresponding system of homogeneous} \\ \textit{equations.}}}$$

The set of solutions to the corresponding system of homogeneous equations in this example forms a two-dimensional subspace of $\mathbf{R}^4$, having basis $(5, 4, 1, 0)$, $(-5, 3, 0, 1)$. This will be a plane through the origin in $\mathbf{R}^4$. The effect of adding the particular solution $(1, 2, 0, 0)$ to every vector in this plane is to slide the plane away from the origin.

Observe that if a number of systems of equations have the same matrix of coefficients their solution will only differ in the particular solution part. The part corresponding to the kernel of the matrix of coefficients will be the same for all the systems.

Example 4*

The analysis of solutions presented in this section is important in the discussion of solutions to linear differential equations. Let V be the vector space of real-valued functions that have derivatives of all orders on some given interval.   Let D denote differentiation and D^k the kth derivative. If $f_0, ..., f_k$ are elements of V, then $f_k D^k + f_{k-1} D^{k-1} + \cdots + f_0$ is called a linear differential operator on V.   It is a linear mapping of V into itself. For a given function g, the equation $(f_k D^k + \cdots + f_0)\, Y = g$ is called a linear differential equation. Y_1 is a solution if and only if it is mapped into g by this linear mapping.   To solve this equation, one determines a specific solution Y_1, called a *particular integral*, and the general solution of $(f_k D^k + \cdots + f_0) Y = 0$, called the *complementary function*.   This is the kernel of the linear mapping.   The set of solutions is the sum of the particular solution and the complementary function.

We illustrate the method for the following differential equation: To solve

$$\frac{d^2 y}{dx^2} - 3\frac{dy}{dx} + 2y = 16 \tag{1}$$

The solution will be the sum of a specific solution and the general solution of

$$\frac{d^2 y}{dx^2} - 3\frac{dy}{dx} + 2y = 0 \tag{2}$$

By inspection it is seen that $y_1 = 8$ is a specific solution.

The theory of differential equations leads one to the general solution of (2), $y = c_1 e^x + c_2 e^{2x}$, where c_1 and c_2 are parameters.

The solution of (1) is

$$y = \underbrace{8}_{\substack{Specific \\ solution}} + \underbrace{c_1 e^x + c_2 e^{2x}}_{\substack{Solutions\ of\ corresponding \\ homogeneous\ equation,\ (2).}}$$

EXERCISES

Consider the following systems of equations.   The solutions are given in each case.   Analyze the solutions by representing each set as the sum of a particular solution and an arbitrary solution of the corresponding system of homogeneous equations.

1. $\begin{aligned} x_1 - x_2 + x_3 &= 1 \\ 2x_1 - 2x_2 + 3x_3 &= 3 \\ x_1 - x_2 - x_3 &= -1 \end{aligned}$ Solutions: $x_1 = x_2$
                                  $x_3 = 1$

2. $\begin{aligned} x_1 + x_2 + x_3 &= 3 \\ 2x_1 + 3x_2 + x_3 &= 5 \\ x_1 - x_2 - 2x_3 &= -5 \end{aligned}$ Solutions: $x_1 = 0$
                                  $x_2 = 1$
                                  $x_3 = 2$

3. $x_1 - 2x_2 + 3x_3 = 1$ Solutions: $x_1 = 1 + x_3$
    $3x_1 - 4x_2 + 5x_3 = 3$ $x_2 = 2x_3$
    $2x_1 - 3x_2 + 4x_3 = 2$

4. $x_1 - x_2 + x_3 = 3$ Solutions: $x_1 = 3 + x_2 - x_3$
   $-2x_1 + 2x_2 - 2x_3 = -6$

5. $x_1 - x_2 - x_3 + 2x_4 = 4$ Solutions: $x_1 = -2 - 4x_3 + 3x_4$
  $2x_1 - x_2 + 3x_3 - x_4 = 2$ $x_2 = -6 - 5x_3 + 5x_4$

6. $x_1 + 2x_2 - x_3 + x_4 + 2x_5 = 0$ Solutions: $x_1 = -19 - 32x_4 - 23x_5$
       $x_2 + 2x_3 - 3x_4 - 3x_5 = 2$ $x_2 = 8 + 13x_4 + 9x_5$
             $x_3 + 5x_4 + 3x_5 = -3$ $x_3 = -3 - 5x_4 - 3x_5$

7. Interpreting the matrix $A = \begin{pmatrix} 1 & 2 & 3 \\ 0 & -1 & 1 \\ 1 & 1 & 4 \end{pmatrix}$ as a mapping of $\mathbf{R}^3$ into

$\mathbf{R}^3$, its range is the set of vectors of the form $\begin{pmatrix} x \\ y \\ x+y \end{pmatrix}$, that is, the plane

$z = x + y$. Use this given information to determine whether solutions exist to the following systems of equations $AX = Y$, Y taking on the various values. (You need not determine the solutions if they exist.)

a) $Y = \begin{pmatrix} 1 \\ 1 \\ 2 \end{pmatrix}$ **b)** $Y = \begin{pmatrix} -1 \\ 2 \\ 3 \end{pmatrix}$ **c)** $Y = \begin{pmatrix} 3 \\ 2 \\ 5 \end{pmatrix}$ **d)** $Y = \begin{pmatrix} 2 \\ 4 \\ 5 \end{pmatrix}$

4–4. MATRIX REPRESENTATIONS OF LINEAR MAPPINGS

In the previous sections we discussed mappings that were defined in terms of matrices. This led to a general definition of a linear mapping. In this section we discuss the fact that any linear mapping between two finite-dimensional vector spaces has a *matrix representation*; that is, there is a matrix that can be used to perform the mapping.

    The following theorem, giving important insights into the behavior of linear mappings, leads up to this concept.

> THEOREM 4-7   *Let U and V be vector spaces, let* $\mathbf{u}_1, \ldots, \mathbf{u}_n$ *be a basis for U, and let f be a linear mapping of U into V. f is completely determined if its value on each base vector is known. The range of f is spanned by* $f(\mathbf{u}_1), \ldots, f(\mathbf{u}_n)$.

    Thus, defining a linear mapping on a set of base vectors automatically defines it on the whole space.

Proof: Let **u** be an element of U. Since $\mathbf{u_1}, ..., \mathbf{u_n}$ is a basis of U, there exist scalars $a_1, ..., a_n$ such that

$$\mathbf{u} = a_1 \mathbf{u_1} + \cdots + a_n \mathbf{u_n}$$

Thus

$$f(\mathbf{u}) = f(a_1 \mathbf{u_1} + \cdots + a_n \mathbf{u_n})$$

Since f is linear,

$$f(\mathbf{u}) = a_1 f(\mathbf{u_1}) + \cdots + a_n f(\mathbf{u_n})$$

Hence $f(\mathbf{u})$ is known if $f(\mathbf{u_1}), ..., f(\mathbf{u_n})$ are known.

Further, $f(\mathbf{u})$ may be taken to be an arbitrary element in the range. It can be expressed as a linear combination of $f(\mathbf{u_1}), ..., f(\mathbf{u_n})$. Thus these vectors span the range of f.

We shall now discuss matrix representations of mappings.

THEOREM 4-8   *Let f be a linear mapping of a vector space U into a vector space V, and let $\mathbf{u_1}, ..., \mathbf{u_n}$ and $\mathbf{v_1}, ..., \mathbf{v_m}$ be bases for U and V, respectively. Let $f(\mathbf{u_i}) = \sum_{j=1}^{m} a_{ij} \mathbf{v_j}$ for $i = 1$ to n. Then the matrix A^t, the transpose of the matrix of coefficients defined by the right sides of these equations, is the* matrix representation *of f.*

Here we have examined the effect of f on each base vector of U, expressing the result in each case as a linear combination of the base vectors of V. The matrix of coefficients must characterize f completely since it indicates the effect of f on each base vector. The theorem states that the transpose of this matrix of coefficients is the matrix that can be used to perform the role of f. We now prove this and in so doing indicate how the matrix is used to define the mapping.

Proof: Let **u** be an arbitrary element of U. It can thus be represented as a linear combination of the vectors $\mathbf{u_1}, ..., \mathbf{u_n}$,

$$\mathbf{u} = \sum_{i=1}^{n} x_i \mathbf{u_i}$$

We have that

$$f(\mathbf{u}) = f\left(\sum_{i=1}^{n} x_i \mathbf{u_i}\right)$$

Since f is linear this can be written

$$f(\mathbf{u}) = \sum_{i=1}^{n} x_i f(\mathbf{u_i}) = \sum_{i=1}^{n} x_i \left(\sum_{j=1}^{m} a_{ij} \mathbf{v_j}\right)$$

The sums being finite, this equation may be rearranged to read

$$f(\mathbf{u}) = \sum_{j=1}^{m}\left(\sum_{i=1}^{n} a_{ij}x_i\right)\mathbf{v}_j$$

$$= \sum_{j=1}^{m} y_j\mathbf{v}_j \quad \text{where} \quad y_j = \sum_{i=1}^{n} a_{ij}x_i \tag{1}$$

Here $x_1, \ldots, x_n$ are the components of $\mathbf{u}$ relative to the basis $\mathbf{u}_1, \ldots, \mathbf{u}_n$ and $y_1, \ldots, y_n$ are the components of $f(\mathbf{u})$ relative to the basis $\mathbf{v}_1, \ldots, \mathbf{v}_m$.

Let $X = \begin{pmatrix} x_1 \\ \vdots \\ x_n \end{pmatrix}$ and $Y = \begin{pmatrix} x_1 \\ \vdots \\ y_m \end{pmatrix}$. Then we have from (1) that

$$Y = A^t X$$

A^t can be used in this manner to map X, the representation of U, into Y, the representation of $f(\mathbf{u})$.

Note that matrix representation does depend upon the bases selected.

Example 1

Consider the linear mapping f of $\mathbf{R}^3$ into itself defined by $f[(a,b,c)] = (a,b,0)$. (This projection mapping was discussed in Section 4-2.) Let us determine a matrix that "does the same job" as f.

In this context we interpret the elements of $\mathbf{R}^3$ as column vectors. Let us use the canonical bases for $\mathbf{R}^3$, $\begin{pmatrix} 1 \\ 0 \\ 0 \end{pmatrix}$, $\begin{pmatrix} 0 \\ 1 \\ 0 \end{pmatrix}$, and $\begin{pmatrix} 0 \\ 0 \\ 1 \end{pmatrix}$.

We determine the effect of f on each of these base vectors:

$$f\left[\begin{pmatrix} 1 \\ 0 \\ 0 \end{pmatrix}\right] = \begin{pmatrix} 1 \\ 0 \\ 0 \end{pmatrix} = 1\begin{pmatrix} 1 \\ 0 \\ 0 \end{pmatrix} + 0\begin{pmatrix} 0 \\ 1 \\ 0 \end{pmatrix} + 0\begin{pmatrix} 0 \\ 0 \\ 1 \end{pmatrix}$$

$$f\left[\begin{pmatrix} 0 \\ 1 \\ 0 \end{pmatrix}\right] = \begin{pmatrix} 0 \\ 1 \\ 0 \end{pmatrix} = 0\begin{pmatrix} 1 \\ 0 \\ 0 \end{pmatrix} + 1\begin{pmatrix} 0 \\ 1 \\ 0 \end{pmatrix} + 0\begin{pmatrix} 0 \\ 0 \\ 1 \end{pmatrix}$$

$$f\left[\begin{pmatrix} 0 \\ 0 \\ 1 \end{pmatrix}\right] = \begin{pmatrix} 0 \\ 0 \\ 0 \end{pmatrix} = 0\begin{pmatrix} 1 \\ 0 \\ 0 \end{pmatrix} + 0\begin{pmatrix} 0 \\ 1 \\ 0 \end{pmatrix} + 0\begin{pmatrix} 0 \\ 0 \\ 1 \end{pmatrix}$$

The matrix of coefficients is $\begin{pmatrix} 1 & 0 & 0 \\ 0 & 1 & 0 \\ 0 & 0 & 0 \end{pmatrix}$ and its transpose is $\begin{pmatrix} 1 & 0 & 0 \\ 0 & 1 & 0 \\ 0 & 0 & 0 \end{pmatrix}$.

The matrix representation of f is thus $\begin{pmatrix} 1 & 0 & 0 \\ 0 & 1 & 0 \\ 0 & 0 & 0 \end{pmatrix}$. Let us show that this matrix does, in fact, do the same job as f.

$$\begin{pmatrix} 1 & 0 & 0 \\ 0 & 1 & 0 \\ 0 & 0 & 0 \end{pmatrix} : \begin{pmatrix} a \\ b \\ c \end{pmatrix} \mapsto \begin{pmatrix} 1 & 0 & 0 \\ 0 & 1 & 0 \\ 0 & 0 & 0 \end{pmatrix} \begin{pmatrix} a \\ b \\ c \end{pmatrix} = \begin{pmatrix} a \\ b \\ 0 \end{pmatrix}$$

Example 2

f is a linear mapping from a vector space U having basis $\mathbf{u_1}, \mathbf{u_2}$ into a vector space V with basis $\mathbf{v_1}, \mathbf{v_2}, \mathbf{v_3}$. f is defined by $f(\mathbf{u_1}) = 2\mathbf{v_1} - \mathbf{v_2} + \mathbf{v_3}$, $f(\mathbf{u_2}) = 3\mathbf{v_1} + 2\mathbf{v_2} - 4\mathbf{v_3}$. Determine the matrix representation of f relative to these bases for U and V. Use this matrix to determine the image of the element $3\mathbf{u_1} + 2\mathbf{u_2}$ under f. Verify your answer by finding the image of $3\mathbf{u_1} + 2\mathbf{u_2}$ directly, using the linear properties of f.

The effect of f on the base vectors of U is given by

$$f(\mathbf{u_1}) = 2\mathbf{v_1} - \mathbf{v_2} + 3\mathbf{v_3}$$

$$f(\mathbf{u_2}) = 3\mathbf{v_1} + 2\mathbf{v_2} - 4\mathbf{v_3}$$

The matrix of coefficients of the right side is $\begin{pmatrix} 2 & -1 & 3 \\ 3 & 2 & -4 \end{pmatrix}$.

The matrix representation of f relative to these bases will be the transpose of this matrix, $\begin{pmatrix} 2 & 3 \\ -1 & 2 \\ 3 & -4 \end{pmatrix}$.

We now find the image of $3\mathbf{u_1} + 2\mathbf{u_2}$ using this matrix. We represent this vector by $\begin{pmatrix} 3 \\ 2 \end{pmatrix}$ and get

$$\begin{pmatrix} 2 & 3 \\ -1 & 2 \\ 3 & 4 \end{pmatrix} \begin{pmatrix} 3 \\ 2 \end{pmatrix} = \begin{pmatrix} 12 \\ 1 \\ 1 \end{pmatrix}$$

The image of the vector $3\mathbf{u_1} + 2\mathbf{u_2}$ has representation $\begin{pmatrix} 12 \\ 1 \\ 1 \end{pmatrix}$ in terms of the basis $\mathbf{v_1}, \mathbf{v_2}, \mathbf{v_3}$ of V. It is the vector $12\mathbf{v_1} + \mathbf{v_2} + \mathbf{v_3}$.

We now check this result by examining the effect of f on $3\mathbf{u_1} + 2\mathbf{u_2}$ directly. We get, using the linearity of f.

$$f(3\mathbf{u_1} + 2\mathbf{u_2}) = f(3\mathbf{u_1}) + f(2\mathbf{u_2})$$

$$= 3f(\mathbf{u_1}) + 2f(\mathbf{u_2})$$

$$= 3(2\mathbf{v_1} - \mathbf{v_2} + 3\mathbf{v_3}) + 2(3\mathbf{v_1} + 2\mathbf{v_2} - 4\mathbf{v_3})$$

$$= 12\mathbf{v_1} + \mathbf{v_2} + \mathbf{v_3}$$

Example 3*

Let V be the vector space of differentiable functions spanned by 1, x, and x^2. Let $D = d/dx$ be the operation of taking the derivative. From a previous section we know that D is linear; thus it has a matrix representation. Let us determine that matrix representation.

We examine the effect of D on the base vectors, 1, x, and x^2.

$$D(1) = 0 = 0(1) + 0(x) + 0(x^2)$$

$$D(x) = 1 = 1(1) + 0(x) + 0(x^2)$$

$$D(x^2) = 2x = 0(1) + 2(x) + 0(x^2)$$

The matrix of coefficients is $\begin{pmatrix} 0 & 0 & 0 \\ 1 & 0 & 0 \\ 0 & 2 & 0 \end{pmatrix}$. The matrix representation

of D is the transpose of this matrix, namely $\begin{pmatrix} 0 & 1 & 0 \\ 0 & 0 & 2 \\ 0 & 0 & 0 \end{pmatrix}$.

We now illustrate the use of this matrix in place of D. Let $a + bx + cx^2$ be an arbitrary element of V (an arbitrary linear combination of the base vectors). We can think of $a + bx + cx^2$ as having representation

$\begin{pmatrix} a \\ b \\ c \end{pmatrix}$ relative to the basis $1, x, x^2$. We find that

$$\begin{pmatrix} 0 & 1 & 0 \\ 0 & 0 & 2 \\ 0 & 0 & 0 \end{pmatrix} \begin{pmatrix} a \\ b \\ c \end{pmatrix} = \begin{pmatrix} b \\ 2c \\ 0 \end{pmatrix}$$

Thus the image of $\begin{pmatrix} a \\ b \\ c \end{pmatrix}$ has representation $\begin{pmatrix} b \\ 2c \\ 0 \end{pmatrix}$ relative to

this basis. It is the element $b + 2cx + 0x^2$. The matrix can be used in this manner to perform the mapping defined by D. As a check we do in fact see that $D(a + bx + cx^2) = b + 2cx$.

Matrix representations of linear mappings play a key role in the use of linear mappings. The representations vary with the bases. When linear mappings arise in analyzing situations, a goal is often to determine a simple representation in order to simplify the problem. This area of linear algebra, called *spectral theory*, involves the use of eigenvalues and eigenvectors (Section 6-5). The technique for determining the simplest representation of a linear mapping is used, for example, in solving systems of differential equations. Readers who desire further knowledge in this area should consult the author's *A Course in Linear Algebra*, Gordon and Breach, Science Publishers, 1972.

EXERCISES

1. Find the matrix representations of the following linear mappings of $\mathbf{R}^3$ into $\mathbf{R}^2$ relative to the canonical bases for these spaces.

   a) $f[(x, y, z)] = (x, z)$ **b)** $f[(x, y, z)] = (x + y, 2x - y)$

2. Determine the matrix representations of the following linear mappings of $\mathbf{R}^3$ into itself relative to the canonical basis of $\mathbf{R}^3$.

   a) $f[(x, y, z)] = (x, y, z)$ **b)** $f[(x, y, z)] = (x, 0, 0)$

   The mapping (b) is also called a *projection mapping*. It projects $\mathbf{R}^3$ onto the one-dimensional subspace, the x axis.

3. Let f be a linear mapping of $\mathbf{R}^3$ into itself defined by its value on the base vectors thus:

$$f\left[\begin{pmatrix} 1 \\ 0 \\ 0 \end{pmatrix}\right] = \begin{pmatrix} 0 \\ 2 \\ 0 \end{pmatrix}, \quad f\left[\begin{pmatrix} 0 \\ 1 \\ 0 \end{pmatrix}\right] = \begin{pmatrix} 3 \\ 0 \\ 0 \end{pmatrix},$$

$$f\left[\begin{pmatrix} 0 \\ 0 \\ 1 \end{pmatrix}\right] = \begin{pmatrix} 0 \\ 0 \\ -1 \end{pmatrix}$$

   By Theorem 4-7 of this section, f is defined completely. Determine the matrix representation of f relative to the canonical basis of $\mathbf{R}^3$ and use it to find the image of the element $\begin{pmatrix} 1 \\ 2 \\ 3 \end{pmatrix}$ under f.

4. Let f be a linear mapping of $\mathbf{R}^3$ into $\mathbf{R}^3$ defined by $f\left[\begin{pmatrix} 1 \\ 0 \\ 0 \end{pmatrix}\right] = \begin{pmatrix} 1 \\ 1 \\ 0 \end{pmatrix}$,

   $f\left[\begin{pmatrix} 0 \\ 1 \\ 0 \end{pmatrix}\right] = \begin{pmatrix} 0 \\ 0 \\ 1 \end{pmatrix}, f\left[\begin{pmatrix} 0 \\ 0 \\ 1 \end{pmatrix}\right] = \begin{pmatrix} 2 \\ 1 \\ 0 \end{pmatrix}$. Find the matrix representation

   of f relative to the canonical basis of $\mathbf{R}^3$. Use this matrix representation to determine the image of the element $\begin{pmatrix} -1 \\ 1 \\ 2 \end{pmatrix}$ of $\mathbf{R}^3$ under f.

5. Let f be a linear mapping from a vector space U with basis $\mathbf{u}_1, \mathbf{u}_2, \mathbf{u}_3, \mathbf{u}_4$ into a vector space V with basis $\mathbf{v}_1, \mathbf{v}_2, \mathbf{v}_3$, defined by $f(\mathbf{u}_1) = \mathbf{v}_1 + \mathbf{v}_2$, $f(\mathbf{u}_2) = 3\mathbf{v}_1 - 2\mathbf{v}_2 + \mathbf{v}_3$, $f(\mathbf{u}_3) = \mathbf{v}_1 + 2\mathbf{v}_2 - \mathbf{v}_3$, $f(\mathbf{u}_4) = 2\mathbf{v}_1 + \mathbf{v}_2 - 2\mathbf{v}_3$. Find the matrix representation of f relative to these bases. Use the matrix representation to determine the image of the element $\mathbf{u}_1 + 2\mathbf{u}_2 - \mathbf{u}_3 + 3\mathbf{u}_4$ under f. Verify your answers by finding the image directly using the linear properties of f.

6. Let f be the linear mapping from a vector space V with basis $\mathbf{v}_1, \mathbf{v}_2, \mathbf{v}_3$ into a vector space W with basis $\mathbf{w}_1, \mathbf{w}_2$, defined by $f(\mathbf{v}_1) = \mathbf{w}_1$, $f(\mathbf{v}_2) = \mathbf{w}_2$,

$f(\mathbf{v_3}) = \mathbf{0}$.   Find the matrix representation of f relative to these bases. Use this matrix representation to determine the image of the element $2\mathbf{v_1} - 3\mathbf{v_2} + 4\mathbf{v_3}$ under f.   Verify your answer by finding the image directly using the linear properties of f.

7. As in Example 3, the functions 1, $2x$, and $3x^2$ can be interpreted as base vectors for a three-dimensional vector space U.   Determine the matrix representation of D, the differential operator relative to this basis.   Note that this vector space is the same vector space as the one in Example 3.   This exercise and Example 3 illustrate how a linear mapping D has two distinct matrix representations relative to two distinct bases.

8. Let V be the vector space generated by the functions $\sin t$ and $\cos t$ defined on the interval $0 \leq t \leq \pi$.   If D is the operation of taking the derivative and D^2 is that of taking the second derivative, find the matrix representation of the linear mapping $D^2 + 2D + 1$ of V into itself relative to the basis $\sin t$, $\cos t$ of V.   What is the image of the element $3 \sin t + \cos t$ under this mapping?

$$5$$

Determinants

Associated with every square matrix is a number called its *determinant*. The determinant is a tool used in many branches of mathematics, science, and engineering. In this chapter the determinant is defined, its properties developed, and its use in proving certain properties of vector operations discussed. In later sections the reader will see its use in developing a formula for the inverse of a matrix. Engineering readers will see how it is employed to determine principal stresses and directions at a point in a body.

5–1. INTRODUCTION TO DETERMINANTS

Consider a 2×2 matrix $\begin{pmatrix} a_{11} & a_{12} \\ a_{21} & a_{22} \end{pmatrix}$. The determinant of such a matrix is defined to be the number $a_{11} a_{22} - a_{12} a_{21}$. The determinant of this matrix is usually denoted $\begin{vmatrix} a_{11} & a_{12} \\ a_{21} & a_{22} \end{vmatrix}$.

Example 1

Find the determinant of the matrix $\begin{pmatrix} -1 & 2 \\ 0 & 1 \end{pmatrix}$.

$$\begin{vmatrix} -1 & 2 \\ 0 & 1 \end{vmatrix} = (-1 \times 1) - (2 \times 0) = -1$$

The definition of the determinant of a 3×3 matrix is given in terms of the determinants of 2×2 matrices, that of a 4×4 matrix in terms of the determinants of 3×3 matrices, etc. For this definition we need the concept of a *minor*. Let A be an $n \times n$ matrix. Associated with an arbitrary element a_{ij} of A we get an $n-1 \times n-1$ matrix obtained by deleting the ith row and jth column. The determinant of this submatrix is called the minor of a_{ij}. There will exist such a minor for every element of the matrix. Denote the minor of the element a_{ij} by A_{ij}.

Example 2

If $A = \begin{pmatrix} 1 & 0 & 3 \\ 4 & -1 & 2 \\ 0 & 1 & 1 \end{pmatrix}$, then the minor of the element a_{11}, the element in the first row and first column, is $A_{11} = \begin{vmatrix} -1 & 2 \\ 1 & 1 \end{vmatrix} = -3$. The minors of the other first row elements are

$$A_{12} = \begin{vmatrix} 4 & 2 \\ 0 & 1 \end{vmatrix} = 4; \quad A_{13} = \begin{vmatrix} 4 & -1 \\ 0 & 1 \end{vmatrix} = 4$$

The minors of the second row elements are

$$A_{21} = \begin{vmatrix} 0 & 3 \\ 1 & 1 \end{vmatrix} = -3; \quad A_{22} = \begin{vmatrix} 1 & 3 \\ 0 & 1 \end{vmatrix} = 1;$$

$$A_{23} = \begin{vmatrix} 1 & 0 \\ 0 & 1 \end{vmatrix} = 1$$

The third row minors are

$$A_{31} = \begin{vmatrix} 0 & 3 \\ -1 & 2 \end{vmatrix} = 3; \quad A_{32} = \begin{vmatrix} 1 & 3 \\ 4 & 2 \end{vmatrix} = -10;$$

$$A_{33} = \begin{vmatrix} 1 & 0 \\ 4 & -1 \end{vmatrix} = -1$$

We now define the determinant of a 3×3 matrix

$$A = \begin{pmatrix} a_{11} & a_{12} & a_{13} \\ a_{21} & a_{22} & a_{23} \\ a_{31} & a_{32} & a_{33} \end{pmatrix}$$

using the first row. The elements of the first row are multiplied by the determinants of their corresponding minors and summed as follows.

$$\begin{vmatrix} a_{11} & a_{12} & a_{13} \\ a_{21} & a_{22} & a_{23} \\ a_{31} & a_{32} & a_{33} \end{vmatrix} = a_{11}A_{11} - a_{12}A_{12} + a_{13}A_{13}$$

Note that the signs alternate.

Example 3

Find the determinant of the matrix $\begin{pmatrix} 1 & 2 & -1 \\ 3 & 0 & 1 \\ 4 & 2 & 1 \end{pmatrix}$.

We know that $a_{11} = 1, a_{12} = 2,$ and $a_{13} = -1$.

$$A_{11} = \begin{vmatrix} 0 & 1 \\ 2 & 1 \end{vmatrix}, \quad A_{12} = \begin{vmatrix} 3 & 1 \\ 4 & 1 \end{vmatrix}, \quad \text{and} \quad A_{13} = \begin{vmatrix} 3 & 0 \\ 4 & 2 \end{vmatrix}$$

Hence

$$\begin{vmatrix} 1 & 2 & -1 \\ 3 & 0 & 1 \\ 4 & 2 & 1 \end{vmatrix} = 1 \begin{vmatrix} 0 & 1 \\ 2 & 1 \end{vmatrix} - 2 \begin{vmatrix} 3 & 1 \\ 4 & 1 \end{vmatrix} + (-1) \begin{vmatrix} 3 & 0 \\ 4 & 2 \end{vmatrix}$$

$$= 1 [(0 \times 1) - (1 \times 2)] - 2 [(3 \times 1) - (1 \times 4)]$$

$$- [(3 \times 2) - (0 \times 4)]$$

$$= 1 (0 - 2) - 2 (3 - 4) - (6 - 0)$$

$$= -2 + 2 - 6$$

$$= -6$$

The determinant of the 3×3 matrix A can be written using sigma notation.

$$\begin{vmatrix} a_{11} & a_{12} & a_{13} \\ a_{21} & a_{22} & a_{23} \\ a_{31} & a_{32} & a_{33} \end{vmatrix} = \sum_{j=1}^{3} (-1)^{1+j} a_{1j} A_{1j}$$

The $(-1)^{1+j}$ ensures that each term has the proper sign.

The determinants of larger matrices are defined similarly in terms of the first row and the minors of the first row. For example, the determinant of a 4×4 matrix is

$$\begin{vmatrix} a_{11} & a_{12} & a_{13} & a_{14} \\ a_{21} & a_{22} & a_{23} & a_{24} \\ a_{31} & a_{32} & a_{33} & a_{34} \\ a_{41} & a_{42} & a_{43} & a_{44} \end{vmatrix} = a_{11} A_{11} - a_{12} A_{12} + a_{13} A_{13} - a_{14} A_{14}$$

Note that the signs always alternate.

Example 4

$$\begin{vmatrix} 3 & 0 & -1 & 2 \\ 4 & 1 & 3 & 2 \\ 6 & -1 & 2 & 0 \\ 1 & 2 & 4 & 1 \end{vmatrix}$$

$$= 3 \begin{vmatrix} 1 & 3 & 2 \\ -1 & 2 & 0 \\ 2 & 4 & 1 \end{vmatrix} - 0 \begin{vmatrix} 4 & 3 & 2 \\ 6 & 2 & 0 \\ 1 & 4 & 1 \end{vmatrix} + (-1) \begin{vmatrix} 4 & 1 & 2 \\ 6 & -1 & 0 \\ 1 & 2 & 1 \end{vmatrix} - 2 \begin{vmatrix} 4 & 1 & 3 \\ 6 & -1 & 2 \\ 1 & 2 & 4 \end{vmatrix}$$

$$= 3\left[1\begin{vmatrix} 2 & 0 \\ 4 & 1 \end{vmatrix} - 3\begin{vmatrix} -1 & 0 \\ 2 & 1 \end{vmatrix} + 2\begin{vmatrix} -1 & 2 \\ 2 & 4 \end{vmatrix}\right]$$

$$- \left[4\begin{vmatrix} -1 & 0 \\ 2 & 1 \end{vmatrix} - 1\begin{vmatrix} 6 & 0 \\ 1 & 1 \end{vmatrix} + 2\begin{vmatrix} 6 & -1 \\ 1 & 2 \end{vmatrix}\right]$$

$$- 2\left[4\begin{vmatrix} -1 & 2 \\ 2 & 4 \end{vmatrix} - 1\begin{vmatrix} 6 & 2 \\ 1 & 4 \end{vmatrix} + 3\begin{vmatrix} 6 & -1 \\ 1 & 2 \end{vmatrix}\right]$$

$$= 3(2 + 3 - 16) - (-4 - 6 + 26)$$

$$- 2(-32 - 22 + 39)$$

$$= 3(-11) - (16) - 2(-15) = -33 - 16 + 30 = -19$$

In terms of sigma notation, the determinant of a 4×4 matrix is

$$\begin{vmatrix} a_{11} & a_{12} & a_{13} & a_{14} \\ a_{21} & a_{22} & a_{23} & a_{24} \\ a_{31} & a_{32} & a_{33} & a_{34} \\ a_{41} & a_{42} & a_{43} & a_{44} \end{vmatrix} = \sum_{j=1}^{4} (-1)^{1+j} a_{1j} A_{1j}$$

The determinant of the $n \times n$ matrix $\begin{pmatrix} a_{11} & \cdots & a_{1n} \\ & \vdots & \\ a_{n1} & \cdots & a_{nn} \end{pmatrix}$ is defined

$$\begin{vmatrix} a_{11} & \cdots & a_{1n} \\ & \vdots & \\ a_{n1} & \cdots & a_{nn} \end{vmatrix} = a_{11} A_{11} - a_{12} A_{12} + a_{13} A_{13} - \cdots (-1)^{1+n} a_{1n} A_{1n} \quad \text{Using}$$

sigma notation, this is written

$$\begin{vmatrix} a_{11} & \cdots & a_{1n} \\ & \vdots & \\ a_{n1} & \cdots & a_{nn} \end{vmatrix} = \sum_{j=1}^{n} (-1)^{1+j} a_{1j} A_{1j}$$

We defined the determinant of a 3×3 matrix in terms of determinants of 2×2 matrices and the determinant of a 4×4 matrix in terms of determinants of 3×3 matrices. The determinant of an $n \times n$ matrix is defined in terms of the determinants of $(n-1) \times (n-1)$ matrices. Since we gave a definition of the determinant of a 2×2 matrix, the determinants of all square matrices of finite order are thus defined. We call such a definition an *inductive definition*. Since a number can be regarded as a 1×1 matrix, we complete this definition by defining the determinant of a 1×1 matrix to be that number.

A square matrix is said to be *singular* if it has a zero determinant; otherwise it is *nonsingular*. These concepts are important in discussing solutions to systems of linear equations, as the following example illustrates.

Example 5

In Example 6, Section 1-2, we discussed the system of equations

$$x - 2y + 3z + w = 0$$

$$2x - 3y + 2z - w = 0$$

$$3x - 5y + 5z = 0$$

$$x - y - z - 2w = 0$$

It was demonstrated that many solutions exist to this system. One solution is $x = 10, y = 7, z = 1, w = 1$; another solution is $x = 0, y = 1, z = 1, w = -1$.

The matrix of coefficients is

$$\begin{pmatrix} 1 & -2 & 3 & 1 \\ 2 & -3 & 2 & -1 \\ 3 & -5 & 5 & 0 \\ 1 & -1 & -1 & -2 \end{pmatrix}$$

It can be verified that the determinant of this matrix is zero,

$$\begin{vmatrix} 1 & -2 & 3 & 1 \\ 2 & -3 & 2 & -1 \\ 3 & -5 & 5 & 0 \\ 1 & -1 & -1 & -2 \end{vmatrix} = 0$$

and therefore the matrix of coefficients is singular.

Later we shall see that the determinant of the matrix of coefficients of certain systems of linear equations is related to the existence and uniqueness of solutions to those equations. For example, if the determinant of the matrix of coefficients is singular, as is the case here, then the system cannot have a single solution; it has either many solutions or none at all.

We have defined determinants in terms of the first row; however, the determinant of a matrix can be found using any row or column.

For example, the determinant of a 3×3 matrix can be found using the third row rather than the first row. The rule is

$$\begin{vmatrix} a_{11} & a_{12} & a_{13} \\ a_{21} & a_{22} & a_{23} \\ a_{31} & a_{32} & a_{33} \end{vmatrix} = a_{31} A_{31} - a_{32} A_{32} + a_{33} A_{33}$$

The elements of the third row play a role in this expansion analogous to the one played by the elements of the first row in the previous expansion.

Example 6

Find the determinant of the matrix $\begin{pmatrix} 1 & 2 & -1 \\ 3 & 0 & 1 \\ 4 & 2 & 1 \end{pmatrix}$ using the third row.

$$\begin{vmatrix} 1 & 2 & -1 \\ 3 & 0 & 1 \\ 4 & 2 & 1 \end{vmatrix} = 4 \begin{vmatrix} 2 & -1 \\ 0 & 1 \end{vmatrix} - 2 \begin{vmatrix} 1 & -1 \\ 3 & 1 \end{vmatrix} + 1 \begin{vmatrix} 1 & 2 \\ 3 & 0 \end{vmatrix}$$

$$= 8 - 8 - 6 = -6$$

Note that we have already evaluated this determinant in terms of the first row in Example 3 of this section and that we got the same number, -6.

Using sigma notation, the expansion of the determinant in terms of the third row is

$$\begin{vmatrix} a_{11} & a_{12} & a_{13} \\ a_{21} & a_{22} & a_{23} \\ a_{31} & a_{32} & a_{33} \end{vmatrix} = \sum_{j=1}^{3} (-1)^{3+j} a_{3j} A_{3j}$$

There is a useful rule that can be memorized for expanding a determinant using any row or column. This rule gives the signs in the expansion—expansions using certain rows and columns begin with a minus sign rather than a plus sign. In any case, they then alternate. The rule is illustrated below.

$$\begin{pmatrix} + & - & + & - & \cdots \\ - & + & - & + & \cdots \\ + & - & + & - & \cdots \\ & & \vdots & & \end{pmatrix}$$

A certain sign, plus or minus, is attached to every element in the matrix. If one expands using the second row or the fourth row, then the signs in the expansion go $-$, $+$, $-$, etc.

Example 7

Find the determinant of the matrix $\begin{pmatrix} 1 & 2 & -1 \\ 3 & 0 & 1 \\ 4 & 2 & 1 \end{pmatrix}$ using the second column.

Using the above rule, the signs involved are

$$\begin{pmatrix} + & - & + \\ - & + & - \\ + & - & + \end{pmatrix}$$

Hence, using the second column, the signs of the terms in the expansion will go $-, +, -$. Thus

$$\begin{vmatrix} 1 & 2 & -1 \\ 3 & 0 & 1 \\ 4 & 2 & 1 \end{vmatrix} = -2 \begin{vmatrix} 3 & 1 \\ 4 & 1 \end{vmatrix} + 0 \begin{vmatrix} 1 & -1 \\ 4 & 1 \end{vmatrix} - 2 \begin{vmatrix} 1 & -1 \\ 3 & 1 \end{vmatrix}$$

$$= 2 + 0 - 8 = -6$$

Note that this matrix is the one in Example 6 and that we get the same value for the determinant.

Let us now give the formal definition of the determinant of a square matrix first in terms of an arbitrary row and then in terms of an arbitrary column.

The determinant of the matrix

$$\begin{pmatrix} a_{11} & \cdots & a_{1n} \\ & \vdots & \\ a_{n1} & \cdots & a_{nn} \end{pmatrix}$$

in terms of the kth row is

$$\begin{vmatrix} a_{11} & \cdots & a_{1n} \\ & \vdots & \\ a_{n1} & \cdots & a_{nn} \end{vmatrix} = \sum_{j=1}^{n} (-1)^{k+j} a_{kj} A_{kj}$$

In terms of the hth column, it is

$$= \sum_{i=1}^{n} (-1)^{i+h} a_{ih} A_{ih}$$

The minor A_{kj} multiplied by $(-1)^{k+j}$ is called the *cofactor* of a_{kj}. Denote this cofactor C_{kj}. Thus

$$C_{kj} = (-1)^{k+j} A_{kj}$$

This method of finding the determinant of a matrix by making use of the definition is not the most efficient way. We present a numerical method in Section 5-3 which is more suitable for programming on a computer. However, as we shall see later, this method is of theoretical importance.

EXERCISES

1. Evaluate the following determinants:

a) $\begin{vmatrix} 1 & 2 \\ 0 & 1 \end{vmatrix}$
b) $\begin{vmatrix} -2 & -1 \\ 3 & 1 \end{vmatrix}$
c) $\begin{vmatrix} -1 & 2 & 3 \\ 0 & 1 & 4 \\ 1 & 1 & 2 \end{vmatrix}$

d) $\begin{vmatrix} 0 & 0 & 1 \\ 2 & 3 & 0 \\ 1 & -1 & 3 \end{vmatrix}$
e) $\begin{vmatrix} -1 & 2 & 3 \\ 0 & 1 & 4 \\ 1 & 2 & 1 \end{vmatrix}$
f) $\begin{vmatrix} 1 & 2 & 3 & -1 \\ 1 & -1 & 0 & 2 \\ 0 & 1 & 0 & 1 \\ 0 & 0 & -1 & 2 \end{vmatrix}$

2. Find out whether the following matrices are singular or nonsingular.

a) $\begin{pmatrix} 1 & 2 \\ 1 & 2 \end{pmatrix}$
b) $\begin{pmatrix} 1 & 2 \\ 0 & 0 \end{pmatrix}$

c) $\begin{pmatrix} 3 & 4 \\ 1 & 2 \end{pmatrix}$
d) $\begin{pmatrix} 1 & 2 & 3 \\ 2 & 4 & 6 \\ 0 & 1 & 2 \end{pmatrix}$

e) $\begin{pmatrix} -1 & 0 & 2 \\ 2 & 1 & -4 \\ 3 & 4 & -6 \end{pmatrix}$
f) $\begin{pmatrix} 1 & 2 & 3 & 4 \\ 0 & -1 & 2 & 3 \\ -1 & -2 & -3 & -4 \\ 0 & 1 & 0 & 1 \end{pmatrix}$

3. Write in sigma notation the determinant of
a) an arbitrary 5×5 matrix
b) an arbitrary 6×6 matrix

4. Evaluate the following determinants in terms of their first columns, then in terms of their second columns.

a) $\begin{vmatrix} -1 & 2 & 3 \\ 0 & 1 & 2 \\ -1 & 1 & 1 \end{vmatrix}$
b) $\begin{vmatrix} 1 & 3 & 0 \\ 4 & 1 & 2 \\ 5 & 0 & 1 \end{vmatrix}$
c) $\begin{vmatrix} -1 & 2 & 5 & 4 \\ 0 & 3 & 2 & 0 \\ 0 & 4 & 1 & -1 \\ 0 & 1 & 1 & 3 \end{vmatrix}$

5. Evaluate the following determinants in terms of their third rows, then in terms of their second columns.

a) $\begin{vmatrix} 1 & 0 & 2 \\ 3 & 1 & 4 \\ 5 & 6 & 1 \end{vmatrix}$
b) $\begin{vmatrix} 2 & 1 & 3 \\ 1 & 1 & 1 \\ 0 & 0 & 1 \end{vmatrix}$
c) $\begin{vmatrix} 3 & 4 & 1 \\ 2 & 1 & 1 \\ 1 & 1 & 1 \end{vmatrix}$

d) $\begin{vmatrix} 2 & 0 & 1 \\ 3 & 0 & 2 \\ 4 & 1 & 1 \end{vmatrix}$
e) $\begin{vmatrix} 1 & 0 & 1 & 2 \\ 4 & 0 & 3 & 4 \\ 5 & 1 & 2 & 3 \\ 0 & 0 & -1 & 2 \end{vmatrix}$

Notice that the arithmetic is simplest when one expands the determinant in terms of the row or column containing the most zeros.

6. Evaluate the determinant $\begin{vmatrix} a & a^2 & a^3 \\ b & b^2 & b^3 \\ c & c^2 & c^3 \end{vmatrix}$ by using the second column.

7. Write out the expansion of the determinant of a general 3×3 matrix in terms of third column, then express your answer in sigma notation.

8. Write out the expansion of the determinant of a general 4×4 matrix in terms of the fourth column, then express your answer in sigma notation.

9. If A is the diagonal matrix

$$\begin{pmatrix} a_{11} & & & 0 \\ & a_{22} & & \\ & & \ddots & \\ 0 & & & a_{nn} \end{pmatrix}$$

prove that $|A| = a_{11} a_{22} \cdots a_{nn}$.

   Hence, for such a matrix A, $|A| = 0$ if and only if one of the diagonal diagonal elements is zero.

10. If A is an upper triangular matrix, a matrix of the type

$$\begin{pmatrix} a_{11} & a_{12} & \cdots & a_{1n} \\ & a_{22} & & \\ & & \ddots & \\ 0 & & & a_{nn} \end{pmatrix}$$

prove that $|A| = a_{11} a_{22} \cdots a_{nn}$.

   Hence, for such a matrix A, $|A| = 0$ if and only if one of the diagonal elements is zero.

5–2. PROPERTIES OF DETERMINANTS

The determinant of a square matrix is a number associated with that matrix. In this section we develop properties of determinants and find that the determinants of various matrices are related.

THEOREM 5-1 *If all the elements of one row or column of a square matrix are zero, then the determinant of that matrix is zero.*

Proof: Let the elements of the kth row of A be zero. If we expand the determinant of A in terms of the kth row,

$$|A| = \sum_{j=1}^{n} (-1)^{k+j} a_{kj} A_{kj}$$

Since the elements a_{kj}, where $j = 1, ..., n$, are zero, $|A| = 0$. Similarly, if all the elements in one column are zero, expanding in terms of that column shows the determinant to be zero.

THEOREM 5-2 *If a matrix B is obtained from A by multiplying the elements of a row or column by a nonzero scalar c, then $|B| = c|A|$.*

Proof: Suppose B is obtained from A by multiplying its kth row by c. Hence the kth row of B is ca_{kj}, where $j = 1, ..., n$, and all other elements of B are the same as the corresponding elements of A. Expand the determinant of B in terms of the kth row:

$$B = \sum_{j=1}^{n} (-1)^{k+j} b_{kj} B_{kj}$$

$$= \sum_{j=1}^{n} (-1)^{k+j} ca_{kj} A_{kj} = c \sum_{j=1}^{n} (-1)^{k+j} a_{kj} A_{kj}$$

$$= c|A|$$

The proof for columns is similar.

Example 1

If $A = \begin{pmatrix} 1 & 2 & 3 \\ 0 & 1 & 1 \\ 4 & -1 & 0 \end{pmatrix}$ and $B = \begin{pmatrix} 1 & 6 & 3 \\ 0 & 3 & 1 \\ 4 & -3 & 0 \end{pmatrix}$, then the second column of B is three times the second column of A. Evaluating the determinants, $|A| = -3$ and $|B| = -9$. Hence $|B| = 3|A|$, illustrating Theorem 5-2.

THEOREM 5-3 *If B is obtained from A by interchanging two rows (columns), then $|B| = -|A|$.*

Proof: The proof is by induction. We see that it holds for 2×2 matrices. Assuming the result holds for $n \times n$ matrices, we shall show that it also holds for $(n+1) \times (n+1)$ matrices, thereby proving by induction that it holds for all square matrices.

Let B be an $(n+1) \times (n+1)$ matrix obtained from A by interchanging two rows. Expand B in terms of a row that is not one of those interchanged, such as the kth row. Then

$$|B| = \sum_{j=1}^{n} (-1)^{k+j} b_{kj} B_{kj}$$

Each b_{kj}, where $j = 1, ..., n$, is identical to the corresponding a_{kj}. Each B_{kj}, where $j = 1, ..., n$, is obtained from the corresponding A_{kj} by interchanging two rows. Hence $b_{kj} = a_{kj}$, and $B_{kj} = -A_{kj}$, where $j = 1, ..., n$. Hence

$$|B| = -\sum_{j=1}^{n} (-1)^{k+j} a_{kj} A_{kj} = -|A|$$

The proof for columns is similar.

Example 2

If $A = \begin{pmatrix} 1 & 2 & -1 \\ 3 & 0 & 1 \\ 4 & 2 & 1 \end{pmatrix}$ and $B = \begin{pmatrix} -1 & 2 & 1 \\ 1 & 0 & 3 \\ 1 & 2 & 4 \end{pmatrix}$, then B can be

obtained from A by interchanging the first and third columns. Evaluating the determinants, we find that $|A| = -6$ and $|B| = 6$, illustrating Theorem 5-3.

THEOREM 5-4   *If two rows (columns) of a square matrix A are proportional, then $|A| = 0$.*

Proof: Let the kth row (column) of a matrix A be c times the hth, c being a nonzero scalar.

Let B be the matrix obtained on dividing every element of the kth row (column) of A by c. Thus $|A| = c|B|$. All other elements of A and B are the same. The matrix B has identical kth and hth rows (columns). On interchanging these identical rows (columns) the determinant remains unchanged. However, according to Theorem 5-3, the sign should be reversed. This is only possible if $|B| = 0$. Hence $|A| = 0$.

Note that this proof demonstrates a special case of this theorem: A matrix with two identical rows (columns) has a zero determinant.

Example 3

The matrix $A = \begin{pmatrix} 1 & 2 & 3 \\ 0 & 1 & 2 \\ 2 & 4 & 6 \end{pmatrix}$ is such that the third row is twice the first.

Evaluating, we find that $|A| = 0$, illustrating Theorem 5-4.

THEOREM 5-5   *If B is obtained from the square matrix A by multiplying the kth row (column) by a nonzero scalar c and adding the result to the ith row (column) using vector addition (where $i \neq k$), then $|B| = |A|$.*

Proof: The element b_{ij} of the ith row of B is of the form $a_{ij} + ca_{kj}$. All elements in rows other than the ith row of B are identical to the corresponding elements in A. Expanding the determinant of B in terms of the ith row,

$$|B| = \sum_{j=1}^{n} (-1)^{i+j} b_{ij} B_{ij}$$

$$= \sum_{j=1}^{n} (-1)^{i+j} (a_{ij} + ca_{kj}) A_{ij}$$

$$= \sum_{j=1}^{n} (-1)^{i+j} a_{ij} A_{ij} + c \sum_{j=1}^{n} (-1)^{i+j} a_{kj} A_{ij}$$

$$= |A| + |A'|$$

where A' is obtained from A by replacing the ith row by the kth row. But since the ith row and the kth row of A' are identical, $|A'| = 0$. Hence $|B| = |A|$.

The proof involving columns is similar.

Example 4

Evaluate the following determinant.

$$|A| = \begin{vmatrix} 0 & 1 & 3 & 0 \\ 4 & 5 & 7 & 2 \\ -2 & -1 & -3 & 0 \\ 4 & 2 & 5 & 0 \end{vmatrix}$$

Multiply the third row by 2 and add the result to the fourth row to get

$$|A| = \begin{vmatrix} 0 & 1 & 3 & 0 \\ 4 & 5 & 7 & 2 \\ -2 & -1 & -3 & 0 \\ 0 & 0 & -1 & 0 \end{vmatrix}$$

Now expand in terms of the last row to get

$$-(-1) \begin{vmatrix} 0 & 1 & 0 \\ 4 & 5 & 2 \\ -2 & -1 & 0 \end{vmatrix} = -1 \begin{vmatrix} 4 & 2 \\ -2 & 0 \end{vmatrix} = -4$$

In this manner, Theorem 5-5 can be used to create zeros in certain locations of the determinant in order to make evaluation easier.

THEOREM 5-6   *If A and B are square matrices, B being the transpose of A, then* $|A| = |B|$.

Proof: We shall proceed by induction. The theorem is true for 1×1 matrices. Assume that the theorem is true for all square matrices of order n. We shall show that it then holds for matrices of order $n+1$ and hence, by induction, that it holds for all square matrices.

Let A and B be $(n+1) \times (n+1)$ matrices. Expand $|A|$ in terms of the first row to get

$$|A| = \sum_{j=1}^{n+1} (-1)^{1+j} a_{1j} A_{1j} \tag{1}$$

Expand $|B|$ in terms of the first column to get

$$|B| = \sum_{i=1}^{n+1} (-1)^{i+1} b_{i1} B_{i1} \tag{2}$$

But since $B = A^t$, the first row of A is identical to the first column of B; $a_{1j} = b_{j1}$, where $j = 1, ..., n+1$. A_{1j} and B_{j1} are both determinants of $n \times n$ matrices. Since $B = A^t$, the one $n \times n$ matrix is the transpose of the other. Hence, by the hypothesis involving determinants of square matrices of order n, $A_{1j} = B_{j1}$ for all $j = 1, ..., n+1$.

The two right-hand sides of identities (1) and (2) are therefore equal, $|A| = |B|$. By induction, the theorem holds for all square matrices.

THEOREM 5-7   *The determinant of the product of two matrices is equal to the product of their determinants.* $|AB| = |A| \, |B|$.

We state this theorem without proof.

We shall now give an alternative definition of determinants that is used mainly for theoretical purposes.† Both definitions are equivalent.

The second definition is in terms of the *permutation symbol*. The permutation symbol is defined as follows.

$$\varepsilon_{i_1 i_2 \cdots i_n} \begin{cases} = +1 \text{ if } i_1, i_2, ..., i_n \text{ is an even permutation} \\ \quad \text{of } 1, 2, ..., n \\ = -1 \text{ if } i_1, i_2, ..., i_n \text{ is an odd permutation} \\ \quad \text{of } 1, 2, ..., n \\ = 0 \text{ otherwise} \end{cases}$$

$i_1, i_2, ..., i_n$ is an even permutation of $1, 2, ..., n$ if it can be rearranged in the form $1, 2, ..., n$ through an even number of interchanges. Otherwise it is an odd permutation.

Example 5

$\varepsilon_{13425} = +1$, because 13425 can be rearranged in the form 12345 in two interchanges:

$$13425 \rightarrow 13245 \rightarrow 12345$$

---

† The remainder of this section is optional.

Example 6 $\varepsilon_{13452} = -1$, because

$$13452 \rightarrow 13425 \rightarrow 13245 \rightarrow 12345$$

An odd number of interchanges is required.

Example 7 $\varepsilon_{13243} = 0$, since two of the indices are identical.

The alternative definition of the determinant in terms of this symbol is

$$|A| = \sum_{p=1}^{2} \cdots \sum_{q=1}^{2} \varepsilon_{p \cdots q} a_{1p} \cdots a_{nq}$$

It can be seen that this definition is equivalent to the previous one for a 2×2 matrix $A = \begin{pmatrix} a_{11} & a_{12} \\ a_{21} & a_{22} \end{pmatrix}$:

$$|A| = \sum_{p=1}^{n} \sum_{q=1}^{n} \varepsilon_{pq} a_{1p} a_{2q} = \varepsilon_{12} a_{11} a_{22} + \varepsilon_{21} a_{12} a_{21}$$

$$= a_{11} a_{22} - a_{12} a_{21}$$

EXERCISES

1. If $A = \begin{pmatrix} 1 & 2 & 3 \\ 0 & 1 & 1 \\ 4 & -1 & 2 \end{pmatrix}$ and $B = \begin{pmatrix} 1 & 2 & -3 \\ 0 & 1 & -1 \\ 4 & -1 & -2 \end{pmatrix}$, prove that $|B| = -|A|$ by expanding the determinants. This illustrates Theorem 5-2.

2. If $A = \begin{pmatrix} 1 & 2 & 3 \\ 1 & 1 & 1 \\ 4 & 1 & 2 \end{pmatrix}$ and $B = \begin{pmatrix} 1 & 2 & 3 \\ 4 & 1 & 2 \\ 1 & 1 & 1 \end{pmatrix}$, prove that $|B| = -|A|$ by expanding the determinants. This illustrates Theorem 5-3.

3. If $A = \begin{pmatrix} 1 & 3 & 1 \\ 2 & 6 & 2 \\ 4 & 1 & 2 \end{pmatrix}$, prove that $|A| = 0$ by expanding the determinant. This illustrates Theorem 5-4.

4. Simplify the following determinants using Theorem 5-5. Then evaluate.

a) $\begin{vmatrix} 1 & 2 & 3 \\ 2 & 4 & 1 \\ 1 & 1 & 1 \end{vmatrix}$

b) $\begin{vmatrix} 0 & 1 & 5 \\ 1 & 1 & 6 \\ 2 & 2 & 7 \end{vmatrix}$

c) $\begin{vmatrix} 2 & 1 & -1 \\ 3 & -1 & 1 \\ 1 & 4 & -4 \end{vmatrix}$

d) $\begin{vmatrix} 3 & -1 & 0 \\ 4 & 2 & 1 \\ 1 & 1 & 2 \end{vmatrix}$

5. If $A = \begin{pmatrix} 1 & 2 & 3 \\ 0 & 1 & 1 \\ 4 & -1 & 2 \end{pmatrix}$ and $B = \begin{pmatrix} 1 & 0 & 4 \\ 2 & 1 & -1 \\ 3 & 1 & 2 \end{pmatrix}$, prove that $|A| = |B|$ by expanding the determinants. Why would you expect this result?

6. Let A be a square matrix whose columns are $\mathbf{a}^1, ..., \mathbf{a}^i + \mathbf{a}'^i, ..., \mathbf{a}^n$; the ith column is interpreted as being the sum of two column vectors. Prove that

$$|A| = |\mathbf{a}^1, ..., \mathbf{a}^i + \mathbf{a}'^i, ..., \mathbf{a}^n|$$

$$= |\mathbf{a}^1, ..., \mathbf{a}^i, ..., \mathbf{a}^n| + |\mathbf{a}^1, ..., \mathbf{a}'^i, ..., \mathbf{a}^n|$$

(*Hint*: Expand the determinant of A using the ith column.)

7. Evaluate the following: $\varepsilon_{12345}, \varepsilon_{32415}, \varepsilon_{324151}$.

8. Give the definition for the determinant of an arbitrary 3×3 matrix in terms of the permutation symbol. Show that this definition is equivalent to the original definition given in terms of the minors.

5–3. THE EVALUATION OF A DETERMINANT

We have introduced the concept of the determinant and given an inductive, theoretical method for its evaluation. Here we give a numerical method that is suitable for programming on the computer.

In the method of Gaussian elimination, to solve a system of linear equations we used certain allowable transformations to change the initial system into another system which was easier to solve. We use a similar approach to compute a determinant. In the previous section we saw that adding a multiple of one row to another left a determinant unchanged and that interchanging two rows negated a determinant. We use these operations to transform a determinant into an upper triangular form. The following theorem tells us that the determinant of such a matrix is the product of its diagonal elements.

THEOREM 5-8 *If A is an upper triangular matrix then its determinant is the product of the diagonal elements.*

Proof: A is an upper triangular matrix, that is a matrix of the type

$$\begin{pmatrix} a_{11} & a_{12} & \cdots & a_{1n} \\ & a_{22} & \cdots & a_{2n} \\ & & \ddots & \\ 0 & & & a_{nn} \end{pmatrix}$$

All the elements below the diagonal are zero.

In evaluating $|A|$ we expand each determinant below in terms of the

273

first column:

$$|A| = \begin{vmatrix} a_{11} & a_{12} & \cdots & a_{1n} \\ & a_{22} & \cdots & a_{2n} \\ & & \ddots & \\ 0 & & & a_{nn} \end{vmatrix} = a_{11} \begin{vmatrix} a_{22} & a_{23} & \cdots & a_{2n} \\ & a_{33} & \cdots & a_{3n} \\ & & \ddots & \\ 0 & & & a_{nn} \end{vmatrix}$$

$$= a_{11} a_{22} \begin{vmatrix} a_{33} & a_{34} & \cdots & a_{3n} \\ & a_{44} & \cdots & a_{4n} \\ & & \ddots & \\ 0 & & & a_{nn} \end{vmatrix}$$

$$= \cdots = (a_{11} a_{22} \cdots a_{nn})$$

proving the theorem.

We now summarize the *method*:   Transform the given determinant into an upper triangular determinant using two types of transformations:

1. A multiple of one row can be added to another. This transformation leaves the determinant unchanged.
2. Rows may be interchanged. This transformation negates the determinant.

The final determinant, in upper triangular form, is the product of the diagonal elements.

The following examples illustrate the method.

Example 1

Evaluate

$$\begin{vmatrix} 1 & 0 & 2 & 1 \\ 2 & -1 & 1 & 0 \\ 1 & 0 & 0 & 3 \\ -1 & 0 & 2 & 1 \end{vmatrix}$$

Using the above transformations,

$$\begin{vmatrix} 1 & 0 & 2 & 1 \\ 2 & -1 & 1 & 0 \\ 1 & 0 & 0 & 3 \\ -1 & 0 & 2 & 1 \end{vmatrix} \underset{\substack{\text{Row } 2-(2)\text{ Row } 1 \\ \text{Row } 3-\text{Row } 1 \\ \text{Row } 4+\text{Row } 1}}{=} \begin{vmatrix} 1 & 0 & 2 & 1 \\ 0 & -1 & -3 & -2 \\ 0 & 0 & -2 & 2 \\ 0 & 0 & 4 & 2 \end{vmatrix}$$

$$\underset{\text{Row } 4+(2)\text{ Row } 3}{=} \begin{vmatrix} 1 & 0 & 2 & 1 \\ 0 & -1 & -3 & -2 \\ 0 & 0 & -2 & 2 \\ 0 & 0 & 0 & 6 \end{vmatrix}$$

Thus the determinant is $1 \times -1 \times -2 \times 6 = 12$.

It sometimes becomes necessary to interchange rows to obtain the upper triangular matrix, as the following example illustrates. Whenever rows are interchanged, remember to negate the determinant.

Example 2

Evaluate

$$\begin{vmatrix} 1 & 2 & 3 & 1 \\ 2 & 4 & 3 & 1 \\ 1 & 3 & 4 & 2 \\ 2 & 5 & 6 & 4 \end{vmatrix}$$

$$\begin{vmatrix} 1 & 2 & 3 & 1 \\ 2 & 4 & 3 & 1 \\ 1 & 3 & 4 & 2 \\ 2 & 5 & 6 & 4 \end{vmatrix} \underset{\substack{\text{Row } 2-(2)\,\text{Row } 1 \\ \text{Row } 3-\text{Row } 1 \\ \text{Row } 4-(2)\,\text{Row } 1}}{=} \begin{vmatrix} 1 & 2 & 3 & 1 \\ 0 & 0 & -3 & -1 \\ 0 & 1 & 1 & 1 \\ 0 & 1 & 0 & 2 \end{vmatrix}$$

At this point we normally use the element in the $(2,2)$ location to create the necessary zeros in the second column. Here we cannot do this since that element is 0. We must interchange the second and third rows.

$$\underset{\substack{\text{interchange} \\ \text{Row } 2 \text{ and Row } 3}}{=} - \begin{vmatrix} 1 & 2 & 3 & 1 \\ 0 & 1 & 1 & 1 \\ 0 & 0 & -3 & -1 \\ 0 & 1 & 0 & 2 \end{vmatrix}$$

$$\underset{\text{Row } 4-\text{Row } 2}{=} - \begin{vmatrix} 1 & 2 & 3 & 1 \\ 0 & 1 & 1 & 1 \\ 0 & 0 & -3 & -1 \\ 0 & 0 & -1 & 1 \end{vmatrix}$$

$$\underset{\text{Row } 4-(1/3)\,\text{Row } 3}{=} - \begin{vmatrix} 1 & 2 & 3 & 1 \\ 0 & 1 & 1 & 1 \\ 0 & 0 & -3 & -1 \\ 0 & 0 & 0 & \frac{4}{3} \end{vmatrix}$$

$$= \quad -(1 \times 1 \times -3 \times \tfrac{4}{3}) = 4$$

If, at any stage, the diagonal element is zero and all elements below it in that column are also zero, then the value of the determinant is zero; it is not necessary to continue to determine the upper triangular matrix.

The following example illustrates this.

Example 3

Evaluate

$$
\begin{vmatrix}
1 & -1 & 0 & 2 \\
-1 & 1 & 2 & 3 \\
2 & -2 & 3 & 4 \\
6 & -6 & 5 & 1
\end{vmatrix}
$$

$$
\begin{vmatrix}
1 & -1 & 0 & 2 \\
-1 & 1 & 2 & 3 \\
2 & -2 & 3 & 4 \\
6 & -6 & 5 & 1
\end{vmatrix}
=
\begin{vmatrix}
1 & -1 & 0 & 2 \\
0 & 0 & 2 & 5 \\
0 & 0 & 3 & 0 \\
0 & 0 & 5 & -11
\end{vmatrix}
= 0
$$

Diagonal element   All elements below the zero
is zero   diagonal element are zero.

We know that the determinant is zero at this point because the zero diagonal element will appear as a diagonal element in the eventual upper triangular matrix; hence on multiplying the diagonal elements of this upper triangular matrix we will get zero.

EXERCISES

Evaluate the following determinants using elementary matrix transformations.

1. $\begin{vmatrix} 1 & 0 & -1 \\ 2 & 1 & 2 \\ -1 & 1 & 1 \end{vmatrix}$

2. $\begin{vmatrix} 2 & 1 & 1 \\ 4 & 0 & 1 \\ -1 & 2 & 0 \end{vmatrix}$

3. $\begin{vmatrix} 1 & -2 & 3 \\ -1 & 2 & 1 \\ 2 & 1 & 3 \end{vmatrix}$

4. $\begin{vmatrix} 2 & -1 & 3 \\ 4 & -2 & 1 \\ 1 & -\frac{1}{2} & 1 \end{vmatrix}$

5. $\begin{vmatrix} -1 & 2 & 0 & 1 \\ 1 & 1 & -1 & 0 \\ 2 & 1 & 1 & 0 \\ -1 & -1 & 0 & 1 \end{vmatrix}$

6. $\begin{vmatrix} 1 & 0 & -2 & 1 \\ 2 & 1 & 0 & 2 \\ -1 & 1 & -2 & 1 \\ 3 & 1 & -1 & 0 \end{vmatrix}$

7. $\begin{vmatrix} -1 & 1 & 2 & 1 \\ 1 & -1 & 3 & -1 \\ 2 & -2 & 3 & 1 \\ 1 & -1 & 0 & 1 \end{vmatrix}$

8. $\begin{vmatrix} 1 & -1 & 0 & 2 \\ -1 & 1 & 0 & 0 \\ 2 & -2 & 0 & 1 \\ 3 & 1 & 5 & -1 \end{vmatrix}$

5–4.*   VECTOR PRODUCTS

In this section we introduce various operations on vectors. These are indispensable tools in many areas of science and engineering; we shall discuss applications in some of the areas.

All vectors are considered to be in three-space. Their representations are thus elements of $\mathbf{R}^3$.

Let $\mathbf{a}$ and $\mathbf{b}$ be the vectors (a_1, a_2, a_3) and (b_1, b_2, b_3), respectively. The vector product of $\mathbf{a}$ and $\mathbf{b}$, denoted by $\mathbf{a} \times \mathbf{b}$, is the vector

$$\mathbf{a} \times \mathbf{b} = (a_2 b_3 - a_3 b_2, \, a_3 b_1 - a_1 b_3, \, a_1 b_2 - a_2 b_1)$$

Example 1

Let $\mathbf{a} = (1, -2, 3)$ and $\mathbf{b} = (0, 1, 3)$. Determine $\mathbf{a} \times \mathbf{b}$.

$$\mathbf{a} \times \mathbf{b} = [(-2) \times 3 - 3 \times 1, \, 3 \times 0 - 1 \times 3, \, 1 \times 1 - (-2) \times 0]$$
$$= (-9, -3, 1)$$

If we write the vector product $\mathbf{a} \times \mathbf{b}$ in the following form

$$\mathbf{a} \times \mathbf{b} = (a_2 b_3 - a_3 b_2)\mathbf{i} + (a_3 b_1 - a_1 b_3)\mathbf{j} + (a_1 b_2 - a_2 b_1)\mathbf{k}$$

it can be easily verified (by expanding the determinant) that

$$\mathbf{a} \times \mathbf{b} = \begin{vmatrix} \mathbf{i} & \mathbf{j} & \mathbf{k} \\ a_1 & a_2 & a_3 \\ b_1 & b_2 & b_3 \end{vmatrix}$$

This is not an ordinary determinant, the elements of the first row being vectors not scalars. However, it has the algebraic properties of an ordinary determinant.

THEOREM 5-9 $\mathbf{a} \times \mathbf{a} = \mathbf{0}$.   *Here $\mathbf{0}$ is the zero vector.*

Proof: $\mathbf{a} \times \mathbf{a} = \begin{vmatrix} \mathbf{i} & \mathbf{j} & \mathbf{k} \\ a_1 & a_2 & a_3 \\ a_1 & a_2 & a_3 \end{vmatrix} \underset{\substack{\text{subtract the third} \\ \text{row from the second}}}{=} \begin{vmatrix} \mathbf{i} & \mathbf{j} & \mathbf{k} \\ 0 & 0 & 0 \\ a_1 & a_2 & a_3 \end{vmatrix}$

$$= 0\mathbf{i} + 0\mathbf{j} + 0\mathbf{k} = \mathbf{0}$$

Thus, in particular, $\mathbf{i} \times \mathbf{i} = \mathbf{j} \times \mathbf{j} = \mathbf{k} \times \mathbf{k} = \mathbf{0}$.

THEOREM 5-10 $\mathbf{a} \times \mathbf{b} = -\mathbf{b} \times \mathbf{a}$.   *The vector $\mathbf{b} \times \mathbf{a}$ is parallel to $\mathbf{a} \times \mathbf{b}$, of the same magnitude as $\mathbf{a} \times \mathbf{b}$, but in a direction opposite to $\mathbf{a} \times \mathbf{b}$.*

Proof: $\mathbf{a} \times \mathbf{b} = \begin{vmatrix} \mathbf{i} & \mathbf{j} & \mathbf{k} \\ a_1 & a_2 & a_3 \\ b_1 & b_2 & b_3 \end{vmatrix} \underset{\substack{\text{interchanging} \\ \text{rows}}}{=} - \begin{vmatrix} \mathbf{i} & \mathbf{j} & \mathbf{k} \\ b_1 & b_2 & b_3 \\ a_1 & a_2 & a_3 \end{vmatrix}$

$$= -\mathbf{b} \times \mathbf{a}$$

THEOREM 5-11 $\mathbf{a} \cdot (\mathbf{a} \times \mathbf{b}) = \mathbf{b} \cdot (\mathbf{a} \times \mathbf{b}) = 0.$ *The vector* $\mathbf{a} \times \mathbf{b}$ *is orthogonal to both* $\mathbf{a}$ *and* $\mathbf{b}$. *Thus it is orthogonal to the vector space spanned by* $\mathbf{a}$ *and* $\mathbf{b}$; *that is, it is orthogonal to the plane containing* $\mathbf{a}$ *and* $\mathbf{b}$ (*Figure* 5-1).

Proof: $\mathbf{a} \cdot (\mathbf{a} \times \mathbf{b}) = (a_1 \mathbf{i} + a_2 \mathbf{j} + a_3 \mathbf{k}) \cdot \begin{vmatrix} \mathbf{i} & \mathbf{j} & \mathbf{k} \\ a_1 & a_2 & a_3 \\ b_1 & b_2 & b_3 \end{vmatrix}$

$$= \begin{vmatrix} a_1 & a_2 & a_3 \\ a_1 & a_2 & a_3 \\ b_1 & b_2 & b_3 \end{vmatrix} = 0$$

It can be proved similarly that $\mathbf{b} \cdot (\mathbf{a} \times \mathbf{b}) = 0.$

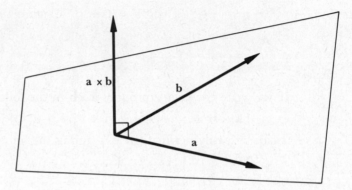

FIGURE 5-1

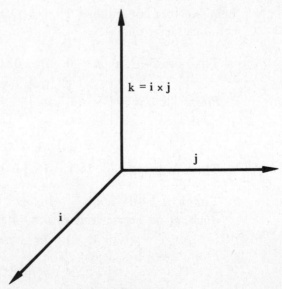

FIGURE 5-2

THEOREM 5-12 $\mathbf{i} \times \mathbf{j} = \mathbf{k}, \mathbf{j} \times \mathbf{k} = \mathbf{i}, \mathbf{k} \times \mathbf{i} = \mathbf{j}$ (*Figure* 5-2).

Proof: $\mathbf{i} \times \mathbf{j} = \begin{vmatrix} \mathbf{i} & \mathbf{j} & \mathbf{k} \\ 1 & 0 & 0 \\ 0 & 1 & 0 \end{vmatrix} = 0\mathbf{i} + 0\mathbf{j} + \mathbf{k} = \mathbf{k}.$

The proofs of the other identities are similar.

THEOREM 5-13 $\|\mathbf{a} \times \mathbf{b}\| = \|\mathbf{a}\| \|\mathbf{b}\| \sin \theta$, *where* θ *is the angle between* $\mathbf{a}$ *and* $\mathbf{b}$.

The reader is asked to prove this identity in the exercises at the end of the section.

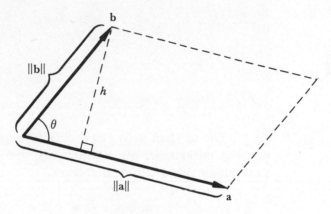

FIGURE 5-3

This leads to an interesting interpretation of the magnitude of the vector $\mathbf{a} \times \mathbf{b}$. Consider the parallelogram in Figure 5-3 defined by the vectors $\mathbf{a}$ and $\mathbf{b}$. Its area is base $\times$ height $= \|\mathbf{a}\| h = \|\mathbf{a}\| \|\mathbf{b}\| \sin \theta$. Thus the magnitude of the vector $\mathbf{a} \times \mathbf{b}$ is the area of the parallelogram defined by the vectors $\mathbf{a}$ and $\mathbf{b}$.

Example 2

The *moment* of a vector about a point, a concept that occurs frequently in mechanics, can be discussed in terms of vector products. The moment of a vector $\mathbf{f}$ about a point O is defined to be the vector $\mathbf{m} = \mathbf{r} \times \mathbf{f}$, where $\mathbf{r}$ is the vector from O to any point of $\mathbf{f}$ (Figure 5-4).

(In Exercise 7 at the end of this section the reader is asked to prove that $\mathbf{m}$ is independent of the choice of $\mathbf{r}$.) Thus $\mathbf{m}$ is a vector in a direction perpendicular to the plane containing $\mathbf{r}$ and $\mathbf{f}$. Its magnitude is

$$\|\mathbf{m}\| = \|\mathbf{r} \times \mathbf{f}\| = \|\mathbf{r}\| \|\mathbf{f}\| \sin \theta = \|\mathbf{f}\| h$$

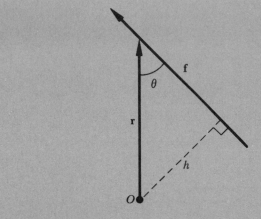

FIGURE 5-4

where h is the perpendicular distance from O to $\mathbf{f}$. If $\mathbf{f}$ is a force, then $\mathbf{r} \times \mathbf{f}$ is called the moment of the force about O. It represents the turning effect of the force about O.

The product $\mathbf{a} \cdot (\mathbf{b} \times \mathbf{c})$ is called the *triple scalar product* of $\mathbf{a}$, $\mathbf{b}$, and $\mathbf{c}$. The parentheses are not actually needed, as the vector operations can only be performed in this order.

THEOREM 5-14 $\mathbf{a} \cdot \mathbf{b} \times \mathbf{c} = \mathbf{a} \times \mathbf{b} \cdot \mathbf{c}$

Proof:

$$\mathbf{a} \cdot \mathbf{b} \times \mathbf{c} = (a_1 \mathbf{i} + a_2 \mathbf{j} + a_3 \mathbf{k}) \cdot \begin{vmatrix} \mathbf{i} & \mathbf{j} & \mathbf{k} \\ b_1 & b_2 & b_3 \\ c_1 & c_2 & c_3 \end{vmatrix} = \begin{vmatrix} a_1 & a_2 & a_3 \\ b_1 & b_2 & b_3 \\ c_1 & c_2 & c_3 \end{vmatrix}$$

and

$$\mathbf{a} \times \mathbf{b} \cdot \mathbf{c} = \begin{vmatrix} \mathbf{i} & \mathbf{j} & \mathbf{k} \\ a_1 & a_2 & a_3 \\ b_1 & b_2 & b_3 \end{vmatrix} \cdot (c_1 \mathbf{i} + c_2 \mathbf{j} + c_3 \mathbf{k})$$

$$= \begin{vmatrix} c_1 & c_2 & c_3 \\ a_1 & a_2 & a_3 \\ b_1 & b_2 & b_3 \end{vmatrix} = \begin{vmatrix} a_1 & a_2 & a_3 \\ b_1 & b_2 & b_3 \\ c_1 & c_2 & c_3 \end{vmatrix}$$

There is an interesting geometrical interpretation of the triple scalar product that leads to an interpretation of 3×3 determinants. Consider the parallelepiped in Figure 5-5 defined by $\mathbf{a}$, $\mathbf{b}$, and $\mathbf{c}$. Its base area is $\|\mathbf{a} \times \mathbf{b}\|$. Its volume is $\|\mathbf{a} \times \mathbf{b}\| h$, where h is the height.

Consider $\left| \dfrac{\mathbf{a} \times \mathbf{b}}{\|\mathbf{a} \times \mathbf{b}\|} \cdot \mathbf{c} \right|$, the absolute value of the scalar projection

of $\mathbf{c}$ onto the vector $\mathbf{a} \times \mathbf{b}$. We can see that $\left| \dfrac{\mathbf{a} \times \mathbf{b}}{\|\mathbf{a} \times \mathbf{b}\|} \cdot \mathbf{c} \right| = h.$

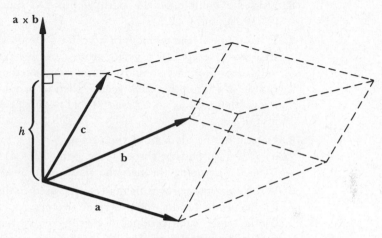

FIGURE 5-5

Thus $|\mathbf{a} \times \mathbf{b} \cdot \mathbf{c}| = \|\mathbf{a} \times \mathbf{b}\| h$; $|\mathbf{a} \times \mathbf{b} \cdot \mathbf{c}|$ is the volume of the parallelepiped defined by $\mathbf{a}$, $\mathbf{b}$, and $\mathbf{c}$. Furthermore, since $|\mathbf{a} \times \mathbf{b} \cdot \mathbf{c}| =$

$\left\| \begin{matrix} a_1 & a_2 & a_3 \\ b_1 & b_2 & b_3 \\ c_1 & c_2 & c_3 \end{matrix} \right\|$ (see above proof), then $\left\| \begin{matrix} a_1 & a_2 & a_3 \\ b_1 & b_2 & b_3 \\ c_1 & c_2 & c_3 \end{matrix} \right\| = $ volume of paral-

lelepiped with edges (a_1, a_2, a_3), (b_1, b_2, b_3), and (c_1, c_2, c_3).

EXERCISES

1. If $\mathbf{a} = (1, 2, 3)$, $\mathbf{b} = (-1, 0, 4)$, and $\mathbf{c} = (1, 2, -1)$, determine
   a) $\mathbf{a} \times \mathbf{b}$ b) $\mathbf{a} \times \mathbf{c}$ c) $(\mathbf{a} \times \mathbf{b}) \times \mathbf{c}$
   d) $\mathbf{a} \times (\mathbf{b} \times \mathbf{c})$ e) $\mathbf{a} \times \mathbf{b} \cdot \mathbf{c}$
   Comment on your results for (c) and (d). $\mathbf{a} \times \mathbf{b} \times \mathbf{c}$ is called the *triple vector product* of $\mathbf{a}$, $\mathbf{b}$, and $\mathbf{c}$.

2. If $\mathbf{a}$, $\mathbf{b}$, and $\mathbf{c}$ are arbitrary vectors in three-space, prove that

   $$\mathbf{a} \times (\mathbf{b} + \mathbf{c}) = (\mathbf{a} \times \mathbf{b}) + (\mathbf{a} \times \mathbf{c})$$

3. If $\mathbf{a}$ and $\mathbf{b}$ are arbitrary vectors in three-space and d is a scalar, prove that

   $$(d\mathbf{a}) \times \mathbf{b} = \mathbf{a} \times (d\mathbf{b}) = d(\mathbf{a} \times \mathbf{b})$$

4. If $\mathbf{a}$ and $\mathbf{b}$ are two vectors in three-space and θ is the angle between them,

prove that $\|\mathbf{a} \times \mathbf{b}\| = \|\mathbf{a}\| \, \|\mathbf{b}\| \sin \theta$. [*Hint*: Let $\mathbf{a} = (a_1, a_2, a_3)$ and $\mathbf{b} = (b_1, b_2, b_3)$. Determine expressions for both sides of the identity. Use $\sin^2 \theta = 1 - \cos^2 \theta$ to find an expression for $\sin \theta$.]

5. Determine the area of the triangle with vertices $(1, 2, 3)$, $(-1, 1, 1)$, and $(1, 0, 4)$.

6. Find the volume of the parallelepiped defined by the vectors $(1, 2, 3)$, $(-1, 0, 4)$, and $(1, 2, 5)$.

7. In Example 2, the moment of a vector $\mathbf{f}$ about a point O was defined to be $\mathbf{m} = \mathbf{r} \times \mathbf{f}$, where $\mathbf{r}$ is a vector from O to any point of $\mathbf{f}$. For this to be a valid definition, we should get the same vector $\mathbf{m}$ for all such vectors $\mathbf{r}$; that is, if $\bar{\mathbf{r}}$ is another such vector, then we would expect $\mathbf{m} = \mathbf{r} \times \mathbf{f} = \bar{\mathbf{r}} \times \mathbf{f}$. Prove that this is so. [*Hint*: $(\mathbf{r} \times \mathbf{f}) - (\bar{\mathbf{r}} \times \mathbf{f}) = (\mathbf{r} - \bar{\mathbf{r}}) \times \mathbf{f}$ by Exercise 2 above.]

8. Determine the moment of the vector
   a) $(1, -1, 4)$ passing through the point $(2, 4, 4)$ about the point $(1, -2, 3)$
   b) $(1, 2, 3)$ passing through the point $(0, 1, 2)$ about the point $(1, 2, -1)$

9. Write computer programs that can be used to determine
   a) the vector product of two given vectors
   b) the triple scalar product of three given vectors
   c) the triple vector product of three given vectors
   Check your answers to Exercise 1 with your programs.

6

Systems of Linear Equations—
A Qualitative Discussion

In this chapter we bring together many of the concepts developed in previous sections. Determinants are used in furthering our understanding of linear systems. The reader will be introduced to the theory of eigenvalues and eigenvectors, important algebraic tools for the mathematician, as well as the social and physical scientist, as we shall see from the examples of applications.

6–1. THE RANK OF A MATRIX

In a qualitative discussion of systems of linear equations, the concept of *rank* plays an important role. Rank enables us to relate vectors to matrices and vice versa. We have seen how a system of linear equations can be characterized by an augmented matrix. Rank enables us to discuss characteristics of systems of equations in terms of the vectors that make up the augmented matrix, particularly the existence and uniqueness of their solutions.

Let A be an $m \times n$ matrix. The rows of A may be viewed as row vectors $\mathbf{a}_1, ..., \mathbf{a}_m$.

DEFINITION 6-1 The *rank* of a matrix A is the maximum number of linearly independent row vectors in the matrix.

If $\mathbf{a}^1, ..., \mathbf{a}^n$ are the column vectors of A, the maximum number of linearly independent column vectors is equal to the maximum number of linearly independent row vectors.   This also gives the rank.

Example 1

Determine the rank of the matrix $\begin{pmatrix} 1 & 2 & 3 \\ 0 & 1 & 2 \\ 2 & 5 & 8 \end{pmatrix}$.

We see by inspection that the third row of this matrix can be expressed as a linear combination of the first two rows,

$$(2,5,8) = 2(1,2,3) + (0,1,2)$$

Hence the three rows of this matrix are linearly dependent, and its rank must be less than 3.

On examining the vectors $(1,2,3)$ and $(0,1,2)$, the identity

$$a(1,2,3) + b(0,1,2) = \mathbf{0}$$

leads to

$$a = b = 0$$

Thus the vectors $(1,2,3)$ and $(0,1,2)$ are linearly independent, and the rank of the matrix is 2.

This method, implied by the definition, is not practical for determining the ranks of matrices larger than 3×3.   We shall now develop a method for determining the rank of a matrix using an echelon form.

THEOREM 6-1   *The rank of a matrix in echelon form is its number of nonzero row vectors.*

Proof: Let E be an $m \times n$ matrix in echelon form with nonzero row vectors $\mathbf{e_1}, \ldots, \mathbf{e_r}$.   Consider the identity

$$a_1 \mathbf{e_1} + a_2 \mathbf{e_2} + \cdots + a_r \mathbf{e_r} = \mathbf{0}$$

where $a_1, \ldots, a_r$ are scalars.

The first nonzero component of $\mathbf{e_1}$ is 1, and it is the only vector with a nonzero component in this slot.   Thus on adding the vectors $a_1 \mathbf{e_1}, \ldots, a_r \mathbf{e_r}$ we get an n-tuple having first component a_1. On equating to zero, $a_1 = 0$.   The identity thus reduces to

$$a_2 \mathbf{e_2} + \cdots + a_r \mathbf{e_r} = \mathbf{0}$$

The first nonzero component of $\mathbf{e_2}$ is 1, and it is the only vector in this identity with a nonzero component in this slot.   Thus $a_2 = 0$. Similarly, $a_3, \ldots, a_r$ are all zero.   The vectors $\mathbf{e_1}, \ldots, \mathbf{e_r}$ are linearly independent.   Since all the other row vectors of E are zero vectors, these are the linearly independent row vectors of E.   Thus the rank of the echelon form is the number of nonzero row vectors.

The following example illustrates the proof.

Example 2

Consider the echelon matrix

$$E = \begin{pmatrix} 1 & 0 & 2 & 1 \\ 0 & 0 & 1 & 3 \\ 0 & 0 & 0 & 1 \\ 0 & 0 & 0 & 0 \end{pmatrix}$$

According to Theorem 6-1, its rank is 3.   Let us prove that this is so.
The row vectors are

$$\mathbf{e_1} = (1, 0, 2, 1)$$

$$\mathbf{e_2} = (0, 0, 1, 3)$$

$$\mathbf{e_3} = (0, 0, 0, 1)$$

$$\mathbf{e_4} = (0, 0, 0, 0)$$

The identity

$$a_1(1, 0, 2, 1) + a_2(0, 0, 1, 3) + a_3(0, 0, 0, 1) = \mathbf{0}$$

gives

$$(a_1, 0, 2a_1, a_1) + (0, 0, a_2, 3a_2) + (0, 0, 0, a_3) = \mathbf{0}$$

$$(a_1, 0, 2a_1 + a_2, a_1 + 3a_2 + a_3) = \mathbf{0}$$

Since the first component must be zero, a_1 is zero.   Since the third component is zero, $a_2 = 0$, and since the fourth component is zero, $a_3 = 0$.   Thus the vectors $\mathbf{e_1}$, $\mathbf{e_2}$, and $\mathbf{e_3}$ are linearly independent.   The vectors $\mathbf{e_1}$, $\mathbf{e_2}$, $\mathbf{e_3}$, and $\mathbf{e_4}$ are linearly dependent, since this set contains the zero vector.   Thus the rank of E is 3.

THEOREM 6-2   *The rank of a matrix remains invariant under elementary matrix transformations.*

Proof: Let A be an $m \times n$ matrix of rank r with rows $\mathbf{a_1}, \ldots, \mathbf{a_i}, \ldots,$ $\mathbf{a_j}, \ldots, \mathbf{a_m}$.   A number r of these vectors are linearly independent, and any $r + 1$ are linearly dependent.

Consider an elementary transformation: the interchanging of two rows such as the ith and the jth rows.   The maximum number of linearly independent vectors in the set $\mathbf{a_1}, \ldots, \mathbf{a_j}, \ldots, \mathbf{a_i}, \ldots, \mathbf{a_m}$ is still r; the rank remains r.

Next let us look at another transformation: the multiplying of the ith row by a nonzero scalar c. The rows are now $\mathbf{a}_1, ..., c\mathbf{a}_i, ..., \mathbf{a}_m$. It can easily be proved that the maximum number of linearly independent row vectors in this set is the same as in the original set of row vectors $\mathbf{a}_1, ..., \mathbf{a}_i, ..., \mathbf{a}_m$. Thus the rank remains unchanged. (See Exercise 12a.)

Finally, consider a third transformation: adding c times the ith row to the jth row. The row vectors are now $\mathbf{a}_1, ..., \mathbf{a}_i, ..., c\mathbf{a}_i + \mathbf{a}_j, ..., \mathbf{a}_m$. The new jth row vector is $c\mathbf{a}_i + \mathbf{a}_j$; the other row vectors are unchanged. The maximum number of linearly independent vectors in this set is again the same as in the set $\mathbf{a}_1, ..., \mathbf{a}_i, ..., \mathbf{a}_j, ..., \mathbf{a}_m$, proving that the transformation leaves the rank unchanged. (See Exercise 12b.)

THEOREM 6-3   *A is an $m \times n$ matrix of rank r if and only if its echelon form has r nonzero row vectors.*

Proof: We know from the previous theorem that the rank of A will be the same as that of its echelon form. But the rank of its echelon form is r if and only if r is the number of nonzero row vectors, proving the theorem.

Example 3

Determine the rank of the matrix

$$A = \begin{pmatrix} 1 & 2 & -1 & 3 & -1 \\ 0 & 1 & 4 & -1 & 0 \\ -1 & 2 & 1 & 1 & 2 \\ 2 & 4 & -2 & 6 & -2 \end{pmatrix}$$

We find that

$$A = \begin{pmatrix} 1 & 2 & -1 & 3 & -1 \\ 0 & 1 & 4 & -1 & 0 \\ -1 & 2 & 1 & 1 & 2 \\ 2 & 4 & -2 & 6 & -2 \end{pmatrix} \cong \begin{pmatrix} 1 & 2 & -1 & 3 & -1 \\ 0 & 1 & 4 & -1 & 0 \\ 0 & 4 & 0 & 4 & 1 \\ 0 & 0 & 0 & 0 & 0 \end{pmatrix}$$

$$\cong \begin{pmatrix} 1 & 2 & -1 & 3 & -1 \\ 0 & 1 & 4 & -1 & 0 \\ 0 & 0 & -16 & 8 & 1 \\ 0 & 0 & 0 & 0 & 0 \end{pmatrix} \cong \begin{pmatrix} 1 & 2 & -1 & 3 & -1 \\ 0 & 1 & 4 & -1 & 0 \\ 0 & 0 & 1 & -\frac{1}{2} & -\frac{1}{16} \\ 0 & 0 & 0 & 0 & 0 \end{pmatrix}$$

A is thus of rank 3.

We complete this section with a geometrical interpretation of rank. From the definition we see that the rank of a matrix is the dimension of the space spanned by its row vectors. It is also equal to the dimension of the space spanned by its column vectors. This latter interpretation leads to the following theorem, by which the rank of a matrix is the dimension of the range of the mapping defined by the matrix.

THEOREM 6-4   *Let an $m \times n$ matrix A define a linear mapping of $\mathbf{R}^n$ into $\mathbf{R}^m$. The columns of A, interpreted as elements of $\mathbf{R}^m$, span the range of this mapping. The rank of A is the dimension of this range.*

Proof: Let Y be an arbitrary element of the range. Thus, there exists an element X of $\mathbf{R}^n$ such that

$$AX = Y$$

Let us express this

$$a_{11} x_1 + \cdots + a_{1n} x_1 = y_1$$
$$\vdots \qquad\qquad \vdots \qquad \vdots$$
$$a_{m1} x_1 + \cdots + a_{mn} x_n = y_m$$

This system can be written

$$x_1 \begin{pmatrix} a_{11} \\ \vdots \\ a_{m1} \end{pmatrix} + \cdots + x_n \begin{pmatrix} a_{1n} \\ \vdots \\ a_{mn} \end{pmatrix} = \begin{pmatrix} y_1 \\ \vdots \\ y_m \end{pmatrix}$$

Thus Y can be expressed as a linear combination of the column vectors that make up A. These column vectors span the range.

Since the rank of A is equal to the maximum number of linearly independent column vectors, it will be the dimension of the range.

EXERCISES

1. Determine the ranks of the following matrices using the definition of rank.

a) $\begin{pmatrix} 1 & 2 \\ 3 & 4 \end{pmatrix}$

b) $\begin{pmatrix} 1 & 2 & 1 \\ 2 & 4 & 2 \\ 1 & 2 & 3 \end{pmatrix}$

c) $\begin{pmatrix} 1 & 0 & 0 \\ 0 & 1 & 0 \\ 0 & 0 & 1 \end{pmatrix}$

d) $\begin{pmatrix} 2 & 1 & 3 \\ 4 & 2 & 6 \\ 2 & 1 & 3 \end{pmatrix}$

Determine the ranks of the following matrices by finding echelon forms.

2. $\begin{pmatrix} 1 & 2 & 0 \\ 0 & 1 & 1 \\ -1 & 2 & 3 \end{pmatrix}$

3. $\begin{pmatrix} 1 & 2 & 3 \\ 0 & -1 & -1 \\ 3 & 4 & 7 \end{pmatrix}$

$$4. \begin{pmatrix} -1 & 0 & 2 \\ 1 & 1 & 1 \\ -1 & 2 & 3 \end{pmatrix} \qquad 5. \begin{pmatrix} 1 & 2 & 3 & 4 \\ -1 & 2 & 0 & 1 \\ 0 & 1 & 0 & 2 \end{pmatrix}$$

$$6. \begin{pmatrix} 1 & 3 & 4 & 1 \\ 2 & 6 & 8 & 2 \\ 0 & 1 & 2 & 1 \end{pmatrix} \qquad 7. \begin{pmatrix} -1 & 2 & 3 & 1 \\ 0 & -2 & -6 & -2 \\ 1 & 2 & 1 & 1 \end{pmatrix}$$

$$8. \begin{pmatrix} -1 & 2 & 3 & 0 \\ 1 & 2 & 0 & 0 \\ -1 & 0 & 1 & 2 \\ 0 & -1 & 1 & 1 \end{pmatrix} \qquad 9. \begin{pmatrix} 1 & 1 & 0 & -1 \\ 2 & 1 & 0 & 0 \\ 3 & 2 & 0 & -1 \\ -1 & 0 & 1 & 1 \end{pmatrix}$$

$$10. \begin{pmatrix} -1 & 2 & 1 & 0 \\ 1 & -2 & -1 & 0 \\ -1 & 0 & 1 & 1 \\ -2 & 0 & 2 & 2 \end{pmatrix}$$

11. If A and B are matrices of the same kind, prove that rank $(A+B) \leqslant$ rank $A+$ rank B.

12. a) Prove that if $\mathbf{a}_1, ..., \mathbf{a}_i, ..., \mathbf{a}_m$ are m vectors with n components each, then the maximum number of linearly independent vectors in this set is the same as in the set $\mathbf{a}_1, ..., c\mathbf{a}_i, ..., \mathbf{a}_m$, where c is any nonzero scalar. This result is used in the proof of Theorem 6-2. (*Hint*: Prove that the spaces spanned by these two sets are identical. The dimension of this space is the maximum number of linearly independent vectors in each set.)

   b) Prove that the maximum number of linearly independent vectors in the sets $\mathbf{a}_1, ..., \mathbf{a}_j, ..., \mathbf{a}_m$ and $\mathbf{a}_1, ..., c\mathbf{a}_i+\mathbf{a}_j, ..., \mathbf{a}_m$ are the same; c is a nonzero scalar. This result is used in the proof of Theorem 6-2.

13. This exercise gives a computational method for deriving a basis for the space spanned by a given set of vectors. Prove that if the given vectors are written as the rows of a matrix and the echelon form computed, then the nonzero vectors of this echelon form are a basis for the space.

   Use the method of Exercise 13 to determine bases for the spaces spanned by the following sets of vectors.

14. $(2, 1, 3, 0, 4)$, $(-1, 2, 3, 1, 0)$, $(3, -1, 0, -1, 4)$.

15. $(1, 2, 3, 4)$, $(0, -1, 2, 3)$, $(2, 3, 8, 11)$, $(2, 3, 6, 8)$.

16. $(-1, 2, 0, 1, 3)$, $(1, 2, 0, 5, 4)$, $(3, 2, 2, 0, -1)$.

17. $(0, -1, 2, 3, -1, 4, 5)$, $(1, 4, 0, -1, 5, 2, 1)$, $(3, 1, 4, 2, 1, 0, 0)$, $(-2, 2, -2, 0, 3, 6, 6)$.

18. This exercise gives a computational method for deriving a basis for the range of a mapping defined by a given matrix A. Prove that the nonzero row vectors of an echelon form of A^t give a basis for this range.

   Use the method of Exercise 18 to determine bases for the ranges of the mappings defined by the following matrices.

19. $\begin{pmatrix} -1 & 2 & -1 & 2 \\ 0 & 1 & 0 & 1 \\ 2 & 0 & 2 & 4 \\ 3 & 1 & 3 & 0 \end{pmatrix}$ 20. $\begin{pmatrix} 2 & -6 & 3 \\ -1 & 3 & -1.5 \\ 3 & -9 & 4.5 \\ 4 & -12 & 6 \end{pmatrix}$

21. Let A be an $m \times n$ matrix. Interpret A as a mapping of $\mathbf{R}^n \to \mathbf{R}^m$ and A^t as a mapping of $\mathbf{R}^m \to \mathbf{R}^n$. Show that the dimension of the range of A is equal to the dimension of the range of A^t.

22. Write a computer program that can be used to determine a basis for the space spanned by a given set of vectors. (Use the result of Exercise 13.)

6–2. EXISTENCE AND UNIQUENESS OF SOLUTIONS

In this section we formulate conditions for existence and uniqueness of solutions to a system of linear equations in terms of the ranks of its matrix of coefficients and augmented matrix. We shall see from examples discussed that the criteria derived are compatible with previous results.

We first discuss existence.

THEOREM 6-5   *A system of m equations in n variables has a solution if and only if the rank of the augmented matrix is equal to the rank of the matrix of coefficients.*

Proof: Consider a system of m equations in n variables,

$$a_{11}x_1 + \cdots + a_{1n}x_n = y_1$$
$$\vdots$$
$$a_{m1}x_1 + \cdots + a_{mn}x_n = y_m$$

Letting $\mathbf{a}^1, \ldots, \mathbf{a}^m$ be the column vectors of the matrix of coefficients A, the system of equations may be written

$$x_1\mathbf{a}^1 + \cdots + x_n\mathbf{a}^n = \mathbf{y}$$

A solution $x_1, \ldots, x_n$ will exist if and only if $\mathbf{y}$ is linearly dependent on $\mathbf{a}^1, \ldots, \mathbf{a}^n$. That is, if and only if the maximum number of linearly independent vectors in the set $\mathbf{a}^1, \ldots, \mathbf{a}^n$ is the same as in the set $\mathbf{a}^1, \ldots, \mathbf{a}^n, \mathbf{y}$, proving the theorem.

Let us now look at uniqueness.

THEOREM 6-6   *Let the matrix of coefficients and augmented matrix of a system of m linear equations in n variables have the same rank r. When $r = n$ the solution is unique; when $r < n$ the solution is not unique.*

289

Proof: Again consider the system in the form

$$x_1 \mathbf{a}^1 + \cdots + x_n \mathbf{a}^n = \mathbf{y}$$

A solution $x_1, \ldots, x_n$ exists since the ranks of the matrix of coefficients and augmented matrix are the same. $\mathbf{y}$ can thus be expressed as a linear combination of $\mathbf{a}^1, \ldots, \mathbf{a}^n$. $\mathbf{y}$ is in the vector space V generated by $\mathbf{a}^1, \ldots, \mathbf{a}^n$.

When $r = n$, the vectors $\mathbf{a}^1, \ldots, \mathbf{a}^n$ are linearly independent and must thus form a basis for the space they generate. $\mathbf{y}$ is thus expressed uniquely as a linear combination of $\mathbf{a}^1, \ldots, \mathbf{a}^n$; the solution is unique.

When $r < n$, the vectors $\mathbf{a}^1, \ldots, \mathbf{a}^n$ are linearly dependent. They span the vector space V that they generate. But any vector, in particular $\mathbf{y}$, can be represented in more than one way as a linear combination of them. The solution is not unique.

We now illustrate the results of this theorem in terms of the systems of equations that were solved in Section 2-2.

Example 1

Discuss the existence and uniqueness of solutions to the system of equations

$$x_1 + x_2 + x_3 = 3$$
$$2x_1 + 3x_2 + x_3 = 5$$
$$x_1 - x_2 - 2x_3 = -5$$

The existence and uniqueness of solutions depend upon the ranks of the matrix of coefficients and the augmented matrix. To determine the rank of the augmented matrix we find the echelon form. The echelon form of the matrix of coefficients is found at the same time. The augmented matrix is

$$\begin{pmatrix} \underbrace{\begin{matrix} 1 & 1 & 1 \\ 2 & 3 & 1 \\ 1 & -1 & -2 \end{matrix}}_{\text{matrix of coefficients}} & \begin{matrix} 3 \\ 5 \\ -5 \end{matrix} \end{pmatrix}$$

matrix of coefficients

$$\cong \begin{pmatrix} 1 & 1 & 1 & 3 \\ 0 & 1 & -1 & -1 \\ 1 & -1 & -2 & -5 \end{pmatrix} \cong \begin{pmatrix} 1 & 1 & 1 & 3 \\ 0 & 1 & -1 & -1 \\ 0 & -2 & -3 & -8 \end{pmatrix}$$

$$\cong \begin{pmatrix} 1 & 1 & 1 & 3 \\ 0 & 1 & -1 & -1 \\ 0 & 0 & -5 & -10 \end{pmatrix}$$

$$\cong \begin{pmatrix} 1 & 1 & 1 & 3 \\ 0 & 1 & -1 & -1 \\ 0 & 0 & 1 & 2 \end{pmatrix} \leftarrow \begin{array}{l} \textit{echelon form of} \\ \textit{augmented matrix} \end{array}$$

$$\underbrace{\phantom{\begin{pmatrix} 1 & 1 & 1 \end{pmatrix}}}_{\begin{array}{c}\textit{echelon form of} \\ \textit{matrix of coefficients}\end{array}}$$

Both echelon forms have three rows of nonzero elements. Thus the ranks of the matrix of coefficients and the augmented matrix are both 3, the number of variables. A solution exists and is unique. This agrees with the discussion of Example 2, Section 2-2.

Example 2

Discuss the existence and uniqueness of the solutions to the system of equations

$$x_1 - 2x_2 + 3x_3 = 1$$
$$3x_1 - 4x_2 + 5x_3 = 3$$
$$2x_1 - 3x_2 + 4x_3 = 2$$

The augmented matrix is

$$\begin{pmatrix} 1 & -2 & 3 & 1 \\ 3 & -4 & 5 & 3 \\ 2 & -3 & 4 & 2 \end{pmatrix}$$

$$\underbrace{\phantom{\begin{pmatrix} 1 & -2 & 3 \end{pmatrix}}}_{\begin{array}{c}\textit{matrix of} \\ \textit{coefficients}\end{array}}$$

In Example 4, Section 2-2 we found that the echelon form was

$$\begin{pmatrix} 1 & -2 & 3 & 1 \\ 0 & 1 & -2 & 0 \\ 0 & 0 & 0 & 0 \end{pmatrix} \leftarrow \begin{array}{l} \textit{echelon form of} \\ \textit{augmented matrix} \end{array}$$

$$\underbrace{\phantom{\begin{pmatrix} 1 & -2 & 3 \end{pmatrix}}}_{\begin{array}{c}\textit{echelon form of} \\ \textit{matrix of coefficients}\end{array}}$$

Thus the ranks of the augmented matrix and the matrix of coefficients are equal, both 2, but less than the number of variables. A solution exists but is not unique.

Example 3

Discuss the existence and uniqueness of the solutions to the system of equations

$$x_1 - x_2 + 2x_3 = 3$$
$$2x_1 - 2x_2 + 5x_3 = 4$$
$$x_1 + 2x_2 - x_3 = -3$$
$$2x_2 + 2x_3 = 1$$

The augmented matrix is

$$\begin{pmatrix} 1 & -1 & 2 & 3 \\ 2 & -2 & 5 & 4 \\ 1 & 2 & -1 & -3 \\ 0 & 2 & 2 & 1 \end{pmatrix}$$

The echelon form was found in Example 6, Section 2-2 to be

$$\begin{pmatrix} 1 & -1 & 2 & 3 \\ 0 & 1 & -1 & -2 \\ 0 & 0 & 1 & -2 \\ 0 & 0 & 0 & 1 \end{pmatrix} \leftarrow \begin{array}{l} \textit{echelon form of} \\ \textit{augmented matrix} \end{array}$$

$$\underbrace{\phantom{\begin{matrix} 1 & -1 & 2 \end{matrix}}}_{\substack{\textit{echelon form of} \\ \textit{matrix of coefficients}}}$$

The rank of the augmented matrix is 4, and the rank of the matrix of coefficients is 3.   A solution does not exist.

THEOREM 6-7   *A solution always exists to a system of $m \times n$ homogeneous linear equations.   A nonzero solution exists if and only if the rank of the matrix of coefficients is less than n.*

Proof: Let

$$a_{11}x_1 + \cdots + a_{1n}x_n = 0$$
$$\vdots$$
$$a_{m1}x_1 + \cdots + a_{mn}x_n = 0$$

be the homogeneous system.

The augmented matrix is

$$\begin{pmatrix} a_{11} & \cdots & a_{1n} & 0 \\ & \vdots & & \\ a_{m1} & \cdots & a_{mn} & 0 \end{pmatrix}$$

Since the last column of the augmented matrix is the zero vector, the matrix of coefficients and augmented matrix have the same number of linearly independent columns and hence the same rank.   Thus a solution always exists.   By inspection we see that $x_1 = 0, \ldots, x_n = 0$ is a solution; this is called the *trivial solution*, or *zero solution*.

A further distinct solution will exist by Theorem 6-6 if and only if the rank of the matrix of coefficients is less than n.

EXERCISES

Discuss the existence and uniqueness of solutions to the following systems of linear equations by examining the ranks of the matrix of coefficients and the augmented matrix for each system.

1. $\begin{aligned} x_1 + 3x_3 &= 0 \\ 3x_1 + 2x_2 + 4x_3 &= 1 \\ 2x_1 + x_2 + 2x_3 &= 2 \end{aligned}$

2. $\begin{aligned} x_1 - 2x_2 &= 1 \\ 2x_1 + 3x_2 &= 4 \\ 3x_1 + 5x_2 &= 2 \end{aligned}$

3. $\begin{aligned} x_1 + 2x_2 + 4x_3 &= 1 \\ 2x_1 + 3x_2 + 2x_3 &= 2 \end{aligned}$

4. $\begin{aligned} 2x_1 + 3x_2 + 2x_3 &= 2 \\ x_1 + 2x_2 + 3x_3 &= 0 \\ 5x_1 + 3x_2 + x_3 &= 0 \end{aligned}$

5. $\begin{aligned} x_1 + 2x_2 + 3x_3 &= 0 \\ 2x_1 + 4x_2 + 6x_3 &= 0 \\ -x_1 - 2x_2 - 3x_3 &= 0 \end{aligned}$

6. $\begin{aligned} -x_1 + 2x_2 + 3x_3 &= 0 \\ x_1 + 2x_2 &= 0 \\ -x_1 + x_3 &= 2 \\ -x_1 + x_2 &= 1 \end{aligned}$

7. $\begin{aligned} x_1 + x_2 &= -1 \\ 2x_1 + x_2 &= 0 \\ 3x_1 + 2x_2 &= -1 \\ -x_1 + x_3 &= 1 \end{aligned}$

8. Give examples of systems of three equations in three variables that
   a) have a unique solution
   b) have many solutions involving one parameter
   c) have many solutions involving two parameters
   d) have no solutions
   (*Hint*: Use systems corresponding to certain echelon forms.)

9. Give examples of systems of six equations in seven variables that
   a) have a unique solution
   b) have many solutions involving one parameter
   c) have many solutions involving two parameters
   d) have no solutions
   (*Hint*: Use systems corresponding to certain echelon forms.)

6-3. CRAMER'S RULE

We have seen how rank plays a role in the analyses of systems of equations. The following theorem relates the determinant of a square matrix to its rank, thus paving the way for the use of determinants in the analyses of certain systems.

THEOREM 6-8 *An $n \times n$ matrix A is nonsingular ($|A| \neq 0$) if and only if it is of rank n.*

Proof: Let A be nonsingular, $|A| \neq 0$. Thus $|A|$ can be transformed into $|B|$, where B is an upper triangular matrix having all nonzero diagonal elements, and $|A| = \pm |B|$. The allowable transformations are the addition of multiples of rows to rows and the interchanging of rows. These are also elementary matrix transformations. Thus B

293

will be of the same rank as A. The rank of B is n; thus the rank of A is n.

We now prove the converse. Let A be a matrix of rank n. Transform $|A|$ into $|C|$ where C is an upper triangular matrix. The rank of C is the same as A, thus C is of rank n. C must have nonzero diagonal elements for otherwise it would have an echelon form having last row zero, violating its rank n. (see Exercise 12.)

Thus $|C| \neq 0$ and since $|A| = \pm |C|$, $|A| \neq 0$.

We are now in a position to relate the determinant of the matrix of coefficients of a system of n equations in n variables to the uniqueness of solutions.

THEOREM 6-9   *A system of n equations in n variables has a unique solution if and only if the matrix of coefficients is nonsingular. If the matrix is singular, solutions may or may not exist.*

Proof: The uniqueness result follows immediately from the previous theorem and Theorem 6-6.

The possibilities of solutions existing or not if the matrix of coefficients is singular are demonstrated by the following two examples.

Example 1

$$x_1 - 2x_2 + 3x_3 = 1$$
$$3x_1 - 4x_2 + 5x_3 = 3$$
$$2x_1 - 3x_2 + 4x_3 = 2$$

The determinant of the matrix of coefficients is 0. Solutions can be shown to be $x_1 = x_3 + 1, x_2 = 2x_3$.

Example 2

$$x_1 + 2x_2 + 3x_3 = 3$$
$$2x_1 + x_2 + 3x_3 = 3$$
$$x_1 + x_2 + 2x_3 = 0$$

The determinant of the matrix of coefficients is 0. It can be shown that there are no solutions to this system.

Consider the following system of n equations in n variables which is assumed to have a nonsingular matrix of coefficients.

$$a_{11} x_1 + a_{12} x_2 + \cdots + a_{1n} x_n = y_1$$
$$\vdots$$
$$a_{n1} x_1 + a_{n2} x_2 + \cdots + a_{nn} x_n = y_n$$

Since the matrix of coefficients is nonsingular, a unique solution to this system exists. Here we give a method for determining the solution called *Cramer's rule*.

THEOREM 6-10   *The above system has a unique solution given by*

$$x_k = \frac{1}{|A|} \begin{vmatrix} a_{11} & \cdots & a_{1(k-1)} & y_1 & a_{1(k+1)} & \cdots & a_{1n} \\ & & \vdots & & & & \vdots \\ a_{n1} & \cdots & a_{n(k-1)} & y_n & a_{n(k+1)} & \cdots & a_{nn} \end{vmatrix}$$

*The kth column of the matrix of coefficients is
replaced by $y_1, ..., y_n$.*

where $k = 1, 2, ..., n$.

Proof: The determinant on the right-hand side is obtained by replacing the kth column of A by $\mathbf{y}$. Expressing this determinant in terms of its columns, we get, using the properties of determinants,

$$|\mathbf{a}^1, ..., \mathbf{y}, ..., \mathbf{a}^n|$$
$$= |\mathbf{a}^1, ..., x_1 \mathbf{a}^1 + \cdots + x_n \mathbf{a}^n, ..., \mathbf{a}^n|$$
$$= |\mathbf{a}^1, ..., x_1 \mathbf{a}^1, ..., \mathbf{a}^n| + \cdots + |\mathbf{a}^1, ..., x_n \mathbf{a}^n, ..., \mathbf{a}^n|$$
$$= x_1 |\mathbf{a}^1, ..., \mathbf{a}^1, ..., \mathbf{a}^n| + \cdots + x_n |\mathbf{a}^1, ..., \mathbf{a}^n, ..., \mathbf{a}^n|$$

The only nonzero determinant in this sum is $|\mathbf{a}^1, ..., \mathbf{a}^k, ..., \mathbf{a}^n|$; all the other determinants having two identical columns. Thus

$$|\mathbf{a}^1, ..., \mathbf{y}, ..., \mathbf{a}^n| = x_k |\mathbf{a}^1, ..., \mathbf{a}^k, ..., \mathbf{a}^n|$$
$$= x_k |A|$$

This proves that the x_k given above is a solution to the system if solutions exist. As has been stated above, a unique solution does exist; hence this must be it.

Cramer's rule is useful in theoretical work involving systems of linear equations because it gives an explicit formula for the solution. However, it is not usually used to obtain numerical solutions to large systems of equations, because the work involved in evaluating large determinants is enormous. We illustrate the rule for a system of three equations in three variables.

Example 3

Solve the following system of equations using Cramer's rule.

$$x_1 - x_2 + x_3 = 2$$
$$x_1 + 2x_2 \qquad = 1$$
$$x_1 \qquad - x_3 = 4$$

The matrix of coefficients is

$$A = \begin{pmatrix} 1 & -1 & 1 \\ 1 & 2 & 0 \\ 1 & 0 & -1 \end{pmatrix} \quad \text{and} \quad \begin{pmatrix} y_1 \\ y_2 \\ y_3 \end{pmatrix} = \begin{pmatrix} 2 \\ 1 \\ 4 \end{pmatrix}$$

$|A| = -5$. Thus a unique solution exists. By Cramer's rule,

$$x_1 = -\tfrac{1}{5} \begin{vmatrix} 2 & -1 & 1 \\ 1 & 2 & 0 \\ 4 & 0 & -1 \end{vmatrix} = \tfrac{13}{5}$$

$$x_2 = -\tfrac{1}{5} \begin{vmatrix} 1 & 2 & 1 \\ 1 & 1 & 0 \\ 1 & 4 & -1 \end{vmatrix} = -\tfrac{4}{5}$$

$$x_3 = -\tfrac{1}{5} \begin{vmatrix} 1 & -1 & 2 \\ 1 & 2 & 1 \\ 1 & 0 & 4 \end{vmatrix} = -\tfrac{7}{5}$$

EXERCISES

Determine whether the following matrices are of rank 3 or less than 3 by finding their determinants.

1. $\begin{pmatrix} 1 & 2 & -3 \\ -1 & 0 & 4 \\ 1 & 6 & -1 \end{pmatrix}$ **2.** $\begin{pmatrix} 0 & 1 & 4 \\ 2 & 1 & 3 \\ 3 & 2 & 6 \end{pmatrix}$

3. $\begin{pmatrix} 1 & 2 & -1 \\ -1 & 1 & 5 \\ 3 & 1 & 7 \end{pmatrix}$ **4.** $\begin{pmatrix} 1 & -1 & 2 \\ 3 & 1 & 1 \\ 5 & -1 & 3 \end{pmatrix}$

5. Apply Theorem 6-9 to determine values of λ for which the system of equations

$$(1 - \lambda)x_1 + \quad 6x_2 = 0$$

$$5x_1 + (2 - \lambda)x_2 = 0$$

can have many solutions. Such λ's are called *eigenvalues* of a matrix of coefficients. Eigenvalues are important in applications, as the reader will see in following sections. Determine the solutions for each such λ.

6. Determine values of λ for which the system of equations

$$(5 - \lambda)x_1 + \quad 4x_2 + \quad 2x_3 = 0$$

$$4x_1 + (5 - \lambda)x_2 + \quad 2x_3 = 0$$

$$2x_1 + \quad 2x_2 + (2 - \lambda)x_3 = 0$$

has many solutions.

7. Prove that the system of equations $AX = \lambda X$ has the possibility of more than one solution if and only if $|A - \lambda I| = 0$. Here I is the appropriate unit matrix.

Solve the following systems of linear equations using Cramer's rule.

8. $\begin{aligned} x_1 + x_2 + x_3 &= 0 \\ 2x_1 - 5x_2 - 3x_3 &= 10 \\ 4x_1 + 8x_2 + 2x_3 &= 4 \end{aligned}$

9. $\begin{aligned} 2x_1 - x_2 + 2x_3 &= 11 \\ x_1 + 2x_2 - x_3 &= -3 \\ 3x_1 - 2x_2 - 3x_3 &= -1 \end{aligned}$

10. $\begin{aligned} x_1 - x_2 + x_3 &= 1 \\ 2x_1 - 2x_2 + 5x_3 &= 0 \\ -x_1 + 3x_2 - 4x_3 &= 3 \end{aligned}$

11. $\begin{aligned} x_1 + 3x_2 - x_3 &= 1 \\ 2x_1 + x_2 + x_3 &= 4 \\ 3x_1 + 4x_2 + 2x_3 &= -1 \end{aligned}$

12. **a)** Let C be the 5×5 upper triangular matrix

$$\begin{pmatrix} 1 & 2 & 4 & 3 & -1 \\ 0 & 0 & 2 & 4 & -8 \\ 0 & 0 & 2 & 1 & -2 \\ 0 & 0 & 0 & -1 & 3 \\ 0 & 0 & 0 & 0 & -3 \end{pmatrix}$$

having a zero diagonal element. Find an echelon form for C. Observe that C is of rank less than 4.

b) Using part (a) as a guideline, generalize the result: Let C be an $n \times n$ upper triangular matrix having a zero diagonal element. Prove that an echelon form of C has last row zero. Thus C is of rank less than n. [This result was used in the proof of Theorem 6-8.]

6–4. A FORMULA FOR THE INVERSE OF A MATRIX

Previously we gave a numerical method for determining the inverse of a matrix. Here we derive a formula for the inverse. As in the case of Cramer's rule, this result if of theoretical importance, but it is not a useful method for actually determining the inverse of a matrix. It gives an explicit formula for the inverse of an arbitrary square matrix. We obtain the formula for the inverse in proving the following useful existence theorem.

THEOREM 6-11   *A square $n \times n$ matrix is invertible if and only if it is nonsingular.*

Proof: First of all, assume that A is invertible. Hence the inverse of A, A^{-1}, exists, and $AA^{-1} = I_n$. Taking the determinant, $|AA^{-1}| = 1$. Thus $|A||A^{-1}| = 1$, proving that $|A| \neq 0$ and that A is nonsingular.

To prove the converse we associate the nonsingular matrix A with a system of linear equations $A\mathbf{x} = \mathbf{y}$, where A is its matrix of coefficients, and we apply Cramer's rule. Here $\mathbf{y}$ is a column vector with n arbitrary components. The system has a unique solution given by

$$x_1 = \frac{1}{|A|} \left| \mathbf{y}, \mathbf{a}^2, \ldots, \mathbf{a}^n \right|, \ldots, x_n = \frac{1}{|A|} \left| \mathbf{a}^1, \ldots, \mathbf{a}^{n-1}, \mathbf{y} \right|$$

Expanding $\left| \mathbf{y}, \mathbf{a}^2, \ldots, \mathbf{a}^n \right|$ in terms of the first column, expanding $\left| \mathbf{a}^1, \mathbf{y}, \ldots, \mathbf{a}^n \right|$ in terms of the second column, etc., the solution may be expressed as

$$x_1 = \frac{1}{|A|} \left[y_1 A_{11} + \cdots + (-1)^{n+1} y_n A_{n1} \right], \ldots,$$

$$x_n = \frac{1}{|A|} \left[(-1)^{1+n} y_1 A_{1n} + \cdots + (-1)^{n+n} y_n A_{nn} \right]$$

where A_{ij} is the minor of the element a_{ij} of A.

This solution may be expressed as a single vector equation

$$\mathbf{x} = B\mathbf{y}$$

where

$$B = \frac{1}{|A|} \begin{pmatrix} A_{11} & \cdots & (-1)^{n+1} A_{n1} \\ & \vdots & \\ (-1)^{1+n} A_{1n} & \cdots & (-1)^{n+n} A_{nn} \end{pmatrix} \tag{1}$$

Combining equations $A\mathbf{x} = \mathbf{y}$ and $\mathbf{x} = B\mathbf{y}$ by inserting the expression for $\mathbf{x}$ from the latter into the former, we get

$$AB\mathbf{y} = \mathbf{y}$$

Similarly, by substituting for $\mathbf{y}$ from the first equation into the second,

$$BA\mathbf{x} = \mathbf{x}$$

Since $\mathbf{x}$ and $\mathbf{y}$ can take all values, these results imply that $AB = BA = I_n$, the unit $n \times n$ matrix, and hence that $B = A^{-1}$.

The proof of this theorem gives us the formula for finding the inverse of a matrix. The inverse of a matrix A is given by B in equation (1). The matrix

$$\begin{pmatrix} A_{11} & \cdots & (-1)^{n+1} A_{n1} \\ & \vdots & \\ (-1)^{1+n} A_{1n} & \cdots & (-1)^{n+n} A_{nn} \end{pmatrix}$$

in (1) is called the *adjoint of A* and denoted adj A.

It is the transpose of the following matrix of *signed minors*:

$$\begin{pmatrix} +A_{11} & -A_{12} & +A_{13} & \cdots & (-1)^{1+n}A_{1n} \\ -A_{21} & +A_{22} & +A_{23} & \cdots & \\ +A_{31} & -A_{32} & +A_{33} & \cdots & \\ & & & \vdots & \\ (-1)^{n+1}A_{n1} & \cdots & & & (-1)^{n+n}A_{nn} \end{pmatrix}$$

Hence

$$A^{-1} = \frac{1}{|A|}\, \text{adj}\, A$$

Example 1

Consider the matrix $A = \begin{pmatrix} 1 & 2 \\ 3 & 4 \end{pmatrix}$.

$|A| = 4 - 6 = -2 \neq 0$.   Hence A^{-1} exists.   We know that $A_{11} = 4$, $A_{12} = 3$, $A_{21} = 2$, and $A_{22} = 1$.   The matrix of signed minors is

$$\begin{pmatrix} 4 & -3 \\ -2 & 1 \end{pmatrix}$$

and the adjoint matrix is the transpose of this matrix.

$$\text{adj}\, A = \begin{pmatrix} 4 & -2 \\ -3 & 1 \end{pmatrix}$$

The inverse of A is thus

$$A^{-1} = -\tfrac{1}{2}\begin{pmatrix} 4 & -2 \\ -3 & 1 \end{pmatrix}$$

As we check, we confirm that

$$-\tfrac{1}{2}\begin{pmatrix} 1 & 2 \\ 3 & 4 \end{pmatrix}\begin{pmatrix} 4 & -2 \\ -3 & 1 \end{pmatrix} = -\tfrac{1}{2}\begin{pmatrix} 4 & -2 \\ -3 & 1 \end{pmatrix}\begin{pmatrix} 1 & 2 \\ 3 & 4 \end{pmatrix} = I$$

Example 2

Consider $A = \begin{pmatrix} 1 & 2 & 0 \\ 2 & 1 & -1 \\ 3 & 1 & 1 \end{pmatrix}$.

$|A| = -8$.   Hence A^{-1} exists.

$$\text{adj}\, A = \begin{pmatrix} \begin{vmatrix} 1 & -1 \\ 1 & 1 \end{vmatrix} & -\begin{vmatrix} 2 & -1 \\ 3 & 1 \end{vmatrix} & \begin{vmatrix} 2 & 1 \\ 3 & 1 \end{vmatrix} \\ -\begin{vmatrix} 2 & 0 \\ 1 & 1 \end{vmatrix} & \begin{vmatrix} 1 & 0 \\ 3 & 1 \end{vmatrix} & -\begin{vmatrix} 1 & 2 \\ 3 & 1 \end{vmatrix} \\ \begin{vmatrix} 2 & 0 \\ 1 & -1 \end{vmatrix} & -\begin{vmatrix} 1 & 0 \\ 2 & -1 \end{vmatrix} & \begin{vmatrix} 1 & 2 \\ 2 & 1 \end{vmatrix} \end{pmatrix}^{t}$$

$$= \begin{pmatrix} 2 & -5 & -1 \\ -2 & 1 & 5 \\ -2 & 1 & -3 \end{pmatrix}^t = \begin{pmatrix} 2 & -2 & -2 \\ -5 & 1 & 1 \\ -1 & 5 & -3 \end{pmatrix}$$

Hence $A^{-1} = -\frac{1}{8} \begin{pmatrix} 2 & -2 & -2 \\ -5 & 1 & 1 \\ -1 & 5 & -3 \end{pmatrix}$.

We complete this section with important results involving orthogonal matrices. These results will, for example, be useful in a discussion of transformations that follows in a later section.

We remind the reader that an orthogonal matrix is a square matrix whose columns form a set of unit, mutually orthogonal vectors (Section 4-1).

THEOREM 6-12   *Let A be an orthogonal matrix.   Then $A^tA = AA^t = I$. Thus $A^{-1} = A^t$, the inverse of an orthogonal matrix is its transpose.   (This result was discussed in Section 4-1.)*

Proof: Let $B = A^t$.   Then

$$A^tA = BA = \left(\sum_k b_{ik} a_{kj} \right) = \left(\sum_k a_{ki} a_{kj} \right) = (\delta_{ij}) = I$$

This proves the first part of the theorem.

Let us now prove that $AA^t = I$.

Since $A^tA = I$, $|A^tA| = 1$, implying that $|A^t| |A| = 1$.   Thus, $|A^t| \neq 0$ and $|A| \neq 0$; $(A^t)^{-1}$ and A^{-1} exist, by Theorem 6-11.

Return to the equation

$$A^tA = I$$

Multiply both sides by (A^t) to get

$$A^tAA^t = A^t$$

Premultiplying both sides by $(A^t)^{-1}$ gives

$$AA^t = I$$

proving the second part of the theorem.

THEOREM 6-13   *Let A be an orthogonal matrix.   Then $|A| = \pm 1$.*

Proof: $AA^t = I$ implies that $|A| |A^t| = 1$.   But $|A| = |A^t|$.   Thus $|A|^2 = 1$, implying that $|A| = \pm 1$.

Example 3

Consider the orthogonal matrix $A = \begin{pmatrix} \cos\theta & -\sin\theta \\ \sin\theta & \cos\theta \end{pmatrix}$. Observe that $A^t = \begin{pmatrix} \cos\theta & \sin\theta \\ -\sin\theta & \cos\theta \end{pmatrix}$ and that

$$A^t A = \begin{pmatrix} \cos\theta & \sin\theta \\ -\sin\theta & \cos\theta \end{pmatrix} \begin{pmatrix} \cos\theta & -\sin\theta \\ \sin\theta & \cos\theta \end{pmatrix} = \begin{pmatrix} 1 & 0 \\ 0 & 1 \end{pmatrix}$$

It can be shown that $AA^t = I$, also.

Thus

$$A^{-1} = A^t = \begin{pmatrix} \cos\theta & \sin\theta \\ -\sin\theta & \cos\theta \end{pmatrix}$$

Further,

$$|A| = \begin{vmatrix} \cos\theta & -\sin\theta \\ \sin\theta & \cos\theta \end{vmatrix} = \cos^2\theta - (-\sin^2\theta) = 1$$

illustrating Theorem 6-13.

Observe that on interchanging the columns of A one gets the orthogonal matrix $B = \begin{pmatrix} -\sin\theta & \cos\theta \\ \cos\theta & \sin\theta \end{pmatrix}$ and $|B| = -1$. In this manner, by interchanging columns, one can always get an orthogonal matrix having determinant -1 from one having determinant $+1$, and conversely. This technique will be useful in a later section on transformations when we are interested in constructing orthogonal matrices having determinants $+1$. There orthogonal matrices will be used to describe rotations of coordinate systems.

EXERCISES

In Exercises 1–8 determine whether or not inverses of the matrices exist. In the cases where inverses exist, determine the inverse using the formula of this section.

1. $\begin{pmatrix} 1 & 4 \\ 3 & 2 \end{pmatrix}$

2. $\begin{pmatrix} -2 & -1 \\ 7 & 3 \end{pmatrix}$

3. $\begin{pmatrix} 1 & 2 \\ 2 & 4 \end{pmatrix}$

4. $\begin{pmatrix} 2 & 1 \\ 4 & 3 \end{pmatrix}$

5. $\begin{pmatrix} 1 & 2 & 3 \\ 0 & 1 & 2 \\ 4 & 5 & 3 \end{pmatrix}$

6. $\begin{pmatrix} 0 & 3 & 3 \\ 1 & 2 & 3 \\ 1 & 4 & 6 \end{pmatrix}$

7. $\begin{pmatrix} 0 & 3 & 3 \\ 1 & 2 & 3 \\ 2 & 4 & 6 \end{pmatrix}$

8. $\begin{pmatrix} 1 & 2 & -1 \\ 2 & 4 & -3 \\ 1 & -2 & 0 \end{pmatrix}$

301

9. If C is a nonsingular matrix, prove that $|C||C^{-1}| = 1$.

10. **a)** Let A and B be invertible matrices of the same kind. Prove that the product matrix AB is invertible with inverse defined by

$$(AB)^{-1} = B^{-1}A^{-1}$$

   b) Let A, B, and C be invertible matrices of the same kind. Prove that ABC is invertible with inverse $(ABC)^{-1} = C^{-1}B^{-1}A^{-1}$.

11. Is the sum of two invertible matrices always invertible?

12. Prove that if A is an invertible matrix then $(A^t)^{-1} = (A^{-1})^t$. Thus the inverse of the transpose of a matrix is the transpose of its inverse.

13. Prove that if A is a symmetric invertible matrix then A^{-1} is also symmetric.

14. Let A be invertible. If B and C are matrices such that $AB = AC$, prove that $B = C$.

15. Use the result $AA^t = I$ to prove that the row vectors of an orthogonal matrix A also form a set of unit, mutually orthogonal vectors, and that A^{-1} is orthogonal.

6–5. EIGENVALUES AND EIGENVECTORS

DEFINITION 6-2   A nonzero vector $\mathbf{x}$ in $\mathbf{R}^n$ is said to be an *eigenvector* of an $n \times n$ matrix A if there exists a scalar λ such that $A\mathbf{x} = \lambda\mathbf{x}$. λ is called the *eigenvalue* corresponding to $\mathbf{x}$.

   Let us look at the geometrical significance of an eigenvector. Let A be a 2×2 matrix with eigenvector $\mathbf{x}$ and corresponding eigenvalue λ; therefore $A\mathbf{x} = \lambda\mathbf{x}$. We can interpret A as a mapping in the plane and $\mathbf{x}$

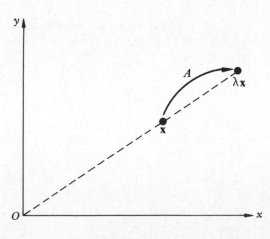

FIGURE 6-1

as the location of a point in the plane. $\lambda\mathbf{x}$ will correspond to a
the plane that is in the same or opposite direction from the or
(depending upon the sign of λ), but that is at a distance λ tin
away. Figure 6-1 illustrates this concept when $\lambda > 1$.

The eigenvectors of A may be interpreted geometrically as the
points in the plane whose directions from O are unchanged or reversed
after mapping. If $\lambda = 1$, then the point remains fixed under the mapping;
the vector is then said to be *invariant* under the matrix operation.

If there exists such a λ for a given $\mathbf{x}$, then it is uniquely determined.
Suppose there exist two such λ's, λ_1 and λ_2. Then $\lambda_1\mathbf{x} = A\mathbf{x} = \lambda_2\mathbf{x}$,
implying that $\lambda_1 = \lambda_2$. The set of all eigenvalues is called the *spectrum*
of A.

In this section we develop a method for determining the eigenvalues and eigenvectors of a matrix, and in following sections we shall
see their uses.

The equation $A\mathbf{x} = \lambda\mathbf{x}$ may be rewritten

$$A\mathbf{x} - \lambda\mathbf{x} = \mathbf{0}$$

giving

$$(A - \lambda I_n)\mathbf{x} = \mathbf{0}$$

This represents the set of homogeneous linear equations

$$\begin{pmatrix} a_{11} - \lambda & \cdots & a_{1n} \\ & \vdots & \\ a_{n1} & \cdots & a_{nn} - \lambda \end{pmatrix} \begin{pmatrix} x_1 \\ \vdots \\ x_n \end{pmatrix} = \mathbf{0}$$

$\mathbf{x} = \mathbf{0}$ is a solution. Since we are looking for eigenvectors, which
have been defined to be nonzero vectors, we are not interested in this
solution. Further solutions can only exist if the matrix of coefficients is
singular (Theorem 6-9, Section 6-3). Hence, solving the equation
$|A - \lambda I_n| = 0$ for λ gives any eigenvalues of A. The polynomial $|A - \lambda I_n|$
is called the *characteristic polynomial* of A, and the equation $|A - \lambda I_n| = 0$
is called the *characteristic equation* of A. The eigenvalues are substituted
back into the equation $(A - \lambda I_n)\mathbf{x} = \mathbf{0}$ to find the corresponding eigenvectors.

Example 1

Find the eigenvalues and eigenvectors of the matrix

$$\begin{pmatrix} 1 & 6 \\ 5 & 2 \end{pmatrix}$$

The characteristic equation $|A - \lambda I_2| = 0$ is

$$\left| \begin{pmatrix} 1 & 6 \\ 5 & 2 \end{pmatrix} - \lambda \begin{pmatrix} 1 & 0 \\ 0 & 1 \end{pmatrix} \right| = 0$$

Thus

$$\begin{vmatrix} 1-\lambda & 6 \\ 5 & 2-\lambda \end{vmatrix} = 0$$

Expanding,

$$\lambda^2 - 3\lambda - 28 = 0$$

$$(\lambda - 7)(\lambda + 4) = 0$$

The roots of this equation are 7 and -4; thus the eigenvalues of A are 7 and -4.

The corresponding eigenvectors are now found. For $\lambda = 7$, $(A - \lambda I_2)\mathbf{x} = \mathbf{0}$ becomes

$$\left[\begin{pmatrix} 1 & 6 \\ 5 & 2 \end{pmatrix} - 7 \begin{pmatrix} 1 & 0 \\ 0 & 1 \end{pmatrix} \right] \begin{pmatrix} x_1 \\ x_2 \end{pmatrix} = \mathbf{0}$$

which reduces to

$$\begin{pmatrix} -6 & 6 \\ 5 & -5 \end{pmatrix} \begin{pmatrix} x_1 \\ x_2 \end{pmatrix} = \mathbf{0}$$

This is the matrix form of the system of equations

$$-6x_1 + 6x_2 = 0$$

$$5x_1 - 5x_2 = 0$$

The solutions are $x_1 = x_2$. Hence any vector of the type $x_2 \begin{pmatrix} 1 \\ 1 \end{pmatrix}$, where x_2 is a nonzero scalar, is an eigenvector corresponding to the eigenvalue 7.

For $\lambda = -4$, $(A - \lambda I_2)\mathbf{x} = \mathbf{0}$ becomes

$$\left[\begin{pmatrix} 1 & 6 \\ 5 & 2 \end{pmatrix} + 4 \begin{pmatrix} 1 & 0 \\ 0 & 1 \end{pmatrix} \right] \begin{pmatrix} x_1 \\ x_2 \end{pmatrix} = \mathbf{0}$$

Simplified,

$$\begin{pmatrix} 5 & 6 \\ 5 & 6 \end{pmatrix} \begin{pmatrix} x_1 \\ x_2 \end{pmatrix} = \mathbf{0}$$

implying that $x_1 = -\frac{6}{5}x_2$. Hence any vector of the type $x_2 \begin{pmatrix} -\frac{6}{5} \\ 1 \end{pmatrix}$, where x_2 is a nonzero scalar, is an eigenvector for $\lambda = -4$.

Here vectors of the forms $\begin{pmatrix} c \\ c \end{pmatrix}$ and $\begin{pmatrix} -\frac{6}{5}c \\ c \end{pmatrix}$, where $c \neq 0$, are

eigenvectors of $\begin{pmatrix} 1 & 6 \\ 5 & 2 \end{pmatrix}$ with eigenvalues 7 and -4, respectively. (For

convenience we use c rather than x_2.) These vectors can be interpreted geometrically as points on the lines OA and OB, respectively.

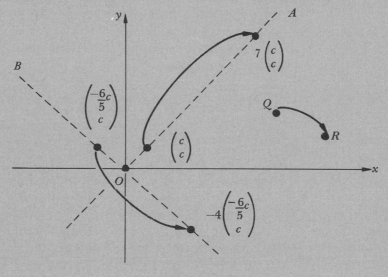

FIGURE 6-2

The effect of matrix multiplication is to map these points into other points on the lines, as shown in Figure 6-2. This only happens for points on these lines. A point such as Q, which does not lie on one of these lines, would be mapped into a point such as R, which docs not lie on OQ.

Note that the eigenvectors in this example can only be determined up to magnitude; if $A\mathbf{x} = \lambda\mathbf{x}$, thcn $cA\mathbf{x} = c\lambda\mathbf{x}$, implying that $A(c\mathbf{x}) = \lambda(c\mathbf{x})$ for any nonzero scalar c. Hence, if $\mathbf{x}$ is an eigenvector for eigenvalue λ, $c\mathbf{x}$ is an eigenvector of this same eigenvalue.

From this discussion we might conjecture the following result.

THEOREM 6-14   *Let A be an $n \times n$ matrix and λ an eigenvalue of A. The subset consisting of all eigenvectors corresponding to λ together with the zero vector is a subspace of $\mathbf{R}^n$ called the eigenspace of the eigenvalue λ.*

Proof: In order to prove that the eigenspace is a subspace, we have to show that it is closed under addition and scalar multiplication.

Let $\mathbf{x}_1$ and $\mathbf{x}_2$ be two elements of the eigenspace corresponding to λ. Then $A\mathbf{x}_1 = \lambda\mathbf{x}_1$ and $A\mathbf{x}_2 = \lambda\mathbf{x}_2$. Hence

$$A\mathbf{x}_1 + A\mathbf{x}_2 = \lambda\mathbf{x}_1 + \lambda\mathbf{x}_2$$

or

$$A(\mathbf{x}_1 + \mathbf{x}_2) = \lambda(\mathbf{x}_1 + \mathbf{x}_2)$$

Thus $\mathbf{x}_1 + \mathbf{x}_2$ is an element of the eigenspace of λ. The eigenspace is closed under addition.

Further, since $A\mathbf{x}_1 = \lambda\mathbf{x}_1$, $cA\mathbf{x}_1 = c\lambda\mathbf{x}_1$, implying that $A(c\mathbf{x}_1) = \lambda(c\mathbf{x}_1)$. $c\mathbf{x}_1$ is an element of the eigenspace.

Hence the eigenspace is closed under scalar multiplication. Thus it is a subspace.

We now illustrate the results of this theorem with an example.

Example 2

Determine the eigenvalues and eigenvectors of the matrix

$$\begin{pmatrix} 5 & 4 & 2 \\ 4 & 5 & 2 \\ 2 & 2 & 2 \end{pmatrix}$$

The characteristic equation is

$$\left| \begin{pmatrix} 5 & 4 & 2 \\ 4 & 5 & 2 \\ 2 & 2 & 2 \end{pmatrix} - \lambda \begin{pmatrix} 1 & 0 & 0 \\ 0 & 1 & 0 \\ 0 & 0 & 1 \end{pmatrix} \right| = 0$$

This reduces to $(\lambda - 10)(\lambda - 1)^2 = 0$, giving the eigenvalues to be 10 and 1.

For $\lambda = 10$, the eigenvectors are given by

$$\begin{pmatrix} -5 & 4 & 2 \\ 4 & -5 & 2 \\ 2 & 2 & -8 \end{pmatrix} \begin{pmatrix} x_1 \\ x_2 \\ x_3 \end{pmatrix} = \mathbf{0}$$

This system of linear equations has solutions $x_1 = 2x_3, x_2 = 2x_3$, so the eigenspace is the one-dimensional space $x_3 \begin{pmatrix} 2 \\ 2 \\ 1 \end{pmatrix}$.

For $\lambda = 1$, the eigenvectors are given by

$$\begin{pmatrix} 4 & 4 & 2 \\ 4 & 4 & 2 \\ 2 & 2 & 1 \end{pmatrix} \begin{pmatrix} x_1 \\ x_2 \\ x_3 \end{pmatrix} = \mathbf{0}$$

This system has solutions $x_1 = -x_2 - \frac{1}{2}x_3$ for arbitrary x_2 and x_3. Hence the eigenvectors corresponding to the eigenvalue 1 are nonzero vectors of the form $\begin{pmatrix} -x_2 - \frac{1}{2}x_3 \\ x_2 \\ x_3 \end{pmatrix}$. They, together with the zero vector, form a two-dimensional subspace of $\mathbf{R}^3$.

We can get two base vectors for this space by letting $x_2 = 1$ and $x_3 = 0$, and then letting $x_2 = 0$ and $x_3 = 1$. The base vectors would be

$$\begin{pmatrix} -1 \\ 1 \\ 0 \end{pmatrix} \text{ and } \begin{pmatrix} -\frac{1}{2} \\ 0 \\ 1 \end{pmatrix}.$$

Eigenvectors and eigenvalues are important tools in linear algebra. Their use is illustrated in the following sections.

EXERCISES

Determine the eigenvalues and corresponding eigenspaces of each of the following matrices. Use the Gram-Schmidt orthogonalization process to determine an orthonormal basis for each eigenspace of the matrices in Exercises 4, 5, and 9.

1. $\begin{pmatrix} 5 & 4 \\ 1 & 2 \end{pmatrix}$
    2. $\begin{pmatrix} 2 & 1 \\ -1 & 4 \end{pmatrix}$

3. $\begin{pmatrix} 3 & 2 & -2 \\ -3 & -1 & 3 \\ 1 & 2 & 0 \end{pmatrix}$
    4. $\begin{pmatrix} 1 & -2 & 2 \\ -2 & 1 & 2 \\ -2 & 0 & 3 \end{pmatrix}$

5. $\begin{pmatrix} 1 & 0 & 0 \\ -2 & 1 & 2 \\ -2 & 0 & 3 \end{pmatrix}$
    6. $\begin{pmatrix} 1 & 0 & 0 \\ -2 & 5 & -2 \\ -2 & 4 & -1 \end{pmatrix}$

7. $\begin{pmatrix} 15 & 7 & -7 \\ -1 & 1 & 1 \\ 13 & 7 & -5 \end{pmatrix}$
    8. $\begin{pmatrix} 5 & -2 & 2 \\ 4 & -3 & 4 \\ 4 & -6 & 7 \end{pmatrix}$

9. $\begin{pmatrix} 4 & 2 & -2 & 2 \\ 1 & 3 & 1 & -1 \\ 0 & 0 & 2 & 0 \\ 1 & 1 & -3 & 5 \end{pmatrix}$
    10. $\begin{pmatrix} 3 & 5 & -5 & 5 \\ 3 & 1 & 3 & -3 \\ -2 & 2 & 0 & 2 \\ 0 & 4 & -6 & 8 \end{pmatrix}$

11. $\begin{pmatrix} 0.5 & 1.5 & -1.5 & 1.5 \\ 3.5 & 3.5 & 3.5 & -3.5 \\ -1 & 1 & 0 & 1 \\ -3 & -1 & -4 & 5 \end{pmatrix}$

Determine the eigenvalues and corresponding eigenspaces of the following matrices. Interpret your results geometrically. (*Hint*: Use Example 3, Section 4-1.)

12. $\begin{pmatrix} 3 & 0 \\ 0 & 3 \end{pmatrix}$

13. $\begin{pmatrix} -5 & 0 & 0 \\ 0 & -5 & 0 \\ 0 & 0 & -5 \end{pmatrix}$

14. Show that the matrix $\begin{pmatrix} 1 & 1 \\ -2 & -1 \end{pmatrix}$ has no real eigenvalues. Readers who have a knowledge of complex numbers will observe that it does have complex eigenvalues. In fact, all matrices have eigenvalues in the complex number system.

15. Prove that the matrix $\begin{pmatrix} 0 & -1 \\ 1 & 0 \end{pmatrix}$ has no real eigenvalues and thus no eigenvectors. Interpret your results geometrically. (*Hint*: Use Example 1, Section 4-1.)

16. Draw a diagram to illustrate the geometrical significance of
   a) a negative eigenvalue   **b)** a fractional eigenvalue

17. Show that if A is an upper triangular matrix

$$\begin{pmatrix} a_{11} & a_{12} & \cdots & a_{1n} \\ 0 & a_{22} & \cdots & a_{2n} \\ & & \vdots & \\ 0 & \cdots & 0 & a_{nn} \end{pmatrix}$$

its eigenvalues are $a_{11}, \ldots, a_{nn}$.

18. Prove that if A is a square matrix, A^t has the same eigenvalues as A.

19. Prove that if $\lambda_1, \ldots, \lambda_n$ are the eigenvalues of A, then $c\lambda_1, \ldots, c\lambda_n$ are the eigenvalues of cA.

20. If A has eigenvalues $\lambda_1, \ldots, \lambda_n$, prove that A^2 has eigenvalues $\lambda_1{}^2, \ldots, \lambda_n{}^2$ and that A^m has eigenvalues $\lambda_1{}^m, \ldots, \lambda_n{}^m$.

21. Prove that a matrix A can have zero as an eigenvalue if and only if it is singular.

6–6.* AN APPLICATION OF EIGENVALUES AND EIGENVECTORS

Eigenvalues and eigenvectors play a central role in Markov processes. We illustrate this role using the population distribution model of Section 1-7 and a weather model.

Assume that the population distribution at a certain time is given by the row vector $\mathbf{x_0}$ and that the stochastic matrix P gives the transition probabilities for one year. We found that the population distribution after one year was $\mathbf{x_1}$, where $\mathbf{x_1} = \mathbf{x_0} P$, and that after n years it was $\mathbf{x_n}$, where $\mathbf{x_n} = \mathbf{x_0} P^n$.

We would like to know if the relative populations become *stable* after a finite number of years—that is, if the number of people in both city and suburbia remains unchanged.   Mathematically this condition of stability would be expressed by the existence of a nonnegative integer n, where $\mathbf{x}_n = \mathbf{x}_{n+1}$.   If this happens at the initial year $(\mathbf{x}_0 = \mathbf{x}_1)$, then $\mathbf{x}_0 = \mathbf{x}_0 P$.   If, after n years, $\mathbf{x}_n = \mathbf{x}_{n+1}$, then

$$\mathbf{x}_{n+1} = \mathbf{x}_0 P^{n+1}$$

$$\mathbf{x}_{n+1} = \mathbf{x}_n P$$

$$\mathbf{x}_n = \mathbf{x}_n P$$

Remembering that $\mathbf{x}_n$ is a vector and P is a matrix, we see that this equation is similar to the type of equation that defines eigenvectors and eigenvalues.   The only differences are that we have row vectors in place of column vectors and that we multiply the matrix on the left rather than on the right. $\mathbf{x}_n$ in this equation is a *left eigenvector* with eigenvalue 1. The eigenvectors previously considered are actually *right eigenvectors*. The theory of left eigenvectors is similar to that of right eigenvectors. For any matrix the eigenvalues are the same set of numbers for both left and right eigenvectors.

Let us return to the equation $\mathbf{x}_n = \mathbf{x}_n P$.   If such an $\mathbf{x}_n$ is going to exist, then P must have eigenvalue 1.   This leads us to investigate the possibility of a stochastic matrix having an eigenvalue 1.   We arrive at the following remarkable result.

THEOREM 6-15   1 *is an eigenvalue for every stochastic matrix.*

Proof: Let $P = (p_{ij})$ be an arbitrary $n \times n$ stochastic matrix.   We know that P will have 1 as an eigenvalue if and only if $|P - 1I_n| = 0$, that is, if and only if

$$\begin{vmatrix} p_{11} - 1 & p_{12} & \cdots & p_{1n} \\ p_{21} & p_{22} - 1 & \cdots & p_{2n} \\ & & \vdots & \\ p_{n1} & p_{n2} & \cdots & p_{nn} - 1 \end{vmatrix} = 0$$

If we add all the remaining columns to the first column, we find that the determinant on the left is equal to

$$\begin{vmatrix} \sum p_{1i} - 1 & p_{12} & \cdots & p_{1n} \\ \sum p_{2i} - 1 & p_{22} - 1 & \cdots & p_{2n} \\ & & \vdots & \\ \sum p_{ni} - 1 & p_{n2} & \cdots & p_{nn} - 1 \end{vmatrix}$$

Since the sum of the elements in each row of a stochastic matrix is 1, each element in the first column of this determinant is zero.   Hence the determinant is zero, proving the result.

We are thus assured that the matrix of transition probabilities P has the necessary unit eigenvalue.

Let us look at the condition $\mathbf{x}_n = \mathbf{x}_n P$ for finite stability. Is there a necessary condition on the matrix P for stability to occur? We find that there is.

THEOREM 6-16   *A necessary condition for finite stability sometime after the first year is that P be singular.*

Proof: We investigate the case of stability after the first year. The condition of immediate stability is that $\mathbf{x}_0$ be an eigenvector with eigenvalue 1.

Assume that $\mathbf{x}_n = \mathbf{x}_n P$ for some $n > 0$ with a given $\mathbf{x}_0$. This may be written $\mathbf{x}_0 P^n = \mathbf{x}_0 P^{n+1}$. Interchanging sides,

$$\mathbf{x}_0 P^{n+1} = \mathbf{x}_0 P^n$$

$$\mathbf{x}_0 (P - I) P^n = \mathbf{0}$$

$$\mathbf{x}_0 (P - I) P^n = 0 \mathbf{x}_0 (P - I)$$

Thus a necessary condition is that P^n have eigenvalue zero [for the left eigenvector $\mathbf{x}_0 (P - I)$].

P^n will have eigenvalue zero if and only if $|P^n| = 0$, that is, if and only if $|P| \cdots |P|$ (n times) $= 0$. Thus a necessary condition for stability in a finite number of steps is that P be singular.

Let us apply the result of this theorem to Example 2 of Section 1-7.

$$P = \begin{pmatrix} 0.96 & 0.04 \\ 0.01 & 0.99 \end{pmatrix}$$

P is nonsingular; hence, if the present trend continues, the situation will not stabilize in a finite number of years.

There are various classes of transition matrices leading to various classes of Markov chains. We now introduce the reader to one of these classes, illustrating the role of eigenvectors.

A transition matrix P of a Markov chain is said to be *regular* if for some power of P all the components are positive. The chain is then called a *regular Markov chain*. The transition matrix of the population movement model, $\begin{pmatrix} 0.96 & 0.04 \\ 0.01 & 0.99 \end{pmatrix}$, is a regular matrix.

The following theorem, which we present without proof, gives the important role of eigenvectors in the theory of regular Markov chains.

THEOREM 6-17   *Consider a regular Markov chain having transition matrix P and initial state* $\mathbf{x_0}$.

a. *The sequence of distributions of states* $\mathbf{x_0}$, $\mathbf{x_1} = \mathbf{x_0}P$, $\mathbf{x_2} = \mathbf{x_0}P^2$, ... *approaches a vector* $\mathbf{x}$ *that satisfies* $\mathbf{x}P = \mathbf{x}$. *This limit vector is thus a left eigenvector of P corresponding to the eigenvalue 1.*
b. *The sequence of matrices* $P, P^2, P^3, ...$ *approaches a stochastic matrix T. The rows of T are all identical, a row being a left eigenvector of P corresponding to the eigenvalue 1.*

The implication of (a) is that no matter what the initial distribution $\mathbf{x_0}$, a regular Markov chain will tend to stabilize as it approaches $\mathbf{x}$. The long term behavior becomes predictable; (b) tells us that, no matter what the initial state, the long term probability of a state S_i occurring is the same.

Let us now illustrate these results for our population movement model. The transition matrix is the regular matrix $P = \begin{pmatrix} 0.96 & 0.04 \\ 0.01 & 0.99 \end{pmatrix}$ and $\mathbf{x_0} = (\ 57{,}633 \quad 71{,}549 \ )$. We find the left eigenvector corresponding to the eigenvalue 1.

$$\mathbf{x}P = \mathbf{x} \text{ gives } \mathbf{x}(P - I) = 0.$$

Let $\mathbf{x} = (c, d)$. Thus

$$(c, d) \begin{pmatrix} 0.96 - 1 & 0.04 \\ 0.01 & 0.99 - 1 \end{pmatrix} = 0,$$

implying that $-0.04c + 0.01d = 0$; $4c = d$. Left eigenvectors are thus of the form $(c, 4c)$.

In our specific problem, let us assume no population increase. Therefore, the sum of the components of $(c, 4c)$ is $57{,}633 + 71{,}549 = 129{,}182$. We have that $5c = 129{,}182$ giving $c = 25{,}836.4$. The limit vector is

$$\begin{array}{cc} \text{city} & \text{suburbia} \\ \mathbf{x} = (\ 25{,}836.4 & 103{,}345.6 \ ) \end{array}$$

The populations of city and suburbia will get closer and closer to this distribution yearly if the conditions remain unchanged. (The units are thousands.)

Further we know that P^n approaches a matrix T. Each row of T is identical, being a vector of the type $(c, 4c)$. Since T is stochastic the sum of the elements in each row is 1. Thus $c + 4c = 1$, giving $c = 0.2$.

$$T = \begin{pmatrix} 0.2 & 0.8 \\ 0.2 & 0.8 \end{pmatrix}$$

We get the matrix sequence

$$
\begin{matrix} P \\ \begin{pmatrix} 0.96 & 0.04 \\ 0.01 & 0.99 \end{pmatrix} \end{matrix},
\begin{matrix} P^2 \\ \begin{pmatrix} 0.922 & 0.078 \\ 0.0195 & 0.9805 \end{pmatrix} \end{matrix},
$$

$$
\begin{matrix} P^3 \\ \begin{pmatrix} 0.8859 & 0.1141 \\ 0.028525 & 0.97147 \end{pmatrix} \end{matrix}, \cdots \rightarrow
\begin{matrix} T \\ \begin{pmatrix} 0.2 & 0.8 \\ 0.2 & 0.8 \end{pmatrix} \end{matrix}
$$

Let us interpret this result. We focus on the $(1, 2)$ element in each matrix—a similar interpretation will apply to the other elements. We get the sequence

$$0.04, 0.078, 0.1141, \cdots \rightarrow 0.8$$

Remembering that the $(1, 2)$ element of P^n gives the probability of going from the state 1 (city) to state 2 (suburbia) in n steps, we see that the probability of moving from city to suburbia increases annually from 0.04 to 0.078 to 0.1141, etc., getting closer and closer to 0.8. The elements of T give the long term probabilities. The long term probability of living in suburbia is 0.8. This is independent of initial location since $t_{12} = t_{22}$. The long term probability of living in a city is 0.2. Here, also, this is independent of initial location, a characteristic of regular Markov chains.

Readers who are interested in the proof of Theorem 6-17 are referred to *Finite Mathematical Structures*, J. G. Kemeny, H. Mirkil, J. L. Snell, and G. L. Thompson, Chapter 6, Prentice-Hall.

We now discuss an interesting application of Markov chains in a model that describes rainfall in Tel Aviv.

Example 1

K. R. Gabriel and J. Neumann have developed "A Markov chain model for daily rainfall occurrence at Tel Aviv", *Quart. J. R. Met. Soc.* **88**, (1962), 90–95. The probabilities used were based on data of daily rainfall in Tel Aviv (Nahami Street) for the 27 years 1923/24–1949/50. Days were classified as wet or dry according to whether or not there had been recorded at least 0.1 mm of precipitation in the 24 hour period from 8 A.M. to 8 A.M. the following day. A Markov chain was constructed for each of the months November through April, these months constituting the rainy season. We discuss the chain developed for November. The model assumes that the probability of rainfall on any day depends only on whether the previous day was wet or dry. The statistics accumulated over the years for November were

Preceding Day	Actual Day Wet
Wet	117 out of 195
Dry	80 out of 615

Thus the probability of a wet day following a wet day is $117/195 = 0.6$. The probability of a wet day following a dry day is $80/615 = 0.13$. The matrix of transition probabilities, P, for the weather pattern in November is thus

$$\text{Preceding Day} \begin{array}{c} \\ \text{Wet} \\ \text{Dry} \end{array} \overset{\begin{array}{c} \textit{Actual Day} \\ \text{Wet} \quad \text{Dry} \end{array}}{\left(\begin{array}{cc} 0.6 & 0.4 \\ 0.13 & 0.87 \end{array} \right)}$$

On any given day one can use P to predict the weather on a future day. For example, if today, a Wednesday in November, is dry, we can determine the probability that Saturday will be wet. Saturday is three days away, the various probabilities will be given by P^3. It can be shown that

$$P^3 = \left(\begin{array}{cc} 0.32 & 0.68 \\ 0.22 & 0.78 \end{array} \right)$$

(We perform all computations to six significant figures in this model but round off to two for ease of reading.) The probability of Saturday being wet is $p_{21}{}^{(3)}$, that is 0.22. Observe that the matrix of transition probabilities, P, is regular. Let us determine a left eigenvector corresponding to the eigenvalue 1.

$$(a, b) \left(\begin{array}{cc} 0.6 & 0.4 \\ 0.13 & 0.87 \end{array} \right) - (a, b)$$

$$(a, b) \left(\begin{array}{cc} -0.4 & 0.4 \\ 0.13 & -0.13 \end{array} \right) = 0$$

$$-0.4a + 0.13b = 0$$

$$a = 0.325b.$$

Thus left eigenvectors corresponding to the eigenvalue 1 are of the form $b(\; 0.325 \quad 1 \;)$.

To construct a left eigenvector, the sum of whose components is 1, we let $b = 1/1.325$. This gives $(\; 0.25 \quad 0.75 \;)$. Thus the sequence

$$P, P^2, P^3, P^4 \dots \text{ approaches the matrix } T = \left(\begin{array}{cc} 0.25 & 0.75 \\ 0.25 & 0.75 \end{array} \right)$$

The convergence is quite rapid as the following sample of powers of P illustrates,

$$\overset{P^2}{\left(\begin{array}{cc} 0.41 & 0.59 \\ 0.19 & 0.81 \end{array} \right)}, \overset{P^4}{\left(\begin{array}{cc} 0.28 & 0.72 \\ 0.23 & 0.77 \end{array} \right)}, \overset{P^6}{\left(\begin{array}{cc} 0.25 & 0.75 \\ 0.24 & 0.76 \end{array} \right)},$$

$$\overset{P^8}{\left(\begin{array}{cc} 0.25 & 0.75 \\ 0.25 & 0.75 \end{array} \right)}$$

We can interpret the limit matrix thus

A Day in the Distant Future

$$\begin{array}{cc} & \text{Wet} \quad \text{Dry} \\ \textit{A Given Day} \begin{array}{c} \text{Wet} \\ \text{Dry} \end{array} & \left(\begin{array}{cc} 0.25 & 0.75 \\ 0.25 & 0.75 \end{array} \right) \end{array}$$

The long term prediction of a future day in November being wet is 0.25 and of being dry is 0.75. These are independent of the current day being either wet or dry.

The model can also be used to discuss wet and dry spells. A wet spell of length k is defined as a sequence of k wet days preceded and followed by dry days. Dry spells are defined analogously. Let us look at a wet spell in terms of a diagram:

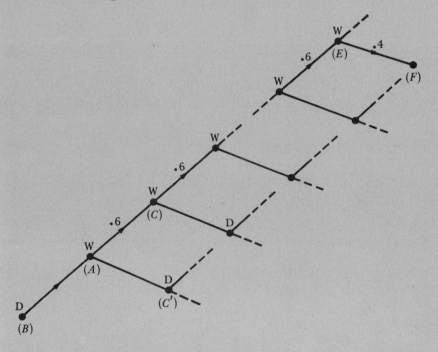

The diagram is to be interpreted from left to right. The current day denoted by the point (A) is wet, W. The previous day (B) was dry, D. At (A) one is interested in going to (C) not (C'); the following day is to be wet. One wants to progress to (E), $k-1$ days later, and finally to (F). The probabilities are marked. It can be seen that the probability of reaching (F) is

$$\underbrace{0.6 \times 0.6 \times \cdots \times 0.6}_{k-1 \text{ times}} \times 0.4$$

Thus, in terms of an arbitrary matrix of transition probabilities, P, the probability of a wet spell of length k is $(p_{11})^{k-1} p_{12}$. Similarly the probability of a dry spell of length k is $(p_{22})^{k-1} p_{21}$.

Using this model, the probability of a wet spell of length 4, for example, is $(0.6)^3 (0.4) = 0.086$.

Gabriel and Neumann in their paper have correlated their predictions to the observed results using statistical methods. The fit is very close, supporting the assumption made for this Markov model that the weather characteristic (wet or dry) on any day depends only on the characteristic of the previous day. They suggest that this is probably connected with a rapid process of "de-coupling", i.e., loss of dependence between some of the meteorological events. Readers who are interested in pursuing this model further are referred to the paper of Gabriel and Neumann.

EXERCISES

1. Prove that if $\mathbf{x}$ is a left eigenvector of a matrix A, $\mathbf{x}^t$ is a right eigenvector of A^t. Prove further that the set of eigenvalues corresponding to left eigenvectors is identical to the set of eigenvalues for right eigenvectors.

2. We return to the model of population flow between metropolitan areas and non-metropolitan areas of Exercise 7, Section 1-7. The populations of metropolitan and non-metropolitan areas in 1971 were 129,182 and 69,723 (in thousands of persons one year old or over). The matrix of transition probabilities was

$$
\begin{array}{cc}
 & \textit{Final} \\
 & \begin{array}{cc} \text{Metro} & \text{Non-metro} \end{array}
\end{array}
$$
$$
\textit{Initial} \begin{array}{c} \text{Metro} \\ \text{Non-metro} \end{array}
\begin{pmatrix}
0.99 & 0.01 \\
0.02 & 0.98
\end{pmatrix}
$$

Determine the figures that the population distributions will approach if conditions remain unchanged.

3. The model of Exercise 8, Section 1-7 was a refinement of the above model in that the metro population was broken down into city and suburbia. In that model the populations of city, suburbia, and non-metro in 1971 were 57,633, 71,549, and 69,723 (in thousands of persons one year old or over), respectively. The stochastic matrix giving the probabilities of moves was

$$
\begin{array}{ccc}
 & \textit{Final} \\
 & \begin{array}{ccc} \text{City} & \text{Suburb} & \text{Non-metro} \end{array}
\end{array}
$$
$$
\begin{array}{c} \text{City} \\ \textit{Initial}\ \text{Suburb} \\ \text{Non-metro} \end{array}
\begin{pmatrix}
0.96 & 0.03 & 0.01 \\
0.01 & 0.98 & 0.01 \\
0.015 & 0.005 & 0.98
\end{pmatrix}
$$

Determine the figures that the population distributions will approach if conditions remain unchanged.

4. We return to the genetic model of Exercise 13, Section 1-7. In that model, offspring of guinea pigs were crossed with hybrids only. The transition matrix P for that model is

$$
\begin{array}{c}
\\
AA \\
Aa \\
aa
\end{array}
\begin{array}{c}
AA\ Aa\ aa \\
\begin{pmatrix}
\frac{1}{2} & \frac{1}{2} & 0 \\
\frac{1}{4} & \frac{1}{2} & \frac{1}{4} \\
0 & \frac{1}{2} & \frac{1}{2}
\end{pmatrix}
\end{array}
$$

Prove that P is regular.

Let $P^n \to T$. Determine T. What information can you get from T?

5. This exercise is based on the weather model of Example 1, this section. The statistics for the month of December in Tel Aviv for the years 1923/24–1949/50 are:

Preceding Day	Actual Day Wet
Wet	213 out of 326
Dry	117 out of 511

Use these statistics to construct a model for the December weather pattern.

a) If a Thursday in December in Tel Aviv is wet, what is the probability that the following Saturday will also be wet?

b) Determine the matrix that describes the long-term prediction of the weather in December.

c) What is the probability of a dry spell of length 5 in December?

d) What is the probability of a dry spell of length m followed by a wet spell of length n?

6. a) Let A be a 2×2 stochastic matrix. Prove that the vectors $c \begin{pmatrix} 1 \\ 1 \end{pmatrix}$ are the only right eigenvectors for the eigenvalue 1; c any scalar.

b) Let A be an $n \times n$ stochastic matrix. Prove that the vectors $c \begin{pmatrix} 1 \\ \vdots \\ 1 \end{pmatrix}$ are eigenvectors for the eigenvalue 1. Can you interpret the result?

7. A psychologist conducts the following experiment. 20 rats are placed at random in a compartment that has been divided into rooms labeled 1, 2 and 3 as follows:

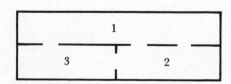

Observe that there are four doors in the arrangement. There are 3 possible states for the rats: they can be in rooms 1, 2 or 3.   Let us assume that the rats move from room to room.   A rat in room 1 has the probabilities $p_{11} = 0$, $p_{12} = \frac{2}{3}$, $p_{13} = \frac{1}{3}$ of moving to the various rooms, based on the distribution of doors.   This approach leads to the following matrix that defines a Markov chain describing the movement of the rats.

$$P = \begin{pmatrix} 0 & \frac{2}{3} & \frac{1}{3} \\ \frac{2}{3} & 0 & \frac{1}{3} \\ \frac{1}{2} & \frac{1}{2} & 0 \end{pmatrix}$$

Predict the distribution of the rats at the end of the experiment.   At the end of the experiment, what is the probability that a given marked rat will be in room 2?

8. 40 rats are placed at random in a compartment having four rooms:

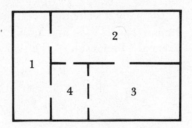

Construct a Markov chain model for describing the movements of the rats between the rooms.   Predict the equilibrium distribution of the rats.   What is the probability that a given marked rat will be in room 4 at the end of the experiment?

9. Two car rental companies A and B are competing for customers at certain airports.   A study has been made of customer satisfaction with the various companies.   These results are expressed in the following matrix Q.

$$Q \quad \begin{matrix} & A & B \\ A & \\ B & \end{matrix} \begin{pmatrix} 75\% & 25\% \\ 20\% & 80\% \end{pmatrix}$$

The elements of Q are to be interpreted as follows.   The first row implies that 75% of those currently using rental company A are satisfied with A and intend to use A next time.   25% are dissatisfied with A and intend using B next time.   The second row implies that 20% of those currently using B are dissatisfied with B and intend switching to A while 80% will remain with B.

Modify the matrix A to obtain transition matrix P that defines a Markov chain model of the rental patterns.   If the current rental trends continue how will the eventual rental distribution settle?   Express the distribution in percentages that use A and B.

10. A market research group has been studying the buying patterns for three competing products I, II and III. The results of the analysis are contained in the following matrix A.

$$A = \begin{matrix} & \\ \text{I} \\ \text{II} \\ \text{III} \end{matrix} \overset{\begin{matrix} \text{I} & \text{II} & \text{III} \end{matrix}}{\begin{pmatrix} 80\% & 5\% & 15\% \\ 20\% & 75\% & 5\% \\ 5\% & 5\% & 90\% \end{pmatrix}}$$

Thus, for example row I implies that of these people currently using product I, 80% will remain with product I while 5% will switch to II and 15% will switch to III, etc.

Construct a stochastic matrix from A that will define a Markov chain model of these trends. If the current patterns continue determine the eventual likely distribution of sales, in terms of percentages.

11. A company developing a certain area of land in Arizona offers two types of houses, ranch style and split level. Residents are polled as to their satisfaction. The result of the poll are contained in the following matrix A.

$$A = \begin{matrix} \\ \text{R} \\ \text{S} \end{matrix} \overset{\begin{matrix} \text{R} & \text{S} \end{matrix}}{\begin{pmatrix} 90\% & 10\% \\ 15\% & 85\% \end{pmatrix}}$$

R indicates ranch style, S split level. The first row of A implies that 90% of those living in ranch style are satisfied while 10% wish they had bought split level. The second row implies that 15% of those currently living in split level wish they had bought ranch style while 85% are satisfied with split level.

Construct a stochastic matrix P from A that defines a Markov chain model of these preferences. Compute an eigenvector X for P, the sum of whose components is 100; the elements then represent percentages. What use can the developers make of X?

12. Write a computer program to illustrate the fact that the sequence of states in the population movement model of this section does indeed approach

$$(\ 25836.4 \quad 103345.6\ ),$$

and that the sequence $P, P^2, P^3, \ldots$ approaches $\begin{pmatrix} 0.2 & 0.8 \\ 0.2 & 0.8 \end{pmatrix}$.

13. Write a computer program for exactly determining the left eigenvector and limit matrix T for the sequence arising from a given 2×2 stochastic matrix. Test your program on the matrix $\begin{pmatrix} 0.96 & 0.04 \\ 0.01 & 0.99 \end{pmatrix}$ of the population movement model, where the vector is $(\ 25836.4 \quad 103345.6\ )$, and the matrix is $\begin{pmatrix} 0.2 & 0.8 \\ 0.2 & 0.8 \end{pmatrix}$. (*Hint:* Let the given stochastic matrix be

$\begin{pmatrix} a & 1-a \\ b & 1-b \end{pmatrix}$, and the left eigenvector be $(\ c \quad d\ )$.   Then

$(\ c \quad d\ ) \begin{pmatrix} a-1 & 1-a \\ b & -b \end{pmatrix} = 0$ leads to algebraic expressions for c and d.)

14. Write a computer program that can be used to give results from the weather model of this section.   Let the program give the probability that any desired future day in November be wet or dry, and the probability of a wet or dry spell of any given length in November.   Using your model of Exercise 5, extend your program to include December, also.

6–7.* EIGENVALUES AND EIGENVECTORS BY ITERATION

Numerical techniques exist for evaluating certain eigenvalues and eigenvectors of various types of matrices.   Here we present an iterative method called the *Power Method*.   It can be used for determining the eigenvalue with the largest absolute value (if it exists) and a corresponding eigenvector† for certain matrices.   Such an eigenvalue is called the *dominant* eigenvalue of the matrix.   In many applications one is only interested in the dominant eigenvalue.

Example 1   Let A be a square matrix with eigenvalues $-5, -2, 1, 3$.   Then -5 is the dominant eigenvalue since $|-5| > |-2|, |-5| > |1|, |-5| > |3|$.

Example 2   Let B be a square matrix with eigenvalues $-4, -2, 1, 4$.   There is no dominant eigenvalue since $|-4| = |4|$.

The power method is based on the following theorem.

THEOREM 6-18   *Let A be an $n \times n$ matrix having n linearly independent eigenvectors and a dominant eigenvalue.   Let $\mathbf{x_0}$ be an arbitrary chosen initial vector such that $A\mathbf{x_0}$ exists.*
The sequence of vectors

$$\mathbf{x_1} = A\mathbf{x_0}, \ \mathbf{x_2} = A\mathbf{x_1}, \ \mathbf{x_3} = A\mathbf{x_2}, ..., \ \mathbf{x_k} = A\mathbf{x_{k-1}}, ...,$$

as k becomes larger, will approach an eigenvector for λ_1, the dominant eigenvalue, if $\mathbf{x_0}$ has a nonzero component in the direction of an eigenvector for λ_1.

Proof: Let the eigenvalues of A be $\lambda_1, ..., \lambda_n$ with corresponding linearly independent eigenvectors $\mathbf{z_1}, ..., \mathbf{z_n}$.   It is assumed that

---

† We shall be working henceforth with right eigenvectors. These are used more than left eigenvectors, and one usually calls them eigenvectors.

the initial guess $\mathbf{x_0}$ has a nonzero component in the direction of $\mathbf{z_1}$. Thus $\mathbf{x_0}$ can be expressed

$$\mathbf{x_0} = a_1 \mathbf{z_1} + \cdots + a_n \mathbf{z_n}$$

where $a_1 \neq 0$.

$$\mathbf{x}_k = A\mathbf{x}_{k-1} = A^2 \mathbf{x}_{k-2} = \cdots = A^k \mathbf{x_0}$$

$$= A^k [a_1 \mathbf{z_1} + \cdots + a_n \mathbf{z_n}]$$

$$= [a_1 A^k \mathbf{z_1} + \cdots + a_n A^k \mathbf{z_n}]$$

$$= [a_1 (\lambda_1)^k \mathbf{z_1} + \cdots + a_n (\lambda_n)^k \mathbf{z_n}]$$

$$= (\lambda_1)^k \left[a_1 \mathbf{z_1} + a_2 \left(\frac{\lambda_2}{\lambda_1} \right)^k \mathbf{z_2} + \cdots + a_n \left(\frac{\lambda_n}{\lambda_1} \right)^k \mathbf{z_n} \right]$$

As k increases, since $|\lambda_j/\lambda_1| < 1$ for $j \geq 2$, $(\lambda_j/\lambda_1)^k$ will approach zero, and the vector $\mathbf{x}_k$ will approach the direction of the vector $\mathbf{z_1}$, an eigenvector for the dominant eigenvalue λ_1.

In the following section we see that symmetric $n \times n$ matrices have n linearly independent eigenvectors, satisfying one requirement of this theorem. The method can thus be tried for a symmetric matrix.

It remains to find an approximation to the dominant eigenvalue. Let λ be an eigenvalue of A and $\mathbf{x}$ a corresponding eigenvector. Then

$$\frac{\mathbf{x} \cdot A\mathbf{x}}{\mathbf{x} \cdot \mathbf{x}} = \frac{\mathbf{x} \cdot \lambda \mathbf{x}}{\mathbf{x} \cdot \mathbf{x}} = \frac{\lambda \mathbf{x} \cdot \mathbf{x}}{\mathbf{x} \cdot \mathbf{x}} = \lambda.$$

This result is used in practice to find an approximation to the dominant eigenvalue. The expression $\dfrac{\mathbf{x}_i \cdot A\mathbf{x}_i}{\mathbf{x}_i \cdot \mathbf{x}_i}$ is computed concurrently with each approximation $\mathbf{x}_i$ to the eigenvector, resulting in a sequence of scalars that approaches the dominant eigenvalue. The method can be continued until successive approximations to the eigenvalue are within the required accuracy. (The ratio $\dfrac{\mathbf{x} \cdot A\mathbf{x}}{\mathbf{x} \cdot \mathbf{x}}$ is called the *Rayleigh quotient*.)

There is one modification that is carried out in practice. The components of the vectors $\mathbf{x_1}, \mathbf{x_2}, \ldots$ may, as the method now stands, become very large, causing significant round-off errors to occur. This problem is overcome by dividing each component of $\mathbf{x}_i$ by the absolute value of its largest component and then using this vector (a vector in the same direction as $\mathbf{x}_i$) in the following iteration.

Example 3

Consider the matrix $A = \begin{pmatrix} 5 & 4 & 2 \\ 4 & 5 & 2 \\ 2 & 2 & 2 \end{pmatrix}$ of Section 6.5.   We already know that a dominant eigenvalue exists and is 10 and that the corresponding eigenvectors are of the form $c \begin{pmatrix} 2 \\ 2 \\ 1 \end{pmatrix}$.   Let us illustrate the method for this known result.   (In the following computations the vectors are actually column vectors—we write them as row vectors for notational convenience.)   Let $\mathbf{x_0} = (-4, 2, 6)$—an arbitrarily selected vector.   We get

Iteration	$A\mathbf{x}$	Adjusted Vector	$(\mathbf{x}.A\mathbf{x})/(\mathbf{x}.\mathbf{x})$
1	$A\mathbf{x_0} = (0, 6, 8)$	$\mathbf{x_1} = (0, 0.75, 1) \left[= \frac{1}{8} A\mathbf{x_0} \right]$	5
2	$A\mathbf{x_1} = (5, 5.75, 3.5)$	$\mathbf{x_2} = (0.869565, 1, 0.608696) \left[= \frac{1}{5.75} A\mathbf{x_1} \right]$	9.88889
3	$A\mathbf{x_2} = (9.56522, 9.69565, 4.95652)$	$\mathbf{x_3} = (0.986547, 1, 0.511211)$	9.99888
4	$A\mathbf{x_3} = (9.95516, 9.96861, 4.99552)$	$\mathbf{x_4} = (0.998651, 1, 0.501125)$	9.99999

Thus, after 4 iterations, an approximation to the dominant eigenvalue is 9.99999 and an approximation to a corresponding eigenvector is $(0.998651, 1, 0.501125)$.

This method has an advantage similar to that of the Gauss-Seidel iterative method discussed earlier.   Any error in computation only means that a new arbitrary vector has been introduced at that stage.   The method is very accurate in that the only round-off errors that occur are those arising from the matrix multiplication carried out during the final iteration and the final computation of $\mathbf{x} \cdot A\mathbf{x}/\mathbf{x} \cdot \mathbf{x}$.

If A is symmetric, a technique called *deflation* can be used for determining further eigenvalues and corresponding eigenvectors.   The method is based on the following theorem which we state without proof.

THEOREM 6-19   *Let A be a symmetric matrix with eigenvalues $\lambda_1, \lambda_2, ...,$ λ_n, where $\lambda_1, \lambda_2, ..., \lambda_n$ are in order according to absolute values, λ_1 being the largest.   Let $\mathbf{x}$ be a unit eigenvector for λ_1, in column form.   Then*

  I. *The matrix $B = A - \lambda_1 \mathbf{x} \mathbf{x}^t$ has eigenvalues $0, \lambda_2, ..., \lambda_n$.   The matrix B is symmetric.*
 II. *If $\mathbf{y}$ is an eigenvector of B corresponding to one of the eigenvalues $\lambda_2, ...,$ λ_n, it is also an eigenvector of A corresponding to the same eigenvalue.*

One determines λ_1 and $\mathbf{x}$ (or good approximations) using the previous technique. B is then found.   Note that, according to the theorem, λ_2 will be the dominant eigenvalue of B (if one exists); hence, on applying the iterative method to B one gets λ_2 and a corresponding eigenvector which is also an eigenvector of λ_2 for A.   In this manner, all the eigenvalues of A and a corresponding eigenvector in each case can be found.   The disadvantage of the method is that the eigenvalues become increasingly inaccurate through compounding of errors.   If at any stage λ_i is not dominant, the method breaks down at that stage.

Readers who are interested in further numerical techniques for determining eigenvalues and eigenvectors will find articles and references in *Methods for Digital Computers*, Vols. 1 and 2, John Wiley & Sons, 1968. Also, in *Numerical Methods*, Germund Dahlquist and Åke Björck, Section 5-8, Prentice-Hall, 1974.

EXERCISES

Using the iterative method in this section determine the dominant eigenvalue and a corresponding eigenvector for each of the following matrices.

1. $\begin{pmatrix} 1 & 7 & -7 \\ -1 & 3 & -1 \\ -1 & -5 & 7 \end{pmatrix}$
2. $\begin{pmatrix} 9 & 4 & -4 \\ -1 & 1 & 1 \\ 7 & 4 & -2 \end{pmatrix}$

3. $\begin{pmatrix} 13 & 7 & -7 \\ -2 & -1 & 2 \\ 12 & 7 & -6 \end{pmatrix}$
4. $\begin{pmatrix} 17 & 8 & -8 \\ 0 & 1 & 0 \\ 16 & 8 & -7 \end{pmatrix}$

5. $\begin{pmatrix} 1 & 1 & -1 & 1 \\ 1 & 1 & 1 & -1 \\ -3 & 3 & 3 & 3 \\ -3 & 3 & 1 & 5 \end{pmatrix}$
6. $\begin{pmatrix} 3 & -1 & 1 & -1 \\ -3 & 1 & -3 & 3 \\ -3 & 3 & 7 & 3 \\ -1 & 5 & 7 & 3 \end{pmatrix}$

7. $\begin{pmatrix} 84 & 5 & -5 & 5 \\ 1 & 0 & 1 & -1 \\ -1 & 1 & 0 & 1 \\ 3 & 5 & -5 & 6 \end{pmatrix}$
8. $\begin{pmatrix} 4.5 & 5.5 & -5.5 & 5.5 \\ 1.5 & 0.5 & 1.5 & -1.5 \\ -1 & 1 & 0 & 1 \\ 3 & 5 & -6 & 7 \end{pmatrix}$

Determine *all* eigenvalues and a corresponding eigenvector in each case for the following *symmetric* matrices using the iterative method of this section.

9. $\begin{pmatrix} 5 & 4 & 2 \\ 4 & 5 & 2 \\ 2 & 2 & 2 \end{pmatrix}$
10. $\begin{pmatrix} 4 & 0 & 2 \\ 0 & 4 & 0 \\ 2 & 0 & 4 \end{pmatrix}$

11. $\begin{pmatrix} 7 & 0 & 5 \\ 0 & 2 & 0 \\ 5 & 0 & 7 \end{pmatrix}$
12. $\begin{pmatrix} 4 & 0 & 0 & 6 \\ 0 & 2 & -4 & 0 \\ 0 & -4 & 2 & 0 \\ 6 & 0 & 0 & 4 \end{pmatrix}$

13. $\begin{pmatrix} 5 & 4 & 1 & 1 \\ 4 & 5 & 1 & 1 \\ 1 & 1 & 4 & 2 \\ 1 & 1 & 2 & 4 \end{pmatrix}$

6–8. SIMILARITY TRANSFORMATIONS—DIAGONALIZATION OF SYMMETRIC MATRICES

In this section and the following one we introduce and see applications of an important class of transformations, similarity transformations. These transformations often enter into a mathematical analysis when a coordinate transformation is involved. Eigenvalues and eigenvectors play a central role in the discussion.

DEFINITION 6-3   If A is a square matrix and C is a nonsingular matrix of the same kind as A, then the transformation of A into $C^{-1}AC$ is called a *similarity transformation*. Two square matrices of the same kind, A and B, are said to be *similar* if there exists a nonsingular matrix C such that $B = C^{-1}AC$.

Example 1

Let $A = \begin{pmatrix} 1 & 2 \\ 3 & 4 \end{pmatrix}$ and $C = \begin{pmatrix} 2 & 5 \\ 1 & 3 \end{pmatrix}$. C is a nonsingular matrix. Its inverse can be shown to be $\begin{pmatrix} 3 & -5 \\ -1 & 2 \end{pmatrix}$. Let us transform A into $C^{-1}AC$.

$$C^{-1}AC = \begin{pmatrix} 3 & -5 \\ -1 & 2 \end{pmatrix} \begin{pmatrix} 1 & 2 \\ 3 & 4 \end{pmatrix} \begin{pmatrix} 2 & 5 \\ 1 & 3 \end{pmatrix}$$

$$= \begin{pmatrix} -12 & -14 \\ 5 & 6 \end{pmatrix} \begin{pmatrix} 2 & 5 \\ 1 & 3 \end{pmatrix} = \begin{pmatrix} -38 & -102 \\ 16 & 43 \end{pmatrix}$$

The matrix $\begin{pmatrix} 1 & 2 \\ 3 & 4 \end{pmatrix}$ has been transformed into the matrix $\begin{pmatrix} -38 & -102 \\ 16 & 43 \end{pmatrix}$ using a similarity transformation involving the matrix $\begin{pmatrix} 2 & 5 \\ 1 & 3 \end{pmatrix}$.

THEOREM 6-20   *Similar matrices have the same eigenvalues.*

Proof: Let A and B be similar matrices. Hence there exists a matrix C such that $B = C^{-1}AC$.

The characteristic polynomial of B is $|B - \lambda I|$.

Substituting for B and using the properties of determinants,

$$|B - \lambda I| = |C^{-1}AC - \lambda I| = |C^{-1}(A - \lambda I)C|$$
$$= |C^{-1}||A - \lambda I||C| = |A - \lambda I|$$

Hence the characteristic polynomials of A and B are identical; their eigenvalues will also be identical.

As the definition implies, any nonsingular matrix C can be used to define a similarity transformation. In practice, one is often interested in transforming a given matrix into a diagonal matrix, if possible. The following theorem tells us when this is possible, and the form that the transforming matrix C takes in such a transformation.

THEOREM 6-21   *An $n \times n$ matrix A can be transformed into a diagonal matrix B if and only if it has n linearly independent eigenvectors. If A is such a matrix, the matrix C, consisting of n linearly independent eigenvectors of A as its columns, can be used to transform A into a diagonal matrix $B = C^{-1}AC$. The diagonal matrix B will then have eigenvalues of A as diagonal elements.*

Proof: First, assume that A has eigenvalues $\lambda_1, ..., \lambda_n$ (which need not be distinct) with corresponding linearly independent eigenvectors $\mathbf{v}_1, ..., \mathbf{v}_n$.

Let C be the matrix with $\mathbf{v}_1, ..., \mathbf{v}_n$ as column vectors. Write $C = (\mathbf{v}_1, ..., \mathbf{v}_n)$. Then, since $A\mathbf{v}_1 = \lambda\mathbf{v}_1, ..., A\mathbf{v}_n = \lambda_n\mathbf{v}_n$, matrix multiplication gives

$$A(\mathbf{v}_1 \cdots \mathbf{v}_n) = (\lambda_1\mathbf{v}_1 \cdots \lambda_n\mathbf{v}_n)$$

$$= (\mathbf{v}_1 \cdots \mathbf{v}_n) \begin{pmatrix} \lambda_1 & & 0 \\ & \ddots & \\ 0 & & \lambda_n \end{pmatrix}$$

Hence

$$(\mathbf{v}_1 \cdots \mathbf{v}_n)^{-1} A(\mathbf{v}_1 \cdots \mathbf{v}_n) = \begin{pmatrix} \lambda_1 & & 0 \\ & \ddots & \\ 0 & & \lambda_n \end{pmatrix}$$

A can be transformed into a diagonal matrix using the matrix C.

Let us now prove the converse. Let C be a non-singular matrix that can be used to transform A into a diagonal form $C^{-1}AC$. Let

$$C^{-1}AC = \begin{pmatrix} \gamma_1 & & 0 \\ & \ddots & \\ 0 & & \gamma_n \end{pmatrix}$$

Thus

$$AC = C \begin{pmatrix} \gamma_1 & & 0 \\ & \ddots & \\ 0 & & \gamma_n \end{pmatrix}$$

Let C have columns $\mathbf{u}_1, \ldots, \mathbf{u}_n$. Then

$$A(\mathbf{u}_1 \cdots \mathbf{u}_n) = (\mathbf{u}_1 \cdots \mathbf{u}_n) \begin{pmatrix} \gamma_1 & & 0 \\ & \ddots & \\ 0 & & \gamma_n \end{pmatrix}$$

This implies that $A\mathbf{u}_1 = \gamma_1 \mathbf{u}_1, \ldots, A_n \mathbf{u}_n = \gamma_n \mathbf{u}_n$. $\mathbf{u}_1, \ldots, \mathbf{u}_n$ are eigenvectors of A with corresponding eigenvalues $\gamma_1, \ldots, \gamma_n$. Since C is non-singular it is of rank n (Theorem 6-8), thus the n eigenvectors $\mathbf{u}_1, \ldots, \mathbf{u}_n$ are linearly independent. This proves that if an $n \times n$ matrix can be transformed into a diagonal matrix using a similarity transformation, it has n linearly independent eigenvectors.

Example 2

Let us transform the matrix $A = \begin{pmatrix} 1 & 0 & 0 \\ -2 & 5 & -2 \\ -2 & 4 & -1 \end{pmatrix}$ into a diagonal matrix using a similarity transformation.

We determine the eigenvalues of A.

$$|A - \lambda I| = 0 \Rightarrow \begin{vmatrix} 1-\lambda & 0 & 0 \\ -2 & 5-\lambda & -2 \\ -2 & 4 & -1-\lambda \end{vmatrix} = 0$$

$$\Rightarrow (1-\lambda) \begin{vmatrix} 5-\lambda & -2 \\ 4 & -1-\lambda \end{vmatrix} = 0$$

$$\Rightarrow (1-\lambda)[(5-\lambda)(-1-\lambda)+8] = 0$$

$$\Rightarrow (1-\lambda)^2(\lambda-3) = 0$$

The eigenvalues are 1, 1, and 3.

We now determine the eigenspaces.

For $\lambda = 1$, $\begin{pmatrix} 0 & 0 & 0 \\ -2 & 4 & -2 \\ -2 & 4 & -2 \end{pmatrix} \begin{pmatrix} x_1 \\ x_2 \\ x_3 \end{pmatrix} = 0$

$\Rightarrow -2x_1 + 4x_2 - 2x_3 = 0$, giving $x_1 = 2x_2 - x_3$

Eigenvectors are $\begin{pmatrix} 2x_2 & -x_3 \\ & x_2 \\ & x_3 \end{pmatrix}$. Two linearly independent eigen-

vectors in this space are $\begin{pmatrix} 2 \\ 1 \\ 0 \end{pmatrix}, \begin{pmatrix} -1 \\ 0 \\ 1 \end{pmatrix}$.

For $\lambda = 3$, $\begin{pmatrix} -2 & 0 & 0 \\ -2 & 2 & -2 \\ -2 & 4 & -4 \end{pmatrix} \begin{pmatrix} x_1 \\ x_2 \\ x_3 \end{pmatrix} = 0 \Rightarrow \begin{cases} -2x_1 & = 0 \\ -2x_1 + 2x_2 - 2x_3 = 0 \\ -2x_1 + 4x_2 - 4x_3 = 0 \end{cases}$

Solving we get $x_1 = 0$, $x_2 = x_3$.   Eigenvectors are $x_2 \begin{pmatrix} 0 \\ 1 \\ 1 \end{pmatrix}$.   Thus A

has three linearly independent eigenvectors $\begin{pmatrix} 2 \\ 1 \\ 0 \end{pmatrix}, \begin{pmatrix} -1 \\ 0 \\ 1 \end{pmatrix}, \begin{pmatrix} 0 \\ 1 \\ 1 \end{pmatrix}$.

Let $C = \begin{pmatrix} 2 & -1 & 0 \\ 1 & 0 & 1 \\ 0 & 1 & 1 \end{pmatrix}$.   Then C^{-1} can be found to be

$\begin{pmatrix} 1 & -1 & 1 \\ 1 & -2 & 2 \\ -1 & 2 & -1 \end{pmatrix}$.   Performing a similarity transformation on A,

$$C^{-1}AC = \begin{pmatrix} 1 & -1 & 1 \\ 1 & -2 & 2 \\ -1 & 2 & -1 \end{pmatrix} \begin{pmatrix} 1 & 0 & 0 \\ -2 & 5 & -2 \\ -2 & 4 & -1 \end{pmatrix} \begin{pmatrix} 2 & -1 & 0 \\ 1 & 0 & 1 \\ 0 & 1 & 1 \end{pmatrix}$$

$$= \begin{pmatrix} 1 & -1 & 1 \\ 1 & -2 & 2 \\ -3 & 6 & -3 \end{pmatrix} \begin{pmatrix} 2 & -1 & 0 \\ 1 & 0 & 1 \\ 0 & 1 & 1 \end{pmatrix} = \begin{pmatrix} 1 & 0 & 0 \\ 0 & 1 & 0 \\ 0 & 0 & 3 \end{pmatrix}$$

This matrix has the eigenvalues of A as its diagonal elements.

Not every matrix has sufficient eigenvectors to be diagonalized. The following matrix is such a one.

Example 3

Determine the eigenvalues and eigenvectors of the matrix

$$\begin{pmatrix} 2 & 1 & 0 \\ 0 & 2 & 0 \\ 0 & 0 & 3 \end{pmatrix}$$

The eigenvalues are given by

$$|A - \lambda I| = 0, \Rightarrow \begin{vmatrix} 2-\lambda & 1 & 0 \\ 0 & 2-\lambda & 0 \\ 0 & 0 & 3-\lambda \end{vmatrix} = 0 \Rightarrow (2-\lambda)^2 (3-\lambda) = 0$$

The eigenvalues are 2, 2, and 3.

Let us determine the corresponding eigenvectors.

For $\lambda = 2$, $\begin{pmatrix} 0 & 1 & 0 \\ 0 & 0 & 0 \\ 0 & 0 & 1 \end{pmatrix} \begin{pmatrix} x_1 \\ x_2 \\ x_3 \end{pmatrix} = 0 \Rightarrow \begin{cases} x_2 = 0 \\ x_3 = 0 \end{cases}$

Eigenvectors are $x_1 \begin{pmatrix} 1 \\ 0 \\ 0 \end{pmatrix}$

For $\lambda = 3$, $\begin{pmatrix} -1 & 1 & 0 \\ 0 & -1 & 0 \\ 0 & 0 & 0 \end{pmatrix} \begin{pmatrix} x_1 \\ x_2 \\ x_3 \end{pmatrix} = 0 \Rightarrow \begin{cases} -x_1 = 0 \\ -x_2 = 0 \end{cases}$

Eigenvectors are $x_3 \begin{pmatrix} 0 \\ 0 \\ 1 \end{pmatrix}$. Thus the 3×3 matrix has two one-dimensional eigenspaces. The 3×3 matrix does not have three linearly independent eigenvectors. It cannot be diagonalized using a similarity transformation.

Let us now look at certain properties of symmetric matrices. We find that all symmetric matrices have sufficient eigenvectors to be diagonalized using a similarity transformation.

The following theorem sums up the properties of eigenspaces of symmetric matrices.

THEOREM 6-22

a. *The eigenvalues of a symmetric matrix are all real numbers.* (*Eigenvalues of a matrix can, in general, be complex numbers.*)

b. *n linearly independent eigenvectors exist for every $n \times n$ symmetric matrix.*

c. *The eigenspaces of a symmetric matrix corresponding to distinct eigenvalues are all orthogonal.* (*Two subspaces are said to be orthogonal if an arbitrary vector in the one subspace is orthogonal to an arbitrary vector in the other subspace.*)

d. *The dimension of an eigenspace of a symmetric matrix is the multiplicity of the eigenvalue as a root of the characteristic equation.*

The reader is asked to prove part (c) in the exercises that follow. The proof of the remainder of the theorem is beyond the reader's scope here.

Thus if A is a symmetric matrix it can be transformed into a diagonal matrix B using a similarity transformation $C^{-1}AC$, where C has linearly independent eigenvectors of A as column vectors. Furthermore, by using eigenvectors that form an orthonormal basis for each eigenspace, the transformation matrix C can be selected to be an orthogonal matrix. It is often desirable to use an orthogonal matrix as transformation matrix— norms and angles are then preserved. Note that since the inverse of an orthogonal matrix is equal to is transpose, the similarity transformation then assumes the simpler form $C^t A C$.

Example 4

Transform the symmetric matrix

$$A = \begin{pmatrix} \frac{3}{2} & -\frac{1}{2} & 0 \\ -\frac{1}{2} & \frac{3}{2} & 0 \\ 0 & 0 & 3 \end{pmatrix}$$

into a diagonal form using an orthogonal similarity transformation.

The eigenvalues of A are 1, 2, and 3 and corresponding eigenvectors are $a(1, 1, 0)$, $b(-1, 1, 0)$, and $c(0, 0, 1)$, respectively, for arbitrary constants a, b, and c. These vectors are already orthogonal, since they are in distinct eigenspaces. Normalizing, we get $(1/\sqrt{2}, 1/\sqrt{2}, 0)$, $(-1/\sqrt{2}, 1/\sqrt{2}, 0)$, and $(0, 0, 1)$. Hence an orthogonal transforming matrix C is

$$C = \begin{pmatrix} \dfrac{1}{\sqrt{2}} & -\dfrac{1}{\sqrt{2}} & 0 \\ \dfrac{1}{\sqrt{2}} & \dfrac{1}{\sqrt{2}} & 0 \\ 0 & 0 & 1 \end{pmatrix}$$

The similarity transformation that will give the diagonal matrix B is

$$B = C^{-1} A C = C^t A C$$

Thus

$$B = \begin{pmatrix} \dfrac{1}{\sqrt{2}} & -\dfrac{1}{\sqrt{2}} & 0 \\ \dfrac{1}{\sqrt{2}} & \dfrac{1}{\sqrt{2}} & 0 \\ 0 & 0 & 1 \end{pmatrix}^t \begin{pmatrix} \dfrac{3}{2} & -\dfrac{1}{2} & 0 \\ -\dfrac{1}{2} & \dfrac{3}{2} & 0 \\ 0 & 0 & 3 \end{pmatrix} \begin{pmatrix} \dfrac{1}{\sqrt{2}} & -\dfrac{1}{\sqrt{2}} & 0 \\ \dfrac{1}{\sqrt{2}} & \dfrac{1}{\sqrt{2}} & 0 \\ 0 & 0 & 1 \end{pmatrix}$$

$$= \begin{pmatrix} \dfrac{1}{\sqrt{2}} & \dfrac{1}{\sqrt{2}} & 0 \\ -\dfrac{1}{\sqrt{2}} & \dfrac{1}{\sqrt{2}} & 0 \\ 0 & 0 & 1 \end{pmatrix} \begin{pmatrix} \dfrac{3}{2} & -\dfrac{1}{2} & 0 \\ -\dfrac{1}{2} & \dfrac{3}{2} & 0 \\ 0 & 0 & 1 \end{pmatrix} \begin{pmatrix} \dfrac{1}{\sqrt{2}} & -\dfrac{1}{\sqrt{2}} & 0 \\ \dfrac{1}{\sqrt{2}} & \dfrac{1}{\sqrt{2}} & 0 \\ 0 & 0 & 1 \end{pmatrix}$$

$$= \begin{pmatrix} 1 & 0 & 0 \\ 0 & 2 & 0 \\ 0 & 0 & 3 \end{pmatrix}$$

Hence A is similar to the diagonal matrix $\begin{pmatrix} 1 & 0 & 0 \\ 0 & 2 & 0 \\ 0 & 0 & 3 \end{pmatrix}$. The diagonal elements of this matrix are indeed the eigenvalues of A. If one is only interested in determining a diagonal matrix, it can be determined immediately from the eigenvalues of A. However, a knowledge of the actual transformation is important in many applications of this theory. Such transformations arise, for example, when coordinate transformations are involved in a mathematical model, as in the next section.

Example 5

Determine a diagonal matrix similar to the symmetric matrix $\begin{pmatrix} 5 & 4 & 2 \\ 4 & 5 & 2 \\ 2 & 2 & 2 \end{pmatrix}$.

We have already determined the eigenvalues of this matrix in a previous section. They are 1, 1, and 10. Hence a diagonal matrix similar to this matrix is

$$\begin{pmatrix} 1 & 0 & 0 \\ 0 & 1 & 0 \\ 0 & 0 & 10 \end{pmatrix}$$

Note that there is a two-dimensional eigenspace corresponding to the repeated eigenvalue 1. The Gram-Schmidt orthogonalization process would have to be used to determine an orthogonal transformation matrix. The reader is asked to determine such a matrix in Exercise 4.

EXERCISES

1. In each of the following, transform A into $C^{-1}AC$ using the given nonsingular matrix C.

   a) $A = \begin{pmatrix} 1 & 2 \\ -1 & 3 \end{pmatrix}$, $C = \begin{pmatrix} 2 & 5 \\ 1 & 3 \end{pmatrix}$

   b) $A = \begin{pmatrix} 0 & 4 \\ 3 & 2 \end{pmatrix}$, $C = \begin{pmatrix} 2 & 1 \\ 7 & 4 \end{pmatrix}$

   c) $A = \begin{pmatrix} -1 & 2 \\ 3 & 1 \end{pmatrix}$, $C = \begin{pmatrix} -4 & 6 \\ -2 & 2 \end{pmatrix}$

   d) $A = \begin{pmatrix} 3 & 2 \\ -1 & 4 \end{pmatrix}$, $C = \begin{pmatrix} 2 & -7 \\ -1 & 4 \end{pmatrix}$

   e) $A = \begin{pmatrix} 1 & 0 & 2 \\ -1 & 3 & 4 \\ 0 & 1 & 3 \end{pmatrix}$, $C = \begin{pmatrix} 0 & 3 & 3 \\ 1 & 2 & 3 \\ 1 & 4 & 6 \end{pmatrix}$

f) $A = \begin{pmatrix} 2 & 1 & 0 \\ 3 & 0 & 2 \\ 4 & 1 & -1 \end{pmatrix}$, $C = \begin{pmatrix} 1 & 2 & 3 \\ 0 & 1 & 2 \\ 4 & 5 & 3 \end{pmatrix}$

g) $A = \begin{pmatrix} -1 & 2 & 1 \\ 0 & 4 & 1 \\ 2 & 0 & 3 \end{pmatrix}$, $C = \begin{pmatrix} 1 & 2 & 0 \\ 2 & 1 & -1 \\ 3 & 1 & 1 \end{pmatrix}$

2. Transform (if possible) each of the following matrices into diagonal form using a similarity transformation involving eigenvectors. Give the transformations.

a) $\begin{pmatrix} 5 & 4 \\ 1 & 2 \end{pmatrix}$ **b)** $\begin{pmatrix} 2 & 1 \\ 2 & 3 \end{pmatrix}$ **c)** $\begin{pmatrix} 4 & -1 \\ 2 & 1 \end{pmatrix}$

d) $\begin{pmatrix} 1 & 1 \\ 0 & 1 \end{pmatrix}$ **e)** $\begin{pmatrix} 15 & 7 & -7 \\ -1 & 1 & 1 \\ 13 & 7 & -5 \end{pmatrix}$ **f)** $\begin{pmatrix} 5 & -2 & 2 \\ 4 & -3 & 4 \\ 4 & -6 & 7 \end{pmatrix}$

g) $\begin{pmatrix} 1 & 0 & 0 \\ -2 & 1 & 2 \\ -2 & 0 & 3 \end{pmatrix}$ **h)** $\begin{pmatrix} 3 & 0 & 0 \\ 1 & 2 & 0 \\ 0 & 0 & -4 \end{pmatrix}$ **i)** $\begin{pmatrix} 1 & -2 & 2 \\ 4 & 5 & -4 \\ 0 & -2 & 3 \end{pmatrix}$

3. Transform each of the following symmetric matrices into diagonal form using an orthogonal similarity transformation. Give the transformation in each case.

a) $\begin{pmatrix} 1 & 2 \\ 2 & 1 \end{pmatrix}$ **b)** $\begin{pmatrix} 11 & 2 \\ 2 & 14 \end{pmatrix}$ **c)** $\begin{pmatrix} 3 & 1 \\ 1 & 3 \end{pmatrix}$

d) $\begin{pmatrix} -1 & -8 \\ -8 & 11 \end{pmatrix}$ **e)** $\begin{pmatrix} \frac{1}{2} & -\frac{3}{2} & 0 \\ -\frac{3}{2} & \frac{1}{2} & 0 \\ 0 & 0 & -2 \end{pmatrix}$ **f)** $\begin{pmatrix} \frac{3}{2} & -\frac{1}{2} & 0 \\ -\frac{1}{2} & \frac{3}{2} & 0 \\ 0 & 0 & 1 \end{pmatrix}$

g) $\begin{pmatrix} 0 & 2 & 0 \\ 2 & 0 & 0 \\ 0 & 0 & 1 \end{pmatrix}$ **h)** $\begin{pmatrix} 9 & -3 & 3 \\ -3 & 6 & -6 \\ 3 & -6 & 6 \end{pmatrix}$ **i)** $\begin{pmatrix} 1 & 2 & -2 \\ 2 & 4 & -4 \\ -2 & -4 & 4 \end{pmatrix}$

4. Determine an orthogonal transformation matrix in Example 5.

5. Prove that if A and B are related through a similarity transformation, then $|A| = |B|$; that is, the determinant of a matrix is invariant under a similarity transformation.

6. If A is a symmetric matrix we know that it is similar to a diagonal matrix. Is such a diagonal matrix unique? (*Hint*: Does the order of the column vectors in the transforming matrix matter?)

7. Let A be a $n \times n$ matrix and C an invertible $n \times n$ matrix. Show that
   a) $C^{-1}A^2C = (C^{-1}AC)^2$
   b) $C^{-1}A^nC = (C^{-1}AC)^n$ for n a positive integer.

8. Two $n \times n$ matrices, A and B, are said to be *orthogonally similar* if there exists an orthogonal matrix C such that $B = C^{-1}AC$. Show that if A is symmetric and if A and B are orthogonally similar then B is symmetric.

9. Show that if A and B are orthogonally similar and B and C are orthogonally similar then A and C are orthogonally similar.

10. Prove that the eigenspaces of a symmetric matrix corresponding to distinct eigenvalues are orthogonal.

11. Let A be an $n \times n$ matrix with eigenvalues $\lambda_1, ..., \lambda_n$ and corresponding orthogonal eigenvectors $\mathbf{v}_1, ..., \mathbf{v}_n$. Let C be the matrix having $\mathbf{v}_1, ..., \mathbf{v}_n$ as column vectors. Prove that

$$C^t AC = \begin{pmatrix} \lambda_1 \|\mathbf{v_1}\|^2 & & 0 \\ & \ddots & \\ 0 & & \lambda_n \|\mathbf{v_n}\|^2 \end{pmatrix}$$

Illustrate this method for the matrix of Example 3.

12. Write a computer program to perform a similarity transformation $C^{-1}AC$, where the matrices A and C are the input. Test your program on Example 1 and use it to check your answers to Exercise 1.

6–9. COORDINATE TRANSFORMATIONS

In geometry a change in coordinate systems is often desirable when it is found that a geometrical figure can be examined more easily in an alternative coordinate system. After formulating a geometric representation of a physical system, one often discovers that there is a more suitable coordinate system, perhaps one that displays certain properties of the situation. In many cases, the change in the coordinates of points from one system to the other (called a *coordinate transformation*) is carried out using matrices, as the following examples illustrate.

Example 1

Consider the two rectangular coordinate systems xy and $x'y'$ in the plane (Figure 6-3). They have a common origin O, the one system being obtained from the other by a rotation through an angle θ. How are the coordinates of the points in the two systems related?

Consider the point A in the plane. The location of A can be given relative to each coordinate system. In the xy coordinate system the x value of A is given by the length OB, and the y value by AB. Thus, in the xy coordinate system, A is the point $\begin{pmatrix} x \\ y \end{pmatrix}$ with $x = OB$ and $y = AB$. In the

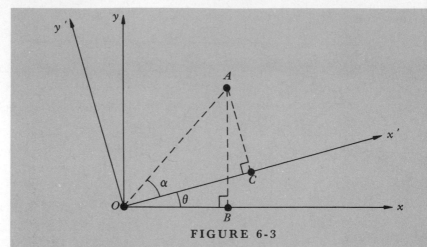

FIGURE 6-3

$x'y'$ coordinate system, the x' value of A is the length of OC, and the y' value the length of AC. In the $x'y'$ coordinate system, A is the point $\begin{pmatrix} x' \\ y' \end{pmatrix}$ with $x' = OC$ and $y' = AC$. We need to know how these two representations of A, $\begin{pmatrix} x \\ y \end{pmatrix}$ and $\begin{pmatrix} x' \\ y' \end{pmatrix}$, are related.

Let $AOC = \alpha$. Then we have

$$x = OB = OA \cos(\alpha + \theta) = OA \cos\alpha \cos\theta - OA \sin\alpha \sin\theta$$

$$= OC \cos\theta - AC \sin\theta = x' \cos\theta - y' \sin\theta$$

$$y = AB = OA \sin(\alpha + \theta) = OA \sin\alpha \cos\theta + OA \cos\alpha \sin\theta$$

$$y = AC \cos\theta + OC \sin\theta = y' \cos\theta + x' \sin\theta$$

$$= x' \sin\theta + y' \cos\theta$$

We can write these equations in matrix form.

$$\begin{pmatrix} x \\ y \end{pmatrix} = \begin{pmatrix} \cos\theta & -\sin\theta \\ \sin\theta & \cos\theta \end{pmatrix} \begin{pmatrix} x' \\ y' \end{pmatrix}$$

This equation defines the coordinate transformation. (Note the similarity between this coordinate transformation and the mapping that defines a rotation of the plane through an angle θ. This arises from the fact that one can interpret a transformation either as moving points in a fixed coordinate system or as changing the coordinate system around fixed points.)

Observe that the column vectors of the matrix that defines the rotation give the directions of the new axes:

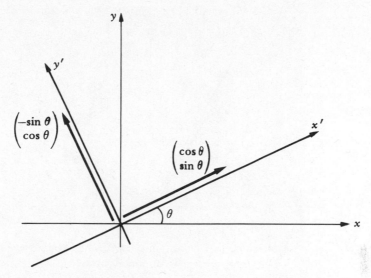

The rotation matrix is an orthogonal matrix having determinant 1.

In general, if C is an orthogonal 2×2 matrix having determinant 1, it can be used to define a coordinate transformation in $\mathbf{R}^2$.†

$$\begin{pmatrix} x \\ y \end{pmatrix} = C \begin{pmatrix} x' \\ y' \end{pmatrix}$$

This will be a rotation about the origin with the columns of C giving the directions of the new axes. The first column of C will define the direction of the x' axis. The second column vector will give the direction of the y' axis.

Example 2

Discuss the coordinate transformation

$$\begin{pmatrix} x \\ y \end{pmatrix} = \begin{pmatrix} \sqrt{3}/2 & -\frac{1}{2} \\ \frac{1}{2} & \sqrt{3}/2 \end{pmatrix} \begin{pmatrix} x' \\ y' \end{pmatrix}$$

The matrix of the transformation is an orthogonal matrix having determinant 1; it thus defines a rotation of coordinates. On comparison with $\begin{pmatrix} \cos \theta & -\sin \theta \\ \sin \theta & \cos \theta \end{pmatrix}$ we see that $\cos \theta = \sqrt{3}/2$ and $\sin \theta = \frac{1}{2}$. θ is thus 30°.

The columns of the rotation matrix give the x' axis to be in the direction defined by the vector $\begin{pmatrix} \sqrt{3}/2 \\ \frac{1}{2} \end{pmatrix}$ and the y' axis in the direction defined by the vector $\begin{pmatrix} -\frac{1}{2} \\ \sqrt{3}/2 \end{pmatrix}$.

† Orthogonal matrices having determinant -1 lead to a reorientation of the axes, that is to new coordinate systems of the type ⟨ x' / y' ⟩. They are not of interest to us.

333

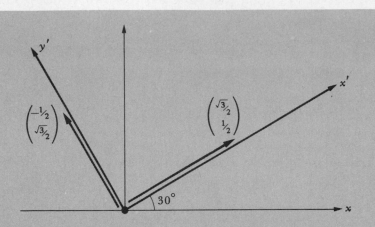

Example 3

Consider two rectangular coordinate systems xy and $x'y'$ in the plane, where one system is obtained from the other by a translation of axes (Figure 6-4). What is the coordinate transformation that relates the systems?

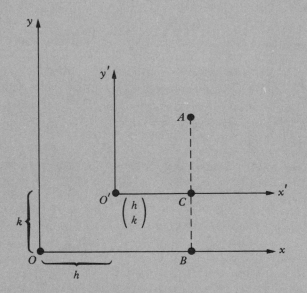

FIGURE 6-4

Let the origins of the coordinate systems be O and O'. Let the coordinates of O' in the xy system be $\begin{pmatrix} h \\ k \end{pmatrix}$. Consider the point A, which has representations $\begin{pmatrix} x \\ y \end{pmatrix}$ and $\begin{pmatrix} x' \\ y' \end{pmatrix}$ in the two coordinate systems.

Thus $OB = x$, $AB = y$, $O'C = x'$, and $AC = y'$. We know that

$$x = OB = O'C + h = x' + h$$

$$y = AB = AC + k = y' + k$$

Thus

$$\begin{pmatrix} x \\ y \end{pmatrix} = \begin{pmatrix} x' \\ y' \end{pmatrix} + \begin{pmatrix} h \\ k \end{pmatrix}$$

Example 4

What is the coordinate transformation that relates two arbitrary rectangular coordinate systems xy and $x'y'$ in the plane?

Let Figure 6-5 illustrate two such systems. The coordinates of the origin O' in the xy system are $\begin{pmatrix} h \\ k \end{pmatrix}$.

Let A be an arbitrary point whose representations are $\begin{pmatrix} x \\ y \end{pmatrix}$ and $\begin{pmatrix} x' \\ y' \end{pmatrix}$ in the two systems. We need to determine how $\begin{pmatrix} x \\ y \end{pmatrix}$ and $\begin{pmatrix} x' \\ y' \end{pmatrix}$ are related. Introduce a new coordinate system $x''y''$ (Figure 6-6), and let the $x'y'$ system be obtained from it by rotation through an angle θ. Let the coordinates of A in $x''y''$ be $\begin{pmatrix} x'' \\ y'' \end{pmatrix}$. Then, using the previous examples,

$$\begin{pmatrix} x \\ y \end{pmatrix} = \begin{pmatrix} x'' \\ y'' \end{pmatrix} + \begin{pmatrix} h \\ k \end{pmatrix}$$

for A in the systems xy and $x''y''$, and

$$\begin{pmatrix} x'' \\ y'' \end{pmatrix} = \begin{pmatrix} \cos\theta & -\sin\theta \\ \sin\theta & \cos\theta \end{pmatrix} \begin{pmatrix} x' \\ y' \end{pmatrix}$$

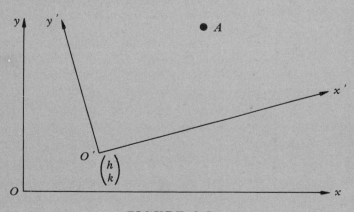

FIGURE 6-5

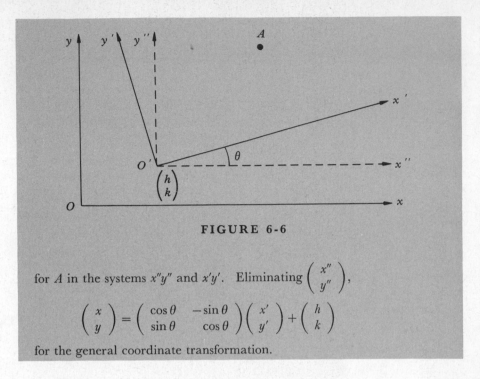

FIGURE 6-6

for A in the systems $x''y''$ and $x'y'$. Eliminating $\begin{pmatrix} x'' \\ y'' \end{pmatrix}$,

$$\begin{pmatrix} x \\ y \end{pmatrix} = \begin{pmatrix} \cos\theta & -\sin\theta \\ \sin\theta & \cos\theta \end{pmatrix} \begin{pmatrix} x' \\ y' \end{pmatrix} + \begin{pmatrix} h \\ k \end{pmatrix}$$

for the general coordinate transformation.

Here we have examined coordinate transformations between rectangular axes in the plane. Rotations and translations of axes also occur in three-dimensional space. In three-dimensional space, 3×3 orthogonal matrices having determinant 1 define coordinate transformations due to rotations of rectangular axes. The coordinate transformation between two arbitrary rectangular coordinate systems xyz and $x'y'z'$ in three dimensions can be written

$$\mathbf{x} = M\mathbf{x}' + \mathbf{k}$$

where M is an orthogonal matrix having determinant 1 that defines a rotation, and $\mathbf{k}$, a column vector, is the coordinates of the origin of the $x'y'z'$ system in the xyz system. This defines the translation part. The column vectors of M will give the directions of the new axes, as in two-dimensional space.

Example 5

Discuss and sketch the curve

$$x^2 + 4y^2 - 6x - 16y + 21 = 0$$

This is a quadratic equation having no xy term. We do not recognize the equation as being in the form of any of the standard curves, such as a circle, parabola, ellipse, or hyperbola.   However, by introducing a second co-ordinate system, we shall find that it is, in fact, the equation of an ellipse.

Rearranging the equation,

$$x^2 - 6x + 4y^2 - 16y + 21 = 0$$

Completing the square,

$$(x-3)^2 - 9 + 4(y-2)^2 - 16 + 21 = 0$$

giving

$$(x-3)^2 + 4(y-2)^2 = 4$$

$$\frac{(x-3)^2}{4} + \frac{(y-2)^2}{1} = 1$$

This resembles the equation of an ellipse.   Performing the coordinate transformation $x' = x - 3$, $y' = y - 2$, we get the equation into the standard form for an ellipse,

$$\frac{x'^2}{4} + \frac{y'^2}{1} = 1$$

We know now that the curve is an ellipse with a major axis of length 2 and a minor axis of length 1 in the $x'y'$ coordinate system (Figure 6-7).

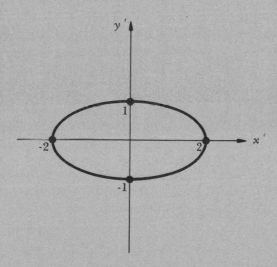

FIGURE 6-7

We must still locate this coordinate system in the original one. The coordinate transformation can be rewritten in the standard form

$x = x' + 3$, $y = y' + 2$. We see that it is a special case of Example 3, a translation of axes with the origin of the $x'y'$ system at the point $\begin{pmatrix} 3 \\ 2 \end{pmatrix}$ of the xy system. In Figure 6-8 we sketch the curve in the original coordinate system.

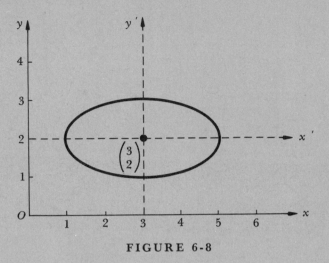

FIGURE 6-8

In this example

1. We found the coordinate system that fits the symmetry of the ellipse, the $x'y'$ coordinate system.
2. We analyzed the equation in this system.
3. We interpreted the analysis in the original system.

An expression of the type

$$ax^2 + bxy + cy^2$$

where a, b, and c are constants is called a *quadratic form*. This expression, which plays an important role in geometry and in applications, can be expressed in the matrix form

$$(x \quad y) \begin{pmatrix} a & b/2 \\ b/2 & c \end{pmatrix} \begin{pmatrix} x \\ y \end{pmatrix}$$

The reader can show that these two expressions are identical by multiplying out the matrices.

The symmetric matrix $\begin{pmatrix} a & b/2 \\ b/2 & c \end{pmatrix}$ associated with the quadratic form is called the *matrix of the quadratic form*.

Example 6

$5x^2 + 6xy - 4y^2$ is a quadratic form.

On comparison with the standard form $ax^2 + bxy + cy^2$ we have that $a = 5$, $b = 6$, and $c = -4$. The matrix of the quadratic form is thus

$$\begin{pmatrix} a & b/2 \\ b/2 & c \end{pmatrix} = \begin{pmatrix} 5 & 3 \\ 3 & -4 \end{pmatrix}$$

The quadratic form can be expressed in terms of matrices

$$\begin{pmatrix} x & y \end{pmatrix} \begin{pmatrix} 5 & 3 \\ 3 & -4 \end{pmatrix} \begin{pmatrix} x \\ y \end{pmatrix}$$

We now illustrate the role of the quadratic form in analyzing equations.

Example 7

Let us discuss the equation

$$6x^2 + 4xy + 9y^2 - 20 = 0$$

This is a quadratic equation having no linear terms in x and y. It incorporates the quadratic form $6x^2 + 4xy + 9y^2$. Let us write the equation in the matrix form

$$\begin{pmatrix} x & y \end{pmatrix} \begin{pmatrix} 6 & 2 \\ 2 & 9 \end{pmatrix} \begin{pmatrix} x \\ y \end{pmatrix} - 20 = 0 \tag{1}$$

Our aim will be to introduce a new coordinate system in which the equation assumes a simpler, recognizable form. The matrix $\begin{pmatrix} 6 & 2 \\ 2 & 9 \end{pmatrix}$ is symmetric; it can thus be diagonalized using a similarity transformation. We shall diagonalize it into the form having eigenvalues on the main diagonal. Let us determine the eigenvalues.

$$\begin{vmatrix} 6-\lambda & 2 \\ 2 & 9-\lambda \end{vmatrix} = 0 \Rightarrow (6-\lambda)(9-\lambda) - 4 = 0$$

$$\Rightarrow \lambda^2 - 15\lambda + 50 = 0$$

$$\text{giving} \quad (\lambda - 10)(\lambda - 5) = 0$$

The eigenvalues are 10 and 5.

We now find the corresponding eigenvectors.

For $\lambda = 10$,

$$\begin{pmatrix} -4 & 2 \\ 2 & -1 \end{pmatrix} \begin{pmatrix} x \\ y \end{pmatrix} = 0 \Rightarrow 2x - y = 0 \Rightarrow 2x = y$$

The eigenvectors are $x \begin{pmatrix} 1 \\ 2 \end{pmatrix}$.

For $\lambda = 5$,

$$\begin{pmatrix} 1 & 2 \\ 2 & 4 \end{pmatrix} \begin{pmatrix} x \\ y \end{pmatrix} = 0 \Rightarrow x + 2y = 0 \Rightarrow x = -2y$$

The eigenvectors are $y \begin{pmatrix} -2 \\ 1 \end{pmatrix}$.

Normalizing these vectors we get unit orthogonal eigenvectors $\begin{pmatrix} 1/\sqrt{5} \\ 2/\sqrt{5} \end{pmatrix}$ and $\begin{pmatrix} -2/\sqrt{5} \\ 1/\sqrt{5} \end{pmatrix}$. Write these vectors as the columns of an orthogonal matrix C having determinant 1:

$$C = \begin{pmatrix} 1/\sqrt{5} & -2/\sqrt{5} \\ 2/\sqrt{5} & 1/\sqrt{5} \end{pmatrix}$$

(The order of selection of the eigenvectors is important, for it decides whether C has determinant $+1$ or -1.)

The matrix $\begin{pmatrix} 6 & 2 \\ 2 & 9 \end{pmatrix}$ can be transformed into the diagonal form $\begin{pmatrix} 10 & 0 \\ 0 & 5 \end{pmatrix}$ using the orthogonal matrix C.

Let us now return to the quadratic equation. We shall rearrange it into a form that suggests a coordinate transformation. Write equation (1) as

$$\mathbf{x}^t A \mathbf{x} - 20 = 0$$

where $\mathbf{x} = (\, x \quad y \,)$ and $A = \begin{pmatrix} 6 & 2 \\ 2 & 9 \end{pmatrix}$. Since C is orthogonal, $C^t C = I$ and this equation can be written

$$\mathbf{x}^t (CC^t) A (CC^t) \mathbf{x} - 20 = 0$$

giving

$$(\mathbf{x}^t C)(C^t A C)(C^t \mathbf{x}) - 20 = 0$$

$$(C^t \mathbf{x})^t \begin{pmatrix} 10 & 0 \\ 0 & 5 \end{pmatrix} (C^t \mathbf{x}) - 20 = 0$$

Introduce a coordinate transformation $\mathbf{x}' = C^t \mathbf{x}$, where $\mathbf{x}' = \begin{pmatrix} x' \\ y' \end{pmatrix}$. In the new coordinate system the equation becomes

$$(\, x' \quad y' \,) \begin{pmatrix} 10 & 0 \\ 0 & 5 \end{pmatrix} \begin{pmatrix} x' \\ y' \end{pmatrix} - 20 = 0$$

giving

$$10(x')^2 + 5(y')^2 - 20 = 0$$

This equation can be written in standard form

$$\frac{(x')^2}{2} + \frac{(y')^2}{4} = 1$$

In the $x'y'$ coordinate system the graph is an ellipse with major axis of length 2 and minor axis of length $\sqrt{2}$.

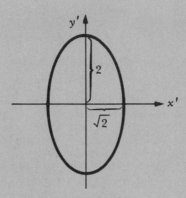

It remains to locate the $x'y'$ coordinate system relative to the original xy system. The coordinate transformation is the rotation

$$\mathbf{x}' = C^t \mathbf{x}$$

or

$$\mathbf{x} = C\mathbf{x}' \qquad \text{(Since } C^{-1} = C^t, C \text{ being orthogonal)}$$

$$\begin{pmatrix} x \\ y \end{pmatrix} = \begin{pmatrix} 1/\sqrt{5} & -2/\sqrt{5} \\ 2/\sqrt{5} & 1/\sqrt{5} \end{pmatrix} \begin{pmatrix} x' \\ y' \end{pmatrix}$$

The x' axis is in the direction $\begin{pmatrix} 1/\sqrt{5} \\ 2/\sqrt{5} \end{pmatrix}$, and the y' axis is in the direction $\begin{pmatrix} -2/\sqrt{5} \\ 1/\sqrt{5} \end{pmatrix}$. Thus the graph in the xy coordinate system is

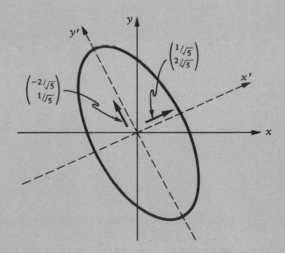

In the last two examples we have discussed the graphs of quadratic equations of the types

$$ax^2 + by^2 + cx + dy + e = 0 \text{ (having no } xy \text{ terms)}$$

and

$$ax^2 + bxy + cy^2 + d = 0 \text{ (having no linear terms in } x \text{ and } y)$$

We found that these could be interpreted by using a translation of coordinates (in the former) and a rotation of coordinates (in the latter). A suitable coordinate system for discussing a general quadratic equation of the form

$$ax^2 + bxy + cy^2 + dx + ey + f = 0$$

having both an xy term and linear terms in x and y can be obtained by first performing a rotation of axes and then a translation. We shall not pursue this general case, but rather introduce the reader to further techniques and applications involving coordinate transformations.

Similarity transformations arise naturally when one considers mappings in various coordinate systems. We now discuss this topic.

Coordinate Transformations and Mappings   Let A be a 2×2 matrix and P be an arbitrary point in the plane with coordinates $\mathbf{p}$ in a coordinate system xy. We have seen how A can be interpreted geometrically as mapping the point P into a point Q with coordinates $\mathbf{q}$ in the system xy, where $\mathbf{q} = A\mathbf{p}$.

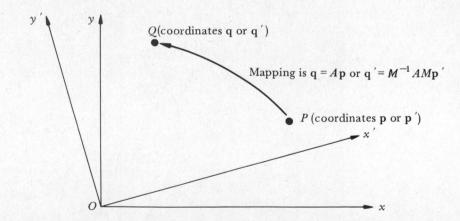

Consider a coordinate rotation in the above figure from the system xy to the system $x'y'$. Let it be defined by the orthogonal matrix M. Let the coordinates of P and Q in this new system be $\mathbf{p}'$ and $\mathbf{q}'$. Then $\mathbf{p} = M\mathbf{p}'$ and $\mathbf{q} = M\mathbf{q}'$. Substituting into the equation, $\mathbf{q} = A\mathbf{p}$ gives

$M\mathbf{q}' = AM\mathbf{p}'$, that is, $\mathbf{q}' = M^{-1}AM\mathbf{p}'$. Thus the relationship between $\mathbf{q}'$ and $\mathbf{p}'$ is specified by the matrix $M^{-1}AM$. The transformation of coordinates has given rise to a similarity transformation.

Example 8

Let the coordinate system $x'y'$ be obtained from xy by a rotation through $\pi/2$. Let $A = \begin{pmatrix} 1 & -1 \\ 0 & 2 \end{pmatrix}$ be a matrix that defines a mapping of the plane itself in the xy coordinate system. What is the equivalent mapping A' in the $x'y'$ coordinate system?

We have seen that the matrix that defines a coordinate rotation through an angle θ is the orthogonal matrix $\begin{pmatrix} \cos\theta & -\sin\theta \\ \sin\theta & \cos\theta \end{pmatrix}$. Thus for $\theta = \pi/2$, the rotation matrix is $\begin{pmatrix} 0 & -1 \\ 1 & 0 \end{pmatrix}$. The matrix A' that defines the mapping in the $x'y'$ coordinate system is therefore

$$A' = \begin{pmatrix} 0 & -1 \\ 1 & 0 \end{pmatrix}^{-1} \begin{pmatrix} 1 & -1 \\ 0 & 2 \end{pmatrix} \begin{pmatrix} 0 & -1 \\ 1 & 0 \end{pmatrix}$$

$$= \begin{pmatrix} 0 & -1 \\ 1 & 0 \end{pmatrix}^{t} \begin{pmatrix} 1 & -1 \\ 0 & 2 \end{pmatrix} \begin{pmatrix} 0 & -1 \\ 1 & 0 \end{pmatrix}$$

$$= \begin{pmatrix} 0 & 1 \\ -1 & 0 \end{pmatrix} \begin{pmatrix} 1 & -1 \\ 0 & 2 \end{pmatrix} \begin{pmatrix} 0 & -1 \\ 1 & 0 \end{pmatrix}$$

$$= \begin{pmatrix} 2 & 0 \\ 1 & 1 \end{pmatrix}$$

Example 9

If the mapping A is a symmetric matrix then the matrix M can be selected to diagonalize A. We illustrate these concepts with the matrix $A = \begin{pmatrix} -1 & -8 \\ -8 & 11 \end{pmatrix}$.

The eigenvalues and corresponding eigenvectors of A can be shown to be 15, -5, and $a\begin{pmatrix} 1 \\ -2 \end{pmatrix}$, $b\begin{pmatrix} 2 \\ 1 \end{pmatrix}$, respectively. Let M be the orthogonal matrix $\begin{pmatrix} 1/\sqrt{5} & 2/\sqrt{5} \\ -2/\sqrt{5} & 1/\sqrt{5} \end{pmatrix}$, having as columns unit orthogonal eigenvectors of A. Thus $M^{-1}AM = \begin{pmatrix} 15 & 0 \\ 0 & -5 \end{pmatrix}$

Consider a rotation of coordinates defined by the matrix M. In the new coordinate system $x'y'$ the mapping is defined by the diagonal matrix $\begin{pmatrix} 15 & 0 \\ 0 & -5 \end{pmatrix}$; we say that the mapping has this "diagonal representation". This is the most suitable coordinate system to discuss this mapping.

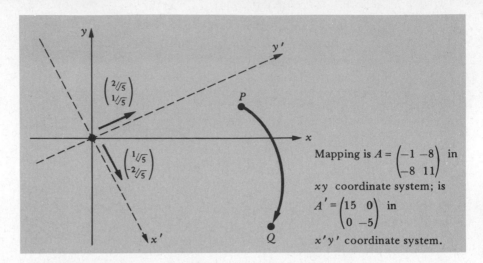

Mapping is $A = \begin{pmatrix} -1 & -8 \\ -8 & 11 \end{pmatrix}$ in xy coordinate system; is $A' = \begin{pmatrix} 15 & 0 \\ 0 & -5 \end{pmatrix}$ in $x'y'$ coordinate system.

Coordinate Transformations and Systems of Equations   Another use of coordinate transformations and similarity transformations is in the solution of certain systems of linear equations.   Let $A\mathbf{x} = \mathbf{y}$ represent a system of n equations in n unknowns.   If C is a nonsingular matrix, let $\mathbf{x} = C\mathbf{x}'$ and $\mathbf{y} = C\mathbf{y}'$.   Substituting for $\mathbf{x}$ and $\mathbf{y}$, the system of equations becomes $AC\mathbf{x}' = C\mathbf{y}'$, that is, $C^{-1}AC\mathbf{x}' = \mathbf{y}'$.   If A is a symmetric matrix, we know that there exists a matrix C such that $C^{-1}AC$ is a diagonal matrix. Using this matrix C, the system $C^{-1}AC\mathbf{x}' = \mathbf{y}'$ is easily solved. $\mathbf{x}$, the required solution, is then obtained from $\mathbf{x} = C\mathbf{x}'$.   If the matrix A is not symmetric, similarity transformations can often be used to obtain a simpler system of equations, although the matrix of coefficients of the resulting system will not always be diagonal.   Similarity transformations can always be chosen so that the matrix of coefficients has only nonzero elements on its main diagonal and immediately above the diagonal, if we work with complex numbers.   Readers who go on to study differential equations will find that this technique is used in solving certain systems of differential equations.

Example 10

Solve the system of equations
$$\begin{pmatrix} \frac{3}{2} & -\frac{1}{2} & 0 \\ -\frac{1}{2} & \frac{3}{2} & 0 \\ 0 & 0 & 3 \end{pmatrix} \begin{pmatrix} x \\ y \\ z \end{pmatrix} = \begin{pmatrix} 4/\sqrt{2} \\ 2/\sqrt{2} \\ 6 \end{pmatrix}$$

using a similarity transformation.

The matrix of coefficients A is symmetric; it can thus be transformed into a diagonal matrix using an orthogonal similarity transformation.   This matrix was discussed in Section 6-8 where it was found that its eigenvalues were 1, 2, and 3 with corresponding eigenvectors $a(1,1,0)$, $b(-1,1,0)$, and $c(0,0,1)$. Normalizing these vectors the transforming orthogonal matrix is

$$C = \begin{pmatrix} 1/\sqrt{2} & -1/\sqrt{2} & 0 \\ 1/\sqrt{2} & 1/\sqrt{2} & 0 \\ 0 & 0 & 1 \end{pmatrix}$$

It will transform the matrix of coefficients into the diagonal form $\begin{pmatrix} 1 & 0 & 0 \\ 0 & 2 & 0 \\ 0 & 0 & 3 \end{pmatrix}$.

In a new coordinate system $x'y'z'$ defined by the rotation $\mathbf{x} = C\mathbf{x}'$, the system of equations becomes $C^{-1}AC\mathbf{x}' = \mathbf{y}'$, that is

$$\begin{pmatrix} 1/\sqrt{2} & -1/\sqrt{2} & 0 \\ 1/\sqrt{2} & 1/\sqrt{2} & 0 \\ 0 & 0 & 1 \end{pmatrix}^{-1} \begin{pmatrix} \frac{3}{2} & -\frac{1}{2} & 0 \\ -\frac{1}{2} & \frac{3}{2} & 0 \\ 0 & 0 & 3 \end{pmatrix}$$

$$\times \begin{pmatrix} 1/\sqrt{2} & -1/\sqrt{2} & 0 \\ 1/\sqrt{2} & 1/\sqrt{2} & 0 \\ 0 & 0 & 1 \end{pmatrix} \begin{pmatrix} x' \\ y' \\ z' \end{pmatrix}$$

$$= \begin{pmatrix} 1/\sqrt{2} & -1/\sqrt{2} & 0 \\ 1/\sqrt{2} & 1/\sqrt{2} & 0 \\ 0 & 0 & 1 \end{pmatrix}^{-1} \begin{pmatrix} 4/\sqrt{2} \\ 4/\sqrt{2} \\ 6 \end{pmatrix}$$

This equation becomes

$$\begin{pmatrix} 1 & 0 & 0 \\ 0 & 2 & 0 \\ 0 & 0 & 3 \end{pmatrix} \begin{pmatrix} x' \\ y' \\ z' \end{pmatrix} = \begin{pmatrix} 1/\sqrt{2} & -1/\sqrt{2} & 0 \\ 1/\sqrt{2} & 1/\sqrt{2} & 0 \\ 0 & 0 & 1 \end{pmatrix}^{t} \begin{pmatrix} 4/\sqrt{2} \\ 2/\sqrt{2} \\ 6 \end{pmatrix}$$

leading to

$$\begin{pmatrix} 1 & 0 & 0 \\ 0 & 2 & 0 \\ 0 & 0 & 3 \end{pmatrix} \begin{pmatrix} x' \\ y' \\ z' \end{pmatrix} = \begin{pmatrix} 3 \\ -1 \\ 6 \end{pmatrix}$$

$$\begin{pmatrix} x' \\ y' \\ z' \end{pmatrix} = \begin{pmatrix} 3 \\ -\frac{1}{2} \\ 2 \end{pmatrix}$$

The solution of the original system is given by $\mathbf{x} = C\mathbf{x}'$,

$$\begin{pmatrix} x \\ y \\ z \end{pmatrix} = \begin{pmatrix} \dfrac{1}{\sqrt{2}} & -\dfrac{1}{\sqrt{2}} & 0 \\ \dfrac{1}{\sqrt{2}} & \dfrac{1}{\sqrt{2}} & 0 \\ 0 & 0 & 1 \end{pmatrix} \begin{pmatrix} x' \\ y' \\ z' \end{pmatrix}$$

or

$$\begin{pmatrix} x \\ y \\ z \end{pmatrix} = \begin{pmatrix} \dfrac{7}{2\sqrt{2}} \\ \dfrac{5}{2\sqrt{2}} \\ 2 \end{pmatrix}$$

Example 11*

In this example we return to the stress analysis of Example 7, Section 1-6. The stress matrix is symmetric—we shall make use of the eigenspace properties of such matrices. We shall also see the significance of using a coordinate transformation that takes us into a special coordinate system, one that displays physical properties of the situation.

Consider a body subject to external forces. Let O be an arbitrary point in the body. We have seen that associated with every plane through O there will be a stress, a force per unit area. This stress is usually discussed in terms of its component perpendicular to the plane, normal stress, and its component parallel to the plane, shearing stress.

Suppose that a plane is oriented in such a way that the shearing stress is zero; the resultant stress at the point is then the normal stress. The plane is then called a *principal plane* at the point, the perpendicular direction is called a *principal direction*, and the stress is called a *principal stress* (Figure 6-9).

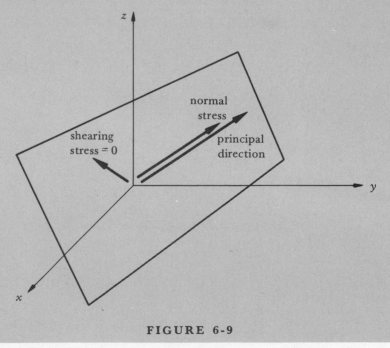

FIGURE 6-9

Determining the principal planes and stresses is an important engineering problem—there is no sliding tendency for such planes, the stress tending only to elongate the body. We shall now see that the problem of determining principal stresses and principal directions is an eigenvalue, eigenvector problem.

Let *xyz* be an arbitrary rectangular coordinate system at *O* and let

$$T = \begin{pmatrix} T_x & T_{xy} & T_{xz} \\ T_{yx} & T_y & T_{xz} \\ T_{zx} & T_{zy} & T_z \end{pmatrix}$$

be the stress matrix relative to the planes defined by this coordinate system. This matrix is a symmetric matrix; thus it has real eigenvalues and three unit orthogonal eigenvectors exist. Let v_1, v_2, and v_3 be such eigenvectors. Use these vectors as the columns of an orthogonal matrix *C*. We are now in a position to consider the situation at *O* in a more suitable coordinate system. Let *x'y'z'* be a rectangular coordinate system defined by the vectors v_1, v_2, and v_3 in the sense that *Ox'* is along v_1, etc. Then the stress matrix at *O*, relative to the planes of this system of coordinates can be shown to be (we do not prove this)

$$C^{-1}TC$$

We know that this will be a diagonal matrix, having eigenvalues on the diagonal,

$$\begin{pmatrix} T_{x'} & 0 & 0 \\ 0 & T_{y'} & 0 \\ 0 & 0 & T_{z'} \end{pmatrix}$$

Observe that all shearing stresses are zero. The new coordinate planes are in fact principal planes, and the new coordinate axes give principal directions. In terms of the original coordinate system the eigenvectors of the stress matrix give principal directions and its eigenvalues principal stresses.

Thus the result that a symmetric matrix can be diagonalized by an orthogonal transformation means here that there always exists an orthogonal set of principal directions for a body subject to external forces.

Let us look at the various possible situations that can arise. There are three possibilities for the roots of the characteristic polynomial of the stress matrix. (σ is usually used in this context for these roots.)

1. If the roots of the characteristic polynomial are distinct, σ_1, σ_2, and σ_3, then these are the principal stresses and there will be three distinct orthogonal eigenvectors giving the corresponding principal directions.

2. If there is a repeated root σ_1 and a single root σ_2 of the characteristic polynomial, then there are just two principal stresses. Corresponding to the repeated root there will be a two-dimensional eigenspace representing a two-dimensional space of principal directions. Corresponding to the single root there will be a principal direction orthogonal to the two-dimensional space.

3. If all three roots of the characteristic polynomial are equal then there is a single principal stress. All vectors are eigenvectors; thus any direction is a principal direction. This corresponds to a state of so-called hydrostatic stress.

Let us consider a specific stress matrix

$$\begin{pmatrix} 5{,}000 & 4{,}000 & 2{,}000 \\ 4{,}000 & 5{,}000 & 2{,}000 \\ 2{,}000 & 2{,}000 & 2{,}000 \end{pmatrix}$$

where the units are pounds per square inch.

The eigenvalues are 1,000 (repeated) and 10,000. 1,000 is a repeated eigenvalue, we have case 2 above.

For $\sigma = 1{,}000$ we have the two-dimensional eigenspace where eigenvectors are of the form $\begin{pmatrix} -x_2 & -\frac{1}{2}x_3 \\ x_2 \\ x_3 \end{pmatrix}$

For $\sigma = 10{,}000$ we have the one-dimensional eigenspace $x_3' \begin{pmatrix} 2 \\ 2 \\ 1 \end{pmatrix}$.

Thus the principal stresses are 1,000 pounds per square inch with principal directions $\begin{pmatrix} -x_2 & -\frac{1}{2}x_3 \\ x_2 \\ x_3 \end{pmatrix}$ and 10,000 pounds per square inch with principal direction $x_3' \begin{pmatrix} 2 \\ 2 \\ 1 \end{pmatrix}$.

We shall not continue any further here with this analysis of the stress matrix; we wish only to show the central role of the concept of eigenvalues and eigenvectors. A further discussion would involve certain invariants that are defined in terms of the coefficients of the characteristic polynomial of the stress matrix. For further reading the reader is referred to *Plasticity: Theory & Practice*, by Alexander Mendelson, Macmillan, 1968. The book does not use eigenvector and eigenvalue terminology, but the mathematics developed is that of eigenvalues and eigenvectors.

Example 12

This example† is intended for readers who have some familiarity with electromagnetic theory.

In a homogeneous crystalline dielectric, the directions of the electric intensity **e** and the electric displacement **d** are not the same, except for certain orientations of the electric field with respect to the crystal. However, any change in the magnitude of the vector **e** changes the magnitude of **d** proportionately without changing the angle α between them. In an arbitrary rectangular coordinate system the vectors **e** and **d** are related as follows.

$$
\begin{pmatrix} d_1 \\ d_2 \\ d_3 \end{pmatrix} = \begin{pmatrix} b_2{}^2 + b_3{}^2 & -b_1 b_2 & -b_1 b_3 \\ -b_1 b_2 & b_3{}^2 + b_1{}^2 & -b_2 b_3 \\ -b_1 b_3 & -b_2 b_3 & b_1{}^2 + b_2{}^2 \end{pmatrix} \begin{pmatrix} e_1 \\ e_2 \\ e_3 \end{pmatrix}
$$

where b_1, b_2, and b_3 are scalars. Since this matrix is symmetric, it is possible to find another coordinate system in which this equation becomes

$$
\begin{pmatrix} d_1' \\ d_2' \\ d_3' \end{pmatrix} = \begin{pmatrix} \varepsilon_1 & 0 & 0 \\ 0 & \varepsilon_2 & 0 \\ 0 & 0 & \varepsilon_3 \end{pmatrix} \begin{pmatrix} e_1' \\ e_2' \\ e_3' \end{pmatrix}
$$

The coordinate transformation will be a similarity transformation. The directions of the axes in this coordinate system are called the *electrical axes of the crystal*. If $\varepsilon_1 = \varepsilon_2 = \varepsilon_3$, the medium is isotropic; if only two of the ε_i are the same, the crystal is said to be *uniaxial*; if all three of the ε_i are different, the crystal is *biaxial*. Since we know that ε_1, ε_2, and ε_3 will be the eigenvalues of the original matrix, these three cases correspond mathematically to the cases of repeated and distinct eigenvalues.

For a more complete discussion of this example refer to the original text.

EXERCISES

1. Consider two rectangular coordinate systems xy and $x'y'$ in the plane with a common origin O. The second system is obtained from the first by a rotation about O through an angle θ in a counterclockwise direction. In each case determine the matrix that defines the coordinate transformation and the coordinates of the point $(1, 1)$ in the new system.
   a) $\theta = 90°$ b) $\theta = 45°$ c) $\theta = 60°$

2. Illustrate the coordinate transformations described by the following matrix equations, sketching the relative locations of the axes.

† Adapted from *Electromagnetic Theory* by Numzio Tralli. Copyright © 1963 by McGraw-Hill, Inc. Used with permission of McGraw-Hill Book Company.

a) $\begin{pmatrix} x \\ y \end{pmatrix} = \begin{pmatrix} x' \\ y' \end{pmatrix} + \begin{pmatrix} 1 \\ 2 \end{pmatrix}$ **b)** $\begin{pmatrix} x \\ y \end{pmatrix} = \begin{pmatrix} x' \\ y' \end{pmatrix} + \begin{pmatrix} -1 \\ 3 \end{pmatrix}$

c) $\begin{pmatrix} x \\ y \end{pmatrix} = \begin{pmatrix} x' \\ y' \end{pmatrix} + \begin{pmatrix} -2 \\ -3 \end{pmatrix}$ **d)** $\begin{pmatrix} x \\ y \end{pmatrix} = \begin{pmatrix} x' \\ y' \end{pmatrix} + \begin{pmatrix} 2 \\ -1 \end{pmatrix}$

e) $\begin{pmatrix} x \\ y \end{pmatrix} = \begin{pmatrix} x' \\ y' \end{pmatrix} + \begin{pmatrix} 0 \\ 2 \end{pmatrix}$ **f)** $\begin{pmatrix} x \\ y \end{pmatrix} = \begin{pmatrix} x' \\ y' \end{pmatrix} + \begin{pmatrix} -3 \\ 0 \end{pmatrix}$

g) $\begin{pmatrix} x \\ y \end{pmatrix} = \begin{pmatrix} 0 & -1 \\ 1 & 0 \end{pmatrix} \begin{pmatrix} x' \\ y' \end{pmatrix}$ **h)** $\begin{pmatrix} x \\ y \end{pmatrix} = \begin{pmatrix} 0 & 1 \\ -1 & 0 \end{pmatrix} \begin{pmatrix} x' \\ y' \end{pmatrix}$

i) $\begin{pmatrix} x \\ y \end{pmatrix} = \begin{pmatrix} 1/\sqrt{2} & -1/\sqrt{2} \\ 1/\sqrt{2} & 1/\sqrt{2} \end{pmatrix} \begin{pmatrix} x' \\ y' \end{pmatrix}$

j) $\begin{pmatrix} x \\ y \end{pmatrix} = \begin{pmatrix} 1/\sqrt{2} & 1/\sqrt{2} \\ -1/\sqrt{2} & 1/\sqrt{2} \end{pmatrix} \begin{pmatrix} x' \\ y' \end{pmatrix}$

k) $\begin{pmatrix} x \\ y \end{pmatrix} = \begin{pmatrix} \sqrt{3}/2 & -\frac{1}{2} \\ \frac{1}{2} & \sqrt{3}/2 \end{pmatrix} \begin{pmatrix} x' \\ y' \end{pmatrix}$

l) $\begin{pmatrix} x \\ y \end{pmatrix} = \begin{pmatrix} -\frac{1}{2} & -\sqrt{3}/2 \\ \sqrt{3}/2 & -\frac{1}{2} \end{pmatrix} \begin{pmatrix} x' \\ y' \end{pmatrix}$

m) $\begin{pmatrix} x \\ y \end{pmatrix} = \begin{pmatrix} -\frac{3}{5} & \frac{4}{5} \\ -\frac{4}{5} & -\frac{3}{5} \end{pmatrix} \begin{pmatrix} x' \\ y' \end{pmatrix}$

n) $\begin{pmatrix} x \\ y \end{pmatrix} = \begin{pmatrix} 1/\sqrt{2} & -1/\sqrt{2} \\ 1/\sqrt{2} & 1/\sqrt{2} \end{pmatrix} \begin{pmatrix} x' \\ y' \end{pmatrix} + \begin{pmatrix} 2 \\ 4 \end{pmatrix}$

o) $\begin{pmatrix} x \\ y \end{pmatrix} = \begin{pmatrix} \sqrt{3}/2 & -\frac{1}{2} \\ \frac{1}{2} & \sqrt{3}/2 \end{pmatrix} \begin{pmatrix} x' \\ y' \end{pmatrix} + \begin{pmatrix} -1 \\ 3 \end{pmatrix}$

p) $\begin{pmatrix} x \\ y \end{pmatrix} = \begin{pmatrix} 0 & -1 \\ 1 & 0 \end{pmatrix} \begin{pmatrix} x' \\ y' \end{pmatrix} + \begin{pmatrix} -1 \\ -2 \end{pmatrix}$

q) $\begin{pmatrix} x \\ y \end{pmatrix} = \begin{pmatrix} \frac{3}{5} & \frac{4}{5} \\ -\frac{4}{5} & \frac{3}{5} \end{pmatrix} \begin{pmatrix} x' \\ y' \end{pmatrix} + \begin{pmatrix} 2 \\ -1 \end{pmatrix}$

3. Discuss and sketch the graphs of each of the following equations. Each graph will be an ellipse, hyperbola, or parabola. (Use a coordinate transformation. Observe that each equation has no term in xy.)

a) $x^2 + 4y^2 - 4x + 3 = 0$
b) $x^2 - 4y^2 - 4x + 3 = 0$
c) $9x^2 + 4y^2 + 18x - 8y - 16 = 0$
d) $x^2 + y^2 - 4x + 6y + 9 = 0$
e) $x^2 - 4y^2 + 2x + 16y - 19 = 0$
f) $y^2 - 4y - 4x = 0$
g) $x^2 + 4y^2 - 6x + 16y + 9 = 0$

4. Express each of the following quadratic forms in terms of matrices.
a) $x^2 + 4xy + 2y^2$

b) $3x^2 + 2xy - 4y^2$

c) $7x^2 - 6xy - y^2$

d) $2x^2 + 5xy + 3y^2$

e) $-3x^2 - 7xy + 4y^2$

5. Discuss and sketch the graphs of each of the following equations. Each graph will be an ellipse, hyperbola, or a pair of straight lines. (Use a coordinate transformation. Observe that each equation has no linear terms in x and y.)

a) $11x^2 + 4xy + 14y^2 - 60 = 0$

b) $3x^2 + 2xy + 3y^2 - 12 = 0$

c) $x^2 - 6xy + y^2 - 8 = 0$

d) $4x^2 + 4xy + 4y^2 - 5 = 0$

e) $-x^2 - 16xy + 11y^2 - 30 = 0$

f) $-7x^2 - 18xy + 17y^2 = 0$

g) $3x^2 - 10xy + 3y^2 = 0$

6. **a)** Discuss the coordinate transformation

$$\begin{pmatrix} x \\ y \end{pmatrix} = \begin{pmatrix} 2\cos\theta & -2\sin\theta \\ 2\sin\theta & 2\cos\theta \end{pmatrix} \begin{pmatrix} x' \\ y' \end{pmatrix}$$

Note that this transformation may be written

$$\begin{pmatrix} x \\ y \end{pmatrix} = \begin{pmatrix} 2 & 0 \\ 0 & 2 \end{pmatrix} \begin{pmatrix} \cos\theta & -\sin\theta \\ \sin\theta & \cos\theta \end{pmatrix} \begin{pmatrix} x' \\ y' \end{pmatrix}$$

b) Discuss coordinate transformations of the type

$$\begin{pmatrix} x \\ y \end{pmatrix} = \begin{pmatrix} a\cos\theta & -a\sin\theta \\ a\sin\theta & a\cos\theta \end{pmatrix} \begin{pmatrix} x' \\ y' \end{pmatrix}$$

for the classes $a > 1$, $0 < a < 1$, $-1 < a < 0$, and $a < -1$.

c) Discuss coordinate transformations of the type

$$\begin{pmatrix} x \\ y \end{pmatrix} = \begin{pmatrix} a\cos\theta & -a\sin\theta \\ b\sin\theta & b\cos\theta \end{pmatrix} \begin{pmatrix} x' \\ y' \end{pmatrix}$$

Note that such a transformation may be written

$$\begin{pmatrix} x \\ y \end{pmatrix} = \begin{pmatrix} a & 0 \\ 0 & b \end{pmatrix} \begin{pmatrix} \cos\theta & -\sin\theta \\ \sin\theta & \cos\theta \end{pmatrix} \begin{pmatrix} x' \\ y' \end{pmatrix}$$

7. Let the coordinate system $x'y'$ be obtained from xy by a rotation through an angle θ. Let A be a 2×2 matrix that defines a mapping of the plane into itself in coordinate system xy. What is the equivalent mapping in the coordinate system $x'y'$ for

a) $\theta = \pi/2$, $A = \begin{pmatrix} 2 & 1 \\ 1 & 0 \end{pmatrix}$

b) $\theta = \pi/4$, $A = \begin{pmatrix} 1 & 2 \\ -1 & 3 \end{pmatrix}$

c) $\theta = \pi/3$, $A = \begin{pmatrix} 2 & 0 \\ 0 & 2 \end{pmatrix}$

8. In each of the following exercises, A is a symmetric matrix that defines a mapping of the plane into itself, in a coordinate system xy. Thus there exists a coordinate system in which this mapping of the plane into itself can be represented by a diagonal matrix. Find this diagonal matrix and the appropriate coordinate system, sketching the location of the axes.

a) $A = \begin{pmatrix} 3 & 1 \\ 1 & 3 \end{pmatrix}$ **b)** $A = \begin{pmatrix} 0 & -2 \\ -2 & 0 \end{pmatrix}$ **c)** $A = \begin{pmatrix} 7 & -1 \\ -1 & 7 \end{pmatrix}$

d) $A = \begin{pmatrix} 7 & -4 \\ -4 & 13 \end{pmatrix}$ **e)** $A = \begin{pmatrix} 11 & 2 \\ 2 & 14 \end{pmatrix}$ **f)** $A = \begin{pmatrix} -1 & -8 \\ -8 & 11 \end{pmatrix}$

9. Solve the following systems of equations using similarity transformations. (The diagonal forms and similarity transformations were determined in the exercises in Section 6-8.)

a) $\begin{pmatrix} 1 & 2 \\ 2 & 1 \end{pmatrix} \begin{pmatrix} x \\ y \end{pmatrix} = \begin{pmatrix} 4 \\ 2 \end{pmatrix}$

b) $\begin{pmatrix} \frac{1}{2} & -\frac{3}{2} & 0 \\ -\frac{3}{2} & \frac{1}{2} & 0 \\ 0 & 0 & -2 \end{pmatrix} \begin{pmatrix} x \\ y \\ z \end{pmatrix} = \begin{pmatrix} 4 \\ 6 \\ 4 \end{pmatrix}$

c) $\begin{pmatrix} \frac{3}{2} & -\frac{1}{2} & 0 \\ -\frac{1}{2} & \frac{3}{2} & 0 \\ 0 & 0 & 1 \end{pmatrix} \begin{pmatrix} x \\ y \\ z \end{pmatrix} = \begin{pmatrix} 1 \\ 2 \\ 2 \end{pmatrix}$

d) $\begin{pmatrix} 1 & -2 & 2 \\ 4 & 5 & -4 \\ 0 & -2 & 3 \end{pmatrix} \begin{pmatrix} x \\ y \\ z \end{pmatrix} = \begin{pmatrix} 1 \\ -1 \\ 1 \end{pmatrix}$

10. In Section 4-4 we discussed the concept of matrix representation of a linear mapping. Let A be an $n \times n$ matrix having n linearly independent eigenvectors. Interpret A as a linear mapping of $\mathbf{R}^n$ into $\mathbf{R}^n$. Prove that A has diagonal matrix representation relative to a basis of eigenvectors. Prove that this representation has eigenvalues of A as diagonal elements.

11. Determine the principal stresses and principal directions for each of the following stress matrices:

a) $\begin{pmatrix} 1,000 & 1,000 & 1,000 \\ 1,000 & 1,000 & 1,000 \\ 1,000 & 1,000 & 1,000 \end{pmatrix}$ **b)** $\begin{pmatrix} 1,000 & 0 & 0 \\ 0 & 2,000 & 1,000 \\ 0 & 1,000 & 2,000 \end{pmatrix}$

c) $\begin{pmatrix} 2,000 & 0 & 0 \\ 0 & 2,000 & 0 \\ 0 & 0 & 2,000 \end{pmatrix}$

12. Write a computer program to determine the mappings in Exercise 7. Use your program to determine the equivalent mappings when

a) $\theta = 62°$, $A = \begin{pmatrix} 1.2 & 3 \\ 1 & 1.7 \end{pmatrix}$ b) $\theta = -21°$, $A = \begin{pmatrix} 2.4 & 0 \\ 0 & 1.6 \end{pmatrix}$

6–10.* NORMAL MODES OF OSCILLATING SYSTEMS

In this section we shall give two examples that illustrate the use of eigenvalues and eigenvectors. The first is from mechanics, and the second from electrical engineering. These examples also illustrate the use of coordinate transformations in such analyses.

Example 1

Consider a horizontal string AB of length $4a$ and negligible mass loaded with three particles, each of mass m. Let the masses be located at fixed distances a, $2a$, and $3a$ from A. The particles are displaced slightly from their equilibrium position and released. We wish to analyze the subsequent motion, assuming that it all takes place in a vertical plane through A and B.

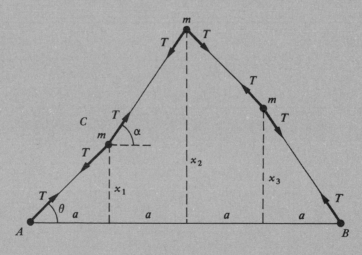

FIGURE 6-10

Let the vertical displacements of the particles at any instant during the subsequent motion be x_1, x_2, and x_3, as illustrated in Figure 6-10. Let T be the tension in the string. Consider the motion of the particle at C. The resultant force on this particle in a vertical direction

353

is $T \sin \alpha - T \sin \theta$.   If we assume that the displacements are small, the motion of each particle can be assumed to be vertical and the tension can be assumed to be unaltered throughout the motion.   Since the angles α and θ will be small, $\sin \alpha$ can be taken to be equal to $\tan \alpha$ and $\sin \theta$ equal to $\tan \theta$.   Thus the resultant vertical force at C is

$$T \tan \alpha - T \tan \theta = T \frac{(x_2 - x_1)}{a} - T \frac{x_1}{a}$$

$$= -\frac{2Tx_1}{a} + \frac{Tx_2}{a}$$

Applying Newton's second law of motion (force = mass × acceleration), we find that the motion of the first particle is described by the equation

$$m\ddot{x}_1 = -\frac{2Tx_1}{a} + \frac{Tx_2}{a}$$

Similarly, the motions of the other two particles are described by the equations

$$m\ddot{x}_2 = \frac{Tx_1}{a} - \frac{2Tx_2}{a} + \frac{Tx_3}{a}$$

and

$$m\ddot{x}_3 = \frac{Tx_2}{a} - \frac{2Tx_3}{a}$$

These equations can be combined into one matrix equation

$$\begin{pmatrix} \ddot{x}_1 \\ \ddot{x}_2 \\ \ddot{x}_3 \end{pmatrix} = \frac{T}{ma} \begin{pmatrix} -2 & 1 & 0 \\ 1 & -2 & 1 \\ 0 & 1 & -2 \end{pmatrix} \begin{pmatrix} x_1 \\ x_2 \\ x_3 \end{pmatrix}$$

We shall now see how the theory of eigenvalues, eigenvectors, and coordinate transformations enables us to solve this matrix equation elegantly, leading to solutions of the three original equations which describe the motions of the three particles.

The matrix $\begin{pmatrix} -2 & 1 & 0 \\ 1 & -2 & 1 \\ 0 & 1 & -2 \end{pmatrix}$ is symmetric; thus it can be transformed into a diagonal form using a similarity transformation.   A matrix of transformation will have linearly independent eigenvectors of this matrix as column vectors.   The eigenvalues of this matrix are -2, $-2 - \sqrt{2}$, and $-2 + \sqrt{2}$.   Corresponding eigenvectors are $(1, 0, -1)$,

$(1, -\sqrt{2}, 1)$, and $(1, \sqrt{2}, 1)$.   Thus a matrix of transformation is

$$\begin{pmatrix} 1 & 1 & 1 \\ 0 & -\sqrt{2} & \sqrt{2} \\ -1 & 1 & 1 \end{pmatrix},$$

and

$$\begin{pmatrix} 1 & 1 & 1 \\ 0 & -\sqrt{2} & \sqrt{2} \\ -1 & 1 & 1 \end{pmatrix}^{-1} \begin{pmatrix} -2 & 1 & 0 \\ 1 & -2 & 1 \\ 0 & 1 & -2 \end{pmatrix}$$

$$\times \begin{pmatrix} 1 & 1 & 1 \\ 0 & -\sqrt{2} & \sqrt{2} \\ -1 & 1 & 1 \end{pmatrix}$$

$$= \begin{pmatrix} -2 & 0 & 0 \\ 0 & -2-\sqrt{2} & 0 \\ 0 & 0 & -2+\sqrt{2} \end{pmatrix},$$

a diagonal matrix where the eigenvalues are diagonal elements.

We introduce this similarity transformation into the matrix equation that describes the motion by rearranging the equations.

$$\begin{pmatrix} 1 & 1 & 1 \\ 0 & -\sqrt{2} & \sqrt{2} \\ -1 & 1 & 1 \end{pmatrix}^{-1} \begin{pmatrix} \ddot{x}_1 \\ \ddot{x}_2 \\ \ddot{x}_3 \end{pmatrix}$$

$$= \frac{T}{ma} \begin{pmatrix} 1 & 1 & 1 \\ 0 & -\sqrt{2} & \sqrt{2} \\ -1 & 1 & 1 \end{pmatrix}^{-1} \begin{pmatrix} -2 & 1 & 0 \\ 1 & -2 & 1 \\ 0 & 1 & -2 \end{pmatrix}$$

$$\underbrace{\times \begin{pmatrix} 1 & 1 & 1 \\ 0 & -\sqrt{2} & \sqrt{2} \\ -1 & 1 & 1 \end{pmatrix}}$$

Similarity transformation that diagonalizes the matrix

$$\times \begin{pmatrix} 1 & 1 & 1 \\ 0 & -\sqrt{2} & \sqrt{2} \\ -1 & 1 & 1 \end{pmatrix}^{-1} \begin{pmatrix} x_1 \\ x_2 \\ x_3 \end{pmatrix}$$

This equation is equivalent to the original matrix equation.   Thus the motion is described by

$$\begin{pmatrix} 1 & 1 & 1 \\ 0 & -\sqrt{2} & \sqrt{2} \\ -1 & 1 & 1 \end{pmatrix}^{-1} \begin{pmatrix} \ddot{x}_1 \\ \ddot{x}_2 \\ \ddot{x}_3 \end{pmatrix}$$

$$= \frac{T}{ma} \begin{pmatrix} -2 & 0 & 0 \\ 0 & -2-\sqrt{2} & 0 \\ 0 & 0 & -2+\sqrt{2} \end{pmatrix} \begin{pmatrix} 0 & 1 & 1 \\ 0 & -\sqrt{2} & \sqrt{2} \\ -1 & 1 & 1 \end{pmatrix}^{-1} \begin{pmatrix} x_1 \\ x_2 \\ x_3 \end{pmatrix}$$

We now introduce new coordinates y_1, y_2, and y_3 defined by the transformation.

$$\begin{pmatrix} y_1 \\ y_2 \\ y_3 \end{pmatrix} = \begin{pmatrix} 1 & 1 & 1 \\ 0 & -\sqrt{2} & \sqrt{2} \\ -1 & 1 & 1 \end{pmatrix}^{-1} \begin{pmatrix} x_1 \\ x_2 \\ x_3 \end{pmatrix}$$

In this new coordinate system, the matrix equation that describes the motion assumes the particularly simple form

$$\begin{pmatrix} \ddot{y}_1 \\ \ddot{y}_2 \\ \ddot{y}_3 \end{pmatrix} = \frac{T}{ma} \begin{pmatrix} -2 & 0 & 0 \\ 0 & -2-\sqrt{2} & 0 \\ 0 & 0 & -2+\sqrt{2} \end{pmatrix} \begin{pmatrix} y_1 \\ y_2 \\ y_3 \end{pmatrix}$$

(Coordinates such as y_1, y_2, y_3 and x_1, x_2, x_3 that in some way describe the configuration of a system are called *generalized coordinates*. They may not be interpretable in terms of rectangular cartesian axes, but since they lead to a description of the configuration, they are coordinates.)

Thus, in terms of the new coordinates y_1, y_2, and y_3, the motion is described by the equations

$$\ddot{y}_1 = -\frac{2T}{ma} y_1$$

$$\ddot{y}_2 = \frac{(-2-\sqrt{2})T}{ma} y_2$$

$$\ddot{y}_3 = \frac{(-2+\sqrt{2})T}{ma} y_3$$

three simple harmonic motions. Solutions of these equations are

$$y_1 = b_1 \cos\left(\sqrt{\frac{2T}{ma}}\, t + \gamma_1\right)$$

$$y_2 = b_2 \cos\left(\sqrt{\frac{(2+\sqrt{2})T}{ma}}\, t + \gamma_2\right)$$

$$y_3 = b_3 \cos\left(\sqrt{\frac{(2-\sqrt{2})T}{ma}}\, t + \gamma_3\right)$$

where b_1, b_2, b_3, γ_1, γ_2, and γ_3 are constants of integration that depend upon the configuration at time $t = 0$. They are not all independent. In standard interpretation of simple harmonic motion b_1, b_2, and b_3 are amplitudes and γ_1, γ_2, and γ_3 phases.

These special coordinates y_1, y_2, and y_3 are called the *normal coordinates* of the motion. The general motion in terms of the original coordinates is described by

$$\begin{pmatrix} x_1 \\ x_2 \\ x_3 \end{pmatrix} = \begin{pmatrix} 1 & 1 & 1 \\ 0 & -\sqrt{2} & \sqrt{2} \\ -1 & 1 & 1 \end{pmatrix} \begin{pmatrix} y_1 \\ y_2 \\ y_3 \end{pmatrix}$$

$$= y_1 \begin{pmatrix} 1 \\ 0 \\ -1 \end{pmatrix} + y_2 \begin{pmatrix} 1 \\ -\sqrt{2} \\ 1 \end{pmatrix} + y_3 \begin{pmatrix} 1 \\ \sqrt{2} \\ 1 \end{pmatrix}$$

giving

$$\begin{pmatrix} x_1 \\ x_2 \\ x_3 \end{pmatrix} = b_1 \cos\left(\sqrt{\frac{2T}{ma}}\,t + \gamma_1\right)\begin{pmatrix} 1 \\ 0 \\ -1 \end{pmatrix}$$

$$+ b_2 \cos\left(\sqrt{\frac{(2+\sqrt{2})T}{ma}}\,t + \gamma_2\right)\begin{pmatrix} 1 \\ -\sqrt{2} \\ 1 \end{pmatrix}$$

$$+ b_3 \cos\left(\sqrt{\frac{(2-\sqrt{2})T}{ma}}\,t + \gamma_3\right)\begin{pmatrix} 1 \\ \sqrt{2} \\ 1 \end{pmatrix}$$

The motion can thus be interpreted as a combination of the three motions

$$\begin{pmatrix} x_1 \\ x_2 \\ x_3 \end{pmatrix} = \cos\left(\sqrt{\frac{2T}{ma}}\,t + \gamma_1\right)\begin{pmatrix} 1 \\ 0 \\ -1 \end{pmatrix}$$

$$\begin{pmatrix} x_1 \\ x_2 \\ x_3 \end{pmatrix} = \cos\left(\sqrt{\frac{(2+\sqrt{2})T}{ma}}\,t + \gamma_2\right)\begin{pmatrix} 1 \\ -\sqrt{2} \\ 1 \end{pmatrix}$$

$$\begin{pmatrix} x_1 \\ x_2 \\ x_3 \end{pmatrix} = \cos\left(\sqrt{\frac{(2-\sqrt{2})T}{ma}}\,t + \gamma_3\right)\begin{pmatrix} 1 \\ \sqrt{2} \\ 1 \end{pmatrix}$$

The actual combination will be a linear combination determined by b_1, b_2, and b_3 that will depend upon the initial configuration.

Each of these motions is a simple harmonic motion of a particle. These modes, called the *normal modes of oscillation*, are illustrated in Figure 6-11.

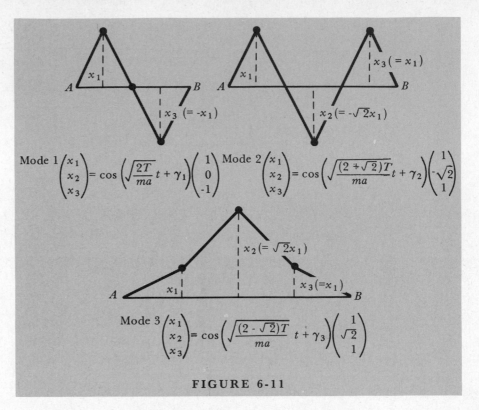

FIGURE 6-11

Before we turn to an analysis of the flow of electricity in two electric circuits in close proximity, we generalize the standard eigenvalue-eigenvector concept.

Let A and B be square matrices of the same kind. A scalar λ and a nonzero vector $\mathbf{x}$ satisfying the identity

$$A\mathbf{x} = \lambda B\mathbf{x}$$

are said to be an eigenvalue and eigenvector of A and B. Note that these reduce to the standard definitions when B is the identity matrix. To determine these eigenvalues and eigenvectors we proceed as previously. Rewriting the identity $A\mathbf{x} - \lambda B\mathbf{x} = \mathbf{0}$ as

$$(A - \lambda B)\mathbf{x} = \mathbf{0} \tag{1}$$

we see that the eigenvalues can be obtained by solving the equation

$$|A - \lambda B| = 0$$

The corresponding eigenvectors are then obtained from identity (1).

In the application of this theory A and B are often $n \times n$ real symmetric matrices. Under these conditions n linearly independent eigenvectors of A and B are known to exist. A matrix C having such eigenvectors as columns can be used to simultaneously diagonalize A and B; $C^t A C = D$ and $C^t B C = E$ where D and E are both diagonal matrices.

To clarify the concept, let us look at a specific situation. We shall determine the eigenvalues and eigenvectors of the matrices $A = \begin{pmatrix} 8 & 2 \\ 2 & 2 \end{pmatrix}$ and $B = \begin{pmatrix} -4 & 0 \\ 0 & -1 \end{pmatrix}$ and diagonalize both matrices simultaneously.

B is already in diagonal form; the theorem then implies that B remains in diagonal form under the transformation that diagonalizes A. We know that the eigenvalues are given by $|A - \lambda B| = 0$. Thus

$$\begin{vmatrix} 8 - \lambda(-4) & 2 \\ 2 & 2 - \lambda(-1) \end{vmatrix} = \begin{vmatrix} 8 + 4\lambda & 2 \\ 2 & 2 + \lambda \end{vmatrix}$$

$$= (8 + 4\lambda)(2 + \lambda) - 4 = 4(\lambda + 1)(\lambda + 3) = 0$$

The eigenvalues are $\lambda_1 = -1, \lambda_2 = -3$; they are distinct. We now find the corresponding eigenvectors.

For $\lambda_1 = -1$, eigenvectors must satisfy

$$\begin{pmatrix} 8 + 4\lambda_1 & 2 \\ 2 & 2 + \lambda_1 \end{pmatrix} \begin{pmatrix} x_1 \\ x_2 \end{pmatrix} = \mathbf{0}$$

That is,

$$\begin{pmatrix} 4 & 2 \\ 2 & 1 \end{pmatrix} \begin{pmatrix} x_1 \\ x_2 \end{pmatrix} = \mathbf{0}$$

The condition on the vector (x_1, x_2) is that $2x_1 + x_2 = 0$; that is, $x_2 = -2x_1$. Thus $(\,1, \quad -2\,)$ is an eigenvector.

For $\lambda_2 = -3$, the eigenvectors are given by

$$\begin{pmatrix} 8 + 4\lambda_2 & 2 \\ 2 & 2 + \lambda_2 \end{pmatrix} \begin{pmatrix} x_1 \\ x_2 \end{pmatrix} = \mathbf{0}$$

That is,

$$\begin{pmatrix} -4 & 2 \\ 2 & -1 \end{pmatrix} \begin{pmatrix} x_1 \\ x_2 \end{pmatrix} = \mathbf{0}$$

$(1, 2)$ is an eigenvector.

Thus a transforming matrix C with linearly independent eigenvectors as column vectors is

$$C = \begin{pmatrix} 1 & 1 \\ -2 & 2 \end{pmatrix}$$

We have that

$$C^t A C = \begin{pmatrix} 1 & -2 \\ 1 & 2 \end{pmatrix} \begin{pmatrix} 8 & 2 \\ 2 & 2 \end{pmatrix} \begin{pmatrix} 1 & 1 \\ -2 & 2 \end{pmatrix} = \begin{pmatrix} 8 & 0 \\ 0 & 24 \end{pmatrix}$$

a diagonal matrix, and

$$C^t BC = \begin{pmatrix} 1 & -2 \\ 1 & 2 \end{pmatrix} \begin{pmatrix} -4 & 0 \\ 0 & -1 \end{pmatrix} \begin{pmatrix} 1 & 1 \\ -2 & 2 \end{pmatrix}$$

$$= \begin{pmatrix} -8 & 0 \\ 0 & -8 \end{pmatrix} = -8 \begin{pmatrix} 1 & 0 \\ 0 & 1 \end{pmatrix}$$

Example 2

Such eigenvalues and eigenvectors can be used to analyze the current flows in the circuits in Figure 6-12.

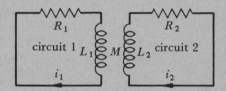

FIGURE 6-12

The two circuits are in close proximity; a change in current in one circuit affects the other. The circuits are said to be *coupled magnetically* by a mutual inductance of M henries. For such circuits, the voltage drop across a resistance is Ri, the voltage drop across an inductance is $L(di/dt)$, and an induced voltage due to the proximity of a changing current is $M(di/dt)$. For currents flowing in the directions shown in Figure 6-12, the second of Kirchhoff's laws gives

Circuit 1, $\qquad L_1 \dfrac{di_1}{dt} + M \dfrac{di_2}{dt} + R_1 i_1 = 0$

Circuit 2, $\qquad M \dfrac{di_1}{dt} + L_2 \dfrac{di_2}{dt} + R_2 i_2 = 0$

These two equations can be combined into one matrix equation

$$\begin{pmatrix} L_1 & M \\ M & L_2 \end{pmatrix} \begin{pmatrix} di_1/dt \\ di_2/dt \end{pmatrix} = \begin{pmatrix} -R_1 & 0 \\ 0 & -R_2 \end{pmatrix} \begin{pmatrix} i_1 \\ i_2 \end{pmatrix}$$

At this stage the technique for diagonalizing matrices simultaneously is used. Let us consider a specific example, where $M = 2$ henries, $L_1 = 8$ henries, $L_2 = 2$ henries, $R_1 = 4$ ohms, and $R_2 = 1$ ohm. The matrix equation that describes the current flows becomes

$$\begin{pmatrix} 8 & 2 \\ 2 & 2 \end{pmatrix} \begin{pmatrix} di_1/dt \\ di_2/dt \end{pmatrix} = \begin{pmatrix} 4 & 0 \\ 0 & -1 \end{pmatrix} \begin{pmatrix} i_1 \\ i_2 \end{pmatrix}$$

The two matrices that enter into this equation are the ones that were simultaneously diagonalized earlier. We can capitalize on this technique by rewriting the above equation

$$\underbrace{C'\begin{pmatrix} 8 & 2 \\ 2 & 2 \end{pmatrix}C}_{or\begin{pmatrix} 8 & 0 \\ 0 & 24 \end{pmatrix}}C^{-1}\begin{pmatrix} di_1/dt \\ di_2/dt \end{pmatrix} = \underbrace{C'\begin{pmatrix} -4 & 0 \\ 0 & -1 \end{pmatrix}C}_{or\,-8\begin{pmatrix} 1 & 0 \\ 0 & 1 \end{pmatrix}}C^{-1}\begin{pmatrix} i_1 \\ i_2 \end{pmatrix}$$

where C is the transforming matrix.

Thus the equation becomes

$$\begin{pmatrix} 8 & 0 \\ 0 & 24 \end{pmatrix}C^{-1}\begin{pmatrix} di_1/dt \\ di_2/dt \end{pmatrix} = -8C^{-1}\begin{pmatrix} i_1 \\ i_2 \end{pmatrix}$$

Introducing new variables j_1 and j_2 (the electrical equivalent of generalized coordinates) defined by

$$\begin{pmatrix} j_1 \\ j_2 \end{pmatrix} = C^{-1}\begin{pmatrix} i_1 \\ i_2 \end{pmatrix}$$

the equation assumes the simple form

$$\begin{pmatrix} 8 & 0 \\ 0 & 24 \end{pmatrix}\begin{pmatrix} dj_1/dt \\ dj_2/dt \end{pmatrix} = -8\begin{pmatrix} j_1 \\ j_2 \end{pmatrix}$$

This gives

$$\frac{dj_1}{dt} = -j_1 \quad \text{and} \quad \frac{dj_2}{dt} = -\tfrac{1}{3}j_2$$

Solving these equations,

$$j_1 = pe^{-t}, \quad j_2 = qe^{-1/3t}$$

where p and q are constants that depend on initial conditions.

Thus

$$\begin{pmatrix} i_1 \\ i_2 \end{pmatrix} = C\begin{pmatrix} j_1 \\ j_2 \end{pmatrix} = \begin{pmatrix} 1 & 1 \\ -2 & 2 \end{pmatrix}\begin{pmatrix} j_1 \\ j_2 \end{pmatrix}$$

$$= j_1\begin{pmatrix} 1 \\ -2 \end{pmatrix} + j_2\begin{pmatrix} 1 \\ 2 \end{pmatrix}$$

$$= pe^{-t}\begin{pmatrix} 1 \\ -2 \end{pmatrix} + qe^{-1/3t}\begin{pmatrix} 1 \\ 2 \end{pmatrix}$$

As in Example 1, the mechanics application, the current flows can be expressed as linear combinations of two normal modes, the actual linear combinations depending upon the initial conditions.

EXERCISES

1. Determine the normal modes of oscillation if the following system is displaced slightly from equilibrium.

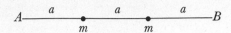

2. The motion of a weight attached to a spring is governed by Hooke's law: tension = $k \times$ extension, where k is a constant for the spring. For oscillations of the system in Figure 6-13, if x_1 and x_2 are the displacements of weights of masses m_1 and m_2 at any instant, the extensions of the two springs are x_1 and $x_2 - x_1$ at that instant. Thus the application of Hooke's law gives the equations

$$m_1 \ddot{x}_1 = -k_1 x_1 + k_2 (x_2 - x_1)$$

and

$$m_2 \ddot{x}_2 = -k_2 (x_2 - x_1)$$

The general motion can be analyzed in terms of normal modes in a manner similar to the one used in Example 1. If $m_1 = m_2 = M$, $k_1 = 3$, and $k_2 = 2$, analyze the motion.

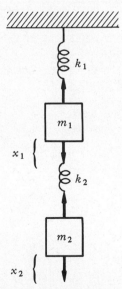

FIGURE 6-13

3. Hooke's law also applies to forces in the extended springs of the system in Figure 6-14. The constants of the springs are 1, 2, and 3, as indicated.

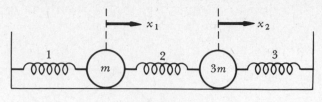

FIGURE 6-14

Applications of the law give the following equations of motion.

$$m\ddot{x}_1 = -x_1 + 2(x_2 - x_1)$$

$$2m\ddot{x}_2 = -2(x_2 - x_1) - 3x_2$$

Analyze this motion in terms of normal modes.

4. Analyze the current flows in the circuits in Figure 6-15, where $M = 2$ henries, $L_1 = 6$ henries, $L_2 = 2$ henries, $R_1 = 2$ ohms, and $R_2 = 1$ ohm.

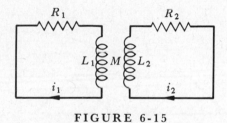

FIGURE 6-15

5. The importance of transformations of the form $C^t A C$ was illustrated in Example 2 of this section. Write a computer program that can be used for performing such transformations. Test your program on the matrices

$$C = \begin{pmatrix} 1 & 1 \\ -2 & 2 \end{pmatrix} \text{ and } A = \begin{pmatrix} 8 & 2 \\ 2 & 2 \end{pmatrix} \text{ in this example.}$$

6–11.* THE CONCEPT OF A COORDINATE SYSTEM

Here we analyze the concepts involved in setting up a coordinate system and the functions of coordinate transformations by means of two examples and an analogy.

Suppose we are asked to describe the framework in Figure 6-16. We might say, "It is made of eight steel girders." We are associating the framework with a sentence in the English language. This association

363

.

defines a mapping from the framework into the set called English language. We can call the mapping *English*.

FIGURE 6-16

This mapping, which gives a "picture" of the framework, is far from unique. Another mapping would be *Welsh*.† The same description in Welsh would be "Mae wedi ei wneud a wyth drawst dur." This gives a second equivalent picture of the framework. The pictures would be related by means of what we call a *translation*; see Figure 6-17. (Give further examples of mappings, descriptions, and translations.) We would call each of these pictures a qualitative description of the framework.

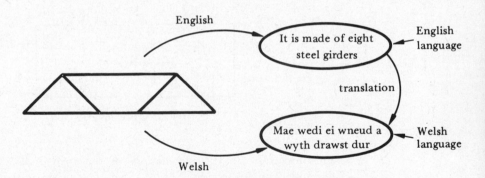

FIGURE 6-17

It is interesting to observe a young bilingual child translate a name for an object from one language into another language. Mentally he will often go from the word in the first language to the object by inverse mapping and then to the word in the other language, rather than translating directly.

In an analogous manner, coordinate systems and transformations are used to give a quantitative picture of the framework, as in the following example.

---

† Welsh is a Celtic language spoken in Wales, one of the countries that make up Great Britain. It is an Indo-European language.

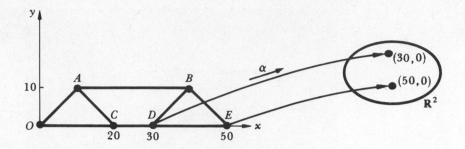

FIGURE 6-18

Example 1

Consider Figure 6-18. We let O be the origin and select x and y axes as shown. Then we associate the coordinates $(0,0)$, $(10,10)$, $(40,10)$, $(20,0)$, $(30,0)$, and $(50,0)$ with O, A, B, C, D, and E, respectively; that is, we map $O, A, ..., E$ onto these elements of $\mathbf{R}^2$ using a mapping α. This gives us a quantitative picture of the framework in $\mathbf{R}^2$. α is a *coordinate system*.

This coordinate system is not unique. We could choose the joint C as the origin and have a coordinate system β (Figure 6-19). O, A, B, C, D, and E would then have coordinates $(-20,0)$, $(-10,10)$, $(20,10)$, $(0,0)$, $(10,0)$, and $(30,0)$, respectively.

A *coordinate transformation* relates the two pictures of the framework in $\mathbf{R}^2$. It is analogous to a translation of a qualitative picture. In one picture, D is the point $(30,0)$, in the second it is $(10,0)$. The coordinate transformation maps $(30,0)$ onto $(10,0)$. Similarly, it maps every point in the α coordinate description of the framework into the corresponding point in the β coordinate description (Figure 6-20).

This type of mathematical analysis lies in the field of *manifold theory*. It is a very powerful area of mathematics using algebra and analysis that is usually presented in graduate courses in mathematics.

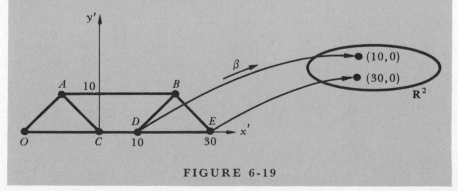

FIGURE 6-19

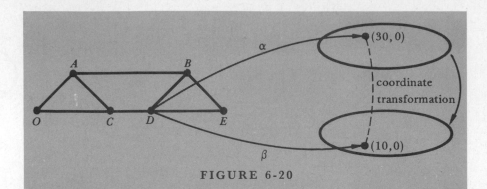

FIGURE 6-20

Example 2   As another example, consider a body moving under certain forces in three-space. If an observer wanted to describe its motion, he might apply Newton's laws of motion in a coordinate system *xyzt*. A second observer moving with uniform velocity relative to the first observer would apply Newton's laws in his own coordinate system *x′y′z′t′* to describe the motion (Figure 6-21).

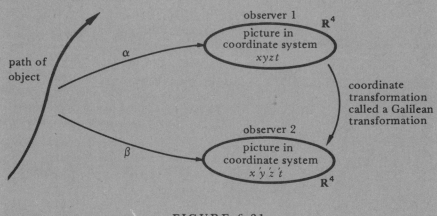

FIGURE 6-21

α and β define the correspondence between the physical situation and each observer's picture. We call them coordinate systems since they actually define *xyzt* and *x′y′z′t′*. The two pictures are related by means of a coordinate transformation called a *Galilean transformation*. These are special transformations that preserve, in some sense, the mathematical structure of Newtonian mechanics.

If the velocity of the second observer is large (in proportion to the velocity of light), then Newtonian mechanics does not lead to an accurate description of the physical situation; we use special relativity.

As we have seen, this involves mapping the physical situation into $\mathbf{R}^4$. Each observer has his own coordinate system, *xyzt* and *x'y'z't'*. Note that here each observer has his own time; the first observer *t*, and the second *t'*.

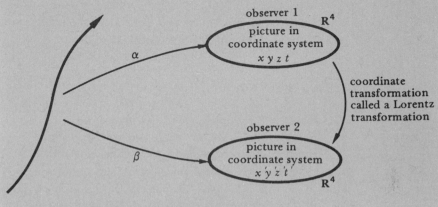

FIGURE 6-22

The two pictures are now related by means of a coordinate transformation called a *Lorentz transformation* (Figure 6-22). These transformations preserve the mathematical structures of special relativity.

Thus, in constructing a mathematical system to describe a physical situation, the mathematician is faced with the following task. He must introduce a coordinate system by mapping the physical situation into $\mathbf{R}^n$ for an appropriate *n* with a mapping α. He must then introduce the necessary mathematics to get Picture 1, a description of the situation. He must do the same for Picture 2, which is a second coordinate description of the situation. The two coordinate pictures must then be related by

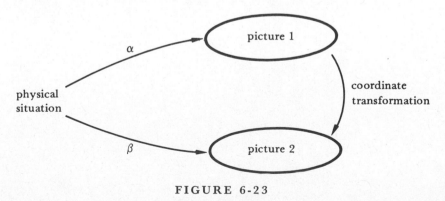

FIGURE 6-23

367

means of a coordinate transformation that preserves the mathematical structures of each picture (Figure 6-23). In linear algebra, the mathematical structure in which we are interested is the vector-space structure. The transformations that preserve this structure are called linear transformations. In both classical mechanics and special relativity, each picture is a vector space and both Galilean transformations and Lorentz transformations are linear mappings.

EXERCISES

1. Relate the analysis of coordinate transformations in this section to the one in Section 6-9. The coordinate transformation introduced in this section is of the type discussed in Example 3 of Section 6-9.
2. Give an example of a third coordinate system that could be used to give a picture in $\mathbf{R}^2$ of the framework in this section. Discuss the specific coordinate transformations between the three pictures you now have.

7^*

Linear Programming

Historically, linear programming was first developed and applied in 1947 by George B. Dantzig, Marshall Wood, and their associates at the U.S. Department of the Air Force. The early applications of linear programming were in the military field, but the emphasis in applications has now moved to the general industrial area. Linear programming is concerned with the efficient use or allocation of limited resources to meet desired objectives.

The 1975 Nobel Prize in Economic Science was awarded to two scientists, Professors Leonid Kantorovich of the Soviet Union and Tjalling C. Koopmans of the United States, for their "contributions to the theory of optimum allocation of resources."

Kantorovich, born in 1912, a winner of the Stalin and Lenin Prizes, is presently with the Mathematics Institute of the Sibirskoje Otdelenie Akademi Nauk, Akademgodorske, in Novosibirsk. He is considered the leading representative of the school in Soviet economic research. Koopmans, 65, who was born in The Netherlands, is currently with the Cowles Foundation for Research in Economics at Yale University in New Haven, Conn. Both economists worked independently on the problem of optimum allocation of scarce resources.

Kantorovich has shown how linear programming can be used to improve economic planning in Russia. He analyzed efficiency conditions for an economy as a whole, demonstrating the connection between the allocation of resources and the price-system. An important element in this analysis was to show that the possibility of decentralizing decisions in

369

a planned economy is dependent on the existence of a rational price system. According to Professor Assar Lindbeck of the International Economics Prize Committee, Kantorovich's work changed the views on planning in the Soviet Union.

Koopmans developed his linear programming theory while working with planning of optimal transportation of ships back and forth across the Atlantic during World War II. Koopmans' work gives new ways of interpreting the relationships between inputs and outputs of a production process. Those are used to clarify the correspondence between efficiency in production and the existence of a system of prices.

We now proceed to develop the mathematical tools that are used in linear programming.

7-1. SYSTEMS OF LINEAR INEQUALITIES

Having discussed systems of equations and having seen the relevance of such mathematical structures, we now turn our attention to a related concept, systems of inequalities. Linear programming problems are described mathematically by systems of inequalities. In this section we introduce these mathematical tools.

We call an expression such as $2x + y = 4$ an equation in the two variables x, y. The graph of an equation is a geometrical representation of the points that satisfy the equation. The graph of this equation consists of the points (x, y) in two-dimensional space that satisfy the condition $2x + y = 4$. Such points form a line of slope -2, y-intercept 4.

We call $2x + y < 4$ an inequality. It also will have a graph, a set of points that satisfy the condition $2x + y < 4$. For example, the origin, $(0, 0)$ will lie on the graph, for $2(0) + 0 = 0$, which is less than 4. The point $(2, 1)$ will not lie on the graph, for $2(2) + 1 = 5$, which is greater than 4. What does the graph of $2x + y < 4$ look like?

Rearrange the equation of the line to read $y = -2x + 4$ and the inequality to read $y < -2x + 4$. Let x_0 be an x value. The point P, $(x_0, -2x_0 + 4)$ lies on the line. The point Q, (x_0, y_I) lies on the graph of the inequality if $y_I < -2x_0 + 4$. Geometrically, this means Q must be below P. The graph of $2x + y < 4$ consists of all points below the line $2x + y = 4$. We call such a region, on one side of a line, a half plane. The graph of $2x + y > 4$ is the half plane above the line $2x + y = 4$.

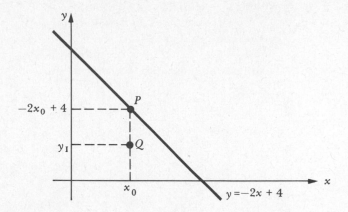

In general, a line $ax + by = c$ divides two-dimensional space into two half planes, $ax + by < c$ and $ax + by > c$.

Example 1

Determine the graph of $x + 3y < 6$.

The graph will be a half plane on one side of the line $x + 3y = 6$. To determine which side, examine the location of a convenient point such as the origin $(0, 0)$. [Any point not on the line will do.] We have that $0 + 3(0) = 0 < 6$. Thus, $(0, 0)$ is on the graph. We now know which half plane is the graph:

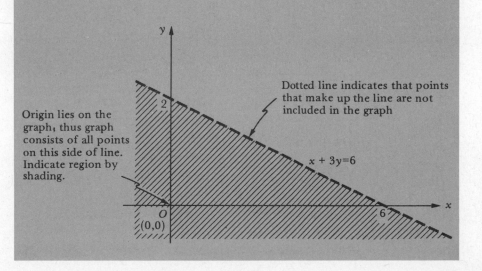

Origin lies on the graph, thus graph consists of all points on this side of line. Indicate region by shading.

Dotted line indicates that points that make up the line are not included in the graph

$x + 3y = 6$

Example 2

Graph the inequality $2x + 3y \geq 6$.

Here we do have possible equality. The graph will consist of all points that lie on one side of the line $2x + 3y = 6$ together with the points that make up this line, for they also satisfy the conditions of the inequality. The point $(3, 3)$ satisfies the inequality; it gives us the relevant side of the line.

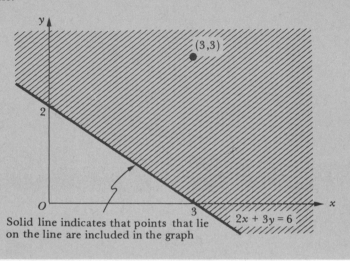

Solid line indicates that points that lie on the line are included in the graph

Let us now proceed to look at solutions to systems of such inequalities. We shall represent these solutions geometrically by regions in the *xy* plane.

Example 3

Describe the solutions to the following system of inequalities geometrically.

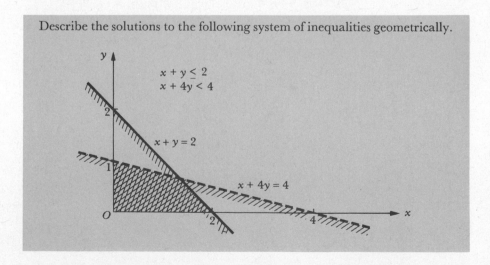

The region shaded ///// is the graph of $x + y \leq 2$; the region shaded ///// is the graph of $x + 4y < 4$. The points that satisfy both inequalities simultaneously are the points that make up the *solution* to the system; these are the points that lie in both regions:

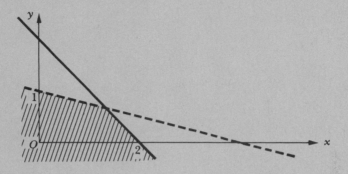

According to this discussion, the origin $(0,0)$ should, for example, be a solution to the system. It can be verified that it is indeed a solution; it does satisfy both inequalities simultaneously.

Example 4

Describe the solutions to the system:

$$-x + y \leq 1$$

$$x + y \leq 3$$

$$x \geq 0$$

We graph the three inequalities. The region common to all three graphs makes up the set of solutions to the system.

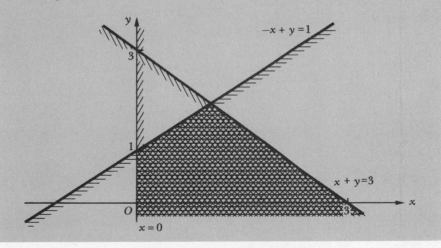

We see that the region common to all three graphs, the solution, is the set of all points in the shaded region:

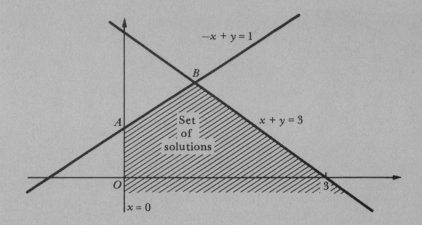

The point $(2, 0)$ for example is in the region. It is indeed seen that $x = 2$, $y = 0$ satisfies all three inequalities. This is a solution. $x = 2$, $y = -1$ would be another solution.

The *boundaries* of the solution set are segments of the lines $x = 0$, $-x+y = 1$, $x+y = 3$. The points A and B are said to be *vertices*. In the following section, when we apply this theory, we shall find that it becomes necessary to determine the vertices of the solution set. The vertices are obtained by solving pairs of simultaneous equations. To find the vertex A, the point of intersection of the lines $x = 0$ and $-x+y = 1$, we solve the system

$$x = 0$$
$$-x + y = 1$$

A is found to be the point $(0, 1)$.

B is the point of intersection of the lines $-x+y = 1$ and $x+y = 3$. Thus, to determine B, we solve the system

$$-x + y = 1$$
$$x + y = 3$$

B is the point $(1, 2)$.

Example 5   Sketch the set of solutions to the system:

$$x + y \leq 3$$
$$-x + 2y \leq 3$$
$$x \geq 0$$
$$y \geq 0$$

The nonnegative restrictions $x \geq 0$ and $y \geq 0$ on the variables often enter into a linear programming problem since these variables usually represent physical quantities. The other inequalities represent restrictions associated with these quantities, limitations such as capital or labor available.

The graph is the shaded region $ABCO$ below:

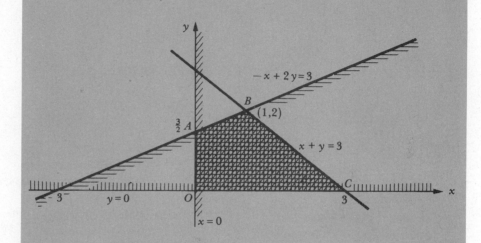

Let us determine the vertices, A, B, C, and O. A is the intersection of the lines $x = 0$ and $-x + 2y = 3$. Substituting $x = 0$ into the equation gives $y = \frac{3}{2}$. Thus A is the point $(0, \frac{3}{2})$. B is the intersection of the lines $-x + 2y = 3$ and $x + y = 3$. Solving the system.

$$-x + 2y = 3$$

$$x + y = 3$$

gives that B is the point $(1, 2)$. C is the intersection of the lines $y = 0$ and $x + y = 3$. Substituting $y = 0$ into the equation gives $x = 3$. Thus C is the point $(3, 0)$. O is of course the origin $(0, 0)$.

EXERCISES

Graph the following inequalities:

1. $x + y \leq 3$ **2.** $2x + y > 4$ **3.** $2x - y < -2$

4. $y \geq 2$ in two dimensional space   **5.** $4x - y \geq 8$

Describe the solutions to the following systems of inequalities geometrically. Determine the vertices to each solution set.

6. $-2x + y < 4$ **7.** $-x + 2y \geq 3$ **8.** $x \geq 0$
$\quad\;\; x + y \geq 3$ $\quad\;\; 2x + y > 2$ $\quad\;\; y < 0$

9. $-x + y \leq 3$
   $x + y \leq 4$
   $x \geq 0$

10. $2x + y \leq 1$
    $-x + y > 4$
    $y \geq 0$

11. $x - y < 2$
    $2x + y \leq 4$
    $4x + y < 6$

12. $2x + y \geq 2$
    $x - 3y \leq 6$
    $-4x + y > -3$

13. $x + y > 1$
    $-x + y \geq 2$
    $5x - y < 4$

14. $-2x + y < 2$
    $3x + y \leq 3$
    $-x + y > -4$
    $x + y \geq -3$

15. $8x + y \geq -10$
    $-2x + y > 6$
    $4x + y < 2$
    $3x - y > -1$

16. $x + y \leq 2$
    $3x + y \leq 6$
    $x \geq 0$
    $y \geq 0$

17. $2x + y \leq 3$
    $5x + 2y \leq 7$
    $x \geq 0$
    $y \geq 0$

18. $-x + y \leq 3$
    $2x + y \leq 6$
    $x \geq 0$
    $y \geq 0$

7–2. LINEAR PROGRAMMING—A GEOMETRICAL INTRODUCTION

We now have the necessary mathematical tools, systems of inequalities, to commence our discussion of linear programming. We motivate the ideas involved with the following example.

Example 1

A company makes two products on separate production lines X and Y. It commands a labor force which is equivalent to 900 hours per week and it has \$2,800 outlay weekly on running costs. It takes 5 hours and 2 hours to produce a single item on X and Y, respectively. The cost of producing a single item on X is \$8 and on Y is \$10. The aim of the company is to maximize its profits. If the profit on each item produced on line X is \$3 and that on each item produced on line Y is \$2, how should the scheduling be arranged to lead to maximum profit?

In this problem there are two restrictions—time and funds available. The aim of the company is to maximize its profits under these restrictions. We can solve the problem in three stages: (1) constructing a mathematical model, that is, describing the situation using mathematics; (2) illustrating the mathematics by means of a graph; and (3) using the graph to determine the solution.

(1) *The mathematical model.* Let x items be produced on production line X and y items on line Y. The weekly profit on line X is thus \$3x$, and on line Y it is \$2y$. The total weekly profit is thus \$(3x + 2y)$.

Let us examine the time constraint. The total time involved in producing x items on line X at the rate of 5 hours per item is $5x$ hours. The time it takes to manufacture y items on line Y at the rate of 2 hours per item is $2y$ hours. Hence the total time involved for both production lines is $5x + 2y$ hours. Since the number of hours available is 900 we must have that $5x + 2y$ is less than or equal to 900. We write this mathematically

$$5x + 2y \leq 900$$

Next we examine the cost involved. The cost of producing x items on line X at \$8 an item is \$8x. The cost for y items on line Y at \$10 an item is \$10y. Hence the total cost is \$8x + \$10y. Since the outlay is \$2,800 we must have that

$$8x + 10y \leq 2,800$$

Finally, we observe that x and y cannot be negative; it would be meaningless for these quantities to be negative as they represent the number of items produced on the production lines. Thus there are the additional restrictions that

$$x \geq 0 \quad \text{and} \quad y \geq 0$$

Therefore, the problem reduces to determining the values of x and y that maximize $3x + 2y$ under the constraints

$$5x + 2y \leq 900$$
$$8x + 10y \leq 2,800$$
$$(x \geq 0, y \geq 0)$$

The function $f = 3x + 2y$ is called the *objective function* of the linear programming problem.

(2) *Graphical representation of the constraints.* The above system of constraints is a system of inequalities. We are interested in the solutions to this system of inequalities that lead to the maximum values of $3x + 2y$. Let us represent the set of solutions to this system geometrically:

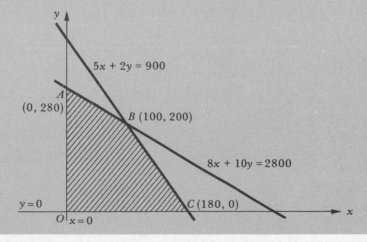

The set of solutions is the shaded region in the above figure. We find the vertices: A is the intersection of the lines $x = 0$ and $8x + 10y = 2{,}800$. Substituting for $x = 0$ into the equation gives $y = 280$. Thus A is the point $(0, 280)$. B is the intersection of the line $8x + 10y = 2{,}800$ and the line $5x + 2y = 900$. Solving the pair of equations

$$8x + 10y = 2800$$

$$5x + 2y = 900$$

gives $x = 100$, $y = 200$. B is the point $(100, 200)$. C is the intersection of the lines $y = 0$, $5x + 2y = 900$. Substituting for $y = 0$ into the equation gives $x = 180$. Thus C is the point $(180, 0)$. O is of course the origin, $(0, 0)$.

Each point in $ABCO$ satisfies all the inequalities. $3x + 2y$ has a value at each of these points. Every one of these points has a chance of being a solution to the problem, that is, a point that leads to a maximum value of $3x + 2y$ under the given constraints. We call such a region, satisfying all the constraints of a linear programming problem, the *feasible region* and a point in the region, a *feasible solution*. Among these points there is one, or possibly many, that gives a maximum value to the objective function; such a point is called an *optimal solution*.

(3) *Determining the optimal solution.* It would be an endless task to examine the value of the objective function $3x + 2y$ at all feasible solutions. General problems of this type have been analyzed, and it has been found that the maximum will occur at a vertex, or at the points that make up one side of $ABCO$ (proved later in this section). That is, in this latter case the maximum value of $3x + 2y$ is attained at all the points that make up one side. Hence, we have only to examine the values of $f = 3x + 2y$ at the vertices A, B, C, and O.

At A,

$$x = 0 \qquad \text{and} \quad y = 280, \qquad \text{thus} \quad f = 3(0) + 2(280) = 560$$

At B,

$$x = 100 \qquad \text{and} \quad y = 200, \qquad \text{thus} \quad f = 3(100) + 2(200) = 700$$

At C,

$$x = 180 \qquad \text{and} \quad y = 0, \qquad \text{thus} \quad f = 3(180) + 2(0) = 540$$

At O,

$$x = 0 \qquad \text{and} \quad y = 0, \qquad \text{giving} \quad f = 0$$

Thus, the maximum value of $3x + 2y$, namely 700, occurs at B, when $x = 100$ and $y = 200$. The interpretation of these results is that the maximum weekly profit that can be achieved under the given constraints is \$700. This will be achieved when production line X turns out 100 items and line Y turns out 200 items.

Example 2

The optimal solution to a linear programming problem need not always be unique as we now illustrate.

Determine the maximum value of the function $5x + 2y$ subject to the constraints:

$$4x + 5y \leq 10$$

$$5x + 2y \leq 10 \qquad (x \geq 0, y \geq 0)$$

$$3x + 8y \leq 12$$

The constraints are represented graphically by the shaded region below:

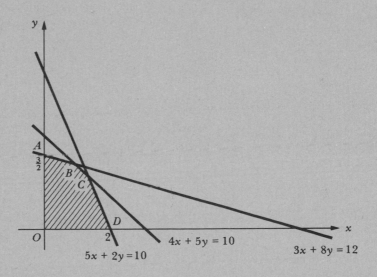

The vertices of the region are found to be $A(0, \frac{3}{2})$, $B(\frac{20}{17}, \frac{18}{17})$, $C(\frac{30}{17}, \frac{10}{17})$, $D(2, 0)$, and $O(0, 0)$. We now determine the values of $5x + 2y$ at each of these vertices. At A, $5x + 2y = 3$; at B, it is 8; at C, 10; and at D, 10.

Thus, the maximum value of $5x + 2y$ is 10, and it occurs at two vertices, C and D. When this happens in a linear programming problem the objective function will have that same value at every point on the edge joining the two vertices. Thus, here $5x + 2y$ has a maximum of 10 subject to these constraints, and this occurs at all points represented by the line CD, that is, at all points that satisfy $5x + 2y = 10$, $0 \leq y \leq \frac{10}{17}$ ($y = 0$ at D and $y = \frac{10}{17}$ at C).

Example 3

It is possible that the constraints in a linear programming problem define an unbounded feasible region and that the objective function is unbounded, having no maximum. The linear programming problem then has no solution. We now illustrate this possibility.

Determine the maximum value of the function $x + 4y$ subject to the constraints

$$-4x + y \leq 2$$
$$2x - y \leq 1 \qquad (x \geq 0, y \geq 0)$$

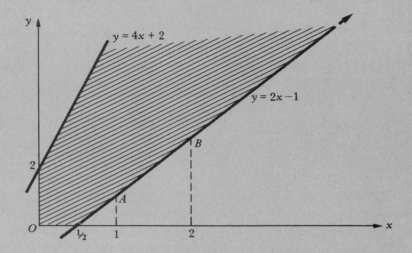

The constraints are represented by the shaded region above. Let us show that the objective function has no maximum value within this region. Consider the values of f at the points along the line $y = 2x - 1$. Along these points

$$f = x + 4y = x + 4(2x - 1) = 9x - 4$$

Thus at A, having x value 1, f is 5. At B, having x value 2, f is 14. As we move in the direction indicated $\rightarrow$, along this line, the x values of the points increase, and f increases without limit.

The objective function has no maximum value within this region, it increases without limit in the direction indicated.

If this situation arises in practice (when one expects a solution), it could be that a constraint has been omitted. For example, suppose the constraint $x + y \leq 7$ had been omitted in the above case, the correct problem being to maximize $x + 4y$ subject to

$$-4x + y \leq 2$$
$$2x - y \leq 1 \qquad (x \geq 0, y \geq 0)$$
$$x + y \leq 7$$

The feasible region is

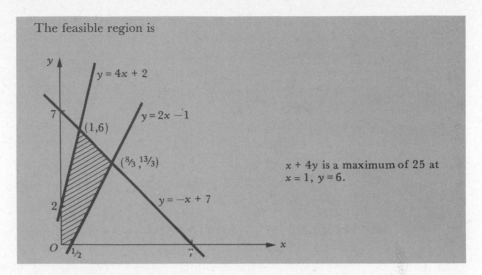

$x + 4y$ is a maximum of 25 at $x = 1$, $y = 6$.

The fact that the feasible region is unbounded does not in itself imply that the objective function has no maximum.   Consider the following example.

Example 4

Maximize the function $3y$ subject to the constraints

$$-x + 2y \leq 4$$
$$y \leq 3 \qquad (x \geq 0, y \geq 0)$$

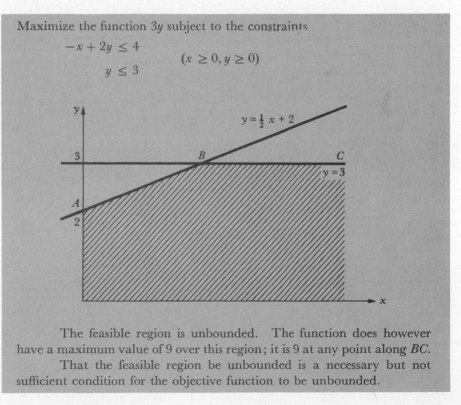

The feasible region is unbounded.   The function does however have a maximum value of 9 over this region; it is 9 at any point along BC.
        That the feasible region be unbounded is a necessary but not sufficient condition for the objective function to be unbounded.

A problem that involves determining a minimum value of a function f can be solved by looking for the maximum value of $-f$, the negative of f, over the same region. Over the given region, the value of f at any point will be the negative of that of $-f$ at that point. *We shall show that the minimum value of f occurs at the point(s) of maximum value of $-f$ and is the negative of that maximum value.*

Denote the value of f at the point A by $f(A)$.

If f has a minimum value at A, in the region, then

$$f(A) \le f(B)$$

where B is any other point in the region. Multiplying both sides by -1,

$$-f(A) \ge -f(B)$$

implying that A is a point of maximum value of $-f$. The converse can also be seen to hold, namely that if A is a point of maximum value of $-f$, then it is a point of minimum value of f. Thus the minimum value of f and maximum value of $-f$ occur at the same point(s).

It can be seen that the minimum value of f, namely $f(A)$ is the negative of the maximum value of $-f$, namely $-f(A)$.

Example 5

Determine the minimum value of the function $2x - 3y$ under the constraints

$$2y + \ x \le 10$$
$$y + 2x \le 11$$

with $x \ge 0$ and $y \ge 0$.

The vertices of the relevant region are $A(0, 5)$, $B(4, 3)$, $C(\frac{11}{2}, 0)$, and $O(0, 0)$. To determine the minimum value of $2x - 3y$ over this region we find the location of the maximum value of $-(2x - 3y)$ in the region. The values of $-(2x - 3y)$ at the vertices are as follows: 15 at A, 1 at B, -11 at C, 0 at O. Thus $-(2x - 3y)$

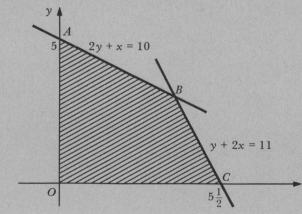

has a maximum of 15 at x = 0, $y = 5$.   The minimum of $(2x - 3y)$ will be -15 at $x = 0, y = 5$.

Notice that each feasible region we have discussed is such that the whole of the segment of a straight line joining any two points within the region lies within that region.   Such a region is called *convex*.   A theorem states that the region satisfying the set of inequalities that represent the constraints in a linear programming problem is always convex.

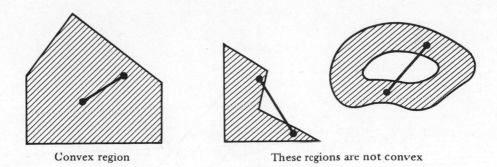

Convex region   These regions are not convex

We now give a geometrical reason for why we expect the maximum value of the objective function to occur at either a vertex or along the side of the feasible region of a linear programming problem in two variables.

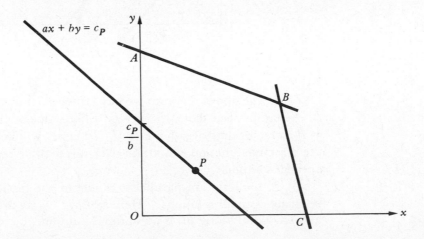

Let the feasible region be *ABCO* above, and let the objective function be $f = ax + by$. f has a value at each point in *ABCO*, at each feasible solution. Let $P(x_p, y_p)$ be an arbitrary feasible solution and let f have value c_p at P. Thus $ax_p + by_p = c_p$. c_p/b has a geometrical interpretation, it is the y intercept of the line $ax + by = c_p$, on which P lies. Thus, associated with each feasible solution there is a line, the y intercept of that line being $1/b$ of the value of the objective function at that feasible solution. The feasible solution leading to the maximum value of the objective function will lie on the line with maximum y intercept since these lines are all parallel.

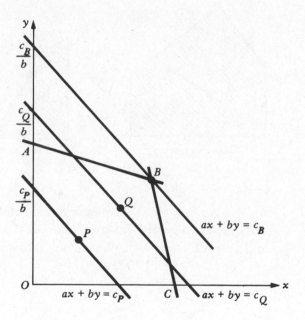

In the above diagram we have three such lines, through P, Q, and B. It can be seen that $c_B > c_Q > c_P$. The value of the objective function at B is greater than at either P or Q. In fact c_B will be the maximum value the objective function can attain over this feasible region. Thus B is the optimal solution.

If the lines are parallel to a side of the feasible region there will be many optimal solutions. Here $ax + by = c_p$ is parallel to *BC*. $c_B = c_C$, and any point along *BC* is an optimal solution.

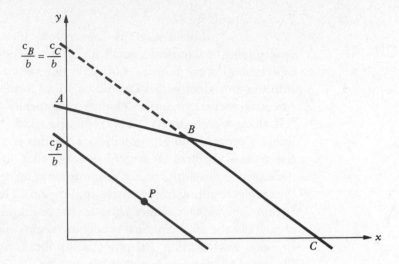

We complete this section by listing examples of problems that are usually solved using linear programming techniques.

Examples of linear programming problems

1. The production-scheduling problem.

    A company manufactures a product on a production line, the rate of which can be varied. The expected sales pattern over a period is known. It is expensive to vary the rate of change of production since cost increases as a result of change. Thus, it is not usual to manufacture the exact amount in each period to satisfy the demand during that period because the changeover cost would be prohibitive. Furthermore, it is expensive to store the product. Linear programming can be used to determine the production schedule that will minimize the sum of the changeover costs and the storage costs.

2. The transportation problem.

    In the transportation problem one is concerned with a product that has to be transported from a number of origins to a number of destinations. The origins could be, for example, production plants, the destinations, distribution warehouses. The quantity of the product available at each origin and the quantity required at each destination are known. The objective is to determine the amount that should be shipped from each origin to each destination in order to minimize total shipping costs.

385

3. The diet problem.

This problem gets its name from its original application—determining economical human diets. In its industrial form it consists of determining the most economical mixture of raw materials that will result in a product with a desired chemical formula. Since the market prices of various materials change periodically, the most economical mix changes accordingly. This repetitive characteristic occurs in many linearly programming problems. That this is so is fortunate because the typical problem is usually so large that the cost of the analysis necessary to develop a linear programming model for it is greater than the savings resulting from a single application. However, if the problem, except for minor changes, has to be re-examined periodically, the changes in the mathematical model are slight, and the cost per solution becomes worthwhile. All practical applications are so involved that the use of a computer becomes mandatory.

EXERCISES

Graph the following linear inequalities, indicating the region that satisfies the inequalities in each case.

1. $2x + y \le 50$
  $4x + 5y \le 160$
  $x \ge 0$
  $y \ge 0$

2. $2x + y \le 60$
  $2x + 3y \le 120$
  $x \ge 0$
  $y \ge 0$

3. $4x + 3y \le 72$
  $4x + 9y \le 144$
  $x \ge 2$
  $y \ge 0$

4. $4x - 3y \ge 60$
  $x + y \le 10$
  $-x + 2y \ge -50$

Solve the following linear programming problems.

5. Maximize $2x + y$
  subject to:
  $4x + y \le 36$
  $4x + 3y \le 60$
  $x \ge 0$
  $y \ge 0$

6. Maximize $x - 4y$
  subject to:
  $x + 2y \le 4$
  $x + 6y \le 8$
  $x \ge 0$
  $y \ge 0$

7. Maximize $2x + y$
  subject to:
  $-3x + y \le 4$
  $x - y \le 2$
  $x \ge 0$
  $y \ge 0$

8. Maximize $4x + 2y$
  subject to:
  $x + 3y \le 15$
  $2x + y \le 10$
  $x \ge 0$
  $y \ge 0$

9. Maximize $3x + y$
  subject to:
  $2x - 3y \ge 10$
  $x \le 8$
  $x \ge 0$
  $y \ge 0$

10. Maximize $x + 5y$
  subject to:
  $x + y \le 10$
  $2x + y \ge 10$
  $x + 2y \ge 10$

11. Maximize $x + 2y$
   subject to:
$$x \geq -2$$
$$x - y \geq -4$$
$$x + 2y \leq 6$$
$$y + 2x \leq 6$$

12. Maximize $-8x + 10y$
   subject to:
$$x \geq -20$$
$$x \leq 5$$
$$y \geq 0$$
$$4x + 3y \leq 40$$
$$-4x + 5y \leq 120$$

13. A manufacturing company makes two products, X and Y, on two machines, I and II. It takes 3 minutes on each machine to produce an X. To produce a Y it takes 1 minute on machine I, and 2 minutes on machine II. The total time available on machine I is 3,000 minutes, and on machine II it is 4,500 minutes. The company realizes a profit of \$15 on each X and \$7 on each Y. How should the manufacturing of the products be arranged in order to obtain the largest possible profit?

14. A company manufactures two types of hand calculators, model $C1$ and model $C2$. It takes 1 hour and 4 hours in labor time to manufacture the $C1$ and $C2$, respectively. The cost of manufacturing the $C1$ is \$30 and that of manufacturing the $C2$ is \$20. The company has 1600 hours per week available in labor and \$18,000 in running costs. The profit on the $C1$ model is \$10, on the $C2$ model it is \$8. What should the weekly production schedule be to ensure maximum profit?

15. A refrigerator company has two plants, at X and at Y. Its refrigerators are sold in a certain town Z. It takes 20 hours (packing, transportation, etc.) to transport a refrigerator from X to Z and 10 hours from Y to Z. It costs \$60 to transport each refrigerator from X to Z and \$10 per refrigerator from Y to Z. 1,200 hours are available (man hours for packing, transportation, etc.) and \$2,400 budgeted for transportation costs. The profit on each refrigerator manufactured at X is \$40 and on each manufactured at Y is \$20 (the plant at X is newer and more efficient than that at Y). How should the company allocate the transportation of refrigerators so as to maximize its profits?

16. A company makes a single product on two separate production lines, X and Y. It commands a labor force which is equivalent to 1000 hours per week and it has a \$3,000 weekly outlay on running costs. It takes 1 hour and 4 hours to produce a single item on X and Y, respectively. The cost of producing a single item on X is \$5 and on Y is \$4. The aim of the company is maximum productivity under these constraints. How can this be achieved?

17. The maximum daily production of an oil refinery is 1,000 barrels. The refinery can produce two types of oil, gasoline and domestic heating oil. A minimum of 200 barrels of heating oil is required daily. The profit is \$3.00 per barrel on gasoline and \$2.00 per barrel on heating oil. What is the maximum profit that can be realized daily, and what quantities of each type of oil are then produced?

18. The organizer of a day conference has \$2,000 available for distribution as expense money to participants. The participants fall into two categories, those having \$30 expenses and those having \$10 expenses. Facilities are available to host 100 participants at the conference. Describe the ways that 100 participants can be invited to the conference.

19. A manufacturer makes two types of fertilizer, X and Y, using chemicals A and B. Fertilizer X is made up of 80% chemical A and 20% chemical B. Fertilizer Y is made up of 60% chemical A and 40% chemical B. The manufacturer requires at least 30 tons of X and at least 50 tons of Y, and has available 100 tons of A and 50 tons of B. The manufacturer wishes to make as much fertilizer as possible. What are the quantities of X and Y that the manufacturer must make to accomplish this?

20. A city has \$600,000 for purchasing cars. Two models, the Arrow and Gazelle, are under consideration, costing \$4,000 and \$5,000, respectively. The estimated annual maintenance cost of the Arrow is \$400 and of the Gazelle it is \$300. The city will allocate \$40,000 for the total annual maintenance of these cars. The Arrow attains 24 miles per gallon and the Gazelle 20 miles per gallon. How many of each model should the city purchase in order to maximize total miles per gallon from these cars?

21. A car dealer imports foreign cars by way of two ports of entry, A and B. The dealer needs 120 cars at city C and 180 at city D. There are 100 cars available at A and 200 at B. It takes 2 hours and 6 hours to transport a car from A and B, respectively, to C, and 4 hours and 3 hours from A and B, respectively, to D. The dealer has 1,030 hours available in driver time to move the cars. The dealer wants to supply as many as possible of the cars for C from B because drivers will not be available for this route in the future. How can this be achieved?

22. A tailor has 80 square yards of cotton material and 120 square yards of woolen material. A suit requires 2 square yards of cotton and 1 square yard of wool, and a dress requires 1 square yard of cotton and 3 square yards of wool. How many of each garment should the tailor make in order to maximize his income if a suit and a dress each sell for \$20? What is the maximum income?

Solve the following linear programming problems which involve determining the minimum value of a function.

23. Minimize $-3x - y$
subject to:
$$x + y \leq 150$$
$$4x + y \leq 450$$
$$x \geq 0$$
$$y \geq 0$$

24. Minimize $-2x + y$
subject to:
$$2x + y \leq 440$$
$$4x + y \leq 680$$
$$x \geq 0$$
$$y \geq 0$$

25. Minimize $-3x - 2y$
  subject to:
  $2x + 3y \leq 260$
  $6x + y \leq 300$
  $x \geq 0$
  $y \geq 0$

26. Minimize $-x + 2y$
  subject to:
  $x + 2y \leq 4$
  $x + 4y \leq 6$
  $x \geq 0$
  $y \geq 0$

7–3. THE SIMPLEX METHOD

The graphical method of solving a linear programming problem has its limitations. The method demonstrated for two variables can be extended to linear programming problems involving three variables, the feasible region being a convex subset of three-dimensional space. However, for problems involving more than three variables the geometrical approach becomes impractical. We now introduce the reader to the simplex method, an algebraic method that can be used for any number of variables. Furthermore, the simplex method has the advantage of being readily programmable for implementation on a computer. The method involves reformulating the constraints in terms of a system of linear equations and then using elementary matrix transformations in a certain sequence in order to arrive at the solution.

We introduce the simplex method by using it to determine the solution for Example 1 of Section 7-2. In that example we wished to determine values of x and y that maximized the objective function $3x + 2y$ subject to the constraints

$$5x + 2y \leq 900$$
$$8x + 10y \leq 2,800 \qquad (x \geq 0, y \geq 0)$$

The simplex method involves reformulating these constraints in terms of a system of linear equations by introducing additional variables called *slack variables*. Let us examine the first constraint $5x + 2y \leq 900$. For each pair of values of x and y that satisfies this condition there will exist a value of a non-negative variable u such that

$$5x + 2y + u = 900$$

The value of u will be the number that has to be added to $5x + 2y$ to bring it up to 900.

Similarly, for each pair of values of x and y that satisfies the second constraint $8x + 10y \leq 2800$ there will exist a value of a nonnegative variable v such that

$$8x + 10y + v = 2800$$

Thus the constraints in the linear programming problem may be represented by the system

$$5x + 2y + u \qquad = 900$$
$$8x + 10y \qquad + v = 2800 \qquad (x \geq 0, y \geq 0, u \geq 0, v \geq 0)$$

The variables u and v are called slack variables because they make up the slack in the original inequalities.

Finally the objective function $f = 3x + 2y$ may be written in the form $-3x - 2y + f = 0$.

The entire problem becomes that of determining the solution to the system of equations.

$$5x + 2y + u \qquad = 900$$
$$8x + 10y \qquad + v \qquad = 2800$$
$$-3x - 2y \qquad + f = \qquad 0$$

such that f is as large as possible with $x \geq 0, y \geq 0, u \geq 0$, and $v \geq 0$. We have thus reformulated the problem in terms of a system of linear equations under certain constraints. The system of equations, consisting of the three equations in the five variables, x, y, u, v, and f will have many solutions in the region defined by $x \geq 0$, $y \geq 0$, $u \geq 0$, and $v \geq 0$. Any such solution is a *feasible solution*. A solution that maximizes f is an *optimal solution*—this is, the solution that we are interested in.

We determine the optimal solution using matrix transformations in a very definite sequence. The method of Gaussian elimination discussed previously involved creating zeros in a very special systematic order. The element that is used to create the zeros in any column when elementary matrix transformations are involved is called the *pivot* for that column. In the simplex method we choose pivots in a manner that differs from Gaussian elimination. Thus, zeros are created in columns in a different order than that of Gaussian elimination.

The augmented matrix of this system, called the *initial simplex tableau*, is

$$\begin{pmatrix} 5 & 2 & 1 & 0 & 0 & 900 \\ 8 & 10 & 0 & 1 & 0 & 2800 \\ -3 & -2 & 0 & 0 & 1 & 0 \end{pmatrix}$$

We now proceed to arrive at a final simplex tableau by following the steps below:

1. Locate the negative entry in the last row, other than the last element, that is largest in magnitude. (If two or more entries share this property,

any one of these can be selected arbitrarily. If all such entries are nonnegative the tableau is in final form.)

2. Divide each positive element in the column containing this negative entry into the corresponding element of the last column.
3. Select as pivot the divisor that yields the smallest quotient.
4. Use this pivot to create a 1 in its location and zeros elsewhere in this column. Zeros are to be created by adding suitable multiples of the pivot row to the other rows.
5. Repeat this sequence until all such negative elements have been eliminated from the last row.

This sequence of transformations ensures that the constraints $x \geq 0, \ldots, v \geq 0$ are not violated at any stage and that we do arrive at an augmented matrix that leads to the solution in an efficient manner.

For this particular tableau we get:

$$
\begin{array}{l}
\text{Pivot} \longrightarrow \\
\text{since} \\
\dfrac{900}{5} < \dfrac{2800}{8}
\end{array}
\left(
\begin{array}{cccccc}
\boxed{5} & 2 & 1 & 0 & 0 & 900 \\
8 & 10 & 0 & 1 & 0 & 2800 \\
-3 & -2 & 0 & 0 & 1 & 0
\end{array}
\right)
$$

$$
\overset{\cong}{(\frac{1}{5})\,\text{Row 1}}
\left(
\begin{array}{cccccc}
\boxed{1} & \frac{2}{5} & \frac{1}{5} & 0 & 0 & 180 \\
8 & 10 & 0 & 1 & 0 & 2800 \\
-3 & -2 & 0 & 0 & 1 & 0
\end{array}
\right)
$$

$$
\begin{array}{l}
\overset{\cong}{} \\
\text{Row 2} - (8)\,\text{Row 1} \\
\text{Row 3} + (3)\,\text{Row 1}
\end{array}
\left(
\begin{array}{cccccc}
1 & \frac{2}{5} & \frac{1}{5} & 0 & 0 & 180 \\
0 & \boxed{\frac{34}{5}} & -\frac{8}{5} & 1 & 0 & 1360 \\
0 & -\frac{4}{5} & \frac{3}{5} & 0 & 1 & 540
\end{array}
\right)
$$

Pivot

$$
\overset{\cong}{(\frac{5}{34})\,\text{Row 2}}
\left(
\begin{array}{cccccc}
1 & \frac{2}{5} & \frac{1}{5} & 0 & 0 & 180 \\
0 & 1 & -\frac{4}{17} & \frac{5}{34} & 0 & 200 \\
0 & -\frac{4}{5} & \frac{3}{5} & 0 & 1 & 540
\end{array}
\right)
$$

$$
\begin{array}{l}
\overset{\cong}{} \\
\text{Row 1} - (\frac{2}{5})\,\text{Row 2} \\
\text{Row 3} + (\frac{4}{5})\,\text{Row 2}
\end{array}
\left(
\begin{array}{cccccc}
1 & 0 & \frac{25}{85} & -\frac{1}{17} & 0 & 100 \\
0 & 1 & -\frac{4}{17} & \frac{5}{34} & 0 & 200 \\
0 & 0 & \frac{7}{17} & \frac{2}{17} & 1 & 700
\end{array}
\right)
$$

This final tableau is equivalent to the system of equations

$$
\begin{aligned}
x + \tfrac{25}{85}u - \tfrac{1}{17}v &= 100 \\
y - \tfrac{4}{17}u + \tfrac{5}{34}v &= 200 \\
\tfrac{7}{17}u + \tfrac{2}{17}v + f &= 700
\end{aligned}
$$

The solutions of this system are, of course, identical to those of the original system. Since $u \geq 0, v \geq 0$, the last equation tells us that f is the maximum of 700 when $u = v = 0$. On substituting these values back into the system we have that $x = 100$ and $y = 200$. Thus, the optimal solution is $x = 100$, $y = 200$, and $f = 700$.

In terms of our original maximal problem this means that the maximum value of the function $3x + 2y$ is 700 and it occurs when $x = 100$ and $y = 200$. (The element in the last row and last column of the final tableau is always the maximum value of f.)

In practice it is usual to omit the next to last column of the simplex tableau, namely $\begin{smallmatrix} 0 \\ 1 \end{smallmatrix}$, as this does not change during the process. Its elimination reduces the computation involved when the method is programmed for the computer. We illustrate this in the following example. The interpretation of the final tableau of this example is not as straightforward as the last one.

Example 1

Determine the maximum value of the function $3x + 5y + 8z$ subject to the constraints:

$$x + y + z \leq 100$$

$$3x + 2y + 4z \leq 200 \qquad (x \geq 0, y \geq 0, z \geq 0)$$

$$x + 2y \qquad \leq 150$$

The corresponding system of equations is:

$$x + y + z + u \qquad = 100$$

$$3x + 2y + 4z \qquad + v \qquad = 200$$

$$x + 2y \qquad + w \quad = 150$$

$$-3x - 5y - 8z \qquad\qquad + f = \quad 0$$

$$(x \geq 0, y \geq 0, z \geq 0, u \geq 0, v \geq 0, w \geq 0)$$

Thus the tableaux are:

$$\text{Pivot} \begin{pmatrix} 1 & 1 & 1 & 1 & 0 & 0 & 0 & 100 \\ 3 & 2 & ④ & 0 & 1 & 0 & 0 & 200 \\ 1 & 2 & 0 & 0 & 0 & 1 & 0 & 150 \\ -3 & -5 & -8 & 0 & 0 & 0 & 1 & 0 \end{pmatrix}$$

$\qquad\qquad\qquad\quad\uparrow\qquad\qquad\qquad\quad\uparrow$

$\qquad\qquad\quad$ *Select this* $\qquad\quad$ *leave this*

$\qquad\qquad\quad$ *column* $\qquad\qquad\quad$ *column out*

$$\underset{(\frac{1}{4}) \text{ Row } 2}{\cong} \begin{pmatrix} 1 & 1 & 1 & 1 & 0 & 0 & 100 \\ \frac{3}{4} & \frac{1}{2} & ① & 0 & \frac{1}{4} & 0 & 50 \\ 1 & 2 & 0 & 0 & 0 & 1 & 150 \\ -3 & -5 & -8 & 0 & 0 & 0 & 0 \end{pmatrix}$$

$$\underset{\substack{\text{Row } 1 - \text{Row } 2 \\ \text{Row } 4 + (8) \text{ Row } 2}}{\cong} \begin{pmatrix} \frac{1}{4} & \frac{1}{2} & 0 & 1 & -\frac{1}{4} & 0 & 50 \\ \frac{3}{4} & \frac{1}{2} & 1 & 0 & \frac{1}{4} & 0 & 50 \\ 1 & ② & 0 & 0 & 0 & 1 & 150 \\ 3 & -1 & 0 & 0 & 2 & 0 & 400 \end{pmatrix}$$

$$\underset{(\frac{1}{2}) \text{ Row } 3}{\cong} \begin{pmatrix} \frac{1}{4} & \frac{1}{2} & 0 & 1 & -\frac{1}{4} & 0 & 50 \\ \frac{3}{4} & \frac{1}{2} & 1 & 0 & \frac{1}{4} & 0 & 50 \\ \frac{1}{2} & ① & 0 & 0 & 0 & \frac{1}{2} & 75 \\ 3 & -1 & 0 & 0 & 2 & 0 & 400 \end{pmatrix}$$

$$\underset{\substack{\text{Row } 1 - (\frac{1}{2}) \text{ Row } 3 \\ \text{Row } 2 - (\frac{1}{2}) \text{ Row } 3 \\ \text{Row } 4 + \text{Row } 3}}{\cong} \begin{pmatrix} 0 & 0 & 0 & 1 & -\frac{1}{4} & -\frac{1}{4} & 12\frac{1}{2} \\ \frac{1}{2} & 0 & 1 & 0 & \frac{1}{4} & -\frac{1}{4} & 12\frac{1}{2} \\ \frac{1}{2} & 1 & 0 & 0 & 0 & \frac{1}{2} & 75 \\ 3\frac{1}{2} & 0 & 0 & 0 & 2 & \frac{1}{2} & 475 \end{pmatrix}$$

$$\uparrow$$

Maximum value
of the function

This is the final tableau. It represents the following system of equations:

$$u - \tfrac{1}{4}v - \tfrac{1}{4}w = 12\tfrac{1}{2}$$
$$\tfrac{1}{2}x + z + \tfrac{1}{4}v - \tfrac{1}{4}w = 12\tfrac{1}{2}$$
$$\tfrac{1}{2}x + y + \tfrac{1}{2}w = 75$$
$$3\tfrac{1}{2}x + 2v + \tfrac{1}{2}w + f = 475$$

(the f in the final equation comes upon remembering about the column that was omitted.) The final equation, on remembering that $x \geq 0$, $v \geq 0, w \geq 0$ gives a maximum value of 475 for f when $x = v = w = 0$.

On substituting these values into the system we get $y = 75, z = 12\tfrac{1}{2}$, $u = 12\tfrac{1}{2}$.

Thus, the maximum value of the function is 475 and it occurs when $x = 0, y = 75$ and $z = 12\tfrac{1}{2}$.

Example 2

We now demonstrate the simplex method for the previous nonunique solution. We were interested in maximizing the function $5x + 2y$ subject to the constraints:

$$4x + 5y \leq 10$$
$$5x + 2y \leq 10 \qquad (x \geq 0, y \geq 0)$$
$$3x + 8y \leq 12$$

The simplex tableaux are:

$$\begin{pmatrix} 4 & 5 & 1 & 0 & 0 & 10 \\ ⑤ & 2 & 0 & 1 & 0 & 10 \\ 3 & 8 & 0 & 0 & 1 & 12 \\ -5 & -2 & 0 & 0 & 0 & 0 \end{pmatrix}$$

$$\cong \begin{pmatrix} 4 & 5 & 1 & 0 & 0 & 10 \\ ① & \frac{2}{5} & 0 & \frac{1}{5} & 0 & 2 \\ 3 & 8 & 0 & 0 & 1 & 12 \\ -5 & -2 & 0 & 0 & 0 & 0 \end{pmatrix}$$

$$\cong \begin{pmatrix} 0 & \frac{17}{5} & 1 & -\frac{4}{5} & 0 & 2 \\ 1 & \frac{2}{5} & 0 & \frac{1}{5} & 0 & 2 \\ 0 & \frac{34}{5} & 0 & -\frac{3}{5} & 1 & 6 \\ 0 & 0 & 0 & 1 & 0 & 10 \end{pmatrix}$$

This final tableau gives the system

$$\frac{17}{5}y + u - \frac{4}{5}v \quad\quad = 2$$
$$x + \frac{2}{5}y \quad\quad + \frac{1}{5}v \quad\quad = 2$$
$$\frac{34}{5}y \quad\quad - \frac{3}{5}v + w = 6$$
$$v \quad\quad + f = 10$$

The last equation gives that the maximum value of f is 10 and that v must be zero.   The system then reduces to:

$$\frac{17}{5}y + u \quad\quad = 2$$
$$x + \frac{2}{5}y \quad\quad = 2 \quad\quad (x \geq 0, y \geq 0, u \geq 0, w \geq 0)$$
$$\frac{34}{5}y \quad\quad + w = 6$$

Any values of x and corresponding y's satisfying these conditions are optimal solutions.   The second equation may be written:

$$5x + 2y = 10$$

Thus the optimal solutions lie on this line.   The maximum value that y can assume according to the first equation, and constraints $y \geq 0, u \geq 0$, is $y = \frac{10}{17}$, when $u = 0$.   The last equation permits $y = \frac{15}{17}$ when $w = 0$. Thus, overall, the maximum value of y is $\frac{10}{17}$.   The minimum value is 0 when $u = 2, w = 6$. y can thus take on values between 0 and $\frac{10}{17}$.   The corresponding values of x (points lie on the line $5x + 2y = 10$) are 2 and $\frac{30}{17}$.   Thus optimal solutions are points on the line $5x + 2y = 10$ from $(2, 0)$ to $(\frac{30}{17}, \frac{10}{17})$, agreeing with the result of our previous geometric approach.

Example 3

We now demonstrate how the simplex method handles a linear programming problem that has no solution, the feasible region being unbounded.

Consider the problem that was discussed in the previous section: Maximize the function $x + 4y$ subject to the constraints:

$$-4x + y \le 2$$

$$2x - y \le 1 \qquad (x \ge 0, y \ge 0)$$

The simplex tableaux are:

$$\begin{pmatrix} -4 & \boxed{1} & 1 & 0 & 2 \\ 2 & -1 & 0 & 1 & 1 \\ -1 & -4 & 0 & 0 & 0 \end{pmatrix} \cong \begin{pmatrix} -4 & \boxed{1} & 1 & 0 & 2 \\ -2 & 0 & 1 & 1 & 3 \\ -1 & -4 & 0 & 0 & 0 \end{pmatrix}$$

$$\cong \begin{pmatrix} -4 & 1 & 1 & 0 & 2 \\ -2 & 0 & 1 & 1 & 3 \\ -17 & 0 & 4 & 0 & 8 \end{pmatrix}$$

It is impossible to proceed further since no pivot exists in the first column because of the negative signs. The simplex method reveals that no maximum exists.

Whenever one arrives at a simplex tableau that contains a column, other than the last column, with a negative last entry and all non-positive elements above it, the feasible region and the objective function are both unbounded.

We now summarize the simplex method as it has been introduced thus far. Since a general linear programming problem can involve many variables, we formulate our results using the notation $x_1, \ldots, x_n$ for variables.

Summary   A *standard linear programming problem* is that of maximizing a function $f = c_1 x_1 + \cdots + c_n x_n$ in n variables subject to m constraints.

$$a_{11} x_1 + \cdots + a_{1n} x_n \le b_1$$
$$\vdots$$
$$a_{m1} x_1 + \cdots + a_{mn} x_n \le b_m$$

with $x_1 \ge 0, \ldots, x_n \ge 0$, where $b_1, \ldots, b_m$ are all non-negative.

We can formulate this problem very compactly in terms of matrices: Maximize the function $f = CX$, subject to the constraints

$$AX \le B$$
$$X \ge 0 \qquad (B \ge 0)$$

Here A is the matrix $\begin{pmatrix} a_{11} & \cdots & a_{1n} \\ & \vdots & \\ a_{m1} & \cdots & a_{mn} \end{pmatrix}$, B is the matrix $\begin{pmatrix} b_1 \\ \vdots \\ b_m \end{pmatrix}$,

C is the matrix $(c_1 \ldots c_n)$ and X is $\begin{pmatrix} x_1 \\ \vdots \\ x_n \end{pmatrix}$. (Note that f is linear.)   A matrix inequality of the type $X \geq 0$ is to be interpreted as $x_1 \geq 0, \ldots, x_n \geq 0$.

The simplex method involves converting the first m inequalities into equations by introducing m slack variables $x_{n+1}, \ldots, x_{n+m}$.   The problem then becomes that of maximizing f subject to

$$
\begin{aligned}
a_{11}x_1 + \cdots + a_{1n}x_n + x_{n+1} &= b_1 \\
a_{21}x_1 + \cdots + a_{2n}x_n \qquad\quad + x_{n+2} &= b_2 \\
\vdots \qquad\qquad\qquad\qquad & \\
a_{m1}x_1 + \cdots + a_{mn}x_n \qquad\qquad\quad + x_{n+m} &= b_m \\
-cx_1 - \cdots - c_n x_n \qquad\qquad\qquad + f &= 0
\end{aligned}
$$

$x_1 \geq 0, \ldots, x_n \geq 0;\ x_{n+1} \geq 0, \ldots, x_{n+m} \geq 0.$

The augmented matrix of this system of equations is called the initial simplex tableau.   One arrives at a final simplex tableau by performing a sequence of elementary matrix transformations, the sequence being selected according to the following rules:

1. Locate the negative entry in the last row, other than the last element, that is largest in magnitude. If two or more entries share this property, any one of these can be selected arbitrarily. If all such entries are non-negative the tableau is already in final form.
2. Divide each positive element in this column into the corresponding element of the last column.
3. Select as pivot the divisor that yields the smallest quotient.
4. Use this pivot to create a 1 in its own location and 0's elsewhere in this column. 0's are to be created by adding suitable multiples of the pivot row to other rows.
5. Repeat this sequence until all such negative elements have been eliminated from the last row.

The final tableau leads to the solution or gives the information that no solution exists.

It is most important to realize the *limitations* of the method that has been stated above. In the above form it can be applied to a standard linear programming problem, that of maximizing a function, where the first set of inequalities are of the type $\leq$, all b_i are nonnegative, and all variables are nonnegative. The simplex method can be applied to other classes of linear programming problems, for example when constraints are a mixture of $\leq$ and $\geq$ types, or some variables are not restricted to being non-negative. In such problems the use of slack variables in setting up the initial tableau differs from the above. Once the initial tableau has been constructed however the sequence of transformations is performed according to the above rules in order to arrive at a final tableau. We shall not pursue these other classes here.

In the class of standard linear programming problems there are two *complications* that can arise:

1. There can be a tie between two or more negative entries in the last row for largest in magnitude. In such a case any of these entries can be selected to provide the pivot column.
2. There can be a tie for pivot. Two or more positive elements in the relevant column on making the divisions into the corresponding elements of the last column can yield equal divisors. Such a case is called *degeneracy*. This will occur for example if some of the b_i are zero. If this occurs any of these elements can be selected as pivot. This technique works for all but a few artificially constructed examples.

We have introduced the reader to the simplex method. Dantzig's discovery of the simplex method ranks high among the achievements of twentieth century applied mathematics.

Readers who desire further knowledge of this useful and interesting area of mathematics are referred to the following texts: *Linear Programming*, by Saul I. Gass, McGraw-Hill Book Company, 1969; and *Linear Programming*, by Robert W. Llewellyn, Holt, Rinehart and Winston, 1964.

EXERCISES

Use the simplex method to maximize (if possible) the following functions under the given restrictions.

1. Maximize $2x + y$
   subject to:
   $$4x + y \leq 36$$
   $$4x + 3y \leq 60$$
   $$x \geq 0$$
   $$y \geq 0$$

2. Maximize $x - 4y$
   subject to:
   $$x + 2y \leq 4$$
   $$x + 6y \leq 8$$
   $$x \geq 0$$
   $$y \geq 0$$

(These are Exercises 5 and 6 of the previous section. Check your answers.)

3. Maximize $4x + 6y$
   subject to:
   $$x + 3y \leq 6$$
   $$3x + y \leq 8$$
   $$x \geq 0$$
   $$y \geq 0$$

4. Maximize $2x + 3y$
   subject to:
   $$-4x + 2y \leq 7$$
   $$2x - 2y \leq 5$$
   $$x \geq 0$$
   $$y \geq 0$$

5. Maximize $10x + 5y$
   subject to:
   $$x + y \leq 180$$
   $$3x + 2y \leq 480$$
   $$x \geq 0$$
   $$y \geq 0$$

6. Maximize $x + 2y + z$
   subject to:
   $$3x + y + z \leq 3$$
   $$x - 10y - 4z \leq 20$$
   $$x \geq 0$$
   $$y \geq 0$$
   $$z \geq 0$$

7. Maximize $100x + 200y + 50z$
   subject to:
   $$5x + 5y + 10z \leq 1000$$
   $$10x + 8y + 5z \leq 2000$$
   $$10x + 5y \leq 500$$
   $$x \geq 0$$
   $$y \geq 0$$
   $$z \geq 0$$

8. Maximize $2x_1 + 4x_2 + x_3$
   subject to:
   $$-x_1 + 2x_2 + 3x_3 \leq 6$$
   $$-x_1 + 4x_2 + 5x_3 \leq 5$$
   $$-x_1 + 5x_2 + 7x_3 \leq 7$$
   $$x_1 \geq 0$$
   $$x_2 \geq 0$$
   $$x_3 \geq 0$$

9. Maximize $2x_1 + x_2 + x_3$
   subject to:
   $$x_1 + 2x_2 + 4x_3 \leq 20$$
   $$2x_1 + 4x_2 + 4x_3 \leq 60$$
   $$3x_1 + 4x_2 + x_3 \leq 90$$
   $$x_1 \geq 0$$
   $$x_2 \geq 0$$
   $$x_3 \geq 0$$

10. Maximize $x_1 + 2x_2 + 4x_3$
    subject to:
    $$8x_1 + 5x_2 - 4x_3 \leq 30$$
    $$-2x_1 + 6x_2 + x_3 \leq 5$$
    $$-2x_1 + 2x_2 + x_3 \leq 15$$
    $$x_1 \geq 0$$
    $$x_2 \geq 0$$
    $$x_3 \geq 0$$

11. Maximize $x_1 + 2x_2 + 4x_3 - x_4$
    subject to:
    $$5x_1 + 4x_3 + 6x_4 \leq 20$$
    $$4x_1 + 2x_2 + 2x_3 + 8x_4 \leq 40$$
    $$x_1 \geq 0$$
    $$x_2 \geq 0$$
    $$x_3 \geq 0$$
    $$x_4 \geq 0$$

12. Maximize $x_1 + 2x_2 - x_3 + 3x_4$
    subject to:
    $$2x_1 + 4x_2 + 5x_3 + 6x_4 \leq 24$$
    $$4x_1 + 4x_2 + 2x_3 + 2x_4 \leq 4$$
    $$x_1 \geq 0$$
    $$x_2 \geq 0$$
    $$x_3 \geq 0$$
    $$x_4 \geq 0$$

7–4. GEOMETRICAL EXPLANATION OF THE SIMPLEX METHOD

We now give an explanation, by means of an example, of the sequence of transformations used in the simplex method.

Example 1

Let us maximize the function $f = 2x + 3y$ subject to the constraints

$$x + 2y \leq 8$$
$$3x + 2y \leq 12$$

$(x \geq 0, y \geq 0)$

The region of interest is $ABCO$, below

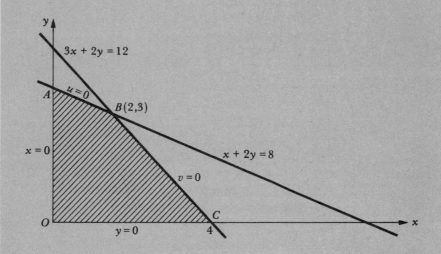

Recall that any point in the region $ABCO$ is called a feasible solution; any such point satisfies the constraints. A feasible solution that gives a maximum value for f is called an optimal solution.

We reformulate the constraints using slack variables u and v,

$$
\begin{aligned}
x + 2y + u \quad\quad &= 8 \\
3x + 2y \quad + v \quad &= 12 \\
-2x - 3y \quad\quad + f &= 0
\end{aligned}
\qquad (x \geq 0, y \geq 0, u \geq 0, v \geq 0) \ (1)
$$

Observe that along AB, $u = 0$ since AB is a segment of the line $x + 2y = 8$. Along BC, $v = 0$ since BC is a segment of the line $3x + 2y = 12$. Furthermore, on OA, $x = 0$ and on OC, $y = 0$. The boundaries of the region of interest are such that one variable is zero along each boundary.

Let us now look at the vertices in terms of the four variables $x, y, u,$ and v:

O is the point $x = 0, y = 0, u = 8, v = 12$. (The u and v values of O are obtained by letting $x = 0$ and $y = 0$ in the constraints.)

A lies on OA and AB. Thus at A, $x = 0$ and $u = 0$. Substituting these values into the constraints gives $y = 4$ and $v = 4$. A is thus the point $x = 0, y = 4, u = 0, v = 4$.

B lies on AB and BC. At B, $x = 2, y = 3, u = 0, v = 0$.

C lies on OC and BC. At C, $y = 0$ and $v = 0$. From the constraints we get that $x = 4$ and $u = 4$. C is the point $x = 4, y = 0, u = 4, v = 0$.

Observe that at each vertex certain variables are nonzero, others are zero. The variables that are not zero are called *basic variables* for that vertex, the remaining variables are called *nonbasic variables*. We summarize the results thus far with the following table

Vertex	Coordinates	Basic variables ($\neq 0$)	Nonbasic variables ($= 0$)
O	$x = 0, y = 0, u = 8, v = 12$	u, v	x, y
A	$x = 0, y = 4, u = 0, v = 4$	y, v	x, u
B	$x = 2, y = 3, u = 0, v = 0$	x, y	u, v
C	$x = 4, y = 0, u = 4, v = 0$	x, u	y, v

The method that we shall develop starts at a vertex feasible solution, O in our case, and then proceeds through a sequence of adjoining vertices, each one giving an increased value of f, until an optimal solution is reached. The initial simplex tableau (in a way to be explained later) corresponds to the situation at the initial feasible solution; further tableaux represent the pictures at other vertices. Each vertex in turn has to be examined to see if it is an optimal solution. If it is not a decision must be made as to which neighboring vertex one moves next. Our aim is to develop the tools to carry out this procedure geometrically and then to translate the geometrical concepts into analogous algebraic ones. The advantage of carrying out the procedure algebraically is that it lends itself to the use of the computer. We shall find that the algebraic procedures that result are those demonstrated previously in the simplex method.

Let us start at O, a vertex feasible solution. We demonstrate whether or not it is necessary to move from O to an adjacent vertex. There are two vertices adjacent to O, A and C. In moving along OC, y is zero, and for each unit increase in x, f will increase by $2 (f = 2x + 3y)$. On the other hand, in moving along OA, x is zero, and for each unit increase in y, f will increase by 3. Because of the larger rate of increase in f along OA this path is selected; A therefore becomes the next feasible solution to be examined. The value of f at A is 12. This is indeed an improvement on the value of f at O. [f value at O is zero.]

We are now confronted with determining whether A is optimal or not. In examining O we expressed f at that point in terms of the non-basic variables x and y, $f = 2x + 3y$. This representation led to the decision to move along the path OA to A. At A we carry out a similar analysis by expressing f in terms of the non-basic variables x and u at this point. To do this we substitute for y from the first constraint in system (1) into $f = 2x + 3y$. We have from the first constraint that $y = 4 - \dfrac{x}{2} - \dfrac{u}{2}$.

Thus in terms of x and u,

$$f = 2x + 3\left(4 - \frac{x}{2} - \frac{u}{2}\right)$$

$$f = 12 + \frac{x}{2} - \frac{3u}{2}$$

In moving along AB from A, $u = 0$ and f increases by $\frac{1}{2}$ for every unit increase in x. In moving along AO from A, $x = 0$ and f decreases by $\frac{3}{2}$ for every unit increase in u (this decrease is to be expected, of course, as we originally moved from O to A in order to increase f). Thus, because of the associated increase in f, we move from A to B. The value of f at B is 13, an increase over its value at A.

Having arrived at B we are confronted with whether this is an optimal solution or not. The non-basic variables at B are u and v. We express f in terms of these variables. From the original form of restrictions

(1) we get that $x = 2 + \frac{u}{2} - \frac{v}{2}$ and $y = 3 - \frac{3u}{4} + \frac{v}{4}$ thus that

$$f = 2x + 3y$$

$$= 2\left(2 + \frac{u}{2} - \frac{v}{2}\right) + 3\left(3 - \frac{3u}{4} + \frac{v}{4}\right)$$

$$= 13 - \frac{5u}{4} - \frac{v}{4}$$

In moving along BC from B, $v = 0$ and f decreases by $\frac{5}{4}$ for every unit increase in u. In moving along BA from B, $u = 0$ and f decreases by $\frac{1}{4}$ for every unit increase in v. Thus f has a maximum value at B. This maximum value is 13.

We have developed the geometrical concepts. We now translate them into algebraic form.

The initial feasible solution was O. The basic variables at O are u and v, having values 8 and 12 respectively. The value of f at O is 0. Let us write an initial tableau (in a more complete form than previously) to reflect this:

$$\begin{array}{cccc} \textit{coefficients of} \\ x & y & u & v \\ \left(\begin{array}{cccc|c} 1 & 2 & 1 & 0 & 8 \\ 3 & 2 & 0 & 1 & 12 \\ -2 & -3 & 0 & 0 & 0 \end{array}\right. & \left.\begin{array}{c} \leftarrow u \\ \leftarrow v \\ \leftarrow f \end{array}\right\} & \textit{Basic variables} \end{array}$$

The last column gives the values of the basic variables and f at O. Knowing the basic variables we also know the non-basic variables. This is the tableau associated with O.

Our geometric discussion told us that O was not an optimal solution and took us to A. The basic variables at O are u and v, at A the basic variables are y and v. In going from O to A, y replaces u as a basic variable. We call u the *departing basic variable*, and y the *entering basic variable*. In moving from O to A, y, the entering variable, corresponded to the largest rate of increase in f, 3. In terms of the above tableau this is reflected by y as the column corresponding to the negative entry in the last row that is largest in magnitude, -3. This rule enables us to select the entering basic variable (if one exists) from any tableau.

The next step is to arrive at the departing variable for the tableau. In going from O to A, the departing variable was u. The first two rows of the tableau correspond to the equations

$$x + 2y + u \quad\;\; = \;\; 8$$

$$3x + 2y \quad\;\; + v = 12$$

In going along OA, x is zero. Thus the equations may be rewritten

$$u = \;\; 8 - 2y$$

$$v = 12 - 2y$$

Since $u \geq 0$ and $v \geq 0$, the maximum value to which y is allowed to increase is 4, when $u = 0$ and $v = 4$ (to the point A). If y goes beyond 4, u would become negative. Thus u becomes a nonbasic variable. Looking at the above two equations we see that the reason u arrives at zero before v is that $\frac{8}{2} < \frac{12}{2}$. In terms of the tableau this corresponds to dividing the elements in the entering variable column into the corresponding elements of the last column. The row containing the numbers that give the smallest result gives the departing variable. Thus

$$\begin{pmatrix} x & y & u & v & \\ 1 & 2 & 1 & 0 & 8 \\ 3 & 2 & 0 & 1 & 12 \\ -2 & -3 & 0 & 0 & 0 \end{pmatrix} \begin{matrix} \\ u \leftarrow \textit{departing variable} \\ v \\ f \end{matrix}$$

$$\underset{\textit{entering variable}}{\uparrow}$$

We now have a method for selecting the entering and departing variables in any tableau.

We must next decide how to transform the above tableau into the tableau that corresponds to the vertex having the new basic variables, y and v. We want the new tableau to have the values of the new basic variables and f at the new vertex as its last column.

The initial constraints, from which the initial tableau was derived were

$$x + 2y + u \qquad\qquad = 8$$

$$3x + 2y \qquad + v \qquad = 12$$

$$-2x - 3y \qquad\qquad + f = 0$$

Observe that the basic variables u and v appear in a single equation each, the coefficient being 1 in each case. It is this fact that causes u and v to assume the values 8 and 12, respectively, at O, on letting the nonbasic variables x and y become zero. Furthermore, the fact that the last equation involves f and only the nonbasic variables causes f to assume the value 0 on the right side of the equation. These are the characteristics we must attempt to obtain for the tableau that represents A, in terms of the basic variables y and v of A. This is achieved by selecting as pivot the element that lies in the entering variable column and departing variable row, creating a 1 in its location and zeros elsewhere in this column. (Zeros are to be created by adding suitable multiples of the pivot row to other rows). Thus the sequence of tableaux becomes

$$
\begin{array}{cccc}
x & y & u & v \\
 & \text{pivot} & &
\end{array}
$$

$$
\begin{pmatrix}
1 & \circled{2} & 1 & 0 & 8 \\
3 & 2 & 0 & 1 & 12 \\
-2 & -3 & 0 & 0 & 0
\end{pmatrix}
\begin{matrix} u \leftarrow \\ v \\ f \end{matrix}
$$

$$
\cong
\begin{pmatrix}
\tfrac{1}{2} & \circled{1} & \tfrac{1}{2} & 0 & 4 \\
3 & 2 & 0 & 1 & 12 \\
-2 & -3 & 0 & 0 & 0
\end{pmatrix} \leftarrow
$$

$$
\begin{array}{cccc}
x & y & u & v
\end{array}
$$

$$
\cong
\begin{pmatrix}
\tfrac{1}{2} & 1 & \tfrac{1}{2} & 0 & 4 \\
2 & 0 & -1 & 1 & 4 \\
-\tfrac{1}{2} & 0 & \tfrac{3}{2} & 0 & 12
\end{pmatrix}
\begin{matrix} y \\ v \\ f \end{matrix}
\begin{matrix} \Big\} \textit{ the new basic variables} \\ \textit{(new value of } f) \end{matrix}
$$

Let us verify that this tableau does indeed correspond to a system of equations that gives $y = 4$, and $f = 12$ in the above order, if we take y and v as basic variables. The corresponding equations are

$$\tfrac{1}{2}x + y + \tfrac{1}{2}u \qquad\qquad = 4$$

$$2x \qquad - u + v \qquad = 4$$

$$-\tfrac{1}{2}x \qquad + \tfrac{3}{2}u \qquad + f = 12$$

Taking x and u as nonbasic variables (they are zero), we do indeed get $y = 4, v = 4, f = 12$ as claimed. This is the tableau for the vertex A.

The analysis is now repeated for this tableau. On account of the negative entry, $-\frac{1}{2}$, in the last row, this value 12 for f is not an optimal solution. The new entering variable is x, on account of the negative sign in this column, the departing variable is v since $\frac{4}{2} < 4/(\frac{1}{2})$.

$$\text{pivot} \longleftarrow \begin{pmatrix} x & y & u & v & \\ \frac{1}{2} & 1 & \frac{1}{2} & 0 & 4 \\ ② & 0 & -1 & 1 & 4 \\ -\frac{1}{2} & 0 & \frac{3}{2} & 0 & 12 \end{pmatrix} \begin{matrix} \\ y \\ v \leftarrow \textit{departing variable} \\ f \end{matrix}$$

$$\underset{\textit{entering variable}}{\uparrow}$$

The entering variable, x, will replace the v in the second row. The sequence of transformations is

$$\begin{pmatrix} x & y & u & v & \\ \frac{1}{2} & 1 & \frac{1}{2} & 0 & 4 \\ ② & 0 & -1 & 1 & 4 \\ -\frac{1}{2} & 0 & \frac{3}{2} & 0 & 12 \end{pmatrix} \begin{matrix} \\ y \\ v \leftarrow \\ f \end{matrix}$$
$$\uparrow$$

$$\cong \begin{pmatrix} x & y & u & v & \\ \frac{1}{2} & 1 & \frac{1}{2} & 0 & 4 \\ ① & 0 & -\frac{1}{2} & \frac{1}{2} & 2 \\ -\frac{1}{2} & 0 & \frac{3}{2} & 0 & 12 \end{pmatrix} \leftarrow$$

$$\cong \begin{pmatrix} x & y & u & v & \\ 0 & 1 & \frac{3}{4} & -\frac{1}{4} & 3 \\ 1 & 0 & -\frac{1}{2} & \frac{1}{2} & 2 \\ 0 & 0 & \frac{5}{4} & \frac{1}{4} & 13 \end{pmatrix} \begin{matrix} \\ y \\ x \\ f \end{matrix}$$

This is the final tableau. The basic variables are x and y, assuming the values 2 and 3, respectively. This is the tableau for the vertex B. The value of f is 13, it is the maximum value possible under the given restrictions.

Example 2

Let us return to Example 2 of the previous section, the linear programming problem that had a nonunique solution.   We now analyze it in terms of entering and departing variables in order to see how the final tableau can be interpreted for such a problem.

The function is $5x + 2y$ and constraints are

$$4x + 5y \leq 10$$

$$5x + 2y = 10 \qquad (x \geq 0, y \geq 0)$$

$$3x + 8y = 12$$

These lead to the introduction of the slack variables u, v, w:

$$4x + 5y + u \qquad\qquad = 10$$

$$5x + 2y \qquad + v \qquad\quad = 10$$

$$3x + 8y \qquad\qquad + w \quad = 12$$

$$-5x - 2y \qquad\qquad\quad + f = \;\; 0$$

$$(x \geq 0, y \geq 0, u \geq 0, v \geq 0, w \geq 0)$$

The simplex tableaux become

$$
\begin{array}{ccccc}
x & y & u & v & w \\
\end{array}
$$

$$
\left(
\begin{array}{ccccc|c}
4 & 5 & 1 & 0 & 0 & 10 \\
\boxed{5} & 2 & 0 & 1 & 0 & 10 \\
3 & 8 & 0 & 0 & 1 & 12 \\
-5 & -2 & 0 & 0 & 0 & 0 \\
\end{array}
\right)
\begin{array}{l}
u \\
v \leftarrow \textit{departing variable} \\
w \\
f \\
\end{array}
$$

$\uparrow$
entering variable

$$
\begin{array}{ccccc}
x & y & u & v & w \\
\end{array}
$$

$$
\cong
\left(
\begin{array}{ccccc|c}
4 & 5 & 1 & 0 & 0 & 10 \\
\boxed{1} & \frac{2}{5} & 0 & \frac{1}{5} & 0 & 2 \\
3 & 8 & 0 & 0 & 1 & 12 \\
-5 & -2 & 0 & 0 & 0 & 0 \\
\end{array}
\right)
\leftarrow
$$

$\uparrow$

$$
\begin{array}{ccccc}
x & y & u & v & w \\
\end{array}
$$

$$
\cong
\left(
\begin{array}{ccccc|c}
0 & \frac{17}{5} & 1 & -\frac{4}{5} & 0 & 2 \\
1 & \frac{2}{5} & 0 & \frac{1}{5} & 0 & 2 \\
0 & \frac{34}{5} & 0 & -\frac{3}{5} & 1 & 6 \\
0 & 0 & 0 & 1 & 0 & 10 \\
\end{array}
\right)
\begin{array}{l}
u \\
x \\
w \\
f \\
\end{array}
$$

This is a final tableau. It leads to a maximum value of 10 for f. It occurs at $x = 2, u = 2, w = 6$. Since y is a nonbasic variable, $y = 0$. Thus we have arrived at the optimal solution $x = 2, y = 0$ (the point D in the figure in Example 2, Section 7-2) with $f = 10$. When the simplex method is applied to a problem that contains many solutions it stops, as here, as soon as it finds one optimal solution. Can it also lead to the other solutions in a manner other than that illustrated in Example 2 of the previous section? Can it lead to the other solutions directly from a tableau? The answer to both questions is yes.

Observe that in the final tableau the coefficient of y, a nonbasic variable, is 0 in the row for f. Each coefficient of a nonbasic variable in this row indicates the rate at which f increases as that variable is increased. Thus, making y an entering variable neither increases or decreases f. We use y as an entering variable, the departing variable is then seen to be u. We get the following sequence:

$$\begin{array}{ccccc} x & y & u & v & w \\ \end{array}$$

$$\begin{pmatrix} 0 & \boxed{\tfrac{17}{5}} & 1 & -\tfrac{4}{5} & 0 & 2 \\ 1 & \tfrac{2}{5} & 0 & \tfrac{1}{5} & 0 & 2 \\ 0 & \tfrac{34}{5} & 0 & -\tfrac{3}{5} & 1 & 6 \\ 0 & 0 & 0 & 1 & 0 & 10 \end{pmatrix} \begin{array}{l} u \leftarrow \textit{departing variable} \\ x \\ w \\ f \end{array}$$

$$\uparrow$$
entering variable

$$\cong \begin{pmatrix} 0 & \boxed{1} & \tfrac{5}{17} & -\tfrac{4}{17} & 0 & \tfrac{10}{17} \\ 1 & \tfrac{2}{5} & 0 & \tfrac{1}{5} & 0 & 2 \\ 0 & \tfrac{34}{5} & 0 & -\tfrac{3}{5} & 1 & 6 \\ 0 & 0 & 0 & 1 & 0 & 10 \end{pmatrix} \leftarrow$$

$$\uparrow$$

$$\begin{array}{ccccc} x & y & u & v & w \\ \end{array}$$

$$\cong \begin{pmatrix} 0 & 1 & \tfrac{5}{17} & \tfrac{4}{17} & 0 & \tfrac{10}{17} \\ 1 & 0 & -\tfrac{2}{17} & \tfrac{5}{17} & 0 & \tfrac{30}{17} \\ 0 & 0 & -2 & 1 & 1 & 2 \\ 0 & 0 & 0 & 1 & 0 & 10 \end{pmatrix} \begin{array}{l} y \\ x \\ w \\ f \end{array}$$

This tableau leads to the optimal solution $x = \tfrac{30}{17}, y = \tfrac{10}{17}$ (the point C in the figure in Example 2, Section 7-2) with f having a maximum value of 10. All points on the line between $(2, 0)$ and $(\tfrac{30}{17}, \tfrac{10}{17})$ would also, as we know, be optimal solutions.

The existence of multiple solutions is indicated in the final tableau by zero coefficients for nonbasic variables in the last row. (There can be numerous vertices that are optimal solutions in a linear programming problem involving many variables.) These solutions can be found in the manner illustrated here. The existence of such multiple solutions could be of interest if factors not included in the model generate a preference for one or more of those solutions over the others. If one is only interested in a single solution the final tableau provides it.

Example 3

We now discuss the geometrical interpretation of the simplex method for the problem having no solution, the objective function being unbounded. Let us return to Example 3 of the previous section.

Maximize the function $x_1 + 4x_2$ subject to the constraints

$$\begin{aligned} -4x_2 + x_2 &\leq 2 \\ 2x_1 - x_2 &\leq 1 \end{aligned} \qquad (x_1 \geq 0, x_2 \geq 0)$$

The simplex tableaux are

$$
\begin{array}{cccc}
x_1 & x_2 & x_3 & x_4 \\
\end{array}
$$

$$
\begin{pmatrix}
-4 & \circled{1} & 1 & 0 & 2 \\
2 & -1 & 0 & 1 & 1 \\
-1 & -4 & 0 & 0 & 0
\end{pmatrix}
\begin{array}{l}
x_3 \leftarrow \\
x_4 \\
f
\end{array}
$$

$$
\cong
\begin{pmatrix}
-4 & 1 & 1 & 0 & 2 \\
-2 & 0 & 1 & 1 & 3 \\
-1 & -4 & 0 & 0 & 0
\end{pmatrix}
$$

$$
\begin{array}{cccc}
x_1 & x_2 & x_3 & x_4 \\
\end{array}
$$

$$
\cong
\begin{pmatrix}
-4 & 1 & 1 & 0 & 2 \\
-2 & 0 & 1 & 1 & 3 \\
-17 & 0 & 4 & 0 & 8
\end{pmatrix}
\begin{array}{l}
x_2 \\
x_4 \\
f
\end{array}
$$

On account of the negative signs in the pivot column, the first column, one cannot choose a pivot; there is no departing variable. What is the geometrical interpretation?

The associated system of equalities is

$$-4x_1 + x_2 + x_3 = 2$$

$$-2x_1 + x_3 + x_4 = 3$$

$$-17x_1 + 4x_3 + f = 8$$

The entering variable should be x_1, that is, the variable that is currently zero and should be increased. It can be seen that x_1 can be increased indefinitely, if x_2, x_4 (the basic variables) and f are increased. There is no limit to f.

Recall that selecting the column containing the negative entry in the last row that is largest in magnitude as pivot column is a convenience that enables us to arrive at a final tableau efficiently. At any stage any nonbasic variable can be selected as the entering variable. In a manner similar to the above, if a tableau appears having a column containing a negative last entry and all nonpositive elements above it, selecting the variable corresponding to that column as entering variable enables the objective function to increase indefinitely—there is no maximum to the objective function.

EXERCISES

Solve the following linear programming problems by means of the simplex method. Determine the basic and nonbasic variables, entering and departing variables for each tableau. Determine the optimal solution and maximum value of the objective function directly from the final tableau. [These problems were given in the previous set of exercises. Use the tableaux that you have already derived, check your previous answers.]

1. Maximize $2x + y$,

   $4x + y \leq 36$

   $4x + 3y \leq 60$

   $x \geq 0, y \geq 0$

   [Exercise 1, Section 7-3]

2. Maximize $x - 4y$,

   $x + 2y \leq 4$

   $x + 6y \leq 8$

   $x \geq 0, y \geq 0$

   [Exercise 2, Section 7-3]

3. Maximize $x + 2y + z$

   $3x + y + z \leq 3$

   $x - 10y - 4z \leq 20$

   $x \geq 0, y \geq 0, z \geq 0$

   [Exercise 6, Section 7-3]

4. Maximize $100x + 200y + 50z$,

   $5x + 5y + 10z \leq 1000$

   $10x + 8y + 5z \leq 2000$

   $10x + 5y \leq 500$

   $x \geq 0, y \geq 0, z \geq 0$

   [Exercise 7, Section 7-3]

5. Maximize $2x_1 + x_2 + x_3$,

   $x_1 + 2x_2 + 4x_3 \leq 20$

   $2x_1 + 4x_2 + 4x_3 \leq 60$

   $3x_1 + 4x_2 + x_3 \leq 90$

   $x_1 \geq 0, x_2 \geq 0, x_3 \geq 0$

   [Exercise 9, Section 7-3]

6. Maximize $x_1 + 2x_2 + 4x_3 - x_4$,

   $5x_1 + 4x_3 + 6x_4 \leq 20$

   $4x_1 + 2x_2 + 2x_3 + 8x_4 \leq 40$

   $x_1 \geq 0, x_2 \geq 0, x_3 \geq 0, x_4 \geq 0$

   [Exercise 11, Section 7-3]

Appendix

Computing

This appendix is intended for readers who will be using the computer in the course. Appendix A provides an introduction to computing while Appendix B is an introduction to the BASIC language. Linear algebra begins in Appendix C. Readers who already have experience in BASIC can commence with Appendix C.

APPENDIX A. INTRODUCTION TO COMPUTING

Charles Babbage (1792–1871), Lucasian Professor of Mathematics at Cambridge University, is recognized by many as being the father of the modern digital computer. Babbage was motivated to construct a calculating machine when a series of mathematical tables prepared for the Royal Astronomical Society proved to be full of errors. His hope was to create a machine that could handle any arithmetic problem. Although the machine was never completed to Babbage's satisfaction, it was the forerunner and model for the successful large computers that began to appear a hundred years later. Babbage's dream might have become an actuality during his own lifetime if technology had been as advanced as his ideas.

The first electromechanical computer, the Mark 1, was developed in 1944 at Harvard University by Howard Aiken. Electromechanical means that the internal computations are performed using electrical relays (rather than mechanical gears). J. P. Eckert and J. W. Mauchly, two scientists at the University of Pennsylvania, are credited with the first electronic digital computer, ENIAC, a thirty-ton machine completed in 1946. In this model, about 18,000 electron vacuum tubes replaced the electrical relays. The UNIVAC 1, which was developed in 1951 by the same scientists, was the first computer to be produced in large quantities. Since that time, computers have become much more sophisticated, capable of performing more and more diverse computations, storing more data, and carrying out operations more rapidly.

In addition to giant computers such as the IBM System/370 and the Burroughs B 6600, there are now mini-computers, which are often the size of a typewriter. It is possible for many users to either "tap" a large computer using special telephone lines or to buy their own relatively inexpensive mini-computer. The mini-computer can be effective if the demands are not too varied in nature and are comparatively light. Mini-computers are, for example, ideal for banks; a bank can have its own mini-computer especially programmed to take care of its accounts. Doctors can subscribe to a local mini-computer service that will keep track of billing procedures.

We have developed a theory of matrices and have seen applications of the theory. Adding and multiplying matrices becomes more time-consuming and tedious, the larger the matrices involved. Matrices used in the analysis of real problems are often very large and many computations are required. The need for a computer was, for example, discussed in the supermarket model in Section 1-4. Here we show how the computer can be used to carry out matrix operations. As the reader progresses through the text, he or she will develop an understanding of the immense potential of the computer and the role that it plays in modern society.

If we are going to use the computer to carry out computations, we have to communicate with it. In communicating with one another we use a language; in our case, English. Many languages have been developed to "talk" to the computer. FORTRAN (*For*mula *Trans*lation), COBOL (*C*ommon *B*usiness *O*riented *L*anguage), and BASIC (*B*eginner's *A*ll-purpose *S*ymbolic *I*nstruction *C*ode) are three such languages. There are many others. Various languages have been constructed to accomplish different purposes. FORTRAN is a scientifically oriented language that is used, for example, in the aerospace industry. COBOL was developed with business applications in mind. BASIC was originally intended as a beginner's language. It has now developed into a very powerful, versatile language used in the scientific community and in business; it is the language that we shall use throughout this text. The reader will learn BASIC through writing *programs*—sets of instructions that tell the computer to perform certain tasks.

There are various ways of feeding a program and data into a computer and getting the results back. The programs duplicated in this book were run on a *teletype* unit. The teletype unit is a typewriter-like terminal hooked up to a computer which may be located some distance away. The program and data are typed in by the user, the computer is given an instruction to execute the program, and the result is either printed out or exhibited on a cathode ray tube (CRT) display device. This type of system is called an *interactive system*. It is ideal for educational purposes and for running short programs, for it permits the user to interact with his running program. Programs can be corrected, modified, and updated at the terminal. This mode of computer usage in which more than one person can use the computer simultaneously is called *time-sharing*. Each user gets the impression that he has complete use of the machine.

We complete this introduction to computing by giving an overall view of how the computer is set up to carry out the instructions that are fed into it.

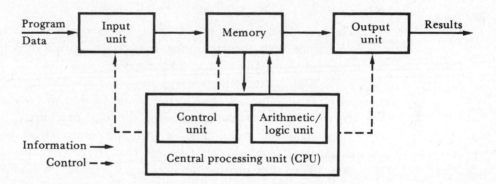

The *input*, the information that is fed into the computer, consists of a program and data. There are many input devices; one has already been mentioned—the teletype. Another common device is a *card reader*. It is designed to read punched cards. The program and data are typed on the cards using a *keypunch machine*. This method is most suitable for large programs. The cards are deposited at the computing center and run on the computer by operators. The results can be picked up later. The time between the moment when the user deposits his cards and when the output is available is called *turnaround time*. Other input devices are *paper tape readers* and *magnetic tape drives*. The paper tape reader functions in a manner similar to the card reader, reading information punched on a paper tape.   The magnetic tape drive is designed to read information that has been recorded magnetically on a coated plastic tape.

The input is stored in the *memory unit* of the machine.   The memory unit works in conjunction with two other components, the *control unit* and the *arithmetic/logic unit*.   These latter two make up the *central processing unit* (CPU) of the computer.   The arithmetic/logic unit performs operations such as those of arithmetic and comparison.   The control unit fetches instructions from memory, interprets these instructions, controls the flow of data between the memory and the arithmetic/logic unit, and directs the arithmetic/logic unit as to the operations it should perform.

Output devices include *card punches*, *high speed printers*, *cathode ray tube display units*, and *magnetic tape units*.   A *plotter* is an output device that can draw curves on paper.

APPENDIX B.   INTRODUCTION TO BASIC

The aim of this section is to give the reader preliminary "hands on" experience with the computer, if necessary, prior to commencing with programs involving vectors and matrices.

Computer systems vary; programs can be implemented through teletypes or punched cards.   It is assumed that the reader has available the necessary instructions on using the local machine.

The user communicates with the computer by giving it a set of instructions in a special language that it understands—BASIC, in our case.   The set of instructions is called a *program*.   A program consists of a sequence of *statements*, each statement corresponding to a sentence in English.   Each statement is written on a separate line and given an appropriate number.   The computer will read the statements in numerical order rather than in the order in which they appear physically in the program.   It is customary to leave gaps between the numbers so that statements can be inserted if necessary.   The numbers also enable reference to be made to specific statements within the program if the need arises.

We now discuss a variety of programs that will enable the reader to get a feel for the computer.   The programs were run on a time-sharing system using a teletype.

Read Statement   The following program reads in a given value, 3, for *x* and prints out that value.   The appropriate passwords are typed in and the terminal is ready to accept the program

Statement 5 is a READ statement. It causes the computer to read in the number 3 from DATA statement 15 as x.

5 READ X

Statement 10 is a PRINT statement that causes the computer to print out this value of x.

10 PRINT X

15 DATA 3

The last statement must always be END.

20 END

RUN causes the program to be executed.

RUN

The output (boldface type will indicate output).

3

The following program shows that the physical locations of the statements in the program are unimportant. The program is identical to the above program from the computer's point of view.

By typing NEW the user prepares the machine for a new program.

NEW

20 END
10 PRINT X
15 DATA 3
5 READ X

RUN

3

It is customary to write data statements at the end of the program. However, the above program would run if the statement had been inserted anywhere in the program before the END. The advantage of placing data statements at the end of the program is that these are statements one invariably desires to update from time to time. When these statements are at the end of the program, they are easy to pick out.

413

Addition   The following program illustrates the operation of addition. The computer reads the values 3.2 and 4.9 for *x* and *y*, respectively, adds them, calls the sum *z*, and then prints out *z*.

The values 3.2 and 4.9 are read in for x and y, respectively. These are added and the sum is called z.

```
5 READ X,Y
10 LET Z=X+Y
```

The computer will print out whatever you enclose between quotation marks.

```
15 PRINT "THE SUM IS"
```

z is not enclosed between quotation marks. The numerical value of z will be printed.

```
17 PRINT Z
```

```
20 DATA 3.2,4.9
25 END
RUN
```

THE SUM IS
8.1

This program can be used to add any two given numbers by updating statement 20.   For example, to add 3.7 and 8.2 we would type the following.

Retyping statement 20 replaces the previous statement 20 with this one.

```
20 DATA 3.7,8.2
```

```
RUN
```

THE SUM IS
11.9

Multiplication   The following program illustrates how multiplication is expressed in BASIC.   The program reads in 5.34 and 6.72, multiplies them, and then prints out the answer.

```
5 READ X,Y
```

* indicates multiplication in BASIC.

```
10 LET Z=X*Y
```

```
15 PRINT "THE PRODUCT IS"
20 PRINT Z
25 DATA 5.34,6.72
30 END

RUN
```

THE PRODUCT IS
35.8848

Division   Determine 4.532/7.634 using the computer.

```
5 READ X,Y
```

*Note the use of / for
division in BASIC.*

```
10 LET Z=X/Y
```

```
15 PRINT "THE QUOTIENT IS"
20 PRINT Z
25 DATA 4.532,7.634
30 END
```

```
RUN
```

THE QUOTIENT IS
0.59366

Exponentiation   Exponentiation can be carried out on the computer using the symbol ↑ or **, depending on what your particular system requires. We also illustrate the technique that is used to insert a statement that has been omitted through oversight.   We type the omitted statement with a suitable number that gives it the desired location in the program.

*On running the
program, there is no
output. A PRINT Y
statement should have
been included.*

```
5 READ X,N
10 LET Y=X↑N
15 DATA 5,4
20 END

RUN
```

*Typing PRINT Y
as statement 17 gives
it the desired location
in the program.*

```
17 PRINT Y
```

```
RUN
```

625

415

*Typing LIST gives
us a clean copy of the
program.*

LIST

5 READ X,N
10 LET Y=X↑N
15 DATA 5,4
17 PRINT Y
20 END

RUN

625

Correcting Mistakes   One inevitably makes mistakes in programming. These can be of the nature of a typing error or an incorrectly written program.   The computer will often tell you in some manner if and where any error occurs in your program.   The diagnostics vary from computer to computer.

In BASIC there are several standard techniques for correcting errors.

1. To correct errors that you become aware of immediately in the current statement, backspace the required number of spaces (the key is probably ←).   This deletes as many of the preceding characters as ←s typed in.
2. Retype the complete statement, in correct form, including the statement number.   This will update the statement in corrected form.
3. If a necessary statement has been omitted, type it in with a suitable number that indicates its location.   This is why we leave a space between statement numbers.   The computer will take the statements in numerical order, not according to their physical location in the program.
4. If a statement is in the wrong place, delete it by typing its statement number, hitting carriage return, and then retyping it with a correct statement number.

A program to multiply three numbers *x*, *y*, and *z* was typed in. The second statement was typed in incorrectly.

*Statement 10
should have read
10 LET P = X*Y*Z.
The computer indicates
the mistake.*

5 READ X,Y,Z
10 LET P=X*Y:Z
ILLEGAL VERB AT LINE 10

416

Retyping statement 10 correctly replaces the incorrect statement with the correct one. The remainder of the program is then typed.

10 LET P=X*Y*Z

15 PRINT "THE PRODUCT IS"
20 PRINT P
25 DATA 3,5,2
30 END

RUN

THE PRODUCT IS
  30

The above program is not easy to follow because of the correction. By typing LIST one can get a clear copy of the correct program.

LIST

5 READ X,Y,Z
10 LET P=X*Y*Z
15 PRINT "THE PRODUCT IS"
20 PRINT P
25 DATA 3,5,2
30 END

RUN

THE PRODUCT IS
  30

EXERCISES

The following programs are to be run on the computer. Predict what the output will be in each case before you run the program.

1. 5 PRINT "COMPUTER"
   10 END

2. 5 READ A
   10 PRINT "A="A
   15 DATA −3
   20 END

3. 5 LET X=8*2/4
   10 PRINT X
   15 END

4. 5 READ A,B,C
   10 LET D=A+B−C
   15 PRINT "D="D
   20 DATA 1,2,3
   25 END

5. 5 LET X=3*4
10 PRINT X
15 END

6. 5 READ P,Q
10 LET R=P*Q
15 PRINT "THE PRODUCT OF
"P"AND"Q"IS"R
20 DATA 3,4
25 END

7. 5 LET X=4+2
10 PRINT X
15 END

8. 5 READ X,Y,Z
10 LET P=(X+Y)/Z
15 PRINT "THE VALUE OF P IS"
17 PRINT P
20 DATA 4,6,5
25 END

9. 5 READ X,N
10 LET Y=X↑N
15 PRINT Y
20 DATA 2,4
25 END

10. 5 READ X,Y,N
10 LET Z=(X+Y)↑N
15 PRINT Z
20 DATA −1,3,4
25 END

11. 5 READ A,B,C,D
10 LET E=(A−B)*(C↑D)
15 PRINT "E="E
20 DATA 3,1,2,3
25 END

Use the computer to evaluate the following.

12. 1.36×5.72

13. 7.345×6.38

14. $9.81 \times 1.36 \times 7.2$

15. $2.26/5.41$

16. $9.23/7.21$

17. $6.23 \times 4.56/9.73$

18. 5^6

19. $(3.2)^{2.1}$

20. $(5.34)^{4.6}/9.1$

Hierarchy of Operations   There is a hierarchy of operations in BASIC: exponentiation, multiplication or division, and then addition or subtraction. We now give examples that illustrate this hierarchy.

Example 1 $x^3 + 2$ can be written X↑3+2 in BASIC. The computer will cube x before performing the addition.

Example 2   Write $3x^2$ in BASIC.
It is written $3 * X↑2$. The computer will perform the X↑2 first and then the $*$.

Example 3   Write $(5x)^{27}$ in BASIC.
This must be written $(5 * X)↑27$ to indicate that multiplication must be performed first. $5 * X↑27$ would have represented $5(x^{27})$.

Example 4   Write $4x^2/(2x+1)$ in BASIC.
This is written $4 * X↑2/(2 * X + 1)$.

When the computer has an apparent choice between multiplication and addition it will perform multiplication first. Exponentiation will be performed before either multiplication or addition. Whenever it becomes necessary to perform these operations in some other sequence, as in Example 3 above, one uses parentheses to indicate the desired sequence of operations. Whenever in doubt, use parentheses to indicate the desired order.

The Pythagorean Theorem   If x and y are the lengths of the short sides of a right triangle, the Pythagorean theorem tells us that the length h of the hypotenuse will be given by

$$h = (x^2 + y^2)^{1/2}$$

The following program can be used to determine the length of the hypotenuse given the sides x and y. It was executed for $x = 4.2, y = 5.6$.

5 READ X,Y

There is no ambiguity as to order of operations here. The computer will always perform exponentiation before addition. The hierarchy of operations is exponentiation, multiplication or division, addition or subtraction.

10 LET H=(X↑2+Y↑2)↑(1/2)

```
15 PRINT H
20 DATA 4.2,5.6
25 END
RUN
 7
20 DATA 3.76,7.83
RUN
 8.68599
```

Many systems have a square root function SQR( ). On such systems statement 10 can be written

10 LET H=SQR(X↑2+Y↑2)

EXERCISES

Write each of the following BASIC expressions in standard algebraic form.

419

1. 2*X−3

2. 3*X↑2−2*X

3. X↑2+3*X

4. 5*X−3/4+X↑−4

5. 3↑X−4*X/3

6. (3*X↑2−4)/(2*X−3)

7. (5/X−2*X)↑(X+2)

8. (4*X−X↑4)/(3*X↑4−3)

Write each of the following algebraic expressions in BASIC.

9. $2x^3 + 3x$

10. $4x^2 − 3x + 2$

11. $(x+2)/(x−3)$

12. $(x^2+3)/(2x^2−4x)$

13. $(4x^4 − 7/x − 3)^2$

14. $(5x^3 + 7x^2 + 2x − 4)^3/(2x^3 − 3x)$

Use the computer to evaluate the following.

15. $\dfrac{9 \times 7.3}{4.6} + \dfrac{1.7}{2.3 \times 1.5}$

16. $\dfrac{5.2 \times 1.6 − 7.2}{5.3}$

17. $\dfrac{(6.5)^3 − 7.1}{2.3}$

18. $\dfrac{4.73 − 5.89}{(1.73)^4} + (2.71)^{1/4}$

19. $(−7.23)^3 − \dfrac{1.57}{7.23} − (5.1)^{1/2}$

20. $3x^2 + 5x$ when $x = 2.1$

21. $5x^3/3 + 7x − 3$ when $x = 1.2$

22. $(x^2+3)/(x+2)$ when $x = 1.7$

23. $(4x^3 + 7x)^3/(2x+1)$ when $x = 1.2$

24. $1.7x + 2.3/x + 4/x^2$ when $x = −1.3$

25. Determine the third side of the triangles in the figure below.

(a)

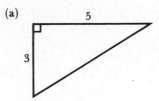

(b)

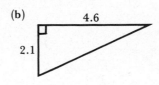

(c)

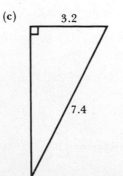

(d)

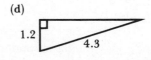

Loops Programs can be written to run through a desired sequence of numbers. The following program illustrates the use of a loop in a program for finding the first eight powers of 2.

This statement reads in the values 2 and 8 for x and n respectively.

```
5 READ X,N
```

Statements 10 to 25 form a loop. i takes on all values between 1 and n in turn. The computer runs through statements 15 and 20 for each value of i.

```
10 FOR I=1 TO N
15 LET Y=X↑I
20 PRINT Y
25 NEXT I
```

```
30 DATA 2,8
35 END
RUN
 2
 4
 8
 16
 32
 64
 128
 256
```

Consider the equation $y = x^2 - 1$. For every value of x there will be a corresponding y. For example, when $x = 1, y = 0$; when $x = 1, y = 3$; and so on. This program uses a loop to print out the values of y corresponding to integer values of x running from 1 through 5.

Statements 5 through 20 form a loop. This enables integer values from 1 to 5 to be fed into $y = x^2 - 1$, the corresponding value of y being printed out in each case. Statement 15 causes the two characters X= to be printed out followed by the current value of x, etc.

```
5 FOR X=1 TO 5
10 LET Y=(X↑2)−1
15 PRINT "X="X,"Y="Y
20 NEXT X
25 END
```

RUN

X= 1 ,Y= 0
X= 2 ,Y= 3
X= 3 ,Y= 8
X= 4 ,Y= 15
X= 5 ,Y= 24

Some systems may require a comma or a semicolon between items to be printed.

15 PRINT "X=";X;",Y=";Y

Input Statement   Time-sharing systems allow the user to interact with his running program.   He can, as we have seen, edit his program while it is running.   Another feature is that he can feed in data while it is running. To do this one uses an input statement.   The program below enables the user to feed any desired value of x into the equation

$$y = \frac{x^2 - 3x + 2}{x - 2}$$

The computer calculates the corresponding value of y.

4 PRINT "GIVE X VALUE"
5 INPUT X
10 LET Y=(X↑2−3*X+2)/(X−2)
15 PRINT "X="X,"Y="Y
20 END

*Input statement 5
causes the computer to
request a value for x.
The user types it in,
3.76 in this case.
The computer then
executes the program
with this x value.*

RUN

GIVE X VALUE
? 3.76
X= 3.76 ,Y= 2.76

RUN

GIVE X VALUE
? 4.6
X= 4.6 ,Y= 3.6

RUN

GIVE X VALUE
? −3.276
X=−3.276 ,Y=−4.276

EXERCISES

The following programs are to be run on the computer. Predict what the output will be in each case before you run the program.

1. 5 READ X,N
   10 FOR K=1 TO N
   15 LET Y=X*K
   20 PRINT Y
   25 NEXT K
   30 DATA 3,4
   35 END

2. 5 READ X
   10 FOR K=1 TO 10
   15 LET Y=X+K
   20 PRINT Y
   25 NEXT K
   30 DATA 1
   35 END

3. 5 FOR X=1 TO 4
   10 LET Y=X↑2
   15 PRINT X",",Y
   20 NEXT X
   25 END

4. 5 FOR X=−2 TO 2
   10 LET Y=2*X−1
   15 PRINT "X=",X",Y="Y
   20 NEXT X
   25 END

5. 5 FOR X=1 TO 5
   10 LET Y=(X/2+1)*2
   15 PRINT X",",Y
   20 NEXT X
   25 END

6. 5 READ X
   10 LET Y=2*X↑3−2
   15 PRINT Y
   20 DATA 3
   25 END

7. 5 READ X
   10 LET Y=(X/2−3)/(5*X−2)
   15 PRINT X",",Y
   20 DATA 8
   25 END

8. 5 READ X
   10 LET Y=2↑X+2/X
   15 PRINT Y
   20 DATA −2
   25 END

9. 5 READ X,Y
   10 LET Z=X*Y+3↑X/2
   15 PRINT Z
   20 DATA 4,2
   25 END

10. 5 READ X,Y
   10 LET Z=X+Y↑2−X/Y+2
   15 PRINT Z
   20 DATA 4,2
   25 END

11. Compute
   a) the first 10 powers of 2.3
   b) the first 12 powers of 1.72

12. Compute the squares of the first five positive integers.

13. Compute the cubes of the first six positive integers.

14. $y = 2x^2 + 1$. Compute the values of y corresponding to integer values of x from -3 to 3.

15. $y = \sqrt{x^3 + 2}$. Compute the values of y corresponding to
   a) integer values of x from 1 to 5
   b) integer values of x from 10 to 15

16. $y = \sqrt{2x^2 + 3x - 1} + \dfrac{2x^2 - 1}{x + 2}$ Use the input statement in a program to compute values of y corresponding to

a) $x = 2$ **b)** $x = 3.5$ **c)** $x = 4.76$ **d)** $x = -2.74$

17. $y = \dfrac{4x^3 - 2x^2 + 3x - 1}{x + 2}$.   Use the input statement in a program that determines values of y corresponding to

a) $x = 1$ **b)** $x = 4.7$ **c)** $x = -3.2$ **d)** $x = 5.7$ **e)** $x = 11.85$

18. Write a program to add the first ten positive integers.

19. Wtite a program to add $\frac{1}{2}, \frac{1}{3}, \frac{1}{4}, \ldots, \frac{1}{9}, \frac{1}{10}, \frac{1}{11}, \frac{1}{12}$.

20. Write a program that will compute the first six powers of the first five positive integers.   (The program will involve two loops, one inside the other.   Such loops are called *nested loops*.)

APPENDIX C.   MATRICES AND VECTORS (SECTION 1–3)

There are various techniques for reading in and printing out matrices.   In this section we discuss these various approaches used in later programs.

MAT READ and MAT PRINT Statements.*†   Many computer systems have built in matrix functions that enable matrices to be read in and printed out as arrays.   The following program reads in and then prints out the matrix

$$\begin{pmatrix} 1 & 2 & 3 \\ 4 & 0 & -4 \end{pmatrix}$$

Statement 5 is an example of a dimension statement. Here it tells the computer to expect an array, called A, consisting of 2 rows and 3 columns.

5 DIM A(2,3)

---

† Some computer systems do not have these built-in matrix commands.   Users who do not have these commands available can modify the programs given in the text by using techniques given in Appendix H.

Statement 10 commands the computer to read in the matrix A. It will scan the remainder of the program until it gets to the data in statement 25, then read in these data as a matrix A.

10 MAT READ A

The computer will print out whatever you enclose between quotation marks. This statement is not essential but makes the printout neater.

15 PRINT "THE MATRIX IS"

Tells the computer to print out A.

20 MAT PRINT A

Note the way the data are typed in; the first row of A is followed by the second row.

25 DATA 1,2,3,4,0,−4

The last statement must always be END.

30 END

You tell it to execute the program by typing RUN.

RUN

The output. In this text, boldface type will be used to indicate output.

THE MATRIX IS
1   2   3
4   0   −4

Using the MAT READ statement as above, the dimension of every matrix read in has to be given in a dimension statement. On many systems one cannot feed variables into the dimension function. Thus the dimension statement may need updating as the program is used for matrices of varying sizes. The following MAT READ statement is useful in that the dimensions of a matrix can be included in a DATA statement.

A is a M × N matrix. M and N are read in from statement 25.

The DATA statements are the only ones that need be updated here.

```
 5 READ M,N
10 MAT READ A(M,N)
15 PRINT "THE MATRIX IS"
20 MAT PRINT A
25 DATA 2,3
30 DATA 1,2,3,4,0,-4
35 END

RUN

THE MATRIX IS
 1     2     3
 4     0    -4
```

MAT INPUT Statement   The MAT INPUT statement enables the user to feed in the desired matrix while the program is running, rather than through DATA statements as in the case of MAT READ.

Very often it is a matter of personal preference which of the two MAT statements one uses. If the data set is large it is usually preferable to use READ. INPUT, on the other hand, allows the user to supply data after seeing partial results.

The MAT INPUT statement causes the computer to request a 2 × 3 matrix. The user types in the elements of the 2 × 3 matrix. The computer then executes the program with this matrix as data. Some systems may require input thus:
? 1, 2, 3
? 4, 0, −4
The semicolon in the MAT PRINT statement will lead to a more compressed output.

```
 5 DIM A (2,3)
10 MAT INPUT A
15 PRINT "THE MATRIX IS"
20 MAT PRINT A;
25 END

RUN

? 1, 2, 3, 4, 0, -4


THE MATRIX IS
 1  2   3
 4  0  -4
```

TAB Function for Matrix Output   If the output matrix has elements consisting of various significant figures the MAT PRINT statement might not, on some systems, lead to a tidy output—the columns of the matrix might not be properly aligned.   When such an output is expected, one can use the following subroutine to print out the matrix.

The TAB (X) function causes the teletype to "tab" to column X for the current value of X and there print the current value of a_{ij}.

```
            5 READ M,N
           10 MAT READ A(M,N)
           15 PRINT "THE MATRIX IS"
          ⎛20 FOR I=1 TO M
          ⎜25 FOR J=1 TO N
Subroutine⎜30 PRINT TAB (J*8−8);A(I,J) ;
          ⎨35 NEXT J
          ⎜40 PRINT
          ⎝45 NEXT I
           50 DATA 2,3
           55 DATA 1.2, 2.3456, 6, 2.7314, 7, 2
           60 END

RUN

THE MATRIX IS
1.2             2.3456          6
2.7314          7               2
```

Vectors   Vectors are row or column matrices.   They can be interpreted as matrices in programs.   They can also be handled as one-dimensional arrays.   We illustrate this approach with the following program that performs scalar multiplication.

Multiply the vector $\mathbf{a}(-1, 2, 3, 0, 4)$ by 2.

```
            5  PRINT "THE SCALAR MULTIPLE IS"
```

This statement limits the program to the scalar multiplication of 5-tuples. If the vector is not a 5-tuple it should be changed accordingly.

```
           10  DIM A(5),B(5)
```

```
           15  READ C
```

These statements read in a component of $\mathbf{a}$, multiply it by the scalar, then move on to the next component and repeat.

```
           20  FOR I=1 TO 5
           25  READ A(I)
           30  LET B(I)= C*A(I)
           35  PRINT B(I);
           40  NEXT I
```

The scalar.

```
           45  DATA 2
```

The vector.

50 DATA −1,2,3,0,4

55 END

RUN

THE SCALAR MULTIPLE IS
−2   4   6   0   8

EXERCISES

The following programs are to be run on the computer. What will the output be in each case?

1. 5 DIM A(2,4)
    10 MAT READ A
    15 PRINT "THE MATRIX A IS"
    20 MAT PRINT A;
    25 DATA 0,1,−2,4,−3,6,7,8
    30 END

2. 5 DIM A(4,1)
    10 MAT READ A
    15 PRINT "A="
    20 MAT PRINT A;
    25 DATA −1,2,3,4
    30 END

3. 5 DIM A(1,3)
    10 MAT READ A
    15 MAT PRINT A;
    20 DATA −1,2,3
    40 END

    5 DIM A (2,3)
    20 DATA −1,2,3,4,5,1

4. Write programs to print out the following matrices.

a) $\begin{pmatrix} 1 & 2 \\ 3 & 4 \end{pmatrix}$

b) $\begin{pmatrix} 1 \\ 2 \\ 3 \end{pmatrix}$

c) $\begin{pmatrix} -1 & 0.3 & 4 & 6 \end{pmatrix}$

d) $\begin{pmatrix} 1.32 & -76 & 2 \\ 0 & 1 & 4 \end{pmatrix}$

e) $\begin{pmatrix} 1 & 2 & -3.6 \\ 7.2 & 8 & 9 \\ 28 & 307 & 21 \end{pmatrix}$

5. Write a program to print out the matrix

$$\begin{pmatrix} 1 & 2 & 3 \\ 4 & 5 & 6 \end{pmatrix}$$

By updating statements in this program print out each of the following matrices in turn.

a) $\begin{pmatrix} 1 & 2.3 & 4 \\ 5 & 0 & 7 \end{pmatrix}$ b) $\begin{pmatrix} 1 & 2 \\ 3 & 4 \\ -1 & 2 \end{pmatrix}$ c) $\begin{pmatrix} 1 & 3 & 5 & 6 \\ 7 & 1 & 2 & 4 \end{pmatrix}$

The following programs contain errors. Find them and correct them. Run each program to investigate if and in what manner your system diagnoses each error. Then correct your error.

6. 5 DIM A(2,3)
10 MAT READ A
15 PRINT "THE MATRIX IS"
20 MAT PRINT A;
25 DATA 1,2,3,4,5
30 END

7. 5 DIM A(2,2
10 MAT READ A
15 MAT PRINT A;
20 DATA 1,2,5,7
25 END

8. 5 DIM A(1,4)
10 PRINT "THE MATRIX IS"
15 MAT PRINT A;
20 DATA 5,−1,3.2,6
25 END

9. 5 DIM A(2,2)
10 DATA 1,2,3,4
15 MAT PRINT A
20 MAT READ A
25 END

10. The statement PRINT A(I, J) prints the specified element of a matrix. For example, PRINT A(2, 3) prints out the element in the second row and third column of A. Write out a program that reads in the matrix $A = \begin{pmatrix} 1 & 2 & 3 \\ 4 & 5 & 6 \end{pmatrix}$ and prints out A(2, 3).

11. Let $A = \begin{pmatrix} 1 & 2 & 4 & 5 \\ 0 & 1 & -1 & 2 \end{pmatrix}$. Write a program to print out A(2, 2) and A(2, 4) on a single line.

APPENDIX D. MULTIPLICATION OF A MATRIX BY A SCALAR (SECTION 1–3)

The following is a program written to perform the scalar multiplication $3\begin{pmatrix} 1 & 2 & 3 \\ -3 & 4 & 8 \end{pmatrix}$.

5 PRINT "THE SCALAR MULTIPLE IS"

We tell the computer that it will be handling two 2 × 3 matrices, called A and B.

10 DIM A(2,3),B(2,3)

We call the scalar c and the matrix that is read in A.

15 READ C
20 MAT READ A

The matrix A is multiplied by the scalar c. The resulting matrix is called B.

25 MAT B=(C)*A

; can be used to compress the matrix output.

30 MAT PRINT B;

All the data could have been placed on one line. The computer would read the first number in the data as c, the remaining numbers as making up A.

35 DATA 3
40 DATA 1,2,3,−3,4,8

45 END

RUN

THE SCALAR MULTIPLE IS
```
 3      6      9
−9     12     24
```

EXERCISES

The following programs are run. What will the output be for each program?

1. 5 DIM A(2,2),B(2,2)
   10 READ C
   15 MAT READ A
   20 MAT B=(C)*A
   25 MAT PRINT B;
   30 DATA 2
   35 DATA 1,−1,2,3
   40 END
2. 5 DIM A(1,2),B(1,2)
   10 READ C
   15 MAT READ A
   20 MAT B=(C)*A
   25 PRINT "THE SCALAR MULTIPLE IS"
   30 MAT PRINT B;
   35 DATA −1,3,2
   40 END

3. Write programs to perform the following multiplications.

a) $2\begin{pmatrix} 1 & -2 \\ 3 & 4 \end{pmatrix}$ 　　 b) $-3\begin{pmatrix} 2 & -4 & 6 \\ 3.2 & 78 & 1 \end{pmatrix}$ 　　 c) $4\begin{pmatrix} 1 \\ 2 \\ -3 \\ 4 \end{pmatrix}$

4. Write a program to execute $4\begin{pmatrix} 1 & 2 \\ 3 & 4 \end{pmatrix}$. Update your program to perform the following operations in turn.

a) $3\begin{pmatrix} -1 & 0 \\ 2 & 4 \end{pmatrix}$ 　　 b) $4\begin{pmatrix} 1 & 2 \\ 3 & 4 \\ 5 & 6 \end{pmatrix}$ 　　 c) $-3(\,1 \quad 2 \quad 4 \quad 1\,)$

5. Modify the program for scalar multiplication given in this section to contain input statements.

6. Write a program that multiplies every element in the third row of the matrix

$$A = \begin{pmatrix} 1 & 2 & 3 \\ 0 & -1 & 2 \\ 2 & 1 & -1 \\ 1 & 2 & -3 \end{pmatrix}$$

by 3 to give a matrix B. All the other rows of B are identical to the corresponding rows of A. We shall use this type of program and the program in Exercise 7 as subroutines in later programs.

7. Write a program that interchanges the third and fifth rows of the matrix

$$A = \begin{pmatrix} 1 & 3 & -1 & 4 \\ 2 & 0 & -1 & 4 \\ 5 & 3 & 2 & -2 \\ 0 & 4 & 1 & 2 \\ 3 & -1 & 2 & 0 \end{pmatrix}$$

Multiplying every element in a row of a matrix by a nonzero scalar and interchanging rows are examples of *elementary matrix transformations*. We shall use these transformations in solving systems of equations.

8. Let

$$A = \begin{pmatrix} 3 & -1 & 2 & 4 \\ 0 & 2 & 2 & 1 \\ 1 & 3 & 4 & 5 \end{pmatrix}$$

a) Use your program from Exercise 6 to divide the second row of A by 2.
b) Use your program from Exercise 7 to interchange the first and third rows of A.

APPENDIX E.   ADDITION OF MATRICES (SECTION 1–3)

We will add the matrices

$$A = \begin{pmatrix} 1 & 3 & 7 \\ 9 & -9 & 0.8 \\ 0 & 5 & -6 \\ 0.5 & -3 & 6 \end{pmatrix} \quad \text{and} \quad B = \begin{pmatrix} 0 & 9 & -7 \\ 9 & 4 & 34 \\ 7 & -6 & 8 \\ 1 & 0 & 8 \end{pmatrix}$$

5 PRINT "THE SUM IS"

We tell the computer to expect to be working with three 4 × 3 matrices A, B, and C.

10 DIM A(4,3),B(4,3),C(4,3)

The computer reads in the matrices A and B from the data statements and calls their sum C.

15 MAT READ A
20 MAT READ B
25 MAT C=A+B

30 MAT PRINT C;
35 DATA 1,3,7,9,−9,.8,0,5,−6,.5,−3,6
40 DATA 0,9,−7,9,4,34,7,−6,8,1,0,8
45 END

RUN

THE SUM IS

1	12	0
18	−5	34.8
7	−1	2
1.5	−3	14

EXERCISES

The following programs are run. Give the output that will be obtained for each program.

1. 5 DIM X(2,1),Y(2,1),Z(2,1)
  10 MAT READ X
  15 MAT READ Y
  20 MAT Z=X+Y

```
25 MAT PRINT Z;
30 DATA 1,2,3,4
35 END
```

2.
```
5 PRINT "C, THE SUM OF A AND B, IS"
10 DIM A(2,3),B(2,3),C(2,3)
15 MAT READ A
20 MAT READ B
25 MAT C=A+B
30 MAT PRINT C;
35 DATA −1,2,3,4,−3,5,7,2,0,1,−3,2
40 END
```

3. Write programs to perform the following matrix additions (if possible) on the computer.

a) $\begin{pmatrix} 1 & 2 \\ 3 & 4 \end{pmatrix} + \begin{pmatrix} 0 & 1 \\ 4 & 6 \end{pmatrix}$ b) $\begin{pmatrix} 1 & 2 & -3 \\ 4 & 1 & 2 \end{pmatrix} + \begin{pmatrix} 0 & 1 & 2 \\ -1 & 3 & 4 \end{pmatrix}$

c) $\begin{pmatrix} 1 \\ 2 \end{pmatrix} + (\ 3 \quad 4\ )$ (Investigate to see how the computer tells you that this addition is not possible.)

d) $\begin{pmatrix} 1 & 3 \\ 2 & 4 \end{pmatrix} + \begin{pmatrix} 3 & 4 \\ -1 & 2 \end{pmatrix} + \begin{pmatrix} 4 & 6 \\ -1 & 3 \end{pmatrix}$

e) $(\ 1 \quad 2 \quad 3 \quad 4\ ) + (\ 0 \quad -1 \quad 4 \quad 6\ ) + (\ 7 \quad 3 \quad 1 \quad 2\ )$

4. Write programs to evaluate the following on the computer.

a) $3\begin{pmatrix} 1 & 2 \\ 3 & 4 \end{pmatrix} + \begin{pmatrix} 1 & 0 \\ -1 & 2 \end{pmatrix}$ b) $4\begin{pmatrix} 3 & 1 \\ 4 & 1 \end{pmatrix} + 5\begin{pmatrix} 1 & 3 \\ 3 & -1 \end{pmatrix}$

c) $-1(\ 1 \quad 3 \quad 4\ ) + 2(\ 3 \quad 5 \quad -1\ ) - 4(\ 2 \quad 4 \quad 6\ )$

5. Modify the given program for adding matrices to use input statements.

6. Write a program to determine a matrix B whose second row is the second row of $A = \begin{pmatrix} 1 & 2 & 3 & -1 \\ 1 & 1 & -1 & 2 \\ 2 & 0 & -2 & 4 \end{pmatrix}$ minus the first row of A. The first and third rows of B are identical to those of A.

   Note that this creates from matrix A a matrix B which has a zero in its $(2, 1)$ location. Write a program to determine a matrix C whose third row is the third row of B minus twice the first row of B. The first and second rows of C are identical to those of B. This creates from B a matrix C which has a zero in its $(3, 1)$ location. In this manner we can use matrix operations (elementary matrix transformations) to create zeros in certain locations in columns. We shall soon see how useful these operations are in solving systems of equations.

7. Let A be a 3×6 matrix and let B be the 3×3 matrix whose columns are the last three columns of A. Write a program that will print out B given A. Test your program on

$$A = \begin{pmatrix} 1 & 2 & 3 & 4 & 1 & 0 \\ 0 & 1 & 2 & -1 & 0 & 4 \\ 3 & 1 & -2 & 4 & 1 & -3 \end{pmatrix}$$

B will be the matrix

$$\begin{pmatrix} 4 & 1 & 0 \\ -1 & 0 & 4 \\ 4 & 1 & -3 \end{pmatrix}$$

We shall use this program later as a subroutine in a program for determining the inverse of a matrix.

8. Let A be a 5×4 matrix. Write a program that will scan the last four elements of the second column to determine the element with the largest absolute value and then interchange the row containing this element with the second row. Test your program on the matrix

$$A = \begin{pmatrix} 1 & 2 & -1 & 3 \\ 0 & 2 & 0 & 4 \\ 1 & 3 & 2 & 1 \\ 2 & -4 & 0 & 1 \\ 3 & 2 & 1 & 1 \end{pmatrix}$$

We shall use this type of program later as a subroutine in a program to evaluate determinants. [BASIC has an absolute value function ABS(X).]

APPENDIX F. MULTIPLICATION OF MATRICES (SECTION 1–4)

We will perform the following matrix multiplication using the computer:

$$\begin{pmatrix} 4 & 7 \\ -6 & -89 \\ 0 & 2 \end{pmatrix} \begin{pmatrix} 4 & -1 \\ 7 & 3 \end{pmatrix}$$

5 PRINT "THE PRODUCT IS"

We are going to call the above matrices A and B. Their product C will be a 3×2 matrix. Here we see the need to be able to predict the shape of the product matrix.

10 DIM A(3,2),B(2,2),C(3,2)

15 MAT READ A
20 MAT READ B

This is the statement for multiplying the matrices A and B and naming their product C.

25 MAT C=A*B

30 MAT PRINT C;
35 DATA 4,7,−6,−89,0,2
40 DATA 4,−1,7,3
45 END

RUN

THE PRODUCT IS

65	17
−647	−261
14	6

EXERCISES

Give the output that would be obtained on running each of the following programs.

1. 5 DIM A(2,1),B(1,3),C(2,3)
10 MAT READ A
15 MAT READ B
20 MAT C=A*B
25 MAT PRINT C;
30 DATA 1,0,−1,2,3
35 END

2. 5 DIM A(2,2),B(2,1),C(2,1)
10 MAT READ A
15 MAT READ B
20 MAT C=A*B
25 PRINT "THE PRODUCT MATRIX IS"
30 MAT PRINT C;
35 DATA 1,2 0,−1
40 DATA 0,1
45 END

3. If $A = \begin{pmatrix} 1 & 2 & -1 \\ 3 & 1 & 0 \\ 4 & 1 & 1 \end{pmatrix}$, $B = \begin{pmatrix} 2 & 0 \\ -1 & 7 \\ -4 & 5 \end{pmatrix}$, $C = \begin{pmatrix} 2 & 3 & 1 \\ -1 & 4 & 2 \\ -5 & 1 & 3 \end{pmatrix}$,

435

write programs to determine

a) AB b) AC c) CA d) BA

e) A^2 f) ACB g) $A^2 BC$

if these products exist.

4. Write a program to evaluate the following.

a) $3\begin{pmatrix} 1 & 2 & 4 \\ -1 & 3 & 2 \end{pmatrix}\begin{pmatrix} 1 & 0 \\ 2 & 1 \\ 3 & 4 \end{pmatrix} + 2\begin{pmatrix} 1 & 2 \\ 3 & 1 \end{pmatrix} + 4\begin{pmatrix} 2 & 0 \\ 1 & 3 \end{pmatrix}$

b) $-1\begin{pmatrix} 2 & 3 \\ 4 & 1 \end{pmatrix}\begin{pmatrix} 1 & 3 \\ 2 & 4 \end{pmatrix}\begin{pmatrix} 1 & 0 \\ 1 & 1 \end{pmatrix}$

c) $2(1 \quad 2) + 3(1 \quad 4) + 5(1 \quad 3)\begin{pmatrix} 1 & 2 \\ 3 & 4 \end{pmatrix}$

5. Write a program that can be used to compute powers of a given square matrix. Use your program to compute the first 20 powers of the matrix $\begin{pmatrix} 1 & 0.2 \\ 3 & -4 \end{pmatrix}$.

6. Write a program that can be used to determine the transpose of a given matrix. Your computer system may have a built-in transpose function TRN(A). Investigate.

Use your program to determine the transpose of the matrix

$\begin{pmatrix} 1 & 2 \\ 3 & 4 \\ -1 & 2 \end{pmatrix}$

7. Write a program to determine the trace of a matrix. Use your program to determine the traces of the following matrices:

a) $\begin{pmatrix} 1 & 2 & 0 \\ 0 & -1 & 3 \\ 4 & 1 & 2 \end{pmatrix}$ b) $\begin{pmatrix} -1 & 2 & 3 & 1 \\ 4 & 2 & 1 & 3 \\ 1 & 2 & 4 & 1 \\ 2 & 1 & 3 & 4 \end{pmatrix}$

APPENDIX G.   SIGNIFICANT FIGURES

There is a limit to the number of digits in a given number that a computer will accept. For example, the Burroughs B 5500 computer at the University of Denver will accept 11 significant figures. Thus, if 7.234568107984 were fed into this machine, it would recognize it as 7.2345681079, cutting the number off at this point. All computations on this machine are rounded off to 11 figures, and any BASIC output is to 7 figures. Thus, if the answer to a computation were 234.76987621, the machine would print this answer as 234.7699. Therefore, when many computations involving numbers of

11 or more significant figures are performed on this machine, errors, called *round-off errors*, can occur.

We illustrate these concepts with the following example.

Example 1

If $A = \begin{pmatrix} 0 & 1 & 2 \\ 0 & 3 & 4 \\ 5 & 0 & 0 \end{pmatrix}$, determine $A^2, A^4, A^8,$ and A^{16}.

The following program was run on a Burroughs B5500.

```
5 DIM A(3,3),M(3,3)
10 MAT READ A
15 FOR I=1 TO 4
20 MAT M=A*A
25 PRINT "POWER:";2↑I
30 MAT PRINT M;
35 MAT A=M
40 NEXT I
45 DATA 0,1,2,0,3,4,5,0,0
50 END
RUN
POWER:   2
   10    3    4
   20    9   12
    0    5   10
POWER:   4
  160   77   116
  380  201   308
  100   95   160
POWER:   8
  66460    38817   60836
 167980    98921  155268
  68100    41995   66460
POWER:  16
  1.508034E+10   8.974402E+9   1.411336E+10
  3.835445E+10   2.282632E+10  3.589761E+10
  1.610617E+10   9.588613E+9   1.508034E+10
```

This loop squares the matrix, substitutes the square for the matrix, then squares that matrix, etc.

This BASIC output is to seven significant figures. However, the computer will accept and perform computations to 11 figures.

$E + 10$ means $\times 10^{10}$. Hence this element is 1.411336×10^{10}.

The computation above is to 11 figures and the output is to 7 figures. The accurate value of A^{16} is in fact

$$\begin{pmatrix} 1.5080342860E+10 & 8.974402097E+9 & 1.4113359076E+10 \\ 3.835445118E+10 & 2.282632356E+10 & 3.589760839E+10 \\ 1.610617210E+10 & 9.588612795E+9 & 1.508034286E+10 \end{pmatrix}$$

437

Compare the elements in the $(1, 1)$ location in both cases. To seven figures, the computer gives 15,080,340,000; the correct number is 15,080,342,860. The computer is in error by 2,860. It can be seen that errors due to round-off occur in each location of A^{16} given by the computer.

Such errors can be crucial—the elements in the above matrix could have been sums of money and the mathematics could have involved a budget for an industrial company or an educational institution! Various techniques have been developed to minimize round-off errors as much as possible. For example, when one has a choice as to the order in which certain operations are to be performed, one can select an order that minimizes round-off errors. Much research has gone into this aspect of computing. The reader should be aware of the limitations of the computer in this respect. Some techniques that are used to cut round-off errors to a minimum are presented in Section 2-4.

EXERCISES

Write a program to read in and print out the following matrix:

$$\begin{pmatrix} 1.44446 & 1.444446 & 1.4444446 & 1.44444446 \\ 1.444444446 & 1.4444444446 & 1.44444444446 & 1.444444444446 \end{pmatrix}$$

Discuss from your output (if you have access to a computer):

a) The number of significant figures to which your computer prints out numbers in BASIC.

b) The number of significant figures to which the computer actually rounds off. The computer will accept a certain number of figures, perform the computations to this number, and then round off to a smaller number in the output. For example, on a computer that prints out to six significant figures and computes to eight significant figures, the output would be

$$\begin{pmatrix} 1.44446 & 1.44445 & 1.44445 & 1.44444 \\ 1.44444 & 1.44444 & 1.44444 & 1.44444 \end{pmatrix}$$

APPENDIX H. PROGRAMS FOR SYSTEMS WITHOUT BUILT-IN MATRIX COMMANDS

Some computer systems do not have the built-in matrix commands that were used in the previous programs. This section is intended for the users of such systems.

Matrix Read-In and Output The following program reads in and prints out the matrix $\begin{pmatrix} 1 & 2 & 3 \\ 4 & 0 & -4 \end{pmatrix}$.

m is number of rows;
n is number of columns.

```
5 READ M,N

10 FOR I=1 TO M
15 FOR J=1 TO N
20 READ A(I,J)
25 NEXT J
30 NEXT I
35 FOR I=1 TO M
40 FOR J=1 TO N
45 PRINT TAB(J*8-8);A(I,J);
50 NEXT J
52 PRINT
55 NEXT I
57 DATA 2,3,1,2,3,4,0,-4
60 END
RUN

1    2    3
4    0   -4
```

2 × 3 matrix. ————————————————————————————————— *Matrix elements.*

Two loops are *nested*. Statements 10 through 30 feed in the matrix, the inside loop scanning each row, the outside loop then leading to the next row. These statements can be used to read in a matrix on a system that does not have MAT READ.

Statements 35 through 55 cause the elements to be printed out in matrix form.

The TAB(X) function causes the teletype to "tab" to column X for the current value of X and there print the current value of a_{ij}. These statements can be used to print out a matrix on a system that does not have MAT PRINT.

Programs in the text will be written using matrix commands. Users who do not have these functions available can modify the programs by using programs provided in this section.

Scalar Multiplication of a Matrix We shall write a program to perform the following scalar multiplication: $3\begin{pmatrix} 1 & 2 & 3 \\ -3 & 4 & 8 \end{pmatrix}$

```
5 READ M,N,C
10 FOR I=1 TO M
15 FOR J=1 TO N
20 READ A(I,J)
```

This statement multiplies the i, jth element of the matrix A by the scalar.

```
25 LET A(I,J)=C*A(I,J)
```

```
25 NEXT J
30 NEXT I
35 FOR I=1 TO M
40 FOR J=1 TO N
45 PRINT TAB(J*8−8);A(I,J);
50 NEXT J
52 PRINT
55 NEXT I
60 DATA 2,3,3,1,2,3,−3,4,8
65 END
```

2 × 3 matrix. ——— *The scalar.*

```
RUN
```

```
 3      6      9
−9     12     24
```

Addition of Matrices   Add the matrices $A = \begin{pmatrix} 1 & 2 & 3 \\ -1 & 2 & 4 \end{pmatrix}$ and

$B = \begin{pmatrix} 0 & -1 & 4 \\ 2 & 3 & 1 \end{pmatrix}$.

```
5 READ M,N
```

The matrix A is fed in.

```
10 FOR I=1 TO M
15 FOR J=1 TO N
20 READ A(I,J)
```

```
25 NEXT J
30 NEXT I
```

```
35 FOR I=1 TO M
40 FOR J=1 TO N
42 READ B(I,J)
```

This statement adds the i, jth elements of A and B; this becomes the i, jth element of C.

```
45 LET C(I,J)=A(I,J)+B(I,J)
```

```
50 NEXT J
55 NEXT I
60 FOR I=1 TO M
65 FOR J=1 TO N
```

```
70 PRINT TAB(J*8−8);C(I,J);
75 NEXT J
80 PRINT
85 NEXT I
90 DATA 2,3,1,2,3,−1,2,4,0,−1,4,2,3,1
95 END
```

2 × 3 matrices. ─────────────

Matrix A   Matrix B

```
RUN

   1    1    7
   1    5    5
```

Multiplication of Matrices   The following program performs the matrix multiplication $\begin{pmatrix} -4 & 7 \\ -6 & -80 \\ 0 & 2 \end{pmatrix} \begin{pmatrix} 4 & -1 \\ 7 & 3 \end{pmatrix}$.

The first matrix is m × n, the second matrix p × q.

```
5 READ M,N,P,Q,

10 FOR I=1 TO M            1,3
15 FOR J=1 TO N            1,3
20 READ A(I,J)
25 NEXT J
30 NEXT I
35 FOR I=1 TO P            1,3
40 FOR J=1 TO Q            1,3
45 READ B(I,J)
50 NEXT J
55 NEXT I
60 FOR I=1 TO M
65 FOR J=1 TO Q
```

These statements determine the i, jth element of the product matrix C by multiplying the ith row of A by the jth column of B in the appropriate manner.

```
66 LET C(I,J)=0
70 FOR K=1 TO N
75 LET C(I,J)=C(I,J)+A(I,K)*B(K,J)
80 NEXT K
85 NEXT J

90 NEXT I
92 PRINT "THE PRODUCT MATRIX IS"
95 FOR I=1 TO M
100 FOR J=1 TO Q
```

441

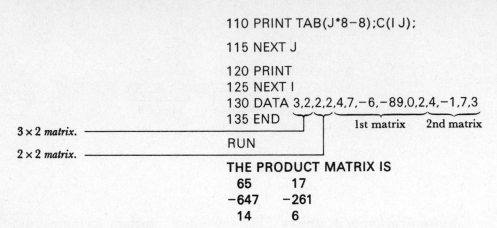

```
110 PRINT TAB(J*8−8);C(I J);
115 NEXT J

120 PRINT
125 NEXT I
130 DATA 3,2,2,2,4,7,−6,−89,0,2,4,−1,7,3
135 END

RUN
```

3 × 2 *matrix.* ⎯⎯⎯⎯⎯⎯⎯⎯⎯⎯
2 × 2 *matrix.* ⎯⎯⎯⎯⎯⎯⎯⎯⎯⎯

1st matrix   2nd matrix

```
THE PRODUCT MATRIX IS
  65      17
−647    −261
  14       6
```

Powers of a Matrix   We shall write a program to determine and print out the first 20 powers of the matrix $\begin{pmatrix} 1 & 0.2 \\ 3 & -4. \end{pmatrix}$.

```
5 READ D,N
10 FOR I=1 TO N
15 FOR J=1 TO N
20 READ A(I,J)
22 LET B(I,J)=A(I,J)
25 NEXT J
30 NEXT I
32 FOR P=1 TO D−1
35 FOR I=1 TO N
40 FOR J=1 TO N

42 LET C(I,J)=0
45 FOR K=1 TO N
50 LET C(I,J)=C(I,J)+A(I,K)*B(K,J)
52 NEXT K
55 NEXT J
60 NEXT I

66 PRINT "A↑";P+1;"="
67 FOR I=1 TO N
68 FOR J=1 TO N
75 PRINT TAB(J*8−8);C(I,J);
76 LET B(I,J)=C(I,J)
80 NEXT J
85 PRINT
90 NEXT I
91 PRINT
```

This loop multiplies A by the kth power of A, B, to get the next higher power.

442

```
92 PRINT
95 NEXT P
100 DATA 20,2,1,.2,3,−4
105 END
```

Powers. ──────────────────────────────── 2×2 *matrix.*

```
A↑ 2=
  1.599999   −.5999999
−9   16.59999

A↑ 3=
−.1999998   2.719997
 40.79998   −68.19995

A↑ 4=
  7.959997   −10.91999
−163.7999   280.9597

A↑ 5=
−24.79997   45.27192
 679.0795   −1156.598
(etc.)
```

APPENDIX I.   ECHELON FORM OF A MATRIX (SECTION 2–2)

Determine an echelon form for the matrix $\begin{pmatrix} 1 & 1 & 1 & 3 \\ 1 & -1 & -2 & -5 \\ 2 & 3 & 1 & 5 \end{pmatrix}$.

```
10 PRINT "ECHELON FORM IS"
15 READ N
20 READ M
25 MAT READ A(N,M)
30 LET L=0
35 FOR K=1 TO N
```

These statements scan the lth column from a_{kl} down to determine a nonzero element. The row containing this element is made into the kth row by interchanging rows. If such a nonzero element does not exist, we start scanning the following column.

```
40 LET L=L+1
45 IF L>M THEN 155
50 IF A(K,L)<>0 THEN 105
55 FOR I=K+1 TO N
60 IF A(I,L)=0 THEN 95
65 FOR J=L TO M
70 LET B=A(K,J)
75 LET A(K,J)=A(I,J)
80 LET A(I,J)=B
85 NEXT J
90 GO TO 105
95 NEXT I
100 GO TO 40
```

*Divide the kth row
by a_{kl} to make the
first nonzero element
in this row unity.*

```
105 IF A(K,L)=1 THEN 125
110 FOR P=L TO M
115 LET A(K,P)=A(K,P)/A(K,L)
120 NEXT P
```

*These statements
make the elements
below the first
nonzero element in
the kth row zero.*

```
125 FOR I=K+1 TO N
126 IF A(I,L)=0 THEN 145
130 FOR J=L TO M
135 LET A(I,J)=A(I,J)−A(I,L)*A(K,J)
140 NEXT J
145 NEXT I
150 NEXT K
155 MAT PRINT A;
```

Kind of matrix.

```
160 DATA 3,4
```

Matrix elements.

```
165 DATA 1,1,1,3,1,−1,−2,−5,2,3,1,5
```

```
170 END
```

```
RUN
```

```
ECHELON FORM IS
 1    1    1    3
 0    1   1.5   4
 0    0    1    2
```

EXERCISES

1. Use the above program to determine echelon forms for the following matrices.
    a) the augmented matrix of the system of equations in Example 2, Section 2-2:

$$\begin{pmatrix} 1 & 1 & 1 & 3 \\ 2 & 3 & 1 & 5 \\ 1 & -1 & -2 & -5 \end{pmatrix}$$

    b) the augmented matrix of the system of equations in Example 4, Section 2-2:

$$\begin{pmatrix} 1 & -2 & 3 & 1 \\ 3 & -4 & 5 & 3 \\ 2 & -3 & 4 & 2 \end{pmatrix}$$

    c) the augmented matrix of the system of equations in Example 6, Section 2-2:

$$\begin{pmatrix} 1 & -1 & 2 & 3 \\ 2 & -2 & 5 & 4 \\ 1 & 2 & -1 & -3 \\ 0 & 2 & 2 & 1 \end{pmatrix}$$

This example checks that the program works when a diagonal element becomes zero.

d) the augmented matrix of the system of equations in Example 7, Section 2-2:

$$\begin{pmatrix} 1 & -1 & 1 & 2 \\ -2 & 2 & -2 & -4 \\ 2 & 1 & 2 & 2 \\ 1 & 1 & 1 & 0 \end{pmatrix}$$

This example checks that the program works when an early row becomes zero.

2. Check your answers to Exercise 5, Section 2-2 using the above program.

3. Modify the above program to get a printout after each elementary matrix transformation with a statement describing the transformation, e.g., ROW 3–(2) ROW 1. The technique of inserting statements to obtain printouts at various locations in a program can be used to analyze programs that are not running correctly, that is, for debugging programs.

4. Modify the above program to give the reduced echelon form of a matrix that is used in the method of Gauss-Jordan Elimination. (Exercise 19 Section 2-2.)

5. Check your answers to Exercises 6–18, Section 2-2 using the computer.

6. Assume that the given matrix is the augmented matrix to a system of linear equations having a unique solution. Extend the echelon form program to print out not only the echelon form but the solution to the system of equations. Use your program to check your answers to Exercises 6, 7, 8, and 10, Section 2-2.

7. Let $A = \begin{pmatrix} 2 & 1 & 3 \\ 1 & 2 & 3 \\ -1 & 3 & 1 \end{pmatrix}$

Write a program that performs the following operations on A, printing out each matrix in the sequence and stating the operations that have been performed.

$$\begin{pmatrix} 2 & 1 & 3 \\ 1 & 2 & 3 \\ -1 & 3 & 1 \end{pmatrix} \xrightarrow{\text{Row} \leftrightarrow \text{Row 2}} \begin{pmatrix} 1 & 2 & 3 \\ 2 & 1 & 3 \\ -1 & 3 & 1 \end{pmatrix}$$

$$\xrightarrow[\substack{R2-2R1 \\ R3+R1}]{} \begin{pmatrix} 1 & 2 & 3 \\ 0 & -3 & -3 \\ 0 & 5 & 4 \end{pmatrix}$$

APPENDIX J. INVERSE OF A MATRIX USING GAUSS-JORDAN ELIMINATION (SECTION 2–5)

Determine the inverse of the matrix $\begin{pmatrix} 0 & 3 & 3 \\ 1 & 2 & 3 \\ 1 & 4 & 6 \end{pmatrix}$.

```
  4 READ N
  5 LET M=2*N
  6 MAT READ B[N,N]
  7 FOR I=1 TO N
  8   FOR J=1 TO M
  9     IF J>N THEN GOTO 12
 10     LET A[I,J]=B[I,J]
 11     GOTO 16
 12     IF J=N+1 THEN GOTO 15
 13     LET A[I,J]=0
 14     GOTO 16
 15     LET A[I,J]=1
 16   NEXT J
 17 NEXT I
 20 FOR K=1 TO N
 25   IF A[K,K]<>0 THEN GOTO 76
 30   FOR I=K+1 TO N
 35     IF A[I,K]=0 THEN GOTO 70
 40     FOR J=K TO M
 45       LET B=A[K,J]
 50       LET A[K,J]=A[I,J]
 55       LET A[I,J]=B
 60     NEXT J
 65     GOTO 76
 70   NEXT I
 75   GOTO 170
 76   IF A[K,K]=1 THEN GOTO 95
 77   LET Y=A[K,K]
 80   FOR P=K TO M
 85     LET A[K,P]=A[K,P]/Y
 90   NEXT P
 95   FOR I=1 TO N
100     IF I=K THEN GOTO 120
102     IF A[I,K]=0 THEN GOTO 120
103     LET Z=A[I,K]
105     FOR J=K TO M
110       LET A[I,J]=A[I,J]-Z*A[K,J]
```

```
115    NEXT J
120    NEXT I
125 NEXT K
130 FOR P=1 TO N
135    FOR Q=N+1 TO M
137       LET R=Q−N
140       LET B[P,R]=A[P,Q]
145    NEXT Q
150 NEXT P
155 PRINT "THE INVERSE IS"
160 MAT PRINT B
165 GOTO 185
170 PRINT "THE INVERSE DOES NOT EXIST"
175 DATA 3
180 DATA 0,3,3,1,2,3,1,4,6
185 END

RUN
```

THE INVERSE IS

0	2	−1
1	1	−1
−.666667	−1	1

EXERCISES

1. Modify the above program to get a printout after each transformation with a statement describing the transformation.
2. Find out whether your system has a built-in matrix inverse function, MAT B = INV(A). Use this function in a program to determine the inverse of the matrix $\begin{pmatrix} 1 & 2 \\ 3 & 4 \end{pmatrix}$.

APPENDIX K. DETERMINANT (SECTION 5–3)

Evaulate $\begin{vmatrix} 1 & 2 & 3 & 1 \\ 2 & 4 & 3 & 1 \\ 1 & 3 & 4 & 2 \\ 2 & 5 & 6 & 4 \end{vmatrix}$.

We use the elimination method.

```
15 READ N
20 MAT READ A[N,N]
```

S will be used as a counter to determine the number of row interchanges.

```
 25 LET S=0
 30 FOR K=1 TO N−1
 35    IF A[K,K]<>0 THEN GOTO 100
 40    FOR I=K+1 TO N
 45      IF A[K,I]=0 THEN GOTO 85
 50      FOR J=K TO N
 55        LET B=A[K,J]
 60        LET A[K,J]=A[I,J]
 65        LET A[I,J]=B
 70      NEXT J
 75      LET S=S+1
 80      GOTO 100
 85    NEXT I
 90    PRINT "THE DETERMINANT IS 0"
 95    GOTO 170
100    FOR I=K+1 TO N
105      IF A[I,K]=0 THEN GOTO 130
110      LET Z=A[I,K]/A[K,K]
115      FOR J=K TO N
120        LET A[I,J]=A[I,J]−Z*A[K,J]
125      NEXT J
130    NEXT I
135 NEXT K
140 LET D=1
145 FOR P=1 TO N
150    LET D=D*A[P,P]
155 NEXT P
157 LET D=D*((−1))↑S
160 PRINT "THE DETERMINANT IS" ;D
165 DATA 4
166 DATA 1,2,3,1,2,4,3,1,1,3,4,2,2,5,6,4
170 END
RUN
```

THE DETERMINANT IS 4

EXERCISES

1. Modify the above program to get a printout after each transformation with a statement describing the transformation.

2. Check your answers to the exercises of Section 5-3 using the computer.

3. Modify the above program to include a pivoting technique that reduces round-off errors.

APPENDIX L. EIGENVALUES AND EIGENVECTORS (SECTION 6–7*)

Determine the dominant eigenvalue and a corresponding eigenvector of the matrix $\begin{pmatrix} 5 & 4 & 2 \\ 4 & 5 & 2 \\ 2 & 2 & 2 \end{pmatrix}$.

Let us use 10 iterations, using the initial vector $(-4, 2, 6)$

```
  5 DIM A[3,3],X[3,1],Y[3,1],Z[3,1]
 10 MAT READ A
 15 MAT READ X
 20 PRINT "THE INITIAL VECTOR IS"
 25 MAT PRINT X
 30 FOR I=1 TO 10
 32   PRINT "ITERATION";I
 35   MAT Y=A*X
 37   MAT PRINT Y
 40   LET K=1
 45   FOR J=2 TO 3
 50     IF ABS(Y[K,1])>=ABS(Y[J,1]) THEN GOTO 0060
 55     LET K=J
 60   NEXT J
 65   MAT Y=(1/ABS(Y[K,1]))*Y
 70   MAT X=Y
 71   PRINT "THE ADJUSTED VECTOR IS"
 72   MAT PRINT X
 73   MAT Z=A*X
 75   LET S=0
 76   LET T=0
 78   FOR L=1 TO 3
 79     LET S=S+X[L,1]*Z[L,1]
 80     LET T=T+X[L,1]*X[L,1]
 81   NEXT L
 82   LET E=S/T
 83   PRINT "APPROX EIGENVALUE";E
 84   PRINT
 85 NEXT I
 90 DATA 5,4,2,4,5,2,2,2,2
 95 DATA −4,2,6
100 END

    RUN

THE INITIAL VECTOR IS
```

−4
2
6
ITERATION 1

0
6
8

THE ADJUSTED VECTOR IS
0
.75
1
APPROX EIGENVALUE 5

ITERATION 2
5
5.75
3.5

THE ADJUSTED VECTOR IS
.869565
1
.608696
APPROX EIGENVALUE 9.88889
⋮
ITERATION 8
9.99999
10
5

THE ADJUSTED VECTOR IS
1
1
.5
APPROX EIGENVALUE 10

ITERATION 9
10
10
5

THE ADJUSTED VECTOR IS
1
1
.5
APPROX EIGENVALUE 10

EXERCISES

1. Use this program to answer Exercises 1–8 of Section 6-7.

2. Extend the program to determine further eigenvalues, and a corresponding eigenvector in each case, for a symmetric matrix (Theorem 2, Section 6-7). Use your program to check your answers to Exercises 9–13, Section 6-7.

3. Determine the dominant eigenvalue and a corresponding eigenvector for each of the following matrices:

a)
$$\begin{pmatrix} 7.2 & -0.6 & -2.4 & 1.8 \\ 3.8 & 5.6 & 0.4 & 4.2 \\ -4.4 & -0.8 & 3.8 & -3.6 \\ -1.4 & 2.2 & 0.8 & 2.4 \end{pmatrix}$$

b)
$$\begin{pmatrix} 9.6 & 3.2 & 0.8 & 2.4 \\ 6 & 5 & -1 & 5 \\ -5.2 & 0.6 & 7.4 & -5.8 \\ -2 & -1 & -3 & 3 \end{pmatrix}$$

c)
$$\begin{pmatrix} 6.4 & 8.8 & 3.2 & 5.6 \\ 9.2 & -2.6 & -5.4 & 3.8 \\ -6.8 & -0.6 & 4.6 & -6.2 \\ 0.4 & -6.2 & -5.8 & -1.4 \end{pmatrix}$$

d)
$$\begin{pmatrix} -2366 & 1470 & 525 & -318 \\ -3498 & 2170 & 783 & -474 \\ -2166 & 1362 & 445 & -270 \\ -1908 & 1200 & 390 & -236 \end{pmatrix}$$

e)
$$\begin{pmatrix} 1990 & -1200 & -495 & 290 \\ 2786 & -1680 & -693 & 406 \\ 2358 & -1422 & -585 & 342 \\ 1940 & -1170 & -480 & 280 \end{pmatrix}$$

APPENDIX M. SIMPLEX METHOD (SECTION 7–3*)

The initial tableau is
$$\begin{pmatrix} 1 & 1 & 1 & 1 & 0 & 0 & 100 \\ 3 & 2 & 4 & 0 & 1 & 0 & 200 \\ 1 & 2 & 0 & 0 & 0 & 1 & 150 \\ -3 & -5 & -8 & 0 & 0 & 0 & 0 \end{pmatrix}.$$

Determine the final tableau. (This is the initial tableau of Example 1, Section 7-3.)

Remarks statements, REM are non-executable; useful in large programs.

```
10 REM SIMPLEX METHOD - - OUTPUT IS FINAL TABLEAU
20 REM DATA STATEMENT 620 IS DIMENSION OF INITIAL TABLEAU
30 REM DATA STATEMENTS 630, 640, 650, 660 ARE INITIAL TABLEAU
```

```
50 READ N
60 READ M
70 MAT READ A[N,M]
80 PRINT "THE INITIAL TABLEAU IS"
```

GOSUB causes computer to begin executing subroutine at statement 550; subroutine for printing out the matrix.

```
90 GOSUB 550
```

Determine the negative entry in the first M-1 locations of the last row that is largest in magnitude.

```
100 LET T=0
110 FOR K=1 TO M−1
120    IF T−A[N,K]<=0 THEN GOTO 150
130    LET T=A[N,K]
140    LET J=K
150 NEXT K
```

Goes to 450 when there is no such negative entry in the last row.

```
160 IF T=0 THEN GOTO 450
```

Check for unbounded function.

```
170 FOR I=1 TO N−1
180    IF A[I,J]<=0 THEN GOTO 200
190    GOTO 220
200 NEXT I
210 GOTO 500
```

Determine the pivot in column j, element a_{sj}. Program will work if pivot $\leq 10^{10}$.

```
220 LET Q=10↑10
230 FOR I=1 TO N−1
240    IF A[I,J]<=0 THEN GOTO 290
250    LET B=A[I,M]/A[I,J]
260    IF B>=Q THEN GOTO 290
270    LET Q=B
280    LET S=I
290 NEXT I
300 IF Q=10↑10 THEN GOTO 530
```

```
310 IF A[S,J]=1 THEN GOTO 360
```

Divide sth row by a_{sj}, the pivot.

```
320 LET Y=A[S,J]
330 FOR P=1 TO M
340    LET A[S,P]=A[S,P]/Y
350 NEXT P
```

Create zeros in the j column.

```
360 FOR I=1 TO N
370    IF I=S THEN GOTO 430
380    IF A[I,J]=0 THEN GOTO 430
400    FOR R=1 TO M
410      LET A[I,R]=A[I,R]-A[(I,J)]*A[S,R]
420    NEXT R
430 NEXT I

440 GOTO 100
450 PRINT
460 PRINT
470 PRINT "THE FINAL TABLEAU IS"
480 GOSUB 550
490 GOTO 670
500 PRINT
510 PRINT "REGION UNBOUNDED AND OBJECTIVE FUNCTION
      UNBOUNDED"
520 GOTO 670
530 PRINT "PROGRAM CANNOT BE USED,ELEMENTS GET TOO LARGE"
540 GOTO 670
```

Subroutine used for printing out the matrix. RETURN transfers control to statement following its GOSUB.

```
550 FOR U=1 TO N
560    FOR V=1 TO M
570      PRINT TAB(V*8-8);A[U,V];
580    NEXT V
590    PRINT
600 NEXT U
610 RETURN
```

Dimension of initial tableau.

```
620 DATA 4,7
```

The initial tableau without the column that remains unchanged.

```
630 DATA 1,1,1,1,0,0,100
640 DATA 3,2,4,0,1,0,200
650 DATA 1,2,0,0,0,1,150
660 DATA -3,-5,-8,0,0,0,0
670 END

RUN
```

THE INITIAL TABLEAU IS

1	1	1	1	0	0	100
3	2	4	0	1	0	200
1	2	0	0	0	1	150
-3	-5	-8	0	0	0	0

THE FINAL TABLEAU IS

0	0	0	1	−.25	−.25	12.5
.5	0	1	0	.25	−.25	12.5
.5	1	0	0	0	.5	75
3.5	0	0	0	2	.5	475

The output matrix is likely to be large here. We use the subroutine consisting of statements 550 to 610 to print out the matrix. The output can be controlled using the TAB function.

EXERCISES

1. Modify the above program to get a printout after each transformation with a statement describing the transformation.

2. Check your answers to the exercises of Section 7-3 using the computer.

Answers to Selected Exercises

CHAPTER 1

Section 1-1

1.

3.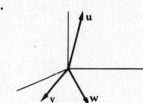

5. a) As points

$-1 \quad 0 \quad 1 \quad 2$

b) As position vectors

$-1 \qquad\qquad\qquad\qquad 2$

0

7. $a = -2, b = -1, c = 3,$

10. b) $a = 1, b = 1, c = 2.$

Section 1-2

1. a) $3(1,4) = (3,12)$

c) $(1,3)$

e) $(-3,6,9)$

g) $(-5,20,-15,10,-25)$

3. $(-2,1)$

5. $(3,5,5)$

7. $(5,2,9,7,4,3)$

9. $(6,11)$

17. a) $\begin{pmatrix} 4 \\ 2 \\ 0 \end{pmatrix}$

c) $\begin{pmatrix} 3 \\ 6 \\ -3 \end{pmatrix}$

20. a) $(-3,2)$

Section 1-3

1. $a_{11} = 1, a_{24} = -1$

2. a) $2A = \begin{pmatrix} 1 & 4 \\ 6 & 0 \end{pmatrix}$

b) $A+B = \begin{pmatrix} 0 & 4 \\ 4 & 1 \end{pmatrix}$, $A+C = \begin{pmatrix} 1 & 3 \\ 4 & 4 \end{pmatrix}$.

c) $A+2B = \begin{pmatrix} -1 & 6 \\ 5 & 2 \end{pmatrix}$, $2A+B-C = \begin{pmatrix} 1 & 5 \\ 6 & -3 \end{pmatrix}$.

3. a) $A-B = \begin{pmatrix} -2 & 3 & 5 \\ -2 & -2 & 1 \end{pmatrix}$.

4. $\begin{pmatrix} a_{12} \\ a_{22} \end{pmatrix}$

8. a) $a = 1, b = 2, c = 2, d = -3$.
   c) $a = -2, b = -1, c = 3, d = 1$.

9. a) $\begin{pmatrix} a+b & a \\ a+b+c & d \end{pmatrix} = \begin{pmatrix} 3 & -4 \\ 6 & 3 \end{pmatrix}$

Section 1-4

1. $AB = \begin{pmatrix} 5 \\ 23 \end{pmatrix}$, $AD = \begin{pmatrix} 2 & 4 & 1 \\ 8 & 18 & 1 \end{pmatrix}$, DB does not exist.

$A^2 = \begin{pmatrix} 2 & 3 \\ 6 & 11 \end{pmatrix}$, $AC+CA = \begin{pmatrix} 8 & 9 \\ 6 & 14 \end{pmatrix}$.

2. AB does not exist, $CB = (12)$, DA does not exist.

$BD = \begin{pmatrix} 1 & 3 \\ 2 & 6 \\ 3 & 9 \end{pmatrix}$.

3. a) 7

5. a) $AB+C$ does not exist   **c)** 4×2 **e)** $2EB+DA$ does not exist.

Section 1-5

1. $\sum_{k=1}^{3} a_{1k} b_{k1}$

3. $a_{21} b_{13} + a_{22} b_{23} + a_{23} b_{33}$

5. $a_{i1} b_{1j} + a_{i2} b_{2j} + a_{i3} b_{3j} + a_{i4} b_{4j} + a_{i5} b_{5j}$

7. $\sum_{k=1}^{5} a_{1k} b_{k1}$

14. a) $B = \begin{pmatrix} -1 & \vdots & -2 \\ \hline 0 & \vdots & 3 \\ 4 & \vdots & 1 \end{pmatrix}$ or $\begin{pmatrix} -1 & -2 \\ \hline 0 & 3 \\ 4 & 1 \end{pmatrix}$

AB is 2×2 or 2×1.

15. a) $AB = \begin{pmatrix} A_1 B \\ \vdots \\ A_n B \end{pmatrix}$, a matrix partitioned into the n rows $A_1 B, \ldots, A_n B$.

c) $AB = \begin{pmatrix} A_1 B^1 & \cdots & A_1 B^n \\ \vdots & & \\ A_n B^1 & \cdots & A_n B^n \end{pmatrix}$, the usual product.

16. a) $\begin{pmatrix} 1 & -R \\ 0 & 1 \end{pmatrix}$ **c)** $\begin{pmatrix} 1 & -R_1 \\ -1/R_2 & 1 + R_1/R_2 \end{pmatrix}$

17. a) $\begin{pmatrix} 1 & -(R_1 + R_2) \\ -1/R_3 & 1 + (R_1 + R_2)/R_3 \end{pmatrix}$.

Section 1-6

1. a) $\begin{pmatrix} -1 & 2 \\ 2 & -3 \end{pmatrix}$, symmetric.   **c)** $\begin{pmatrix} 3 & 2 \\ -1 & 4 \end{pmatrix}$, not symmetric.

e) $\begin{pmatrix} 2 & 4 & 7 \\ -1 & 5 & 8 \\ 3 & 6 & 9 \end{pmatrix}$, not symmetric

g) $\begin{pmatrix} 1 & -1 & 3 \\ -1 & 2 & 0 \\ 3 & 0 & 4 \end{pmatrix}$, symmetric.

i) $(\ 1 \quad 2 \quad 3\ )$, not symmetric.

3. a) 4   **c)** 0

17. a) $A = \begin{pmatrix} 0 & 1 & 0 & 0 \\ 0 & 0 & 1 & 0 \\ 0 & 0 & 0 & 1 \\ 0 & 0 & 0 & 0 \end{pmatrix}$.

18. a) a_{24}

Section 1-7

1. a) Stochastic   **b)** No

4. a) 0.02   **c)** Low density residential

6. a) 0.078   **9. a)** 0.2

10. b) (10,400   600)

15. a) (i) $(\frac{1}{3}, \frac{2}{3})$ **b) (ii)** $\begin{pmatrix} \frac{1}{3} & \frac{2}{3} \\ \frac{3}{7} & \frac{4}{7} \end{pmatrix}$

16. $(\begin{array}{cc} 59488 & 30512 \end{array})$

Section 1-8

1. a) $\begin{pmatrix} 0 & 1 & 0 \\ 0 & 0 & 1 \\ 1 & 0 & 0 \end{pmatrix}$

2.

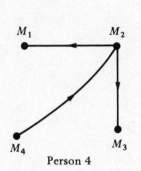

9.

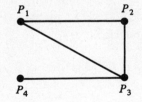

13.

M_1 M_2

M_4 M_3

Person 4

20. Hawaii has the greater connectivity.
Santa Clara and Camagüey are cities of greatest connectivity in Cuba.
All cities have same connectivity in Hawaii.

22. P_1 P_2

P_4 P_3

CHAPTER 2

Section 2-1

1. $x = \frac{3}{2}, y = -\frac{1}{2}$ **3.** No solution. **4.** Many solutions, $(x, 2 - x/2)$.

5. $x = 11, y = -4.$ **7.** $x = \frac{6}{7}, y = \frac{1}{7}.$

Section 2-2

1. a) $\begin{pmatrix} 1 & -1 \\ 2 & 1 \end{pmatrix}$ and $\begin{pmatrix} 1 & -1 & 1 \\ 2 & 1 & 3 \end{pmatrix}$

2. a) $\begin{aligned} x_1 + 2x_2 &= 1 \\ -x_1 \qquad &= 3 \end{aligned}$

3. a) $x_1 = 1, \ x_2 = 1$ **b)** $x_1 = -1, \ x_2 = -3, \ x_3 = 2$

    c) No solution.

    d) $x_1 = -x_2, \ x_2$ any value, $x_3 = 0, \ x_4 = 0$.

      i.e. $(-x_2, x_2, 0, 0)$

    f) $(3 - 2x_3, 2 - 2x_3, x_3)$.

4. a) $\begin{pmatrix} 1 & 2 & 3 & 4 \\ 0 & 1 & 4 & 2 \\ 0 & 0 & 1 & 3 \end{pmatrix}$ **c)** $\begin{pmatrix} 1 & 2 & 3 & 4 & 1 \\ 0 & 1 & 4 & 2 & 1 \end{pmatrix}$

e) $\begin{pmatrix} 1 & 2 & 3 \\ 0 & 0 & 2 \end{pmatrix}$.

5. a) In echelon form

b) $\begin{pmatrix} 1 & 0 & 1 \\ -1 & 1 & -1 \\ 0 & 2 & 2 \end{pmatrix} \cong \cdots \cong \begin{pmatrix} 1 & 0 & 1 \\ 0 & 1 & 0 \\ 0 & 0 & 1 \end{pmatrix}$

d) $\begin{pmatrix} 1 & 2 & -1 & 0 \\ 2 & 4 & 1 & 2 \\ 0 & 1 & 2 & 3 \end{pmatrix} \cong \cdots \cong \begin{pmatrix} 1 & 2 & -1 & 0 \\ 0 & 1 & 2 & 3 \\ 0 & 0 & 1 & \frac{2}{3} \end{pmatrix}$

f) $\begin{pmatrix} 1 & 2 & -1 & 1 \\ 0 & 0 & 0 & 1 \\ 0 & 1 & 1 & 2 \\ 0 & -1 & 0 & 1 \end{pmatrix} \cong \cdots \cong \begin{pmatrix} 1 & 2 & -1 & 1 \\ 0 & 1 & 1 & 2 \\ 0 & 0 & 1 & 3 \\ 0 & 0 & 0 & 1 \end{pmatrix}$

h) $\begin{pmatrix} 1 & -1 & 1 & 2 & 1 \\ 2 & -1 & 0 & 3 & 0 \\ -1 & 1 & 1 & 1 & -1 \\ 0 & 1 & 0 & 1 & 1 \end{pmatrix} \cong \cdots \cong \begin{pmatrix} 1 & -1 & 1 & 2 & 1 \\ 0 & 1 & -2 & -1 & -2 \\ 0 & 0 & 1 & \frac{3}{2} & 0 \\ 0 & 0 & 0 & 1 & -3 \end{pmatrix}$

6. $x_1 = 1, \ x_2 = 2, \ x_3 = 3$.

8. $x_1 = 2, \ x_2 = 1, \ x_3 = 1$.

9. No solution.

11. $x_1 = x_2, \ x_3 = 1$.

13. $x_1 = x_2 = 0, \ x_3 = 1, \ x_4 = 2$.

15. $x_1 = -4x_3 + 3x_4 - 2, \ x_2 = -5x_3 + 5x_4 - 6$

17. No solution.

19. a) $x_1 = 1, \ x_2 = 3, \ x_3 = 2$.

    c) $x_1 = x_3 + 3, \ x_2 = -2x_3 + 1, \ x_4 = -1$.

Section 2-3

1. a) 12, 4, 4, 4 and 8 amps.

c) 1, 1, 0 amps.

e) 2, 2, 0 amps.

2. a) 3, 2, 1 amps.

3. Minimum flow along AB is 150 v.p.h.

5. a) $P = 4, S = 3, D = 3$

c) $P_1 = 1, P_2 = 2, D_1 = S_1 = 2, D_2 = S_2 = 9.$

Section 2-4

1. $x_1 = 6, x_2 = -2, x_3 = 1.$

3. Let $y_1 = x_1, y_2 = 0.001x_2, y_3 = x_3.$

Properly scaled system is

$$
\begin{aligned}
-y_1 + 2y_2 \qquad &= \quad 0 \\
y_1 \qquad + 2y_3 &= -2 \\
y_2 + \ y_3 &= \quad 1
\end{aligned}
$$

Solution does not exist.

5. Let $y_1 = x_1, y_2 = x_2, y_3 = 0.1x_3.$

Multiply second equation by 100.

Properly scaled system is

$$
\begin{aligned}
y_2 - y_3 &= 2 \\
2y_1 + \ y_2 \qquad &= 1 \\
y_1 - 4y_2 + y_3 &= 2.
\end{aligned}
$$

$\Rightarrow x_1 = 1, \ x_2 = -1, \ x_3 = -30.$

7. Let $y_1 = x_1, y_2 = 0.001x_2, y_3 = x_3.$

Multiply second equation by 10 and third equation by 10.

Properly scaled system is

$$
\begin{aligned}
-y_1 \qquad + 2y_3 &= 1 \\
y_1 + \ y_2 \qquad &= 1 \\
y_1 + 2y_2 \qquad &= 2.
\end{aligned}
$$

$\Rightarrow x_1 = 0, \ x_2 = 1000, \ x_3 = 0.5.$

9. $x_1 = 0.1667, x_2 = -1.333, x_3 = -3.167$

11. Properly scaled system is

$$
\begin{aligned}
y_1 - 2y_2 + \ y_3 &= 2 \\
y_1 - 3y_2 + 7y_3 &= 1 \\
y_1 + \ y_2 - \ y_3 &= 4.
\end{aligned}
$$

$\Rightarrow x_1 = 3.313, \ x_2 = 0.625, \ x_3 = -0.0625.$

Section 2-5

1. $\begin{pmatrix} 1 & 0 \\ -2 & 1 \end{pmatrix}$ **3.** $\begin{pmatrix} \frac{1}{2} & -3 \\ \frac{1}{2} & 0 \end{pmatrix}$ **5.** $\begin{pmatrix} \frac{5}{6} & \frac{2}{3} & -2 \\ \frac{1}{3} & \frac{2}{3} & -1 \\ -\frac{1}{6} & -\frac{1}{3} & 1 \end{pmatrix}$

7. Inverse does not exist.

9. $\begin{pmatrix} \frac{3}{2} & -\frac{1}{2} & \frac{1}{2} \\ \frac{3}{4} & -\frac{1}{4} & -\frac{1}{4} \\ 2 & -1 & 0 \end{pmatrix}$

11. Inverse does not exist.

12. $\begin{pmatrix} x_1 \\ x_2 \end{pmatrix} = \begin{pmatrix} -\frac{1}{5} & \frac{3}{5} \\ \frac{2}{5} & -\frac{1}{5} \end{pmatrix} \begin{pmatrix} 5 \\ 10 \end{pmatrix} = \begin{pmatrix} 5 \\ 0 \end{pmatrix}$

14. $\begin{pmatrix} x_1 \\ x_2 \\ x_3 \end{pmatrix} = \begin{pmatrix} \frac{3}{4} & -\frac{1}{4} & \frac{1}{2} \\ -\frac{1}{4} & -\frac{1}{4} & \frac{1}{2} \\ -\frac{1}{4} & \frac{3}{4} & -\frac{1}{2} \end{pmatrix} \begin{pmatrix} 1 \\ 2 \\ 0 \end{pmatrix} = \begin{pmatrix} \frac{1}{4} \\ -\frac{3}{4} \\ \frac{5}{4} \end{pmatrix}$

16. $\begin{pmatrix} 1 & 2 & -1 \\ 1 & 1 & 2 \\ 1 & -1 & -1 \end{pmatrix}^{-1} = \begin{pmatrix} \frac{1}{9} & \frac{1}{3} & \frac{5}{9} \\ \frac{1}{3} & 0 & -\frac{1}{3} \\ -\frac{2}{9} & \frac{1}{3} & -\frac{1}{9} \end{pmatrix}$

Solutions are $\begin{pmatrix} \frac{22}{9} \\ -\frac{6}{9} \\ \frac{1}{9} \end{pmatrix}, \begin{pmatrix} \frac{23}{9} \\ -\frac{12}{9} \\ -\frac{1}{9} \end{pmatrix}, \begin{pmatrix} \frac{26}{9} \\ \frac{6}{9} \\ -\frac{7}{9} \end{pmatrix}.$

20. $A = \begin{pmatrix} 1 & -\frac{1}{2} \\ -2 & \frac{3}{2} \end{pmatrix}.$

Section 2-6

1. a) $a_{32} = 0.25$ **c)** electrical industry.

2. Output levels: $\begin{pmatrix} 60 \\ 40 \end{pmatrix}, \begin{pmatrix} 22\frac{1}{2} \\ 16\frac{2}{3} \end{pmatrix}, \begin{pmatrix} 15 \\ 20 \end{pmatrix}.$

5. $\begin{pmatrix} 15 \\ 24 \\ 32 \end{pmatrix}, \begin{pmatrix} 15 \\ 32 \\ 56 \end{pmatrix}, \begin{pmatrix} 30 \\ 56 \\ 48 \end{pmatrix}$

7. $D = (I - A)X = \begin{pmatrix} 0.80 & -0.40 \\ -0.50 & 0.90 \end{pmatrix} \begin{pmatrix} 8 \\ 10 \end{pmatrix} = \begin{pmatrix} 2.4 \\ 5 \end{pmatrix}$

Section 2-7

1. $x = 2.4, y = 0.4, z = 2$
3. $x = 4.51, y = 5.25, z = 2.69$
5. $x = 7, y = -0.43, z = 2.29$

CHAPTER 3

Section 3-1

1. XZ plane.   **2.** Y axis   **3.** Plane $\perp$ to xy plane through line $y = 2x$.
5. No   **7.** Yes   **9.** No   **11.** No

Section 3-2

1. $(-1, 7) = -3(1, -1) + (2, 4)$
3. $(6, 22) = 4(2, 3) + 2(-1, 5)$

5. Not a combination

7. $(2, 2, -2) = (2 - 2a_3)(1, 1, -1) - a_3(2, 1, 3) + a_3(4, 3, 1)$

9. Not a combination.

11. $(x, y) = \dfrac{(-x + 3y)}{5}(1, 2) + \dfrac{(2x - y)}{5}(3, 1)$

13. $(x, y) = -y(3, -1) + 0(2, 3) + \dfrac{(x + 3y)}{4}(4, 0)$

15. No.

17. $(x, y, z) = \dfrac{(3x + z - 2y)}{4}(1, -1, -1)$

$$+ \dfrac{(7x - 2y - 3z)}{4}(0, 1, 2) + \dfrac{(z - x - 2y)}{4}(1, 2, 1)$$

Section 3-3

10. $\mathbf{v} = a(2, 1, 0) + b(-1, 2, 0)$

Section 3-4

9. $(1, 2, 0)$ and $(2, 3, 0)$ are linearly independent, thus they form a basis for this subspace. It is of dimension 2. It is XY plane.

16. No: $(1, 2, -1) = x(1, -1, 0) + y(3, -1, 2)$ has no solution.

17. $(a, -b, 3a) = a(1, 0, 3) + b(0, -1, 0)$. Basis is $(1, 0, 3)$, $(0, -1, 0)$.

Section 3-5

1. 6   **3.** $\sqrt{5}, 5$ **5.** $\dfrac{1}{\sqrt{13}}(2, 3)$, $\dfrac{1}{\sqrt{5}}(-1, 2)$

9. $45°$ **11.** $\cos \theta = \dfrac{\sqrt{2}}{\sqrt{5}} \Rightarrow \theta = 50.7685°$

15. Any vector of type $(2b + 3c, b, c)$ **20. a)** 5   **c)** 8.3066   **e)** 6.9282

22. a) vector   **c)** no sense   **e)** no sense   **i)** scalar   **k)** vector

Section 3-6

1. $\dfrac{12}{\sqrt{29}}, \dfrac{12}{29}(2, 5)$

3. $\sqrt{5}, (1, 2, 0)$

5. $0, (0, 0, 0, 0)$. Vectors are orthogonal.

8. $(1/\sqrt{2}, b)$ for any b.

9. a) $\frac{1}{\sqrt{5}}(1,0,2),\ \frac{1}{\sqrt{5}}(-2,0,1).$

10. a) $\frac{1}{\sqrt{30}}(1,2,3,4),\ \frac{1}{\sqrt{78}}(-7,4,-3,2)$

15. a) $3/\sqrt{2}$ **c)** 2.

17. a) $10/\sqrt{3},\ 5,\ 5/\sqrt{3}.$ **c)** $5\sqrt{2},\ 5\sqrt{2},\ 0$

Section 3-7

1. Arbitrary vector: $\mathbf{v} = p + qx + rx^2.$
Two distinct sets of bases:
$$\{1, x - x^2, x^2\} \quad \text{and} \quad \{1 + x, x, x^2\}.$$

Section 3-8

1. Duration 12 years. Speed 0.8 speed of light.

3. 820.98 years.

CHAPTER 4

Section 4-1

1. a) $\begin{pmatrix} 1/\sqrt{2} & 1/\sqrt{2} \\ -1/\sqrt{2} & 1/\sqrt{2} \end{pmatrix},\ \begin{pmatrix} 3/\sqrt{2} \\ -1/\sqrt{2} \end{pmatrix}.$

    c) $\begin{pmatrix} -1 & 0 \\ 0 & -1 \end{pmatrix},\ \begin{pmatrix} -2 \\ -1 \end{pmatrix}$

3. $\begin{pmatrix} 1 & 0 \\ 0 & -1 \end{pmatrix}$

5. a) $\begin{pmatrix} 2 & 0 \\ 0 & 2 \end{pmatrix}$

7. $\begin{pmatrix} c\cos\theta & -c\sin\theta \\ c\sin\theta & c\cos\theta \end{pmatrix}$

10. $\begin{pmatrix} \cos\theta & -\sin\theta & 0 \\ \sin\theta & \cos\theta & 0 \\ 0 & 0 & 1 \end{pmatrix}$

15. $\begin{pmatrix} 5 \\ -8 \end{pmatrix} \to \begin{pmatrix} 2 \\ -4 \end{pmatrix},\ \begin{pmatrix} -1 \\ 2 \end{pmatrix} \to \begin{pmatrix} 0 \\ 2 \end{pmatrix}.$

Section 4-2

1. a) $\begin{pmatrix} -1 \\ 1 \end{pmatrix} \to \begin{pmatrix} 1 \\ 4 \\ 1 \end{pmatrix}$ **c)** $\begin{pmatrix} 1 \\ 4 \end{pmatrix} \to \begin{pmatrix} 9 \\ 11 \\ 9 \end{pmatrix}.$

3. Kernel is XZ plane, Range is y axis.

5. $(1, 2) \rightarrow (5, 4, -1)$ and $(2, -1) \rightarrow (5, -2, 3)$.

9. a) Kernel is $\begin{pmatrix} 0 \\ 0 \end{pmatrix}$, range is R^2.

    c) Kernel is $\left\{ \begin{pmatrix} -2y \\ y \end{pmatrix} \right\}$, Range is $\left\{ \begin{pmatrix} x \\ 2x \end{pmatrix} \right\}$.

    e) Kernel is $\left\{ \begin{pmatrix} z \\ -2z \\ z \end{pmatrix} \right\}$, Range is R^2.

    g) Kernel is X axis, Range is $\left\{ \begin{pmatrix} x \\ 2x \\ z \end{pmatrix} \right\}$.

16. a) Kernel is YZ plane, Range is x axis.

    c) Kernel is plane $\{(x, y, -x-y)\}$, Range is R.

Section 4-3

1. $(0, 0, 1) + x_1(1, 1, 0)$

3. $(1, 0, 0) + x_3(1, 2, 1)$

5. $(-2, -6, 0, 0) + x_3(-4, -5, 1, 0) + x_4(3, 5, 0, 1)$

Section 4-4

1. a) $\begin{pmatrix} 1 & 0 & 0 \\ 0 & 0 & 1 \end{pmatrix}$

2. b) $\begin{pmatrix} 1 & 0 & 0 \\ 0 & 0 & 0 \\ 0 & 0 & 0 \end{pmatrix}$

3. $\begin{pmatrix} 0 & 3 & 0 \\ 2 & 0 & 0 \\ 0 & 0 & -1 \end{pmatrix} \cdot \begin{pmatrix} 1 \\ 2 \\ 3 \end{pmatrix} \rightarrow \begin{pmatrix} 6 \\ 2 \\ -3 \end{pmatrix}$

5. $\begin{pmatrix} 1 & 3 & 1 & 2 \\ 1 & -2 & 2 & 1 \\ 0 & 1 & -1 & -2 \end{pmatrix} \cdot u_1 + 2u_2 - u_3 + 3u_4 \rightarrow 12v_1 - 2v_2 - 3v_3$

7. $\begin{pmatrix} 0 & 2 & 0 \\ 0 & 0 & 3 \\ 0 & 0 & 0 \end{pmatrix}$

CHAPTER 5

Section 5-1

1. a) 1   **c)** 7   **e)** 12

2. a) Singular   **c)** Non-singular   **e)** Singular

3. a) $\displaystyle\sum_{j=1}^{5}(-1)^{i+j}a_{ij}A_{ij}$

4. a) 0   **c)** 14

5. a) 3   **c)** 3   **e)** 6

7. $|A| = a_{13}A_{13} - a_{23}A_{23} + a_{33}A_{33} = \displaystyle\sum_{i=1}^{3}(-1)^{i+3}a_{i3}A_{i3}$.

Section 5-2

1. $|A| = -1$ **2.** $|A| = -4$

4. a) -5 **c)** 0

7. $1, 1, 0$

Section 5-3

1. -4 **3.** -20 **5.** -9 **7.** 0

Section 5-4

1. a) $(8, -7, 2)$ **c)** $(3, 10, 23)$ **e)** -8

5. $\frac{1}{2}\sqrt{45}$ **6.** 4

8. a) $(25, -3, -7)$

CHAPTER 6

Section 6-1

1. a) 2   **c)** 3

2. 3   **3.** 2   **5.** 3   **7.** 3   **9.** 3

14. e.g. $\{(1, \frac{1}{2}, \frac{3}{2}, 0, 2), (0, 1, \frac{9}{5}, \frac{2}{5}, \frac{4}{5})\}$.

15. e.g. $\{(1, 2, 3, 4), (0, 1, -2, -3), (0, 0, 1, \frac{3}{2})\}$.

17. e.g. $\{(1, 4, 0, -1, 5, 2, 1), (0, 1, -2, -3, 1, -4, -5), (0, 0, 18, 28, 3, 50, 58)\}$.

19. e.g. $\{(1, 0, -2, -3), (0, 1, 4, 7), (0, 0, 4, -1)\}$.

Section 6-2

1. Rank of augmented matrix = rank of matrix of coefficients.
$$= 3 \quad = \text{number of variables.}$$

Solution exists and is unique.

3. Rank of augmented matrix = rank of matrix of coefficients.
$$= 2 < \text{number of variables.}$$
Solution exists, not unique.

5. Ranks = 1. Solution exists, not unique.

7. Ranks = 3. Solution exists, is unique.

Section 6-3

1. Det = 0. Rank < 3. **3.** Det = 50. Rank = 3.

5. $\lambda = 7, x_1 = -x_2.$
$\lambda = -4, x_1 = -\frac{6}{5}x_2.$

8. $x_1 = 2, x_2 = 0, x_3 = -2.$

9. $x_1 = 2, x_2 = -1, x_3 = 3.$

11. $x_1 = -\frac{13}{5}, x_2 = -\frac{11}{5}, x_3 = -30.$

Section 6-4

1. $\begin{pmatrix} -\frac{1}{5} & \frac{2}{5} \\ \frac{3}{10} & -\frac{1}{10} \end{pmatrix}$ **3.** Inverse does not exist.

5. $\begin{pmatrix} \frac{7}{3} & -3 & -\frac{1}{3} \\ -\frac{8}{3} & 3 & \frac{2}{3} \\ \frac{4}{3} & -1 & -\frac{1}{3} \end{pmatrix}$

7. Inverse does not exist.

Section 6-5

1. $\lambda = 1, c\begin{pmatrix} 1 \\ -1 \end{pmatrix}.$ $\lambda = 6, c\begin{pmatrix} 4 \\ 1 \end{pmatrix}.$

3. $\lambda = -1, c\begin{pmatrix} 1 \\ -1 \\ 1 \end{pmatrix}.$ $\lambda = 2, c\begin{pmatrix} 0 \\ 1 \\ 1 \end{pmatrix}.$ $\lambda = 1, c\begin{pmatrix} 1 \\ 0 \\ 1 \end{pmatrix}.$

5. $\lambda = 3, c\begin{pmatrix} 0 \\ 1 \\ 1 \end{pmatrix}.$ $\lambda = 1, 1 \text{ (repeated)}, \begin{pmatrix} x_1 \\ x_2 \\ x_1 \end{pmatrix}.$

Orthonormal basis $\frac{1}{\sqrt{2}}\begin{pmatrix} 1 \\ 0 \\ 1 \end{pmatrix}, \begin{pmatrix} 0 \\ 1 \\ 0 \end{pmatrix}.$

7. $\lambda = 1, c\begin{pmatrix} 1 \\ -1 \\ 1 \end{pmatrix}.$ $\lambda = 2, c\begin{pmatrix} 0 \\ 1 \\ 1 \end{pmatrix}.$ $\lambda = 8, c\begin{pmatrix} 1 \\ 0 \\ 1 \end{pmatrix}.$

9. $\lambda = 4,\ c\begin{pmatrix} 0 \\ 1 \\ 0 \\ 1 \end{pmatrix}.\quad \lambda = 6,\ c\begin{pmatrix} 1 \\ 0 \\ 0 \\ 1 \end{pmatrix}.\quad \lambda = 2, 2 \text{ (repeated)},\ \begin{pmatrix} x_1 \\ -x_1 \\ x_3 \\ x_3 \end{pmatrix}.$

13. $\lambda = -5.$ $\mathbf{R}^3$ is eigenspace.

Section 6-6

2. $(132{,}603.34 \quad 66{,}301.67)$
   $\uparrow \qquad\qquad \uparrow$
   metro   nonmetro

4. $T = \begin{pmatrix} \frac{1}{4} & \frac{1}{2} & \frac{1}{4} \\ \frac{1}{4} & \frac{1}{2} & \frac{1}{4} \\ \frac{1}{4} & \frac{1}{2} & \frac{1}{4} \end{pmatrix}$

7. $3:3:2.\quad p = \frac{3}{8}$ **9.** $A\ 44.4\%,\ B\ 55.6\%$

Section 6-7

1. $\lambda = 8,\ c\begin{pmatrix} 1 \\ 0 \\ -1 \end{pmatrix}.$

3. $\lambda = 6,\ c\begin{pmatrix} 1 \\ 0 \\ 1 \end{pmatrix}.$

5. $\lambda = 6,\ c\begin{pmatrix} 0 \\ 0 \\ 1 \\ 1 \end{pmatrix}.$

7. $\lambda = 84.311,\ c\begin{pmatrix} 1 \\ 0.0112559 \\ -0.0112559 \\ 0.039746 \end{pmatrix}$

9. $\lambda = 10,\ c\begin{pmatrix} 2 \\ 2 \\ 1 \end{pmatrix}.\quad \lambda = 1,\ c\begin{pmatrix} -1 \\ 1 \\ 0 \end{pmatrix}.$

11. $\lambda = 12,\ c\begin{pmatrix} 1 \\ 0 \\ 1 \end{pmatrix}.\quad \lambda = 2, 2 \text{ (repeated)},\ \begin{pmatrix} x_1 \\ x_2 \\ -x_1 \end{pmatrix}.$

13. $\lambda = 10,\ c\begin{pmatrix} 2 \\ 2 \\ 1 \\ 1 \end{pmatrix}.\quad \lambda = 5,\ c\begin{pmatrix} 1 \\ 1 \\ -2 \\ -2 \end{pmatrix}.\quad \lambda = 2,\ c\begin{pmatrix} 0 \\ 0 \\ 1 \\ -1 \end{pmatrix}.\quad \lambda = 1,\ c\begin{pmatrix} 1 \\ -1 \\ 0 \\ 0 \end{pmatrix}.$

Section 6-8

1. a) $\begin{pmatrix} 7 & 13 \\ -2 & -3 \end{pmatrix}$ 　　　　　 **c)** $\begin{pmatrix} 21 & -31 \\ 14 & -21 \end{pmatrix}$

e) $\begin{pmatrix} 10 & 24 & 39 \\ 5 & 16 & 24 \\ -\frac{13}{3} & -\frac{37}{3} & -19 \end{pmatrix}$ 　　 **g)** $-\frac{1}{8}\begin{pmatrix} -32 & -22 & -2 \\ -8 & 7 & 5 \\ 16 & 3 & -23 \end{pmatrix}$

2. a) $C = \begin{pmatrix} 1 & 4 \\ -1 & 1 \end{pmatrix}, \begin{pmatrix} 1 & 0 \\ 0 & 6 \end{pmatrix}$

c) $C = \begin{pmatrix} 1 & 1 \\ -2 & 1 \end{pmatrix}, \begin{pmatrix} 2 & 0 \\ 0 & 3 \end{pmatrix}.$

e) $C = \begin{pmatrix} 1 & 0 & 1 \\ -1 & 1 & 0 \\ 1 & 1 & 1 \end{pmatrix}, \begin{pmatrix} 1 & 0 & 0 \\ 0 & 2 & 0 \\ 0 & 0 & 8 \end{pmatrix}.$

g) $C = \begin{pmatrix} 1 & 0 & 0 \\ 0 & 1 & 1 \\ 1 & 0 & 1 \end{pmatrix}, \begin{pmatrix} 1 & 0 & 0 \\ 0 & 1 & 0 \\ 0 & 0 & 3 \end{pmatrix}.$

i) $C = \begin{pmatrix} 0 & 1 & 1 \\ 1 & 0 & -1 \\ 1 & 1 & 1 \end{pmatrix}, \begin{pmatrix} 1 & 0 & 0 \\ 0 & 3 & 0 \\ 0 & 0 & 5 \end{pmatrix}.$

3. a) $C = \begin{pmatrix} 1/\sqrt{2} & 1/\sqrt{2} \\ -1/\sqrt{2} & 1/\sqrt{2} \end{pmatrix}, \begin{pmatrix} -1 & 0 \\ 0 & 3 \end{pmatrix}$

c) $C = \begin{pmatrix} 1/\sqrt{2} & 1/\sqrt{2} \\ -1/\sqrt{2} & 1/\sqrt{2} \end{pmatrix}, \begin{pmatrix} 2 & 0 \\ 0 & 4 \end{pmatrix}$

e) $C = \begin{pmatrix} 1 & 1/\sqrt{2} & 1/\sqrt{2} \\ 0 & 1/\sqrt{2} & -1/\sqrt{2} \\ 1 & 0 & 0 \end{pmatrix}, \begin{pmatrix} -2 & 0 & 0 \\ 0 & -1 & 0 \\ 0 & 0 & 2 \end{pmatrix}.$

g) $C = \begin{pmatrix} 1/\sqrt{2} & 0 & 1/\sqrt{2} \\ -1/\sqrt{2} & 0 & 1/\sqrt{2} \\ 0 & 1 & 0 \end{pmatrix}, \begin{pmatrix} -2 & 0 & 0 \\ 0 & 1 & 0 \\ 0 & 0 & 2 \end{pmatrix}.$

Section 6-9

1. a) $\begin{pmatrix} 0 & -1 \\ 1 & 0 \end{pmatrix}, \begin{pmatrix} 1 \\ -1 \end{pmatrix}.$

c) $\begin{pmatrix} \frac{1}{2} & -\frac{\sqrt{3}}{2} \\ \frac{\sqrt{3}}{2} & \frac{1}{2} \end{pmatrix}, \begin{pmatrix} \dfrac{1-\sqrt{3}}{2} \\ \dfrac{1+\sqrt{3}}{2} \end{pmatrix}$

2. a)

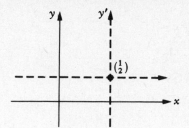

e)

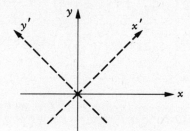

i)

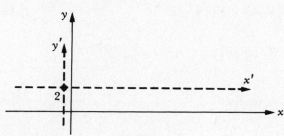

o)

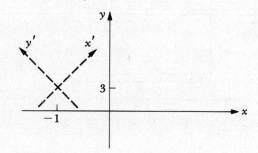

3. a) $\dfrac{(x')^2}{1} + \dfrac{(y')^2}{(\frac{1}{2})^2} = 1$ **d)** $(x')^2 + (y')^2 = 2^2$

e) $\dfrac{(x')^2}{4} - \dfrac{(y')^2}{1} = 1$ **g)** $\dfrac{(x')^2}{16} + \dfrac{(y')^2}{4} = 1.$

4. a) $(\ x\ \ y\ )\begin{pmatrix} 1 & 2 \\ 2 & 2 \end{pmatrix}\begin{pmatrix} x \\ y \end{pmatrix}$ **c)** $(\ x\ \ y\ )\begin{pmatrix} 7 & -3 \\ -3 & -1 \end{pmatrix}\begin{pmatrix} x \\ y \end{pmatrix}$

5. a) $\dfrac{(x')^2}{4} + \dfrac{(y')^2}{6} = 1.$ **c)** $\dfrac{-(x')^2}{4} + \dfrac{(y')^2}{2} = 1.$

e) $\dfrac{(x')^2}{2} - \dfrac{(y')^2}{6} = 1.$ **g)** $x' = \pm 2y'.$

7. a) $\begin{pmatrix} 0 & -1 \\ -1 & 2 \end{pmatrix}$

8. a) $\begin{pmatrix} 2 & 0 \\ 0 & 4 \end{pmatrix}$ **c)** $\begin{pmatrix} 6 & 0 \\ 0 & 8 \end{pmatrix}.$ **e)** $\begin{pmatrix} 15 & 0 \\ 0 & 10 \end{pmatrix}.$

9. a) $x = 0, y = 2.$ **c)** $x = \frac{5}{4}, y = \frac{7}{4}, z = 2.$

11. a) $0, \begin{pmatrix} -x_2 & -x_3 \\ & x_2 \\ & x_3 \end{pmatrix};$ $3000, \begin{pmatrix} 1 \\ 1 \\ 1 \end{pmatrix}.$

Section 6-10

1. $\begin{pmatrix} x_1 \\ x_2 \end{pmatrix} = b_1 \cos\left(\sqrt{\dfrac{3T}{ma}}\,t + \gamma_1\right)\begin{pmatrix} 1 \\ -1 \end{pmatrix} + b_2 \cos\left(\sqrt{\dfrac{T}{ma}}\,t + \gamma_2\right)\begin{pmatrix} 1 \\ 1 \end{pmatrix}$

3. $\begin{pmatrix} x_1 \\ x_2 \end{pmatrix} = b_1 \cos\left(\sqrt{\dfrac{11}{3m}}\,t + \gamma_1\right)\begin{pmatrix} -3 \\ 1 \end{pmatrix} + b_2 \cos\left(\sqrt{\dfrac{1}{m}}\,t + \gamma_2\right)\begin{pmatrix} 1 \\ 1 \end{pmatrix}$

Section 6-11

1.
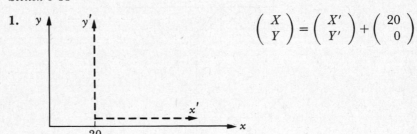

$$\begin{pmatrix} X \\ Y \end{pmatrix} = \begin{pmatrix} X' \\ Y' \end{pmatrix} + \begin{pmatrix} 20 \\ 0 \end{pmatrix}$$

CHAPTER 7

Section 7-1

1.
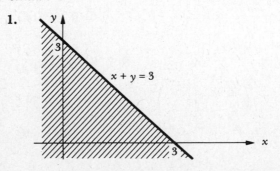

3.

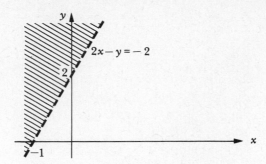

$2x - y = -2$

2

-1

5.

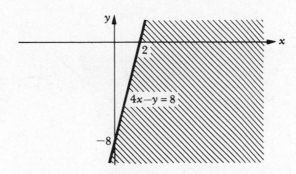

2

$4x - y = 8$

-8

7.

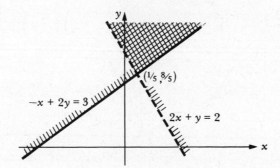

$(\frac{1}{5}, \frac{8}{5})$

$-x + 2y = 3$

$2x + y = 2$

9.

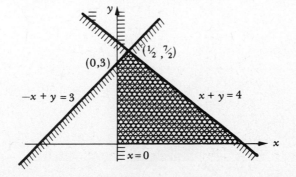

$(\frac{1}{2}, \frac{7}{2})$

$(0,3)$

$-x + y = 3$

$x + y = 4$

$x = 0$

11.

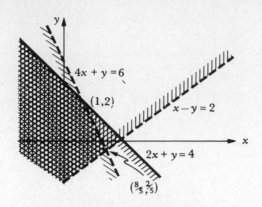

$4x + y = 6$

$(1,2)$

$x - y = 2$

$2x + y = 4$

$\left(\frac{8}{5}, \frac{2}{5}\right)$

13.

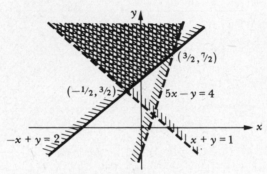

$(3/2, 7/2)$

$(-1/2, 3/2)$

$5x - y = 4$

$-x + y = 2$

$x + y = 1$

15.

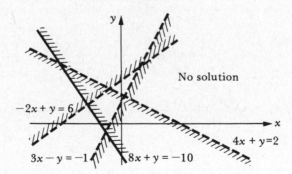

No solution

$-2x + y = 6$

$4x + y = 2$

$3x - y = -1$

$8x + y = -10$

17.

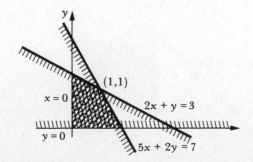

$(1,1)$

$x = 0$

$2x + y = 3$

$y = 0$

$5x + 2y = 7$

Section 7-2

1.

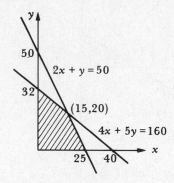

3.

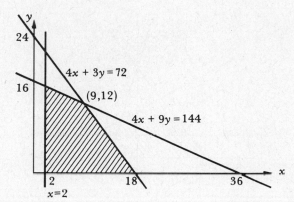

5.

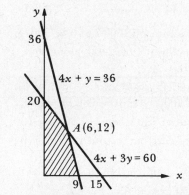

$2x + y$ is a maximum of 24 at
A, $(6,12)$.

7. $\begin{matrix} -3x + y = 4 \\ x - y = 2 \end{matrix}\Big\}$ Intersect at
$(-3,-5)$.

Thus no solution, feasible region
is empty.

9.

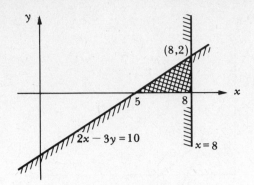

$3x + y$ is a maximum of 26 at (8,2).

11.

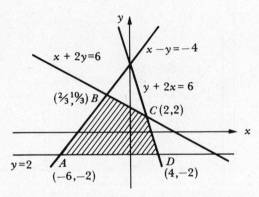

$x + 2y$ is a maximum of 6 along BC.

13. $15x + 7y$ is a maximum of 18000 at $x = 500$, $y = 1500$.

15. $40x + 20y$ is a maximum of 2400 along line $20x + 10y = 1200$, between (0, 120) and (30, 60).

17. Let x and y be barrels of automobile and heating oil respectively. $3x + 2y$ is a maximum of 2800 when $x = 800$, $y = 200$.

19. $x + y$ is a maximum of 150 when $x = 50$, $y = 100$.

21. Let distribution of cars be: x from A to C, y from B to C, p from A to D, q from B to D.
Maximum y is 54 when $q = 146$, $x = 66$, $p = 34$.

23. Minimum of -350 at $x = 100$, $y = 50$.

25. Minimum of -240 at $x = 40$, $y = 60$.

Section 7-3

1. Maximum of 24 at $x = 6$, $y = 12$.

3. 16.5 at $x = 2.25$, $y = 1.25$.

5. 1600 at $x = 160$, $y = 0$.

7. 22,500 at $x = 0$, $y = 100$, $z = 50$.

9. 40 at $x_1 = 20$, $x_2 = x_3 = 0$.

11. 50 at $x_1 = 0$, $x_2 = 15$, $x_3 = 5$, $x_4 = 0$.

Section 7-4

1. Maximum of 24 at $x = 6$, $y = 12$.

3. 6 at $x = 0$, $y = 3$.

5. 40 at $x_1 = 20$, $x_2 = x_3 = 0$.

Index

Absorbing state, 69
Addition. *See* sum
Adjoint matrix, 298
Admissable chronological ordering, 88
Alpha Centauri, 214
Amplitude of simple harmonic motion, 357
Angles between vectors, 190
Archaeology, 53, 87
Associative property, 15, 22, 40
Augmented matrix, 107

Back substitution, 106
BASIC, 412
Basic variables, 400
Basis, 173, 177, 210
   canonical, 177
   orthogonal, 192
   orthonormal, 192
Biaxial crystal, 349
Black box, 40
Bridge structures, 202

Cancer, constellation of, 218
Canonical basis, 177
Capella, 217
Chain, 76
Characteristic equation of a matrix, 303
Characteristic polynomial of a matrix, 303
Class score analysis, 12
Clock paradox, in relativity theory, 213
"Closed," defined, 163
COBOL, 410
Cofactor, 265
Color-matching experiment, 181, 218
Column vector, 11
Communication network, 73

Commutative property, 15, 22
Competition graph, 90
Complementary function, 250
Components of a vector, 3, 198, 201
Composite mapping, 243
Computer programs, 409
Connectivity, 81
Convex region, 383
Coordinate transformation, 331
Coordinates, 363
   generalized, 356
Cost of living analysis, 35
Cramer's rule, 293
Currents, 120

Deflation, 321
Deformation, 225
Degeneracy, 397
Degrees of freedom, 112
Departing basic variable, 402
Dependence, linear, 174
Determinant, 259
Diagonal matrix, 51
Diagonalization of matrices, 324
Digraph, 75
Diet problem, 386
Differential geometry, 217
Dilatation, 222
Dimension, 178
Distance, 194
Distributive property, 45
Doctor/patient model, 24
Domain, 234
Dominant eigenvalue, 319
Dominant factors, 68
Dot product, 186
Doubly stochastic matrix, 69

DuPont, 75

Echelon form, 109
Ecology, 90
Economic geography, 81
Edward III, 84
Eigenspace, 302
Eigenvalue, 302
Eigenvector, 302
Einstein, Albert, 27, 213
Elasticity, 225
Electrical axes of a crystal, 349
Electrical networks, 40, 120, 361
Elementary matrix transformation, 107, 108
Elementary transformation, 105
Elizabeth of York, 85
Energy production, 23
Entering basic variable, 402
Equality
    of matrices, 20
    of vectors, 4
Equilibrium model, 127
Equilibrium solution, 128
Equivalent matrices, 108
Equivalent systems of equations, 105
European Common Market, 36
Event, 214
Existence of solutions, 289
Expansion matrix, 222

Feasible solution, 378
Finite dimension, 178
Flatland, 217
Food price comparison, 30
Food web, 90
Forces, 9
FORTRAN, 410
Function space, 207

Galilean transformation, 366
Gas station model, 129
Gauss-Jordan method for equations, 118
    for inverse of a matrix, 144
Gauss-Seidel method, 158
Gaussian elimination, 109
Genealogical table, 84
General relativity, 27
Generate, 170
Generalized coordinates, 356
Genetics, 68
Genotype, 68
Geometrical interpetation of a matrix as a
    mapping, 219
Geometrical significance of an eigenvector,
    302
Gram-Schmidt orthogonalization process, 200

Graph, 74
Graph theory, 73
Grave analysis, 53
Gravity model, 18
Group relationships, 73
Guinea pig model, 68

Henry VII, 84
Henry VIII, 85
Heredity factors, 68
Hilbert space, 211
History model, 84
Homogeneous system of equations, 11, 115,
    244
Hooke's law, 362
Houses of York and Lancaster, 84
Hyades, 218
Hybrids, 68
Hydrostatic stress, 348

Image, 226, 232
Infinite dimension, 178
Inner product, 186
Input coefficients, 152
Interactive system, 411
Intergenerational mobility, 67
Interval graph, 86
Invariant vectors, 303
Inverse mapping, 229
Inverse of a matrix, 143, 297
Invertible matrix, 143
Isotropic medium, 349
Iterative methods, 157, 319

Kernel, 234
Kirchhoff's laws, 120
Kronecker delta, 52

Lake District network, 81
Left eigenvector, 309
Length of a vector, 187
Leontief input-output model in economics, 19,
    151
Light-year, 214
Linear approximations, 132
Linear combinations, 16, 167
Linear dependence, 174
Linear differential equation, 249
Linear differential operator, 240, 250
Linear equation, 99
Linear independence, 174
Linear inequality, 370
Linear mapping, 232, 234
Linear model, 132
Linear programming, 369

Loop, 421
Lorentz transformation, 367

Magnitude of a vector, 187
Main diagonal, 51
Manifold theory, 365
Mapping, 219, 222
Markov, chain, 66, 310
MAT statement, 425
Matrix, 17
  adjoint, 298
  augmented, 107
  determinant of, 259
  diagonal, 51
  inverse of, 143, 297
  nonsingular, 263
  of coefficients, 107
  of transportation probabilities, 63
  orthogonal, 227, 300
  product of, 27
  rank of, 283
  representation, 251
  scalar multiplication of, 20
  singular, 263
  spectrum, 303
  square, 17
  sum of, 20
  symmetric, 50
  trace of, 52
  transmission, 41
  transpose of, 48
Mileage chart, 13
Minkowski space, 215
Minor, 260
Moment of a vector, 279
Multidigraph, 75
Multiplication. See product
Multiplicative inverse of a matrix, 143
n-chain, 77
n-step transition matrix, 65
Newtonian mechanics, 19, 213, 366
Newton's laws, 366
Nonbasic variables, 400
Nonlinear system, 132
Nonsingular matrix, 263
Norm, 187
Normal coordinates, 357
Normal modes of oscillation, 353, 357
Normal stress, 56, 346
Normalizing, 189

Objective function, 377
Ohm's law, 41
Open sector, 153
Optimal solution, 378
Origin-destination analysis, 18
Orthogonal matrix, 227, 300

Orthogonal vectors, 192
Orthonormal basis, 192
Oscillating systems, 353

Parallel vectors, 193
Parallelogram of forces, 9
Parameter, 127
Particular integral, 250
Partitioning of matrices, 43
Path, 76
Permutation symbol, 271
Perpendicular vectors, 192
Petrie Flanders, 53, 88
Pivot, 136, 390
Plasticity, 225
Pleiades, 217
Pointwise addition, 208
Pointwise scalar multiplication, 208
Population movement model, 60
Position vector, 1
Pottery analysis, 53
Power method, 319
Praesepe, 218
Principal direction, 346
Principal plane, 346
Principal stress, 346
Print statement, 413
Probability, 60
Product of matrices, 27
Production-scheduling problem, 385
Projection, 198
Projection mapping, 236
Properly scaling, 138

Quadratic form, 338
Quantum mechanics, 211

R^n, 3
Rainfall model, 312
Range, 234
Rank of a matrix, 283
Rayleigh quotient, 320
Recessive, 68
Reduced echelon form, 118
Redundant chain, 77
Regular Markov chain, 310
Resistance, 120
Resultant force, 9
Riemannian geometry, 217
Road network, 76, 81
Rotation, 221, 335
Round-off errors, 138
Row vector, 11

Scalar, 4

Scalar multiple of a vector, 5
Scalar product, 5
Scalar projection, 198
Scaling of equations, 138
Schwarz inequality, 197
Sequence dating, 53
Seriation, 53
Shearing stress, 56, 346
Shortest-chain matrix, 81
Sigma notation, 37
Significant figures, 136, 437
Similar matrices, 323
Similarity transformation, 323
Simple harmonic motion, 356
Simplex method, 389
Simplex tableau, 390
Simultaneous diagonalization of matrices, 359
Singular matrix, 263
Sirius, 217
Slack variables, 389
Sociology, 80
Solutions of linear equations, 99
Space, vector, 163, 208
Space-time diagram, 215
Spanning set, 169
Special relativity, 213
Spectral theory, 255
Spectrum of a matrix, 303
Square matrix, 17
Stable population, 308
Stochastic matrix, 60, 61, 309
Stochastic vector, 61
Stress, 56, 202
Stress matrix, 55, 346
Sub-matrix, 43
Subspace, 163, 164
Sum
   of matrices, 20
   of vectors, 5
Supermarket model, 30
Supply and demand model in economics, 19,
   127
Symmetric matrix, 50, 327
Symmetry property of inner product, 186, 218

Systems of linear equations, 99

Taurus, 218
Teletype, 411
Time-sharing, 411
Trace, 52
Trade figures, 23
Traffic flow model, 123
Transformation, coordinate, 331, 365
Translation of axes, 334
Transmission matrix, 41
Transportation problem, 385
Transportation science, 18
Transpose of a matrix, 48
Triangular matrix, 59
Triple scalar product, 280
Triple vector product, 281
Trivial solution, 292
Two-port, 40

Uniaxial crystal, 349
Uniqueness of solutions, 289
Unit matrix, 51
Unit vector, 188
Upper triangular matrix, 59, 273

Vectors, 1, 163, 208
   components of, 3, 198
   orthogonal, 192
   product, 277
   projection of, 198
   sum of, 5
Vector space, 5, 163, 212

Weather model, 312

Zero solution, 292
Zero vector, 10, 16